Human–Centered System Design for Electronic Governance

Saqib Saeed
Bahria University Islamabad, Pakistan

Christopher G. Reddick
University of Texas at San Antonio, USA

T0338600

Managing Director:	Lindsay Johnston
Editorial Director:	Joel Gamon
Book Production Manager:	Jennifer Yoder
Publishing Systems Analyst:	Adrienne Freeland
Development Editor:	Austin DeMarco
Assistant Acquisitions Editor:	Kayla Wolfe
Typesetter:	Christy Fic
Cover Design:	Jason Mull

Published in the United States of America by
Information Science Reference (an imprint of IGI Global)
701 E. Chocolate Avenue
Hershey PA 17033
Tel: 717-533-8845
Fax: 717-533-8661
E-mail: cust@igi-global.com
Web site: http://www.igi-global.com

Library of Congress Cataloging-in-Publication Data

Human-centered system design for electronic governance / Saqib Saeed, Christopher G. Reddick, Editors.
  pages cm
 Includes bibliographical references and index.
 Summary: "This book provides special attention to the most successful practices for implementing e-government technologies, highlighting the benefits of well designed systems in this field, while investigating the implications of poor practices and designs"-- Provided by publisher.
  ISBN 978-1-4666-3640-8 (hardcover) -- ISBN (invalid) 978-1-4666-3641-5 (ebook) -- ISBN 978-1-4666-3642-2 (print & perpetual access) 1. Internet in public administration--Technological innovations. I. Saeed, Saqib, 1970- editor of compilation. II. Reddick, Christopher G., editor of compilation.
 JF1525.A8H86 2013
 352.3'802854678--dc23
                              2012045322

British Cataloguing in Publication Data
A Cataloguing in Publication record for this book is available from the British Library.

All work contributed to this book is new, previously-unpublished material. The views expressed in this book are those of the authors, but not necessarily of the publisher.

Table of Contents

Section 2
Human-Centered E-Government: Effectiveness and Organizations

Section 3
Case Studies on Human-Centered E-Government

Detailed Table of Contents

Section 1
User-Centric E-Government

This chapter focuses on the information and communication technologies (ICTs) adoption by governments in various countries. Theoretical models related to information systems and technology adoption are presented in order to understand the various constructs of importance from the adoption and diffusion of innovations perspective. Moreover, this chapter highlights the drivers and barriers to ICT adoption from the government perspective. Furthermore, this chapter provides important information of ICT adoption in different world regions by governments. Future implications and conclusions are provided.

As social media has become integrated into the public's everyday lives, local governments have started to take advantage of the power of social media as another governance tool to both inform and involve the public in local government. This new tool also introduces a new responsibility for government to monitor and analyze the actions taken on municipal social media sites. For this to be achieved, municipalities must implement a social media policy that addresses the abundant concerns inherent when engaging in social media use. This research indicates the areas that local governments must address in social media policy and offers a best practice approach to completing the task of policy development.

This chapter focuses on dynamic taxonomies, a semantic model for the transparent, guided, user-centric exploration of complex information bases. Although this model has an extremely wide application range, it is especially interesting in the context of e-government because it provides a single framework for the access and exploration of all e-government information and, differently from mainstream research, is citizen-centric, i.e., intended for the direct use of end-users rather than for programmatic or agent-mediated access. This chapter provides an example of interaction and discusses the application of the model to many diverse e-government areas, going from e-services to disaster planning and risk mitigation.

Chapter 4

Nonprofit organizations are an important sector of society working to support underprivileged citizens. The operations of nonprofit organizations differ from their organizational size, scope, and application domain. Modern computer systems are quite effective in managing organizational tasks, but the nonprofit sector lacks in technological systems concerning organizational settings. In order to foster a successful use of electronic services, it is vital that computer systems are appropriate according to user needs. The diversity of users and their work practices in nonprofit organizations make it difficult for standardized infrastructure to work optimally in diverse organizational settings. In this chapter, the authors discuss the issues and complexities associated with system design for nonprofit organizations. They analyze important open issues that need to be explored for appropriated technology design in this domain.

Chapter 5

Websites connect businesses with customers. They are an important medium that facilitates online transactions, a necessity for businesses. The design and usability of an Electronic Commerce (EC) website play an important role in achieving its objectives (Kumar, Smith, & Bannerjee, 2004; Marcus, 2005; Nielsen, 2003; 2005; Krug, 2006; Cappel & Huang, 2007). Recognizing their importance, design and usability aspects of EC websites have been widely researched in both applied and academic research (Lecerof & Paterno, 1998; Lohse & Spiller, 1999; Nielsen, 2000; Cao, Zhang, & Seydel, 2005; Flavian & Guinaliu, 2006; Nathan, Yeow, & Murugesan, 2008; Nathan & Yeow, 2009; Robins & Holmes, 2008). This chapter discusses the recent work with web design and electronic commerce. The importance of usability and user-centered web designs are highlighted. Usability to specific target groups and industries, such as airlines, government, and services portals, are also discussed. Altogether, design guidelines are given for web industries, and recommendations are made for better usability in designing websites.

Chapter 6

The e-government paradigm became an essential path for governments to reach citizens and businesses and to improve service and public performance. One of the important tools used in political and administrative venues is e-voting, where ICT tools are used to facilitate the process of voting for electing representatives and making decisions. The integrity and image of such applications won't be maintained unless strict measures on security and authenticity are applied. This chapter explores the e-voting process, reviews the authentication techniques and methods that are used in this process and proposed in the literature, and demonstrates few cases of applying e-voting systems from different countries in the world. Conclusions and proposed future work are stated at the end of the chapter.

Section 2
Human-Centered E-Government: Effectiveness and Organizations

Chapter 7

In modern times, people and their governments have struggled to find easy, cheap, and effective ways to run countries. The use of Information and Communication Technologies is gaining ground as a means of streamlining public service provision by shifting tasks from the government to its citizens, resulting in reduced government costs, increased public revenues, and greater government transparency and accountability. The new buzzword is e-Government: the use of ICTs by government, civil society, and political institutions to engage citizens through dialogue to promote greater participation of citizens in the process of institutional governance. However, the implementation of such projects is complicated by the reality that while developmental problems in these countries are many, the resources available to tackle them are scarce. In attempting to investigate the interaction between new technologies, information flows, and the complexities of public administration reform in the developing world, this chapter examines not only the interplay of local contingencies and external influences acting upon the project's implementation but also aims to offer an insight into disjunctions in these relationships that inhibit the effective exploitation of ICTs in the given context.

Chapter 8

Port authorities constitute very active organizations that frequently interact with citizens as well as public and private organizations. The employees and administration of port authorities require effective e-government services in order to implement their tasks. The required services should provide effective information flow and collaboration to improve decision making, governance, and integration of all sectors. In this chapter, the authors briefly outline issues concerning the usefulness of intranets in organizations and corresponding services provided to organization employees. They briefly present key aspects of certain recent approaches concerning e-governance and intranets in ports. The authors also present a case study involving the e-government services implemented for Patras's Port Authority in Greece. The specific port authority has a lot of workload because the corresponding port is the third largest in Greece and a main gate to countries abroad. The case study combined Internet-based technologies with e-learning technologies. E-learning services assist employees in acquainting themselves with newly introduced e-government services. Therefore, e-learning may contribute in the successful realization of e-government projects.

Chapter 9

Evaluating e-government systems is a difficult task involving multi-faceted perspectives. Although a review of the literature discovers several e-government evaluation frameworks, numerous shortcomings still exist. The objective of this chapter is to propose a formative and holistic framework to remedy the current research gaps. The formative position of the evaluation framework ensures the evaluation objective achievement, and the holistic approach ensures completeness and continuity of the evaluation process. The framework can be used as a template for researchers and practitioners to assess e-government projects. The authors demonstrate the applicability and practicability of the framework by applying it to the Korean Government-for-Citizen (G4C) project.

Chapter 10

In the era of economic liberalisation, institutions of higher education in the government sector, particularly universities, are facing tremendous challenges in terms of academic, general, and financial administration, which need effective governance. Recently, some of the universities are trying to adopt e-governance as a platform for such a purpose. However, the design of such a system is very much important, as it has to cater to the needs of various stakeholders in the public system. In this context, the effectiveness measurement of such an e-governance system is really necessary either to improve its performance level by re-aligning its organisational culture or by providing inputs for re-designing the system in order to make it more effective. Hence, the performance of such a system can be known if a human-centric approach with multiple criteria of evaluation is considered in the governance environment. This chapter attempts to determine those criteria by multiple factor analyses carried out for the purpose of considering multiple stakeholders. Analytic hierarchical processes as well as fuzzy analytic hierarchical processes have been then employed to measure the effectiveness of e-governance systems along those criteria, taking an Indian university as a case study.

The authors present a literature review, carried out by searching through conference proceedings, journal articles, and other secondary sources for papers focusing on the usability of electronic voting (e-voting) systems and related aspects such as ballot design and verifiability. They include both user studies and usability reviews carried out by HCI experts and/or researchers, and analyze the literature specifically for lessons on designing e-voting system interfaces, carrying out user studies in e-voting and applying usability criteria. From these lessons learned, the authors deduce recommendations addressing the same three aspects. In addition, they identify for future research open questions that are not answered in the literature. The recommendations hold for e-voting systems in general, but this chapter especially focuses on remote e-voting systems providing cryptographic verifiability, as the authors consider these forms as most promising for the future.

Section 3
Case Studies on Human-Centered E-Government

Chapter 12

Charles C. Hinnant, Florida State University, USA
Jisue Lee, Florida State University, USA
Lorri Mon, Florida State University, USA

For public organizations, the ability to harness web-based Information and Communication Technologies (ICT) to make information and services directly available to the public has become an important goal. Simultaneously, the use of volunteers by public organizations has become a crucial component of service delivery within the US. Court Appointed Special Advocate (CASA) programs rely heavily upon volunteers to advocate for neglected children. While there is no doubt variation exists across specific CASA programs, their generally ubiquitous reliance on volunteers indicates a need for recruitment, training, and coordination to successfully achieve program goals. While the discussion of User-Centered Design (UCD) factors illustrates issues for consideration, the case study of Florida's Guardian ad Litem (GAL) program more concretely illustrates how a state-level CASA can begin to harness online ICT to achieve programmatic goals. This chapter discusses key information design characteristics needed for online systems to effectively deliver required information to both volunteers and staff.

Chapter 13

Charlie E. Cabotaje, Center for Leadership, Citizenship and Democracy (CLCD), University of the Philippines, Philippines
Erwin A. Alampay, Center for Leadership, Citizenship and Democracy (CLCD), University of the Philippines, Philippines

Increased access and the convenience of participation to and through the internet encourage connectivity among citizens. These new and enhanced connections are no longer dependent on real-life, face-to-face interactions, and are less restricted by the boundaries of time and space (Frissen, 2005). In this chapter, two cases from the Philippines are documented and assessed in order to look at online citizen engagement. The first case looks at how people participate in promoting tourism in the Philippines through social media. The second case involves their use of social media for disaster response. Previous studies on ICTs and participation in the Philippines have looked at the role of intermediaries (see Alampay, 2002). Since then, the role of social media, in particular that of Facebook and Twitter, has grown dramatically and at times completely circumvents traditional notions of intermediation. The role of Facebook, in particular, will be highlighted in this chapter, and the authors will analyze its effectiveness, vis-à-vis traditional government channels for communication and delivery of similar services. By looking at these two cases and assessing the abovementioned aspects, it is hoped that the use of social media can be seen as an integral part of e-governance especially in engaging citizens to participate in local and national governance.

This chapter focuses on a change effort for introduction of an e-governance innovation in the operating room management of a medium-sized Italian hospital, which led to higher levels of efficiency and effectiveness at once. The innovative project has made all the stages of the surgical process transparent, highlighting where there is an opportunity to improve overall performance via the introduction of organizational and process innovations. New techniques implemented and the specific factors that led to the hospital's success in achieving improved outcomes at lower costs are discussed. The chapter concludes by highlighting that low cost and human-centricity are amongst the key characteristics of success of this innovation.

Local governments are increasingly embracing Web 2.0 technologies to encourage the use of means of bidirectional communication to change how they interact with stakeholders, thus providing the greater accountability demanded. Nonetheless, to make Web 2.0 tools efficient, there must be qualified people to operate and supervise the Web 2.0 and social network technologies implemented by local governments. These people, called "Community Managers," play a key role in the implementation of social networks in local government, successfully or otherwise. In this chapter, the authors analyse whether the training and education of community managers in Spanish local governments is associated with the successful use of social networks by these local governments in their interaction with the public. Their empirical study of local government in Spain shows that the position of community manager is mostly held by men who are aged 25-45 years and have a university degree in journalism, performing in addition, tasks such as updating the municipal website or running the press office.

Chapter 16

An e-Participation ecology is composed of five elements—actors, contents, traditional culture of participation, existing media skills and practices, and discourses in conflicts (establishment vs. antagonists)—and three macro-dimensions—cultural/traditional, political, and socio-technological—with which the five elements are interacting (Cavallo, 2010). Game theory can be used to understand how a certain actor or a group of actors can develop a successful strategy in/for each one of the three dimensions. Therefore, the concept of Nash equilibrium (Nash Jr., 1950), developed in physics and successfully applied in economy and other fields of study, can be borrowed also by e-Participation analysts/project managers to develop "Win-Win" scenarios in order to increase e-Participation projects' chances of success and consequently reduce e-Participation's "risk of failures," especially in developing countries where they usually occur more frequently (Heeks, 2002). The Kenyan e-Participation platform, Ushahidi, generated a techno-discourse about the rise of African Cyberdemocracy and the power of crowd-sourcing that is probably more relevant than the real impact that these e-Participation platforms had or will have on the lives of normal citizens and media activists.

Research on e-government is taking a new phase nowadays, with researchers focusing more to evaluate the continued usage intention by the citizens rather than the initial intention. Continuance intention is defined as a person's intention to continue using, or long term usage intention of a technology. Unlike initial acceptance decision, continuance intention depends on various factors that affect the individual's decision to continue using a particular system, with trust being one for the most important factors. Therefore, this case study aims to examine the role of trust, particularly trust in the system, on continuance usage intention of an e-filing system by taxpayers in Malaysia. The primary discussion in this case study concerns the e-filing system in Malaysia, followed by the strategies for successful adoption of e-government services and the benefits of e-government adoption, concluding with future research directions.

This chapter aims to better understand what citizens think regarding the currently available e-government public services in Egypt. This is done through an analysis of a public opinion survey of Egyptian citizens, examining citizens' use and associated issues with usage of e-government portals. This chapter is different from existing research in that most of the studies that examine e-government and citizens focus on developed countries. This study focuses on a developing country, Egypt, as an emerging democracy, which has very unique and important challenges in the delivery of public services to its citizens. The results revealed that only gender, daily use of the internet, and the desire to convert all of the services to electronic ones were important factors that affected the use of the Egyptian e-government portal. On the other hand, age, education, trust in information confidentiality on the internet, and believing in e-government did not play any role in using e-government.

Preface

Electronic government or e-government projects are a huge undertaking and require serious political commitment, clear vision, and a robust long-term strategy. The successful realization of such projects in practice requires appropriately designed technology infrastructure. Higher failure rate in technology adoption has highlighted that human factors are an important aspect when designing and implementing e-government projects. In order to foster successful usage, technologies need to be consistent with human practices. Failure to design e-government technologies properly can lead to project failure, which can represent a substantial cost for governments and a loss of confidence by users in these systems. Public sector organizations need to deal with a vast number of stakeholders, which makes it especially important that these projects are designed with the user or citizen in mind.

The edited book, *Human-Centered System Design for Electronic Governance* examines the impact of human factors on the development of e-government. The objectives of the book are twofold. First this book provides innovative ideas, suggestions, and recommendations examining the inherent issues, technology design implications, user experiences, and guidelines for technology in government. Second this book provides case studies on best practices employed by organizations in the field of e-government. The aim of the second objective is to provide opportunities for discussion of implications and dissemination of best practices that will be useful, or of interest to academics and practitioners from a range of fields including information systems, human computer interaction, organizational science, public administration, and political science. The book will be helpful for students and researchers working in the domain of information systems. Furthermore, government officials working to devise IT strategy and solutions for e-governance are also potential readers. This book will be helpful for those that want to learn about the design of information systems in the context of e-government, from a human-centric approach.

The book is timely since the focus here is on how e-government has influenced human behavior, something relatively unexplored in the literature. Most of the existing research handbooks examine the adoption of e-government, or various aspects of technological innovations in government; this book focuses specifically on the human element. This book also provides a broad array of chapters examining both developed and developing countries and their experiences. There are many different methods used in this book ranging from case studies, surveys, interviews with public officials, and comprehensive literature reviews. The end results and findings of this book is that human factors are indeed important, especially as governments are spending more money on the development of e-government projects in these tough economic times.

There are three parts to this book. Section 1 deals with user-centric e-government. In this part, there are chapters on the impact of citizens and their impact on the development of e-government. Section 2 of this book examines the human-centered e-government effectiveness and organizations. The chapters in this part of this book focus on evaluating the use of e-government projects that focus on human factors. Section 3 of this book provides selected case studies on human-centered e-government. In this part, there are case studies of not only successful experiences with e-government, but challenges that governments have faced.

Within the three parts of the book, in Section 1, there is first a chapter by Adapa that examines the adoption of e-government globally focusing on its benefits and barriers. Examining adoption in both developed and developing countries shows that there are vast differences in what determines success. The author here notes that developed countries could set the tone for developing countries. Important challenges in developing countries are issues of leadership and decision-making, which needs to be improved for more successful adoption. Education and training are essential for successful adoption. Therefore, the importance of this chapter is that technology adoption must take into account human factors.

Chapter 2, by Foster and Chen, examines the adoption of social media in local governments in the U.S. Social media is one important tool that governments can use to engage their citizens in public service delivery. As the authors mention, social media applications are an example of human-centered design that benefits local governments and their citizens. Their research is different since social media is so new; there is little literature on social media in local government. According to Foster and Chen, through social media governments can make the governance process more inclusive and transparent, which may stimulate citizen trust in government. The authors of this chapter found that since social media is new for local governments, policies on its use have not kept pace; therefore, governments should spend more time thinking about appropriate policies.

In Chapter 3, Sacco argues that user-centric access to complex information is critical for the development of e-government. This author discusses dynamic taxonomies; which are a data management model to make better sense of information placed online. According to Sacco, some of the applications for e-government are seen through laws and regulations where dynamic taxonomies can provide guided browsing and personalized exploration of a complex set of government regulations and laws. Dynamic taxonomies also provide a way for citizens to better access useful information from government websites. The contribution of this chapter is to show the importance of the semantic web and its application to government.

Saeed and Rohde, in Chapter 4, examine the issue of technology design for nonprofit organizations. This is an important contribution to the book since it deals with a topic that has not received much scholarly attention in e-government research. As these authors point out the composition and operation of nonprofit organizations are very different from public sector organizations. These authors also argue that nonprofits lack funding to take on risky and uncertain technology projects. The findings from this study indicated that lack of funding to support technology infrastructure appeared to be a major challenge for technology adoption.

Chapter 5, by Nathan and Suki, examines the design of e-commerce websites that connect businesses with their customers. The simply but important message from these authors is that "one size does not fit all" in prevalent e-commerce websites. One of the recommendations these authors propose is that designing user-centered websites will humanize the web experience for users. These authors argue that web designers need to constantly learn about the changing business environment, and give customers what they want in terms of features.

The last chapter in Section 1 is Chapter 6, which deals with authentication for electronic voting or e-voting. E-voting is the use of Information and Communications Technologies (ICT) for voting. As Abu-Shanab, Khasawneh, and Alsmadi, the authors of this chapter, state, the main issue for the legitimacy of the voting is security and privacy. Their chapter discusses the various techniques and methods used to gain a higher level of security and authentication of the voter. The authors argue that biometric methods will increase e-voting systems accuracy and mixed methods are also critical to employ.

Section 2 of this book examines human centered e-government effectiveness and organizations. Chapter 7 Virkar examines the design and implementation of e-governance projects. This chapter discusses the most important benefits of e-government projects such as cost reductions and efficiency gains, quality of service delivery, transparency, anticorruption, and accountability, improvements in decision making, and increased capacity of government. Some of the challenges for e-government projects are the lack of ICT infrastructure, organizational and management, human capital, information, and general management. The author makes the reader aware of the complexity of implementing e-government projects, which really need to account for the human element.

In Chapter 8, by Prentzas, Derekenaris, and Tsakalidis, the authors examine e-government system design in port authorities. They argue that e-government projects involving ports require careful design because of the complex services provided. Automation and efficient data exchange are among the functionalities required for port authorities since they deal with a lot of traffic. Coordination is also important and requires that systems in place must be carefully designed to take into account human factors. The authors of this chapter especially focus on the Intranet and some of the functionalities of this system for the port authority.

In Chapter 9, Alalwan and Thomas argue that e-government projects are very expensive, risky, and difficult to successful accomplish. These authors provide a framework to evaluate the efficiency of e-government systems. Their framework seeks to classify e-government systems into four quadrants. This framework can be used by researchers and practitioners to assess e-government projects in the public sector. The contribution of this paper is to provide more research on the evaluation of e-government projects.

In Chapter 10, Mangaraj and Aparajita discuss the measurement of the effectiveness of e-government systems using a human-centered approach. There is a necessity for the improvement of the design of e-government in order to make it more user-friendly. Effectiveness can be explained by examining the dimensions of price, content, availability, usability, quality, and communicability. These authors argue that their methodology can be used to evaluate the effectiveness of e-government projects.

In Chapter 11, Olembo and Volkamer examine the usability of e-voting systems. The results from this chapter are useful for those designing e-voting systems to focus more on human factors. Direct Recording Electronic (DRE) such as electronic touch screens and traditional voting methods presently have a lot of research; however, research looking at the usability of verifiable e-voting systems needs further study.

In Section 3, there is an examination of human-centered e-government case studies. Chapter 12, by Hinnant, Lee, and Mon, examines a case study on the adoption of ICT in U.S. Court Appointed Special Advocate (CASA) programs that rely heavily on volunteers for neglected children. This chapter examines ICT systems to delivery information online to both volunteers and staff. The results of their case study show that websites and social media sites can be useful to deliver communication between program staff and stakeholders.

In Chapter 13, Cabotaje and Alampay provide a case study of social media and citizen engagement using a case study of the Philippines. There are two case studies presented: one case is on using social media to promote tourism, and the second examines social media for disaster response. The findings of this case study show that social media can be used to harness greater citizen engagement in government.

In Chapter 14, Padovani, Orelli, Agnoletti, and Buccioli examine human-centered health care service delivery. The results of the case study show that the use of ICT can create more efficiency and effectiveness in the delivery of medical care. This technology can be successfully used by clinicians and managers to better management health care delivery.

In Chapter 15, Bolívar, Pérez, Hernández examines the impact of social media implementation on local governments in Spain. Their case study showed that social networks by improving communication with the public create greater transparency and accountability. The key contribution of their study is to show the impact of social media on public service delivery at the local level.

Chapter 16, by Cavallo, examines electronic participation or e-participation in Kenya. E-participation investigates how ICT can be used to improve citizen interaction with government. Their case study uses a game theoretical approach and found that in Kenya e-participation has inherent risks and failures.

Chapter 17, by Santhanamery and Ramayah, provides a case study of e-filing for income taxes in Malaysia. Their major finding is that trust in the e-filing system leads to greater use of these online systems. This paper supports the existing literature that argues that trust and confidence in e-government leads to greater adoption.

In Chapter 18, by Abdelsalam, Reddick Hatem Elkadi, and Gamal, the final chapter in this book examines through a survey of citizens the issues associated with the use of e-government portals. The results showed that demographic factors had an influence on the use of web portals. In addition, there is a digital divide in Egypt with face-to-face communications being the most preferable way to contact government entities.

Saqib Saeed
Bahria University Islamabad, Pakistan

Christopher G. Reddick
University of Texas at San Antonio, USA

Section 1
User-Centric E-Government

Chapter 1
Government ICT Adoption:
Global Trends, Drivers, and Barriers

Sujana Adapa
University of New England, Australia

ABSTRACT

This chapter focuses on the information and communication technologies (ICTs) adoption by governments in various countries. Theoretical models related to information systems and technology adoption are presented in order to understand the various constructs of importance from the adoption and diffusion of innovations perspective. Moreover, this chapter highlights the drivers and barriers to ICT adoption from the government perspective. Furthermore, this chapter provides important information of ICT adoption in different world regions by governments. Future implications and conclusions are provided.

INTRODUCTION

Governments as public sector organisations throughout the world are the largest users of ICT (Information and Communication Technologies) and often play a lead role in the adoption and diffusion of ICTs (Affisco & Soliman, 2006). The impacts associated with the adoption/non-adoption of ICTs are huge on both governments and citizens. The key benefits associated with government adoption of ICTs include enhanced performance of the government organisations, modernisation of the public sector from the traditional models, improved interactions with the businesses and citizens, reduced transaction costs, increased connections between different levels of the government (federal/state/local), decreasing the levels of business processes and improved cooperation and communication at various levels.

As the citizens and various businesses operate in an integrated manner in a society, the factors driving government ICT adoption from the regional perspective relate to the economic status of the country, prevailing political situation, technological readiness of the country, socio-cultural situation and prevailing laws and regulations in a country. Similarly, citizens and organisations heav-

DOI: 10.4018/978-1-4666-3640-8.ch001

ily rely on their knowledge and awareness, service delivery, efficiency, quality and implementation of the government ICT adoption. Government ICT adoption is evident to the businesses and citizens in the form of enhanced public sector processes (e-Administration), maintaining better connections with the citizens (e-Services) and building interactions with the external stakeholders (e-Society) (West, 2005).

ICTs have been adopted phenomenally by the commercial sectors in many countries. The adoption of ICTs by citizens also increased in a phenomenal manner. Therefore, citizens and other businesses expectation about the adoption of ICTs by their governments have emerged (Nelou, 2004). Today, citizens, businesses, employees and other interest groups if any, expect increase in service provision by the governments, call for better administration practices and tend to create an innovation climate by fostering e-Society (Burne, 2002). Governments in many countries have realised the importance of embracing ICTs in their day today operations as they associate the ICT developments with the immediate benefits of economic development, upgrading infrastructure and productivity gains.

Governments adopting ICTs are often referred to as eGovernment in the extant literature. The concept of e-government enhances the process of decentralisation by bringing the decision-making aspects closer to the citizens, businesses, employees and other governments. Thus e-government focuses on delivering the relevant information and services online through the internet or other digital means for its users (Muir and Oppenheim, 2002). These users are identified to be citizens, businesses, employees and other governments. Furthermore ICTs are creating networked economies where by businesses involving different stakeholders such as suppliers, manufacturers, customers etc., are linked with the business processes in an integrated manner in the real time in order to create more value.

THEORETICAL MODELS ON THE ACCEPTANCE OF TECHNOLOGY

Several businesses are utilising innovative technological advancements in order to make their services more accessible to consumers as well as to improve their business performance and increase their productivity (Winch & Joyce 2006; Reid 2008). However, the correlation between technological advancements and increase in business productivity is feasible only if they are accepted by the intended users (Venkatesh *et al.* 2003). The objective of most research on ICT adoption has been to analyse the process of acceptance, intention to adopt or adoption of the new technology and compare it with the other technologies that are in place. Nevertheless, in recent years ICT adoption has shown increasing uptake by several commercial users and there are more and more citizens willing to adopt ICTs. Recently, there has been an upsurge in the public sector to embrace ICTs in their day to day operations for better service provision to its users. Therefore, governments throughout the world are trying to provide enhanced eService and eAdministration through their web portals in order to create an eSociety.

Thus, from the existing research, possible theoretical models that provide a comprehensive understanding of user acceptance of innovations come from disciplines such as information systems, psychology and sociology. The present study proposes the application of two main theoretical models that of an integrated technology model of consumer adoption and diffusion of innovations model, in order to explain the factors that might have a significant impact on the usage patterns of ICTs offered by eGovernments in several countries.

Four theories are discussed in detail which might explain the perceptions consumers have towards their use of internet banking. The theories that are reviewed include:

1. Theory of Reasoned Action (TRA) predicts the determinants of intended behaviour of individuals (Fishbein & Ajzen 1975).
2. Theory of Planned Behaviour (TPB) is an extension of TRA by the addition of perceived behavioural control (Ajzen 1985).
3. Technology Acceptance Model (TAM) determines the adoption and usage patterns with regard to the general acceptance of the technology (Davis 1989).
4. Diffusion of Innovations (DOI) proposes factors that facilitate the decision to adopt an innovation (Rogers 1995).

The Theory of Reasoned Action (TRA) is a widely validated intention model and has been proved to be successful in predicting and explaining behaviour across a wide variety of domains (Fishbein & Ajzen 1975). Due to its limitation in terms of volitional control, the additional construct of perceived behavioural control, which predicts behavioural intentions and behaviour, has been included and the extended model is known as the Theory of Planned Behaviour (TPB) (Ajzen 1985). Studies conducted by Mathieson (1991); Taylor & Todd (1995) and Venkatesh *et al.* (2000) provide valid empirical support to these two theories for studying the determinants of ICT behaviour.

The theoretical foundation for identifying the dimensions of user behaviour, adoption and diffusion stem from intention models in social psychology (Swanson 1982). In predicting and explaining behaviour across a wide variety of domains, TRA is a widely validated intention model (Fishbein & Ajzen 1975). TRA has been applied to explain behaviour related to the acceptance of technology and includes the four general concepts of behavioural attitudes, subjective norms, intention to use and actual use. Individuals or businesses in general evaluate the consequences of a particular behaviour and develop intentions to act that are consistent with their evaluations. More specifically, TRA states that behaviour is predicated on their attitudes and subjective norms. Attitudes

can be predicted from beliefs about the consequences of their behaviour. Attitude is defined as an individual's positive or negative feelings about performing a specific behaviour and is determined by one's beliefs that the behaviour would lead to various consequences multiplied by the subjective evaluation of those consequences (Davis *et al.* 1989). Subjective norms can be predicted from the beliefs they generate about what others think (Fishbein & Ajzen 1975). The TRA model emphasises the fact that behavioural intention is the only antecedent of actual behaviour. Bagozzi (1982) praises the TRA model because it is intuitive, parsimonious and insightful in its ability to explain behaviour from a theoretical perspective.

Davis *et al.* (1989) reported that TRA is a general model and is not able to specify the beliefs that are responsible for a particular behaviour. Also, TRA deals with the prediction of behaviour rather than the actual outcome of the behaviour (Szmigin & Foxall 1998). As per the TRA model, behaviour is determined by the behavioural intention, which limits the predictability of the model to situations where intention and behaviour are highly correlated. Moreover, according to TRA, behaviour must be under the volitional control of the consumer and is not suitable to predict situations in which individuals have low levels of volitional control (Ajzen 1991). Therefore, including another construct called perceived behavioural control that predicts both behavioural intention and behaviour has extended the TRA model. This extended model is called the Theory of Planned Behaviour (TPB).

The TRA model is the predecessor to TPB. TPB has the added construct of perceived behavioural control (PBC) to the antecedents identified by the TRA (Ajzen 1991). Thus the construct of behavioural intention is formed by one's attitude, which is a combination of the social norm and the perceived behavioural control. Such a model is an improvement on TRA because it reflects perceptions of both internal and external constraints on behaviour simultaneously. Perceived behavioural control is defined as a perception of the difficulty of performing a particular behaviour. By including PBC, the TPB model accounted for more variance in intention (44.50%) than the TRA model (37.27%) (Hagger *et al.* 2002; Ajzen & Madden 1986).

Liao *et al.* (2000) empirically supported the effect of perceived behavioural control. However, the exact nature of the relationships between attitudes, subjective norm and perceived behavioural control were still unclear. Conceptualisation of PBC has been controversial and measurement of salient beliefs underlying the model remains a problem, thus making it difficult to operationalise the TPB model (Kraft *et al.* 2005). Moreover, the model suggests that behavioural intention is an antecedent to the actual behaviour that is predicting the adoption behaviour.

Similar to TRA, TPB also assumes proximity between intention and behaviour. Thus TPB also emphasises behavioural intention as an antecedent to actual behaviour (Szmigin & Foxall 1998). Only one exogenous variable, that of perceived behavioural control, is introduced in the model where evidence suggests that other factors such as personal norms and affective evaluations significantly add to the models predictive power (Manstead & Parker 1995). Also existing studies indicate a problem with measuring the perceived behavioural control construct as existing studies indicated the complexity associated with the conceptualisation of the perceived behavioural construct (Manstead & Parker 1995).

The similarities between TRA and TPB are that both models stress that behavioural intention is an antecedent to the actual behaviour. However, by considering the control-related variable, TRA assumes that the behaviour being studied is under total volitional control of the performer (Madden *et al.* 1992). On the contrast, TPB when compared to TRA extends to more goal-directed actions. The main difference between TRA and TPB is that TPB has an additional exogenous variable, perceived behavioural control that has both a direct

and indirect effect on actual behaviour through intention. The indirect effect of perceived behavioural control on actual behaviour through intentions is based on the assumption that perceived behavioural control has motivational implications for behavioural intentions. The direct effect is assumed to reflect the actual control an individual has over performing the behaviour (Ajzen 1985).

In certain applications, it may be found that only the attitude toward the behaviour has a significant impact on behavioural intention, in others, both attitude toward the behaviour and perceived behavioural control may be significant and, in still others, the attitude toward the behaviour, subjective norms and perceived behavioural control may predict the behavioural intention (Ajzen 1985). The ability of perceived behavioural control and behavioural intention to predict actual behaviour varies across behaviours and situations. Furthermore, several researchers claimed that TPB has a better prediction power of behaviour than TRA because it is so contextually grounded (Madden *et al.* 1992; Cheung *et al.* 2003).

The decomposed TPB model proposed by Taylor & Todd (1995) is an alternative version of the TPB model with decomposed belief structures. In this model, attitudinal, normative and control beliefs are decomposed into multidimensional belief constructs. The decomposed TPB model produced mixed responses from researchers. The model is considered more complex by some researchers as it included a large number of factors (Hsu & Chiu 2004); on the other hand, it is often considered as more useful than other models and is supported by some researchers (Ok & Shon 2006; Jaruwachirathanakul & Fink 2005). Thus, TRA, TPB and DTPB models discussed so far, are focused to measure either behavioural intention to adopt or their adoption.

The Technology Acceptance Model (TAM) has been developed with an objective of providing a better understanding of the determinants of user behaviour across a broad range of end-user technologies, thus offering both parsimonious and theoretical justifications (Davis 1986). The main purpose of TAM is to provide a basis for identifying the impact of external factors on internal beliefs, attitudes and intentions. Perceived usefulness is defined as 'the degree to which using a particular system enhances the performance of the user' (Davis *et al.* 1989, p.89). On the other hand, perceived ease of use is defined as 'the degree to which using a particular system would be free of effort' (Davis *et al.* 1989, p.90). Attitude formation triggers the behavioural intention to use. Attitude is defined as a positive or negative consequence towards an intended behaviour.

Subjective and objective measures of TAM exhibited little similarity. The correlation between the subjective measures and intention is relatively high when compared to actual usage, thus questioning its validity (Szajna 1996; Straub *et al.* 2004). TAM focuses only on the determinants of intention such as perceived usefulness and perceived ease of use and does not provide valid reasons as how users' perceptions towards acceptance and usage are actually formed. There are a wider array of studies which used the TAM model for predicting the acceptance and use of information systems (Cheng *et al.* 2006). In the past decade, TAM has been established as a robust, powerful and parsimonious model for predicting behavioural intention to use. A significant body of research supports the role of perceived usefulness as a strong factor that influences intention behaviour over a period of time (Venkatesh & Davis 1996; Taylor & Todd 1995). Chau (1996) studied the impact of perceived near-term and perceived long-term usefulness and its subsequent influence on behavioural intention to use a particular technology. However, the role of perceived ease of use in the TAM has been reported differently in previous studies (Gefen & Straub 2003).

Perceived ease of use often deals with the motivation that is based on the assessment of ease of use and ease of learning (Luarn & Lin 2004). Vijayasarathy (2004) has reported that the two measures 'perceived usefulness' and 'perceived ease of use' significantly determine the adoption of online banking. Results obtained from statistical meta-analysis of TAM, as applied in eighty-eight published studies, indicated that TAM is a highly reliable, valid and robust predictive model that may be used in a variety of contexts (King & He 2006).

Later the TAM2 model was developed to include a number of determinants of the perceived usefulness construct. It is a theoretical extension of the technology acceptance model that explains perceived usefulness and usage intentions in terms of social influence processes such as subjective norms, voluntariness and image, and cognitive instrumental processes such as job relevance, output quality, result demonstrability and perceived ease of use (Igbaria *et al.* 1995). Longitudinal data were collected from four different organisations that spanned a range of industries, organisational contexts, functional areas and types of systems being used. The results indicated that all the above mentioned social influences and cognitive instrumental processes significantly influenced user acceptance of the systems.

An innovation is a new concept, object, technology or system presented to a target audience for adoption. The form of an innovation may vary depending upon the product, service, process or management system under consideration (Lorente *et al.* 1999), and the success of an innovation is based on the competitive advantage, profit maximisation, cost reduction and organisation's strategic position (Johansen *et al.* 1999). Diffusion of innovations (DOI) theory emerged from sociology and has been used since the 1960s. Since its emergence, the theory has been used widely to study a variety of innovations ranging from agricultural tools to organisation innovations (Venkatesh *et al.* 2003; Tornatzky & Klein 1982). DOI is the concept that explains how diffusion of innovations takes place in a social system (Rogers 1995). According to DOI theory, individuals develop certain perceptions towards an innovation and, based on these perceptions, an individual makes a decision whether to accept or reject an innovation (Agarwal

& Prasad 1997; Moore & Benbasat 1991). An innovation is more likely to be adopted based on the innovation characteristics of relative advantage, compatibility, complexity, trialability and observability which are critical for potential adopters' perceptions (Rogers 1995). Moore & Benbasat (1996) refined a set of constructs that represents characteristics of innovations that are present in innovation diffusion theory. These constructs of relative advantage, compatibility, complexity, trialability and observability, are widely used to predict technology acceptance (Plouffe *et al.* 2001; Karahanna *et al.* 1999; Agarwal & Prasad 1998; Moore & Benbasat 1991).

The DOI model has been refined to develop an instrument that can be used across a variety of innovation domains related to technology (Moore & Benbasat 1991). The developed model is intended to tap a variety of perceptions related to innovations. Two new constructs, 'image' and 'voluntariness', were added. Image is the degree to which an individual believes that the adoption of a technology enhances their prestige in the community and voluntariness is the degree to which an innovation adoption is perceived to be under the adopter's control (Moore & Benbasat 1991). The extended model received very little empirical attention with all of its constructs (Agarwal & Prasad 1998; Plouffe *et al.* 2001). Relative advantage, visibility, compatibility, trialability and result demonstrability were identified as significant predictors of intention to adopt an innovation (Agarwal & Prasad 1997). Whereas Tan & Teo (2000) found relative advantage, complexity, compatibility and trialability, Chin & Gopal (1995) found compatibility, Karahanna *et al.* (1999) found voluntariness, Taylor & Todd (1995) found relative advantage and ease of use and Chan & Lu (2004) identified image and result demonstrability as significant in affecting an individual's intention to adopt an innovation. From a meta-analysis of innovation adoption studies, relative advantage, compatibility and complexity were identified as being consistently related to adoption of innovations (Tornatzky & Klein 1982).

Later the Unified Theory of Acceptance and Use of Technology (UTAUT) has been framed by Venkatesh *et al.* (2003) that combines eight behavioural models of technology adoption such as the theory of reasoned action, the theory of planned behaviour, the technology acceptance model, the combines technology acceptance model and the theory of planned behaviour, the motivational model, the model of personal computer utilisation, diffusion of innovations and the social cognitive theory. Three constructs effort expectancy, performance expectancy and social influence were found to have direct effect on usage intentions and the fourth construct facilitating conditions was identified to have direct impact on actual usage. Thus, the behavioural intention models and diffusion of innovations model discussed so far, outline the relative importance and significance of the various constructs included in the aforementioned models and their predictive ability of the ICTs pre-adoption, adoption and post-adoption behaviour by the governments.

Government ICT Adoption in the USA

The internet boom that was started in the 1990s enabled private businesses to largely embrace internet technologies in their day today business offerings and the successful launch of websites to communicate their offerings to the public. However, there has been a significant lag in the adoption and diffusion of ICTs by the government sector in the USA since the identified internet boom (Dutton, 1996). It was only in 2001, that the e-government task force was officially launched in the USA. The US government mainly focused on offering its services to citizens, businesses and other governments. Therefore the e-government activities were appropriately classified as government to citizen (G2C), government to business (G2B) and government to government (G2G) services (Garson, 2006). Furthermore, the US government took a step forward and successfully introduced IEE (Internal Efficiency and Effectiveness) in order to successful best practices to government operations thereby enhancing the efficiency gains. All of the aforesaid has resulted in the launch of e-government act in 2002. Also, the US government developed USA.gov that acts as a central portal for its citizens in accessing various government-based services. From the extant literature it is evident that to date the US government's adoption of ICTs followed an evolutionary approach and has attracted significant number of users from the public and business sectors. However, the adoption of the ICTs by regional governments and diffusion of government service offerings to regional populations is lagging behind the urban-based adoption and diffusion practices.

Government ICT Adoption in Europe

The Commission of the European Communities in 2002 has implemented 'e-Europe Action Plan: An information Society for All' in order to enhance the information technology handling skills of the citizens. Later 'i2012 – A European Information Society for Growth and Employment' plan was also initiated to encourage the use of ICTs by European citizens. Literature emphasises on the fact that there are more e-government users in Denmark, Finland, Sweden and the United Kingdom than those in other European Union member states. Estonia is the first country to hold online local and parliamentary elections in 2005 (Jaeger, 2003). Similarly, Finland has shown significant advances in regional and local adoption of ICTs and increased e-participation of its citizens and businesses. Finland has a network of citizen centres to promote the usage of ICTs and enhance the delivery of the government services in the country (Kostopoulos, 2004). However the state of the Balkans with regards to the adoption and diffusion of ICTs by the governments, citizens and businesses is not that promising. In Bulgaria, e-government activity started in 2007

with the launch of a website that communicates to its citizens about the government services. In Romania, a transactional platform incorporating online public services was launched in 2003. The Ministry of Public Administration was set up in Slovenia in 2004 with an aim of incorporating all the government departments under a single leadership. The Government Commissions office introduced e-government program in Hungary in 2001. The Occupational Program for the Information Society was initiated in Greece in this decade in order to promote ICTs in the public sector.

Government ICT Adoption in Australia and New Zealand

On the basis of the economic growth and infrastructure of the countries, literature indicates that Australia and New Zealand have the developed status. In late 1990s and early 2000, the governments in both these countries were successful in the launch of websites that mainly showcase the activities undertaken and services offered by the government. As a result, a majority of the government services are currently offered via online and a majority of the citizens in both these countries were also receptive to this change. Several businesses also have embraced significant changes since then. For example, all the accounting businesses file their taxes through the tax department by being paperless and accessing ICTs (Heeks, 2001). The government departments at federal, state and regional levels are well connected through ICTs. The Australian government has taken a step forward by launching National Broadband Network (NBN) services in a few regional areas on a pilot basis. It is anticipated that NBN network would enhance the broadband connectivity, increase the speed levels and focuses on offering online health related advice by experts to the regional communities. It is also anticipated that NBN will be introduced in many regional areas towards the end of 2012 with an aim of potentially serving regional communities with better offerings.

Government ICT Adoption in the Middle East

There has been a significant increase in the usage of ICTs by citizens in the Middle East during the past six years (King *et al.* 1994). However, the adoption of ICTs by the government sector in the Middle East was largely hindered by political leadership, inertia towards the usage of ICTs, religious and cultural conditions, gender inequality and lack of sufficient e-participation by the citizens. Egypt invested heavily in offering basic services to its citizens through provision of e-information, e-consultation and e-decision making in the past decade. The main focus of Egypt has been the development of conducive infrastructure to promote government-based service offerings. Saudi Arabia has taken the initiative to enhance the public sector organisation's productivity by advertising e-government services through Yesser program and the Ministry of Information and Communications Technology (MICT). Jordon has developed a national plan that focuses on increasing the rates of adoption and diffusion of ICTs across key government departments and levels (Themistocleous & Irani, 2001).

Government ICT Adoption in Africa

The government adoption rates of ICTs in the African countries are far behind the strategies implemented by governments of other countries. Strategic plans to incorporate ICTs in the government sector were initiated by Senegal, Kenya, South Africa and Mozambique (Berman & Tettey, 2001). African Information Society Initiative (1996) focuses on the development and implementation of national policies and the promotion of ICTs in the government sector. The key barrier identified for the successful implementation, adoption and diffusion of ICTs by governments in African countries relate to citizen engagement. The bureaucratic nature of the government and the authoritarian type of leadership styles were identified to be critical factors that hinder the adoption of ICTs by government sectors in Africa (Shung & Seddon, 2000).

Government ICT Adoption in Asia

Within Asia, ICTs adoption by government could be delineated by the economic growth of the countries. Singapore, Malaysia, Hong Kong and South Korea were identified to be the newly industrialised countries and governments in these countries are also proactive in their approach with the adoption of ICTs. The Malaysian Administrative Modernisation and Management Planning Unit (MAMPU) has implemented a comprehensive plan that focuses on leadership styles and political conditions (Clay, 2001). The Republic of Korea Supreme Prosecutor's Office and the Seoul District Prosecutor's Office have introduced computer crime investigation departments. The success of these departments soon motivated the Korean government to establish teams in local prosecutor's office throughout the nation. Similarly the Korean government has introduced National Tax Service of the Republic of Korea that focused on the development of an integrated and computerised tax system. Other Asian countries followed an incremental approach with the adoption of ICTs by their governments. In 2001, Thai customs department introduced the computerised system, thus eliminating all of the possible manual errors whist processing imports and exports. The government of Philippines focused on posting of budget related information on its website through the Department of Budget and Management in order to increase the transparency levels of government-based practices. The government of Taipei focused on introducing 'one-window' service that integrates intranet and internet facilities and provides added advantages to its citizens.

The government of India has introduced the Computer-aided Administration of Registration Departments (CARA) in 1998, whereby 214 registration offices were completely computerised (Sharma & Gupta, 2003). Similarly, the government of India has computerised the national tax departments; provision of videoconferencing facilities between the state leaders and the general public, computerisation of the central, state and local banks etc., Pakistan has focused on tax department restructuring in order to decrease the contact time between the tax collectors and tax payers. The government of Sri Lanka through its Ministry of Finance and Planning designed and developed an integrated system to serve the needs of all of its departments and enhance its service provision to the citizens.

DRIVERS OF GOVERNMENT ICT ADOPTION

It is evident that successful adoption of ICTs by governments and further diffusion are possible only through better infrastructure, strong leadership, e-readiness (governments, businesses and citizens), organisation capacity, increased technical capacity, financial resources and appropriate decision making skills. Adoption of ICTs by government departments foster many benefits such as better communications, 24/7 service delivery, decreased delays, fast and convenient service provision (King *et al.* 1994). Governments and its stakeholders whether citizens, businesses or other government departments need to understand that the whole process will enhance the transparency, accountability and governance of the service offerings in the long-run.

Barriers to Government ICT Adoption

Significant barriers to government ICT adoption relates to the general reluctance or inertia of various stakeholders such as governments, businesses and citizens; poor infrastructure; poverty levels in a country; economic status of the country; bureaucratic style of leadership and cultural and/or social divide (King *et al.* 1994). Certain cultural and social factors such as gender inequality, religious conditions, age gap etc., may not allow governments in countries to adopt ICTs in a meaningful manner.

DISCUSSION

Although some of the technology acceptance theories relate to the individual's adoption of information and communication technologies, a majority of the constructs in these theories seem to have a profound influence on the adoption practices of ICT's in government organisations. Moreover, the adoption of ICT's by governments needs to be aligned with relevant user acceptance strategies. Therefore the theoretical models discussed and explained in this chapter provide important implications from the organisation as well as the individual perspectives. Critical variables of importance that influence adoption, diffusion and acceptance variations within organisations and citizens are of prime focus of this chapter. This chapter also touch bases on the readiness levels of the citizens as well the governments to embrace ICTs in their day to day practices. Similarly, towards the end of this chapter relevant drivers and barriers to government ICT adoption have been identified. Governments need to focus on strategies to up skill their ICT adoption practices and need to look at identified barriers as opportunities in promoting government ICT adoption for the benefit of its citizens as well as the society.

CONCLUSION

In comparison to the barriers and challenges associated with the adoption of ICTs by governments, the benefits expected are many more. The e-governments that are successful from developed countries could set a model for the developing and least developed countries. Leadership and decision-making styles need to be improved at the top most level in the hierarchy to offer better services to various stakeholders involved with the government. Citizens and businesses need to be offered sufficient education and training programmes that elicit the use of ICTs. It is also worthy to publish credible information about the benefits associated with the use of ICTs and disseminate this information to citizens, businesses and other governments through trustworthy sources. Governments need to take initiative to integrate various services and departments and levels to provide its service offerings in an efficient and effective manner. It is also equally important for governments to educate its stakeholders that investment in ICTs is for a worthy cause that enhances the transparency, accountability and governance issues between all of the parties involved.

REFERENCES

Affisco, J., & Soliman, K. (2006). E-government: A strategic operations management framework for service delivery. *Business Process Management Journal, 12,* 13–21. doi:10.1108/14637150610643724

Agarwal, R., & Prasad, J. (1997). The role of innovation characteristics and perceived voluntariness in the acceptance of information technologies. *Decision Sciences, 28*(93), 557–582. doi:10.1111/j.1540-5915.1997.tb01322.x

Agarwal, R., & Prasad, J. (1998). A conceptual and operational definition of personal innovativeness in the domain of information technology. *Information Systems Research, 9*(2), 204–215. doi:10.1287/isre.9.2.204

Ajzen, I. (1985). From intentions to actions: A theory of planned behaviour. In Kuhl, J., & Backmann, J. (Eds.), *Action control: From cognition to behaviour* (pp. 11–39). New York, NY: Springer-Verlag. doi:10.1007/978-3-642-69746-3_2

Ajzen, I. (1991). The theory of planned behaviour. *Organizational Behavior and Human Decision Processes, 50*(2), 179–221. doi:10.1016/0749-5978(91)90020-T

Ajzen, I., & Fishbein, M. (1980). *Understanding attitudes and predicting social behaviour.* Upper Saddle River, NJ: Prentice-Hall.

Ajzen, I., & Madden, T. J. (1986). Predication of goal-directed behaviour: Attitude, intentions, and perceived behavioural control. *Journal of Experimental Social Psychology, 22,* 453–474. doi:10.1016/0022-1031(86)90045-4

Bagozzi, R. P. (1982). A field investigation of causal relations among cognitions, affects, intentions, and behaviour. *JMR, Journal of Marketing Research, 19*(4), 562–584. doi:10.2307/3151727

Berman, B. J., & Tettey, W. J. (2001). African states, bureaucratic culture and computer fixes. *Public Administration and Development, 21*(1), 1–13. doi:10.1002/pad.166

Burne, J. (2002). *Better public services through e-government.* London, UK: The National Audit Office.

Chan, M., & Chung, W. (2002). A framework to develop an enterprise information portal for contract manufacturing. *International Journal of Production Economics, 75*(1), 113–126. doi:10.1016/S0925-5273(01)00185-2

Chan, S., & Lu, M. (2004). Understanding internet banking adoption and use behaviour: A Hong Kong perspective. *Journal of Global Information Management, 12*(3), 21–43. doi:10.4018/jgim.2004070102

Chau, P. Y. K. (1996). An empirical assessment of a modified technology acceptance model. *Journal of Management Information Systems, 13*(2), 185–204.

Cheng, T. C. E., David, Y. C. L., & Yeung, A. C. L. (2006). Adoption of internet banking: An empirical study in Hong Kong. *Decision Support Systems*, *42*(3), 1558–1572. doi:10.1016/j.dss.2006.01.002

Cheung, C. M. K., Zhu, L., Kwong, T., Chan, G. W. W., & Limayem, M. (2003). Online consumer behaviour: A review and agenda for future research. *16th Bled eCommerce Conference*, Bled, Slovenia, June 9-11, (pp. 194-218).

Chin, W. W., & Gopal, A. (1995). Adoption intention in GSS: Importance of beliefs. *Data Base Advances*, *26*, 42–64. doi:10.1145/217278.217285

Clay, G. W. (2001). E-government in the Asia-Pacific region. *Asian Journal of Political Science*, *9*(2), 1–26.

Davis, F. D. (1986). *A technology acceptance model for empirically testing new end-user information systems: Theory and results*. Doctoral dissertation, Sloan school of management, Massachusetts Institute of Technology.

Davis, F. D. (1989). Perceived usefulness, perceived ease of use, and user acceptance of information technology. *Management Information Systems Quarterly*, *13*(3), 319–339. doi:10.2307/249008

Davis, F. D., Bagozzi, R. P., & Warshaw, P. R. (1989). User acceptance of computer technology: A comparison of two theoretical models. *Management Science*, *35*(8), 982–1003. doi:10.1287/mnsc.35.8.982

Dutton, W. H. (1996). *Information and communication technologies: Visions and realities*. London, UK: Oxford University Press.

Fishbein, M., & Ajzen, I. (1975). *Belief, attitude, intention and behaviour: An introduction to theory and research*. Reading, MA: Addison-Wesley.

Garson, G. D. (2006). *Public information technology and e-government managing the virtual state*. Sudbury, MA: Jones & Bartlett Publishers.

Gefen, D., & Straub, D. W. (2003). The relative importance of perceived ease of use in IS adoption: A study of e-commerce adoption. *Journal of the Association for Information Systems*, *1*(8), 1–28.

Heeks, R. (2001). *Understanding e-governance for development*. Manchester, UK: Institute for Development Policy Management.

Hsu, M. H., & Chiu, C. M. (2004). Internet self-efficacy and electronic service acceptance. *Decision Support Systems*, *38*(3), 369–381. doi:10.1016/j.dss.2003.08.001

Igbaria, M., Guimaraes, T., & Davis, G. B. (1995). Testing the determinants of microcomputer usage via a structural equation model. *Journal of Management Information Systems*, *11*(4), 87–114.

Jaeger, P. T. (2003). The endless wire: E-government as global phenomenon. *Government Information Quarterly*, *20*(4), 323–331. doi:10.1016/j.giq.2003.08.003

Jaruchirathanakul, B., & Fink, D. (2005). Internet banking adoption strategies for a developing country: The case of Thailand. *Internet Research*, *15*(3), 295–311. doi:10.1108/10662240510602708

Johansen, J. A., Olalsen, J., & Olsen, B. (1999). Strategic use of information technology for increased innovation and performance. *Information Management & Computer Security*, *7*(1), 5–22. doi:10.1108/09685229910255133

Karahanna, E., Straub, D. W., & Chervany, N. L. (1999). Information technology adoption across time: A cross-sectional comparison of pre-adoption and post-adoption beliefs. *Management Information Systems Quarterly*, *23*(2), 183–213. doi:10.2307/249751

King, J. L., Gurbaxani, V., Kraemer, K. L., Mc-Farlan, F. W., Raman, K. S., & Yap, C. S. (1994). Institutional factors in information technology innovations. *Information Systems Research*, 5(21), 139–169. doi:10.1287/isre.5.2.139

King, W. R., & He, J. (2006). A meta-analysis of the technology acceptance model. *Information & Management*, 43, 740–755. doi:10.1016/j.im.2006.05.003

Kostopoulos, G. K. (2004). E-government in the Arabian Gulf: A vision toward reality. *Electronic Government: An International Journal*, 1(3), 293–299. doi:10.1504/EG.2004.005553

Kraft, P., Rise, J., Sutton, S., & Roysamb, E. (2005). Perceived difficulty in the theory of planned behaviour: Perceived behavioural control or affective attitude? *The British Journal of Social Psychology*, 44, 479–496. doi:10.1348/014466604X17533

Liao, S., Shao, Y. P., Wang, H., & Chen, A. (1999). The adoption of virtual banking: An empirical study. *International Journal of Information Management*, 19, 63–74. doi:10.1016/S0268-4012(98)00047-4

Lorente, A. R. M., Dewhust, F., & Dale, B. G. (1999). TQM and business innovation. *European Journal of Innovation Management*, 2(1), 12–19. doi:10.1108/14601069910248847

Luarn, P., & Lin, L. H. (2004). Towards an understanding of the behavioural intention to use mobile banking. *Computers in Human Behavior*, 21(6), 1–19.

Madden, T. J., Ellen, P. S., & Ajzen, I. (1992). A comparison of the theory of planned behaviour and the theory of reasoned action. *Personality and Social Psychology Bulletin*, 18(1), 3–9. doi:10.1177/0146167292181001

Manstead, A. S. R., & Parker, D. (1995). Evaluating and extending the theory of planned behaviour. In Stroebe, W., & Hewstone, M. (Eds.), *European review of social psychology* (pp. 69–96). Chichester, UK: John Wiley & Sons. doi:10.1080/14792779443000012

Mathieson, K. (1991). Predicting user intentions: Comparing the technology acceptance model with the theory of planned behaviour. *Information Systems Research*, 2(3), 173–191. doi:10.1287/isre.2.3.173

Moore, G. C., & Benbasat, I. (1991). Development of an instrument to measure the perceptions of adopting an information technology. *Information Systems Research*, 2(3), 173–191. doi:10.1287/isre.2.3.192

Moore, G. C., & Benbasat, I. (1996). Integrating diffusion of innovations and theory of reasoned action models to predict utilisation of information technology by end-users. In Kautz, K., & Prier-Hege, J. (Eds.), *Diffusion and adoption of information technology* (pp. 132–146). London, UK: Chapman & Hall.

Nelou, V. (2004). E-government for developing countries: Opportunities and challenges. *The Electronic Journal of Information Systems in Developing Countries*, 18(1), 1–24.

Ok, S. J., & Shon, J. H. (2006). The determinants of internet banking usage behaviour in Korea: A comparison of two theoretical models. [[th] December, Adelaide.]. *CollECTor*, 06, 9.

Plouffe, C. R., Hulland, J. S., & Vandenbosch, M. (2001). Richness versus parsimony in modelling technology adoption decisions - Understanding merchant adoption of a smart card-based payment system. *Information Systems Research*, 12(2), 208–222. doi:10.1287/isre.12.2.208.9697

Reid, M., & Levy, Y. (2008). Integrating trust and computer self-efficacy with technology acceptance model: An empirical assessment of customers' acceptance of banking information systems (BIS) in Jamaica. *Journal of Internet Banking and Commerce, 12*(3). Retrieved 15th April, 2009, from www.arraydev.com/commerce/jibc

Rogers, E. M. (1995). *Diffusion of innovations* (4th ed.). New York, NY: The Free Press.

Sharma, S., & Gupta, J. (2003). Building blocks of an e-government: A framework. *Journal of Electronic Commerce in Organizations, 1*(4), 1–15. doi:10.4018/jeco.2003100103

Shung, S., & Seddon, P. (2000). *A comprehensive framework for classifying the benefits of ERP systems.* Paper presented at American Conference on Information Systems, Dallas, Texas.

Straub, D., Boudreau, M. C., & Gefen, D. (2004). Validation guidelines for IS positivist research. *Communications of AIS, 13*, 380–427.

Swanson, E. B. (1982). Measuring user attitudes in MIS research: A review. *Omega, 10*(2), 157–165. doi:10.1016/0305-0483(82)90050-0

Szajna, B. (1996). Empirical evaluation of the revised technology acceptance model. *Management Science, 42*(1), 85–92. doi:10.1287/mnsc.42.1.85

Szmigin, I., & Foxall, G. (1998). Three forms of innovation resistance: The case of retail payment methods. *Technovation, 18*(6/7), 459–468. doi:10.1016/S0166-4972(98)00030-3

Tan, M., & Teo, T. S. H. (2000). Factors influencing the adoption of internet banking. *Journal of the Association for Information Systems, 1*, 1–42.

Taylor, S., & Todd, P. A. (1995). Understanding information technology usage: A test of competing models. *Information Systems Research, 6*(2), 144–176. doi:10.1287/isre.6.2.144

Themistocleous, M., & Irani, Z. (2001). Benchmarking the benefits and barriers of application integration. *Benchmarking: An International Journal, 8*(4), 317–331. doi:10.1108/14635770110403828

Tornatzky, L. G., & Klein, K. J. (1982). Innovation characteristics and innovation adoption implementation: A meta-analysis of findings. *IEEE Transactions on Engineering Management, 29*(1), 28–45.

Venkatesh, V., & Davis, F. D. (1996). A model of the antecedents of perceived ease of use: Development and test. *Decision Sciences, 27*, 451–481. doi:10.1111/j.1540-5915.1996.tb01822.x

Venkatesh, V., Morris, M. G., & Ackerman, P. L. (2000). A longitudinal field investigation of gender differences in individual technology adoption decision making processes. *Organizational Behavior and Human Decision Processes, 83*, 33–60. doi:10.1006/obhd.2000.2896

Venkatesh, V., Morris, M. G., Davis, G. B., & Davis, F. D. (2003). User acceptance of information technology: Toward a unified view. *Management Information Systems Quarterly, 27*(3), 425–478.

Vijayasarathy, L. R. (2004). Predicting consumer intentions to use on-line shopping: The case for an augmented technology acceptance model. *Information & Management, 41*(6), 747–762. doi:10.1016/j.im.2003.08.011

West, D. M. (2005). *Digital government: Technology and public sector performance.* Princeton, NJ: Princeton University Press.

Winch, G., & Joyce, P. (2006). Exploring the dynamics of building, and losing, consumer trust in B2C eBusiness. *International Journal of Retail and Distribution Management, 34*(7), 541–555. doi:10.1108/09590550610673617

KEY TERMS AND DEFINITIONS

Adoption: The rate of embracing information and communication technologies in government-based organisations.

Barriers: Obstacles that hinder the eGovernment adoption practices.

Drivers: Factors that promote the eGovernment adoption by the public sector.

eGovernment: The use of information and communication technologies by the public sector organisations to improve their service delivery and service efficiency.

eReadiness: Factor that determines the usage rate of eGovernment by citizens, businesses and other governments.

ICTs: Focuses on the usage of the Internet, wireless communications, social media, broadband etc., to facilitate convenience to the users in accessing various services.

Trends: A general tendency of inclination towards the adoption of eGovernment practices.

Chapter 2
Municipal Social Media Policy:
A Best Practice Model

Melissa Foster
Northern Illinois University, USA

Yu-Che Chen
Northern Illinois University, USA

ABSTRACT

As social media has become integrated into the public's everyday lives, local governments have started to take advantage of the power of social media as another governance tool to both inform and involve the public in local government. This new tool also introduces a new responsibility for government to monitor and analyze the actions taken on municipal social media sites. For this to be achieved, municipalities must implement a social media policy that addresses the abundant concerns inherent when engaging in social media use. This research indicates the areas that local governments must address in social media policy and offers a best practice approach to completing the task of policy development.

INTRODUCTION

Communications have changed drastically throughout the world as a result of the increased use of social media for social and business interactions. Individuals, businesses, and governments are now able to communicate with each other in a way that was not possible before. Government relations with residents, visitors, and businesses must evolve with the changing digital and social landscape. Social media, as well as other Web 2.0 applications, is an example of a human-centered system design that benefits local governments and their citizens. Social media can be utilized as an e-governance tool that allows for increased dialogue with citizens in an environment where citizens drive discussion. As Web 2.0 application

DOI: 10.4018/978-1-4666-3640-8.ch002

use becomes more widespread, governments can utilize these applications to increase citizen inclusion in the governance process.

Social media is a relatively new phenomenon. Although utilization of "community development" tools began in 1997, social media as we know it today did not become a mainstream phenomenon until 2002-2003 with the launch of Friendster and MySpace, both of which are no longer mainly utilized for social networking but for gaming and music respectively (Boyd & Ellison, 2008, pp. 214-216). Facebook, the most commonly used social media website today, was founded in 2004 and has grown to include millions of users including individuals, businesses, non-profits, and governments. According to the site's administrators, Facebook's purpose as a social media outlet is to "give people the power to share and make the world more open and connected" (Facebook, 2012). Although there is no clear, definitive definition, social media generally refers to an online application that centers around social interaction online (Bryer & Zavatarro, 2011). This term encompasses a wide variety of applications, including social networking sites (such as Facebook, MySpace or LinkedIn), blogs/mini-blogs (such as Twitter), media sharing sites (such as YouTube or Flickr), and a myriad of other applications (Boyd & Ellison, 2008). Social media provides a medium where people and/or entities can easily provide information to a large group of people with the use of few resources.

As social media has become integrated further into the public's everyday lives, local governments have started to take advantage of the power of social media as another governance tool to both inform and involve the public in local government. Governments can use these tools to inform citizens with up-to-date information and allow citizens to comment on issues and events occurring in their communities. Because social media has only been utilized for the past decade, there is little literature from which to study social media as a tool for local governments.

Although social media use is an important e-governance tool for all levels of government, local government social media use should be studied. Local governments are vast in number, which allows for a larger, more diverse sample from which to examine social media use. Local governments also have a closer relationship with citizens as they directly impact citizens on a daily basis through municipal service delivery. Residents can easily communicate with local government officials, whereas it may be more difficult to communicate with state or federal government agencies. Local governments also maintain other avenues for citizen participation, including mandatory public meetings, which create other avenues through which citizens may participate in government. Through the use of social media, governments can make the governance process more inclusive and transparent, which may stimulate citizen trust in government (Tolbert & Mossberger, 2006).

As discussed above, there is a varying array of social media options available to the public. Local governments most commonly use Facebook, Twitter and YouTube as their main social media tools (ICMA, 2011). Local governments can also take advantage of sites specifically designed for local governments, such as patch.com. Because communities are all different, each local government can use the social media outlet or tool that most directly meets its needs. Depending on location, size, demographics, community involvement, and any number of other factors, different organizations will use different types of social media outlets. To illustrate this point, the city of Chicago utilizes multiple Facebook, Flickr, Twitter, Tumblr, and YouTube sites and also utilizes foursquare, an application that lets citizens/visitors "check in" to various locations in the city. Chicago clearly utilizes social media to inform citizens in a variety of ways about all different issues, including topics and departments such as the city bicycle program, the public library, police department, transportation, and special events (City of Chicago, 2011). Chicago is a large city that must address these and other complex and continually developing issues that citizens may need to be informed about, and thus, it reasons that the city should employ a variety of social media outlets. Compare this to small municipalities in Illinois who may utilize one Facebook page for informational purposes. Social media can serve the purpose that each community needs.

As local governments begin to take advantage of evolving social media, it is necessary to form a policy dictating the government's use of social media because of the many issues that could develop for the government as a result. This may include the need to adhere to laws, address content issues, and establish the purpose of social media use. State and federal laws, including sunshine laws and laws that protect employees, may affect how governments use social media. Local governments must be cognizant of the content included on their social media sites, including information published by both the government itself and the comments of citizens/users of the site as well. Social media sites can be used for different local government purposes. A municipality can use social media technology for informational purposes or to increase citizen participation. There could also be issues depending on who is using the site on behalf of the government, including the administrative employees or the elected officials.

It is the intent of this chapter to create a best practice social media policy for municipalities. Although each municipality may use social media in different ways, the issues that are inherent in using social media will be the same for every municipality. Municipalities using social media must create and implement a policy that serves the purpose for which the municipality intends and that also adheres to state and federal law, privacy concerns, and acceptable conduct in the context of local government.

SOCIAL MEDIA AND ELECTRONIC GOVERNANCE

Challenges and Opportunities of Social Media for E-Governance

Over the last twenty years, information technology has been gaining more and more importance for local governments. As technology becomes more ingrained into public life, local governments have begun to take advantage of it as a way to communicate more readily with the public (Streib & Willoughby, 2005). Governments have taken advantage of the internet as a tool to improve service through better policy and services, increased citizen participation, and more informed decision-making (Dawes, 2008). According to a 2011 ICMA survey, 97% of municipalities with a population of 10,000 or more have a municipal website (ICMA, 2011).

Many local governments have scarce resources causing a lack of funding for more innovative information technology or e-governance tools (Streib & Willoughby, 2005). Lack of financial funding has been cited as the most common and most difficult barrier to e-government utilization in municipalities across the United States (ICMA, 2011). Many social media tools are free and easy to use, which affords these communities an opportunity to reach their citizens in a new way without having to spend limited funds on expensive software or IT personnel.

However, many concerns have followed the increased use of technology in local government. Issues with privacy and security must be addressed when utilizing these new technologies (Dawes, 2008). Local governments must strive to protect the individual privacy of the public using their online services, and all private information must be kept secure at all times (Streib & Willoughby, 2005). All personal information and user information must be kept secure (Bertot et al., 2010). Because social media sites are third-party providers, governments may not have complete control over the information included on and transacted on the social media sites that they utilize. As citizens "friend" or join a government group on a social media site, the government is then able to access those citizens' personal social media pages, which could potentially contain personal information (Bryer & Zavattaro, 2011).

Citizens benefit from the increased use of e-governance tools. Citizens are afforded another avenue to receive information about their government, as well as become more engaged by giving feedback to local government through the website or social media channels (Streib & Willoughby, 2005). Social media can be used as a tool for citizen engagement by involving the public in "participatory dialog" with the government (Bertot et al., 2010, p. 2). Higher levels of citizen participation beyond simply informing or consulting citizens (Arnstein, 1969) can be reached through social media (Bertot et al., 2010). Social media can be used to create an environment where governments partner with the public to improve services. Government can utilize social media to be more open and transparent with citizens and increase the level of accountability to the public (Bertot et al., 2010). However, government must employ social media in a way that allows and invites participation, or citizens will not want to participate (Bryer & Zavatarro, 2011). Although it is possible for government to use social media as a citizen engagement tool, most governments have only been using it for one-way communication to citizens (Hand & Ching, 2011). This is demonstrated through the results of the 2011 ICMA e-government survey in which only a small minority of municipalities indicated that they use e-governance tools for mainly interactive purposes (ICMA, 2011).

One problem inherent in the use of social media is that not all citizens have access to the internet and therefore do not have access to social media. According to studies by both Dawes (2008) and Leighninger (2011), any information only posted on a social media site or website will not be accessible to all citizens. According to the PEW Research Center, 78 percent of the adult population are internet users (PEW, 2011), and 82 percent of these users look for information on government websites (Smith, 2010). According to the PEW Research Center Internet and American Life project, 65% of adult internet users utilize social networking sites (PEW, 2011). According to another study, these users not only look for information but also want to share their opinions on government (Leighninger, 2011). Multiple studies have indicated that younger people are more likely to use social networking sites, and minorities and underserved populations have expressed the importance of governments' use of social media to inform citizens (Smith, 2010). This indicates that social media may be a way to reach historically underserved populations of the community

The integration of social media and the internet into citizens' everyday lives means that people are more knowledgeable and more aware of issues than previous generations. With the widespread use of the internet, there is a greater ability to find information and connect with other individuals in a community who care about the same issues (Leighninger, 2011). Currently, most local governments are using social media as another outlet for information dissemination instead of as an outlet for citizens to directly engage with the government (ICMA, 2011). According to the 2011 ICMA survey, a majority of municipalities using social media did so as a one-way communication source from the municipality to citizens with only a small percentage indicating that the goal is two-way communication with citizens (ICMA, 2011). This has given citizens the ability to discuss government with other citizens and groups but not the government. Local governments must also contend with incorrect information, disparaging sites, and fake municipal sites that may affect their ability to communicate the government's real message to citizens.

The future of local government must include enhanced e-governance through the use of interactive social media that allows for two-way communication between the government and citizens. Those that do not will be labeled as "undemocratic" because they are ignoring the opportunity of fully engaging citizens. In order to reach all different types of citizens, municipalities must use a variety of avenues to reach citizens, including different types of social media (Leighninger, 2011).

A Policy Framework for Social Media Use: Issues and Concerns

For a government to successfully embrace and utilize e-governance tools, it must create a policy framework to legitimize the e-governance process and protect both the government and citizens in the ever changing digital environment (Dawes, 2008). Before a government entity decides to engage in social media, it must first create and implement a social media policy to mitigate the potential problems that are accompanied by the use of social media (Hrdinova, Helbig & Stollar Peters, 2010). According to the 2011 ICMA survey, 67% of municipalities with a population of over 10,000 use social media, and 63% of those have a policy or procedure regarding social media use (ICMA, 2011).

A report by the Center for Technology in Government at SUNY University at Albany establishes eight elements that should be present in any government social media policy. This analysis was based on a collection of government social media policies and interviews with government professionals who have used or were planning to use social media. The elements suggested in this study for inclusion in government social media policy include: Employee Access, Account Management, Acceptable Use, Employee Conduct, Content, Security, Legal Issues, and Citizen Conduct (Hrdinova, Helbig, & Stollar Peters, 2010, p.2). This study was conducted using samples from different levels of government, including federal, state, county, and municipal policies, and the results are general guidelines for all government; therefore, some of the findings may not apply specifically to municipal policy.

Each government must decide which employees should have access to social media sites and who should have the capacity to manage the municipal site. Many municipalities already have established acceptable use policies that may be included in the social media policy to outline employee restrictions on the use of social media. An employee code of conduct must be established in regards to social media use to ensure that all behavior is professional and respectful. Acceptable content must also be addressed by each government depending on the purpose of the social media use. Agencies may prefer to let all employees post or may task one employee with posting official government messages. The accuracy of information that is perceived as official government communication is a concern when engaging in social media use (Hrdinova, Helbig, & Stollar Peters, 2010).

As has been discussed previously, security is a major issue that must be addressed by municipal social media policy. Government information, as well as citizen/user information, must be kept secure. Governments must be vigilant in their assessment of information posted on social media sites to ensure that private citizen information is not inadvertently posted to a public website. Government must also protect against viruses that may be exchanged through links shared on the government's site or file sharing that may take place over social media (Hrdinova, Helbig, & Stollar Peters, 2010).

Although governments must contend with these types of logistical problems when managing a site, they must also be aware of laws and regulations that can impact the use of social media in the municipal setting. There are many state and federal laws that apply to different aspects of municipal social media use. Aspects of the Illinois Open Meetings Act (5 ILCS 120) and Freedom of Information Act (5 ILCS 140) both apply to municipal social media use. The Illinois Open Meetings Act is intended to ensure that "public business is conducted in public view" by prohibiting private government considerations on public matters (Madigan, 2004, p.3). The Illinois Open Meetings Act applies to "any gathering whether in person or by video or audio conference, telephone call, electronic means (such as electronic mail, electronic chat, and instant messaging) or other means of contemporaneous interactive communication, of a majority of a quorum of the members of a public body held for the purpose of discussing public business" (Madigan, Public Access Counselor Annual Report 2010, 2011, p.19). Therefore, the Illinois Open Meetings Act would apply to any municipal social media site where a majority of the quorum of the members of the government's board or governing body is present. The government must also give public notice of the date, time, and location of a public meeting. In terms of social media, the public would have to be notified that a meeting was going to be held on the social media site at least 48 hours prior to the conversation occurring (Madigan, Public Access Counselor Annual Report 2010, 2011).

The Freedom of Information Act (FOIA) gives citizens the right to access government records and documents. The law enables citizens (including groups, associations, corporations, firms or organizations) to ask a public entity for copies of specific records or subjects, and the public entity must provide the records unless they are exempt for some reason. According to the State of Illinois, public records include "all records, reports, forms, writings, letters, memoranda, books, papers, maps, photographs, microfilms, cards, tapes, recordings, electronic data, processing records, electronic communications, recorded information and all other documentary materials pertaining to the transaction of public business" (State of Illinois, 2010). Any type of electronic communication is subject to FOIA requests, including social networking communications and pictures, recordings

or postings made on social media sites. According to a Public Access Opinion issued by the Office of the Attorney General of the State of Illinois, private correspondence from government elected officials using electronic communications (email and social media) that involve public matters are subject to FOIA law requirements as well (Madigan, Public Access Opinion No. 11-006, 2011). Because all interaction occurs online at different times and on different areas of sites, it may be difficult for governments to create adequate paper records of all business conducted on social media sites (Bryer & Zavatarro, 2011).

Federal laws may also constrain governments in relation to their use of social media. The First Amendment protects free speech, but there are limitations to this protection. As decided in Chaplinksy v. New Hampshire, "fighting" words are not protected under the First Amendment. These are words that have a "direct tendency to cause acts of violence" and include offensive words used towards another person in a public place (Chaplinksy v. New Hampshire, 1942). Sexually obscene speech is also not under the protection of the First Amendment. Hate speech cannot be prohibited by government on the basis of the hate content. In R.A.V v. St. Paul, the court decided that limiting only hate speech is invalid because this is prohibiting speech based solely on content, and government cannot limit speech on the basis of the speech being an unpopular sentiment. The government can prohibit "fighting" words but cannot prohibit only those based on race, religion, or gender (R.A.V. v. St. Paul, 1992).

Cases concerning the First Amendment also define a designated or limited public forum. A limited public forum is one that is set aside by government for communicative purposes. In defining a limited public forum, a government can adopt reasonable restrictions on who can use the forum, but content based restrictions in a limited public forum are subject to strict scrutiny under federal law (Southeastern Promotions, Ltd. v. Conrad et al., 1974). Governments must take all of these aspects of social media use into consideration when determining if social media should be utilized.

Because the use of social media is so new for governments, many simply use their internet use policy as a social media policy, but these policies do not address the specific concerns of social media use. Other governments simply do not have any policy created for their use of social media. Those that do have a policy may not adhere to it, which could potentially cause problems for the organization. The following study will address the practical application of social media policy in relation to the issues currently facing municipal use of social media through a case study of Illinois municipalities. The research contained in this chapter documents a study of a sample of Illinois municipalities that utilize social media, their policies, and the adherence to those policies on the social media sites utilized by each municipality. Using a combination of the previous literature and analysis of the problems identified from the sample municipalities, a "best practice" social media policy for Illinois municipalities will be developed with emphasis on implementation of the policy. This best practice model can be applied to any municipality by taking state law differences into consideration.

RESEARCH DESIGN AND DATA COLLECTION

The state of Illinois has the largest number of local governments with taxing authority in the United States (Simpson & Moll, 1994) with 1367 municipalities and is also the 5th most populous state, according to the 2010 United States Census (State of Illinois, 2012). Illinois also includes a variety of types of municipalities, including cities and metropolitan areas, suburbs, and rural communities. This large variety of municipalities allows for an understanding of social media use across the spectrum of different types and sizes of municipalities. By focusing on a single state, the entire sample can be judged by the same criteria, including state law that may affect municipal social media use.

To complete this research, a sample of sixty (60) Illinois municipalities was randomly selected to be included in the study. The results were analyzed to ensure the selected municipalities were representative of all state of Illinois municipalities. This allows the findings of the study to be generalized for the entire state of Illinois as the sampling represents all types of municipalities from all areas of the state. The sampling was divided into categories based on population and location. The random sampling was devised from the State of Illinois 2000 census population comparison, which included 1,313 municipalities (State of Illinois, n.d.). From this list, it was determined that the following number of municipalities per population count was needed to create a representative sample.

According to the State of Illinois, 624 Illinois municipalities have a population less than 1,000. For the sample size of sixty (60) municipalities included in this study, twenty-eight (28) municipalities were included to represent this portion of the municipalities. Sixteen (16) municipalities were selected to represent the 368 municipalities with a population of 1,000-4,999. Five (5) municipalities were selected to represent the 116 Illinois municipalities with a population of 5,000-9,999. Four (4) municipalities were selected to represent the ninety (90) municipalities with a population of 10,000-19,999. Four (4) municipalities represent the eighty-nine (89) municipalities with a population of 20,000-49,999. Two (2) municipalities were chosen to represent the nineteen (19) municipalities with a population of 50,000-99,999. One (1) municipality was chosen to represent the seven (7) municipalities with a population of 100,000 or more. See Table 1 for a list of municipalities included in this study (State of Illinois, n.d.).

The sample was also assessed to verify that the sample was representative of the entire state instead of one centralized location. To complete this, the state of Illinois was portioned into six (6) sections, titled Northeast, Northwest, Central East, Central West, Southeast, and Southwest. Each population category sample included municipalities from each quadrant, except those categories that have less than six (6) samples.

After verifying that the sample was representative of the entire population of Illinois municipalities, the selected municipalities were analyzed in terms of their social media presence. This process was carried out in a series of phases described below:

Phase 1: Determine municipalities with a social media presence.

This phase was completed by first determining if each municipality had its own municipal website and if that website allowed commenting from site visitors/citizens. If the municipality had a link to any social media site, this was noted. After determining the status of each municipal website, various social media websites were searched to determine if each municipality had a page(s). If a link was provided on the municipality's website to a social media site, this was used to locate the municipality's social media site. Included in the social media site search were Facebook, Twitter, and YouTube. If a municipality had a link to another social media site not included on this list, that site's municipal page was also researched and included in the study.

Phase 2: Determine municipalities with a social media policy.

After determining the municipalities with a social media presence, a search was then conducted of the municipalities' website/social media sites to determine if the municipality had instituted a social media policy.

Phase 3: Analyze municipal social media policies.

Table 1. Illinois Municipality List

Municipality	Population	Region
Golconda city	726	S2
Spaulding village	559	C1
Tonica village	685	N2
Equality village	721	S2
Parkersburg village	234	S2
Dahlgren village	514	S2
Garrett village	198	C2
De Witt village	188	C2
El Dara village	89	C1
New Minden village	204	S1
Walnut Hill village	109	S1
Royal village	279	C2
Sauget village	249	S1
New Canton town	417	C1
Brocton village	322	C2
Millington village	458	N2
Gladstone village	284	N1
Ursa village	595	C1
Magnolia village	279	N1
Rose Hill village	79	C2
Capron village	961	N2
Spillertown village	220	S1
Prairie Grove village	960	N2
Columbus village	112	C1
Cordova village	633	N1
Keithsburg city	714	N1
Ransom village	409	N2
Calhoun village	222	S2
Atwood village	1,290	C2
Macon city	1,213	C2
Bunker Hill city	1,801	C1
Windsor city	1,125	N1
Albion city	1,933	S2
Savanna city	3,542	N1
Momence city	3,171	N2
McLeansboro city	2,945	S2
Harristown village	1,338	C2
Palos Park village	4,689	N2
Fox River Grove village	4,862	N2

continued on following page

Table 1. Continued

Municipality	Population	Region
Atkinson town	1,001	N1
Carthage city	2,725	C1
Okawville village	1,355	S1
North Pekin village	1,574	C1
Marissa village	2,141	S1
Fox Lake village	9,178	N2
Spring Valley city	5,398	N2
Metropolis city	6,482	S2
Hoopeston city	5,965	C2
Monmouth city	9,841	C1
Grayslake village	18,506	N2
Macomb city	18,558	C1
Cahokia village	16,391	S1
Sterling city	15,451	N1
Downers Grove village	48,724	N2
Collinsville city	24,707	S1
Rock Island city	39,684	N2
Charleston city	21,039	C2
Oak Park village	52,524	N2
Decatur city	81,860	C2
Aurora city	142,990	N2

Key:
N1: Northwest Illinois
N2: Northeast Illinois
C1: West Central Illinois
C2: East Central Illinois
S1: Southwest Illinois
S2: Southeast Illinois

Each of the policies collected in Phase 2 was analyzed by the same factors. To analyze the collected social media policies in a consistent manner, a framework was created from the main themes discussed in the literature review, focusing heavily on the features discussed in the Center for Technology in Government Social Media study (Hrdinova, Helbig & Stollar Peters, 2010). Each social media policy was analyzed using the framework below:

- Purpose of Social Media

- Acceptable Use/Employee Practices
- Content Restrictions
- Legal Issues
- Management of Social Media
- Consistency of Policy Across Social Media
- Security of Social Media (Hrdinova, Helbig & Stollar Peters, 2010, p.2)

Phase 4: Collect observations of conduct on social media sites.

After completing the analysis of the social media policies collected, all municipal social media sites were observed to determine the conduct of both administrators and users of the site.

Phase 5: Comparison of policy document and website observations.

After compiling observations from municipal social media websites, the observations from each site were compared to the organizations' social media policies, as well as the framework outlined in Phase 3. If the organization did not have a policy, the observations were compared to the framework outlined in Phase 3.

Phase 6: Creation of Social Media Policy Best Practice.

The information collected in the previous steps, along with the literature review, was used to form the basis of the social media policy best practices outlined in the Model Social Media Policy section of this chapter.

SOCIAL MEDIA PRESENCE AND POLICY

Sixty (60) Illinois municipalities were surveyed regarding their use of social media. Of the sixty municipalities included in Table 1, twenty-nine (29) (48%) of the municipalities had a municipal website, but none of these websites allowed for commenting by citizens/visitors. Of the sixty municipalities, eleven (11), or 18%, of the municipalities used some form of social media, as outlined below.

Facebook

- Eleven (11) municipalities (18%) have a municipal Facebook page, including Aurora, Decatur, Oak Park, Rock Island, Collinsville, Downers Grove, Sterling, Grayslake, Monmouth, the Hoopeston Fire Department, and Atwood village.
- All the municipalities that use Facebook had a population of over 5,000 except one, the village of Atwood, which will be discussed in greater depth later.
- Facebook users are spread throughout the entire state with six (6) northern municipalities, four (4) central municipalities, and one (1) southern municipality.

YouTube

- Four (4) municipalities (7%) have a YouTube page, including Decatur, Oak Park, Rock Island, and Downers Grove.
- Municipalities that use YouTube had a population of over 15,000.
- All YouTube users also have a Facebook page.
- Three of the four (75%) YouTube users are northern municipalities with the remaining being a central municipality.

Twitter

- Seven (7) municipalities (12%) use a Twitter fccd, including Aurora, Dccatur, Oak Park, Rock Island Fire Department, Collinsville, Downers Grove, and Grayslake.
- The municipalities that use Twitter all had a population of over 35,000.

- All Twitter users also have a Facebook page.
- Five (5) Twitter users are northern municipalities, one (1) is a central municipality, and one (1) is a southern municipality.

Of the eleven municipalities that use social media, only six (55%) had any form of social media policy outlined on their website or social media site. Each of these six policies was posted on the municipality's official Facebook page, except one, which was on the YouTube site. These policies fall into two categories: a basic content statement and a comprehensive policy. The features of each policy are outlined below ranging from the most minimal policy to the most comprehensive.

Village of Grayslake

The Village of Grayslake includes a short statement on the municipal Facebook page in the "Info" section that indicates the Village will remove any comments that are deemed inappropriate (Village of Grayslake, 2012).

City of Sterling

The City of Sterling includes "Public Commenting Guidelines" in the "Notes" section of the municipal Facebook page. This policy includes a purpose and commenting guidelines. The purpose of the city's Facebook page is to share information with citizens. The city attempts to respond to all user comments within 24-72 hours on normal business days. The city limits user commenting on the site and does not allow any offensive language, personal attacks, discriminatory speech, spam, off-topic comments, illegal activity, political speech, or copyright infringement (City of Sterling, 2010). The city also adds that comments posted on the site "do not reflect the opinions or positions of the city of Sterling or its officials and employees" and any questions about the policy should be directed to the city manager (City of Sterling, 2010).

Village of Downers Grove

The Village includes its social network policy on the municipal Facebook page under "General Information" on the "Info" section of the site. The policy outlines only content restrictions. The Village asserts that it will delete postings that contain obscene language, personal attacks, discriminatory remarks of any kind, spam comments, off-topic comments, illegal activity and copyright infringements. The city also includes a note on the policy that comments expressed on the site "do not reflect the opinions and position of the Village of Downers Grove, its officers and employees" and that all comments are considered public record and are therefore subject to the Illinois Freedom of Information Act (Village of Downers Grove, 2012).

City of Collinsville

The City of Collinsville has its Facebook Terms of Use posted in the "Notes" section of the city's official Facebook page. The policy includes the purpose of the Facebook page and posting restrictions. According to the city, the purpose of social media use is to both provide citizens with information and to engage in dialogue with citizens and provide a platform for them to interact with each other (City of Collinsville, 2010).

The policy also outlines the criteria for posting on the city's Facebook page. Citizens are encouraged to post about issues that affect residents, including diverse viewpoints; however, certain activity is not allowed on the site, including adult content, hate speech, alcohol/firearms/tobacco content, threats of violence, discriminatory speech, illegal activity, sexual content, false advertising, obscene language, cruelty to animals, gambling, political campaign speech, solicitations, or advertisements (City of Collinsville, 2010).

City of Aurora

The City of Aurora has a social media policy that is posted in the "Info" section of its official Facebook page. The policy is not included on the city's Twitter page. The policy includes a purpose and definitions, which includes content restrictions. The City of Aurora describes the purpose of social media as a way to communicate with residents, the media, and other stakeholders. This communication should make communications between the city and stakeholders "a rich and robust experience" that increases transparency (City of Aurora, 2012). The city plans to use the site "to publish news releases, highlight events, ordinances and positive media coverage as well as other information that supports the goals and mission of the City" (City of Aurora, 2012). The city establishes that its social media sites are "limited public forums" that are not to be used for unlimited discussion on any topic but are to be used for topics limited by the city (City of Aurora, 2012). The city also established content rules that must be followed by all users of the site. The city reserves the right to remove content that it deems unsuitable because it is either not related to the subjects on the site, or it violates one of the commenting restrictions outlined in the policy. The types of restricted content include: slanderous and offensive language, discriminatory speech, sexual content, solicitations, personal information, political activity, illegal activity, information that may cause a safety or security concern, and content that "violates a legal ownership interest of any other party" (City of Aurora, 2012). Aurora Police Department also operates its own Facebook page. The only policy included on this page is the following: "We reserve the right to remove any comments that are vulgar, disrespectful or disparaging" (City of Aurora Police Department, 2012).

City of Rock Island

The City of Rock Island has its e-government policy posted on its official Facebook page in the "Notes" section as a post. The policy includes the purpose, scope of application, and policy. According to the City of Rock Island, the purpose of the city website and social media sites is to provide information of the city's choosing to users. The policy applies to the website and any other webpage, including social media pages, that represents the City of Rock Island or any of its departments. The policy section of the document includes statements on content, public records law, linking, and social networking rules (City of Rock Island, 2010).

In regards to content, the city lays out the following policies. All content is posted by an authorized member of the city staff. The city manager develops the content guidelines. Content should promote awareness of city programs and services, promote tourism, and promote the "economic welfare" of the city. The city has the right to restrict or remove content that is in violation of the website policy. The following content is unacceptable: profane language, slander, political speech, discriminatory content, sexual content, solicitations, illegal activity, and content that causes safety or security concerns (City of Rock Island, 2010). All city employees are expected to follow the policy.

The policy acknowledges that the city is subject to State of Illinois public records laws in regards to electronic communications. The city expresses that relevant records will be maintained in accordance with Illinois law. Federal laws and the Public Records Act will also be obeyed with regard to electronic communication and email addresses. The policy clearly states that the city's intention is to be transparent in all e-government applications. When providing links to other websites, it is the city's intention to provide these links only for non-profit or civic groups that promote tourism and industry within the City of Rock Island.

The City specifically addresses social networking sites, including Facebook and Twitter. Social network sites are used to "have conversations with the public in relation to the governance of the City" (City of Rock Island, 2010). The city uses this form of communication to solicit comments from the public with the understanding that it is a "limited public forum" (City of Rock Island, 2010). The city includes specific social networking rules in the policy. All pages and sites must be approved by the website coordinator and city manager. The site moderator allows public comments if they are related to posts unless they violate the content rules contained in the policy. Site moderators must check the sites at least three times a day during normal business hours to ensure all comments are in compliance with the content portion of the policy. Anonymous posting is not allowed. Videos, photos and written word may be posted on social media sites, which are to be maintained by the website coordinator (City of Rock Island, 2010).

ANALYSIS OF SOCIAL MEDIA POLICY: COMMON FEATURES AND IMPLEMENTATION

Common Features of Municipal Social Media Policy

The social media policies established by each of the municipalities contain common features. Even the simplest policy includes restrictions on commenting and content. The social media policies were analyzed using the framework described in the Research Design and Data Collection section (p.15) of this chapter. The analysis below is a combination of that framework and prominent features that emerged during the research process. Consistency of policy, which was included in the framework, was not a common feature in the research results and has not been included below, but it is discussed case-by-case in the Analysis of

Social Media section of this chapter. The "Statement Liability" category was added below as it was a common feature in the social media policies but was not a part of the original framework. Each feature is described in detail below.

Purpose of Social Media

This section of the social media policy outlines the purpose of the municipality's social media use. Each policy that includes a purpose includes an informational aspect as part of that purpose. In addition to providing information for citizens, many policies indicate that social media can also be used for dialogue or conversation between government and citizens and among citizens themselves in relation to government issues or topics. For example, the City of Aurora asserts that social media should be used to communicate with stakeholders and increase the transparency of government (City of Aurora, 2012).

Acceptable Use/Employee Practices/ Management of Social Media

A portion of the policies examined name the position responsible for the social media use of the organization or which employees are allowed administrative access to the sites. In the case of Rock Island, the website coordinator and city manager are responsible for the site, but "site monitors" have access to the site in order to enforce content restrictions or add information to the site (Rock Island, 2010).

Content Restrictions

Each policy specifies that the government will not allow any content that is not suitable for the site, whether it is simply deemed "inappropriate" as in the Village of Grayslake policy or if specific rules for content are established, as in the City of Rock Island and City of Aurora policies. Most policies indicate that the sites are a limited public forum

and are therefore subject to the scrutiny of the municipality. Citizens' First Amendment right to free speech is protected, but the government can delete activity that is not consistent with its policy.

Statement Liability Clause

Municipalities can also include a statement explaining that comments made on their municipal social media pages are not the opinion of the municipality. Municipalities may want to include this in their policy so that citizen opinions or statements are not taken as official municipal positions on issues.

Legal Issues

A policy may also acknowledge that the site is subject to state and federal laws that affect the purpose, content, and recordkeeping of the site. For example, the Village of Downers Grove explicitly states in its policy that any comments made are subject to the Illinois Freedom of Information Act and are therefore public record. Users are informed that their comments are going to be saved and may be accessed by others at a later date.

Security

Many municipalities include provisions that address the safety of information posted to the social media site. These are usually contained in the content portion of the policy, where it is established that users should not post personal information or information detrimental to the security or safety of any persons or entities.

Implementation of Municipal Social Media Policy

Not all aspects of the implementation of municipal social media policy are able to be discerned from studying the social media sites of the municipality. It is not possible to know if the municipality is following the Illinois Freedom of Information Act (FOIA) unless one were to file a request for comments made on the site. Most of these sites are still new, and therefore all commenting is still available on the social media site and a FOIA request would be unnecessary. It is also not possible to know the municipal employee who was responsible for every piece of information available on the social media site, especially in larger municipalities that have multiple employees contributing to a site. However, there are two aspects of the social media policy that are able to be examined in depth: purpose and content. These will be discussed below at length for each municipality with a social media policy.

Village of Grayslake

The Village of Grayslake policy only includes the following statement: "The Village of Grayslake, Illinois reserves the right to remove any posts, comments, or general content that is deemed inappropriate" (Village of Grayslake, 2012). The Grayslake Facebook page has only two comments from citizens at the time of this study and neither is inappropriate.

City of Sterling

Social media used at the City of Sterling is for purely informational purposes, according to the city's policy. In this regard, social media serves its purpose as the city continually posts information, pictures, and reminders on the municipal Facebook page.

Most of the comments on the city's Facebook page conform to the restrictions outlined in the city's policy. However, the city should consider clarifying the content restrictions. One comment on the site indicated that the campaign to elect an individual for County Recorder would be at an event the city was hosting, but the policy indicates that the city can delete comments that "promote or oppose particular services, products, or political

organizations and candidates". It is unclear if this statement: "The Campaign to elect [candidate] will be there!" constitutes a promotion of a political candidate on the site, and this should be addressed in the policy. The city also indicates that it will delete any postings that are "clearly off-topic" yet on a posting for renovations to a bridge, a comment is included advertising a "Ladies' Day Out" charity event that has no connection to the bridge topic.

Village of Downers Grove

As discussed previously, the Village of Downers Grove only establishes content rules in their social media policy. The village's Facebook page has only recently been established (February 1, 2012); therefore, there is only a limited amount of postings on the site at this time. At this time, there are no violations of the policy, but it has not been in place a sufficient amount of time to be certain this will remain the case.

City of Collinsville

As with the other examples already discussed, the purpose of the city's social media is to both provide information and engage in dialogue with citizens. The city does use the site for informational purposes of issues that may be concerning citizens, emergency situations, and event announcements. However, there seems to be a lack of real discussion on the site. The city rarely responds to citizen comments or questions on the Facebook site. To illustrate this point, from January 1, 2012 to January 31, 2012, there were 13 instances where citizens made comments on city postings that posed a question or comment that should have been addressed by the city, but the city only responded to 5 of those postings (38%). Citizen trust in social media as a medium for conversation with the local government may wane if the city does not make a concerted effort to increase the response to citizen comments.

There is no evidence of blatant violations of the conduct restrictions on the city's Facebook page; however, the policy may be confusing as to the language actually allowed on the site, as is evidenced by an example of a comment from one resident. Under a posting of a picture of a building that was scheduled to be demolished, one resident commented that the structure looked like a "crack house" to which another poster took offense by indicating that they had lived at the premises previously, and the structure was never a "crack house". It is unclear as to whether this would fall under content that is restricted as the policy indicates that the site will not allow "nudity/pornography, adult and/or hate language, tobacco, alcohol or similar information" or "illegal or inappropriate use of firearms or drugs, or any illegal activity" (City of Collinsville, 2010). None of the commentators is condoning drug use, but there are allusions to it, and it is unclear if the language should be removed.

City of Aurora

According to the City of Aurora policy, the purpose of social media is to communicate with residents, the media and other stakeholder in a "rich" experience that increases transparency. The city plans to use the site to publish news reports and other information that the city feels supports its mission. In reviewing the municipal social media sites, the city periodically (daily during business days) posts news articles, pictures, and important information for residents on the municipal Facebook page. For example, on January 16, 2012, the city posted a notice that downtown street lights would be under construction in preparation for bridge construction that would shut down two bridges for nine months. This is pertinent information that residents may need to function on a daily basis and serves the purpose of informing citizens. This posting also serves to exemplify the second purpose of the social media site, which is to create an experience between residents and the government. In response to this post, a resident of the city posted a question concerning the one-way streets in downtown Aurora to which the site monitor responded by saying that the question would be forwarded to the "decision makers". Another resident then posted under the same topic the benefits to one-way traffic in that area of the city. This example shows that the social media site does serve the purpose that the city intends, at least when citizens choose to engage in this type of dialogue. From the beginning of the city's site in September 2011 through February 14, 2012, there were four instances where citizens engaged in this type of dialogue. Each instance was in regard to a physical problem occurring in the city; three of these occurrences were in reference to snow plowing/winter storm issues, and the other occurrence was in regard to the construction example described above. In each of these instances, the city responded to citizen questions and comments and directed these concerns to the appropriate channel (city department or elected official). Although the policy applies to the city's Facebook page, it does not apply in both applications to the city's Twitter feed. The Twitter feed is only used for informational purposes as there is no evidence that any type of dialogue is established through this medium.

In regards to the implementation of the content restrictions instituted in the city's policy, this is easily monitored through the social media site. The city allows commenting, "liking", and sharing of municipal postings on the official Facebook page. There was no evidence of any kind of posting that violates the city's policy in the entire history of postings on the city's Facebook page. The content restrictions are being enforced as evidenced by a city posting in response to a citizen question posted under the traffic construction posting discussed earlier. The city posted the following comment under its traffic construction posting: "Jill, we received your question but took it down from this post discussion because it was not on topic. If you can provide your email address to publicinfo@aurora-il.org, we will check on the information and get back in touch with you. Thanks (City of Aurora, 2012)." There is no evidence of any violations of the policy on the city's Twitter page.

City of Rock Island

Similar to Aurora, the purpose of Rock Island's social media use is for both informational purposes and also to engage in conversation with citizens in relation to the governance process. The city accomplishes the informational aspect of its policy through the posting of news releases, public meeting announcements, and issues relating to the everyday operations of the city. Since the beginning of the postings on the municipal Facebook page, there have been numerous instances where the city engages in discussions with citizens on various topics, including operations, economic development, and the funding of projects. Through these discussions, the city shows a clear commitment to involving citizens through their social media use.

There is no evidence of conduct restriction violations on the city's Facebook page. The city does not limit discourse to comments that are only complementary in nature. Comments that are critical of the city are included on the site. The city allows other Rock Island organizations to post their advertisements on the Rock Island Facebook page. For example, there are several postings for Circa 21, a theater business located in Rock Island. In the policy, it asserts that linking will be allowed for non-profit and civic entities that address economic advancement in Rock Island. In allowing this business to post on their site, the city is violating its policy. This could also set a precedent for the site in which other businesses may want to post their information on the city's site. The policies appear to only apply to the city's Facebook page as commenting is disabled for videos on the city's YouTube page.

Lack of Implementation of Social Media Policy

Perhaps even more valuable is the examination of municipal social media sites that do not have any type of social media policy. As discussed previously, eleven (11) municipalities use some form of social media but only six (6) had any type of policy. Municipalities without a policy may encounter problems due to their lack of policy, including inappropriate content or use of the site for unintended purposes. With no formal policy on recordkeeping or management of the site, the municipality may not be able to conform to Freedom of Information Act requests as a result of mismanagement of the site. The City of Decatur has no policy for social media yet it allows commenting on its Facebook and YouTube sites. As a result, content on the sites may not be monitored. As an example, on Decatur's AccessDecatur YouTube Channel, the Decatur police department posted a video concerning motorized bicycles in May 2011, and there was one comment posted by a user that included obscene language and derogatory remarks toward the Police Department. If the city had a policy that included content restrictions and procedures for maintaining these social media sites, these comments would most likely be deleted by the city.

The case of the Village of Atwood perfectly illustrates the need for a social media policy. The Village website contains a link to the Village's Facebook page, but after following the link to the Facebook page, it is unclear if this is indeed the official municipal Facebook page. From the beginning of the site (postings begin in April 2009), users post random comments not related to any Village business. There are spam/advertisement postings throughout the site's wall page. The Atwood Armory posted advertisements and pictures of guns on the municipal page. One posting included derogatory comments toward Village Board members. Although this behavior is rampant on the site, it seems as though it is the official page because there are postings on official Village business as well as a reminder that the purpose of the website and Facebook site is to provide information to residents. If these activities had been occurring on a site with a social media policy, the advertisements, spam, and off-topic and derogatory comments would not be allowed on the site, which could then focus on actual Village business.

These cases showcase the need for social media policy. Although these showcase only a few examples of the need for social media policy on municipal social media websites, municipalities should be prepared for any possibility when communicating with the public using this still evolving technology.

MODEL SOCIAL MEDIA POLICY: A RECOMMENDATION

After reviewing the current literature and the results of this study, a "best practice" social media policy has been devised to address the concerns uncovered thus far. A framework was devised from the literature review, mainly focusing on the features outlined in the Center for Technology in Government report, which included acceptable use and employee conduct, legal issues, management, security, and site content (Hrdinova, Helbig & Stollar Peters, 2010). The 2011 ICMA survey indicated that municipalities use social media for different purposes and to different degrees (ICMA, 2011). A main focus of the study was determining if social media policy addresses the real purpose of the site, as each municipality that uses social media may fall at different points along the spectrum from total one-way communication to a dynamic dialogue with citizens through social media.

Each municipal social media policy, as well as the implementation of the policy and behavior on the municipal social media sites, was analyzed. Through a synthesis of the literature review and

the conclusions of the aforementioned study, a model social media policy has been created. As asserted by multiple authors, including both Hand and Ching (2011) and Bertot, Jaeger, Munson and Glaisyer (2010), municipalities can decide if the use of social media is for informational purposes only or if it will be used to create a deeper relationship and include conversation with citizens. If the purpose is only informational, the municipality can disengage the ability of citizens to comment and participate, and therefore, many of the issues uncovered through this study and the literature review would not apply, including citizen content and behavior, and the municipality could address social media sites similar to the municipal website (Hrdinova, Helbig & Stollar Peters, 2010). However, if the purpose of social media use is to engage in discourse with citizens to create a more dynamic relationship between citizens and the government, the municipality will need a comprehensive policy that will address the issues currently known about municipal social media use.

Each section that should be included in the social media policy is outlined below with a sample example that can be formatted for each individual municipality:

- **Purpose:** The purpose of the social media site should be outlined. The municipality should explicitly state why it is using the site: for informational purposes, to create a conversation with residents on municipal issues, or another reason. From this section, users should understand that the site is for municipal business only and is not a social/business avenue.

The purpose of the Village/City of _____ social media page is for both informational and communication purposes. The Village/City will provide useful information to residents, businesses, and visitors that facilitate the provision of Village/City services or furthers the objectives and goals of the Village/City. The Village/City site should also be used as an avenue to increase meaningful interaction between the Village/City and citizens, as well as among citizens themselves, in regards to issues affecting the citizens of _____.

- **User Content/Behavior:** The policy should clearly and concisely explain in detail the behavior and content that is not allowed on the municipal site. The rules described in this section are at the discretion of the municipality but should include content descriptions, advertising/business rules, and anonymous posting restrictions.

The Village/City of _____ reserves the right to restrict or remove any content that is deemed a violation of this policy or any applicable law. This includes but is not limited to:

- *Profane or offensive language or content*
- *Personal attacks, threats or slander*
- *Content that promotes, fosters, or perpetuates discrimination on the basis of race, creed, color, age, religion, gender, marital status, status with regard to public assistance, national origin, physical or mental disability or sexual orientation*
- *Sexual content*
- *Solicitations or advertisements*
- *Illegal activity*
- *Spam or links to other websites*
- *Off-topic comments or content not related to Village/City issues*
- *Content that will compromise the safety and/or security of the public*
- *Political links, promotions or endorsements*
- *Programs and events not sanctioned by the Village/City*
- *Personally identifiable information, such as address, phone number, or social security number*

This notice informs users that this social media site is a limited public forum and any content in violation of this policy will be removed. Anonymous posts are not allowed and will be removed from the site. All comments must represent the person posting the comment.

- **Linkage Policy:** The municipality must create a policy determining whether to allow links to other sites or organizations or not. The municipality can choose to allow no linkages, linking to non-profit or government agencies, linking for certain purposes, or allows all links.

The Village/City will allow links to other organizations to be posted on the social media site provided that the organization is a non-profit or government organization within the Village/City that promotes the goals and objectives of the Village/City, and all content conforms to this policy.

- **Legal Disclosures:** The municipality must state that it will comply with all federal, state, and local laws and ordinances that may be affected by the use of social media, including records retention, the First Amendment of the United States Constitution, the Fourteenth Amendment of the United States Constitution, the Open Meetings Act, the Freedom of Information Act, copyright laws, and safety/security measures. The municipality should additionally inform citizens that, as the municipal social media site is subject to the Illinois Freedom of Information Act, any comments made on the site by any person are subject as well. The municipality should also include a clause that indicates the personal opinion of commenters does not reflect the views of the municipality.

Comments expressed on this site, other than those posted by the City/Village, do not reflect the opinions and/or positions of the Village/City of _____, its officers and/or employees. The Village/City will adhere to all applicable federal, state, county and Village/City laws, ordinances, and regulations, including but not limited to the Freedom of Information Act, the Illinois Open Meetings Act, the First and Fourteenth Amendments of the United States Constitution, copyright laws, and privacy and information security laws. Under the Illinois Freedom of Information Act, comments posted to this site are considered public record and may be requested at any time.

- **Employee Access/Site Monitoring:** The policy should name the individual responsible for enforcement of the policy, the employee(s) that will have administrative access to the site, and the employee(s) who are responsible for monitoring and posting to the site. The process of monitoring/updating the site should also be included in this section. This will be specific to the organization.

The City/Village Administrator has sole discretion over the content of this site. The City/Village monitors and assesses the site daily during regular business hours for policy compliance and informational updates. The City/Village attempts to answer every question or comment within 72 hours of a comment being posted.

- **Security:** The security of information is an important aspect of the policy that must be addressed. No private information should be posted on the site by either users or the municipality. Passwords must be protected so that unknown users are not allowed access to the site.

The Village/City does not allow any personal information to be posted on this site. The passwords for Village social media should be highly guarded

and changed quarterly so as to guard against unintended users from gaining access to the site.

CONCLUSION AND FUTURE RESEARCH DIRECTIONS

Municipal use of social media is a relatively new phenomenon, and as such, municipalities are still developing strategies and policies regarding their use of social media. Although there may be additional challenges in the future concerning social media use, it is clear by the review of the literature on the subject, as well as the findings of this study, that now is the time for municipalities using social media to implement policy regulating the use of social media. As has been described throughout this study, there is a need for policy to standardize the use of social media in response to a number of different criteria, including purpose, user content and behavior, legal disclosures, employee access and site monitoring, linkage policy, and security. The model social media policy outlined above serves as a guide for municipal social media policy development.

The social media policy is an important aspect of social media use for municipalities, but it must be implemented and enforced in order to protect the municipality, as well as citizens. As both the literature and this study have shown, this can be achieved if the municipality has an organized system with the appropriate measures in place to monitor social media use in local government (Hrdinova, Helbig, & Stollar Peters, 2010). Each municipality has the choice on how to utilize social media for the benefit of citizens, but the rules and purpose must be clearly articulated and implemented for social media to serve the purposes the government intends. In order for citizens to have confidence in the government's use of social media, and in the government itself, the policy must be enforced and the purpose must be met. Policy and implementation must also be consistent throughout all social media use, if the municipality utilizes multiple social media sites or applications.

Social media use is starting to gain popularity with local governments throughout the United States (ICMA, 2011). This increase in social media use will be accompanied by new problems and issues that governments will have to address, but with the implementation of a social media policy, the challenges facing local governments concerning social media use will be diminished.

Although social media is a relatively new application being utilized by local governments, it will continue to evolve as the needs of the public and municipalities grow. Even now, new trends are beginning to take form within the local government sphere in terms of social media use. Local governments are starting to utilize customized mobile phone applications and mobile social media applications for government use. This may create new concerns about privacy and security of information, which remains an issue as the use of new technological applications continues.

Research opportunities in this area are immense and will continue to grow as the use of social media continues to grow. Other states may be examined and compared to the findings of this study. Implementation of social media policy can be tracked over a period of time to determine long-term implementation problems and solutions. A comparison of local governments with a sound, comprehensive policy and governments without a policy can be conducted to determine the effects of policy on social media use. Social media use should be studied to determine the long-term impact of successful social media policy implementation. Study of other levels of government, including federal, state, and county levels, can be conducted to identify the problems and solutions for each of these levels of government in terms of social media use. These results can then be compared to the results of this study, as well as other state studies, to determine if a comprehensive policy for all governments can be created.

REFERENCES

Arnstein, S. R. (1969). A ladder of citizen participation. *Journal of the American Institute of Planners*, *35*(1), 216–224. doi:10.1080/01944366908977225

Bertot, J. C., Jaeger, P. T., Munson, S., & Glaisyer, T. (2010). Engaging the public in open government: Social media technology and policy for government transparency. *U.S. National Science Foundation Workshop Draft Report*, (pp. 1-18).

Boyd, d. m., & Ellison, N. B. (2008). Social network sites: Definition, history, and scholarship. *Journal of Computer-Mediated Communication,13*(1), 210-230.

Bryer, T., & Zavattaro, S. M. (2011). Social media and public administration: Theoretical dimensions and introduction to the symposium. *Administrative Theory & Praxis*, *33*(3), 325–340. doi:10.2753/ATP1084-1806330301

Center for Technology in Government. (2009). *Exploratory social media project*. The Research Foundation of State University of New York, 2009.

Chaplinksy v. New Hampshire. (March 9, 1942). 315 U.S. 568; 62 S. Ct. 766; 86 L. Ed. 1031; 1942 U.S. Lexis 851.

City of Aurora. (2012). *City of Aurora, IL info*. Retrieved January 20, 2012, from http://www.facebook.com/cityofaurorail#!/cityofaurorail?sk=info

City of Aurora Police Department. (2012). *City of Aurora Police Department Facebook page*. Retrieved January 20, 2012, from http://www.facebook.com/AuroraPolice?sk=notes

City of Chicago. (2011). *Connect with the City of Chicago vis social media*. Retrieved November 15, 2011, from htt://www.cityofchicago.org/content/city/en/narr/misc/social_media.html

City of Collinsville. (March 1, 2010). *City of Collinsville notes*. Retrieved January 15, 2012, from http://www.facebook.com/cityofcollinsville#!/cityofcollinsville?sk=notes

City of Rock Island. (February 10, 2010). *City of Rock Island notes*. Retrieved January 10, 2012, from http://www.facebook.com/rockislandil?sk=info#!/rockislandil?sk=notes

City of Sterling. (November 10, 2010). *City of Sterling notes*. Retrieved January 14, 2012, from http://www.facebook.com/sterling.il?sk=notes

Dawes, S. (2008). The evolution and continuing challenges of e-governance. *Public Administration Review*, *68*(6), S86–S102. doi:10.1111/j.1540-6210.2008.00981.x

Facebook. (2012). *Facebook basic information*. Retrieved January 4, 2012, from http://www.facebook.com/#!/facebook?sk=info

Hand, L. C., & Ching, B. D. (2011). You have one friend request: An exploration of power and citizen engagement in local governments' use of social media. *Administrative Theory and Praxis*, *33*(3), 362–382. doi:10.2753/ATP1084-1806330303

Hrdinová, J., & Helbig, N. (2011). *Designing social media policy for government* (pp. 1–9). Issues in Technology Innovation.

Hrdinova, J., Helbig, N., & Stollar Peters, C. (2010). *Designing social media policy for government: Eight essential elements. Center for Technology in Government*. The Research Foundation of State University of New York.

ICMA. (2011). *E-government 2011 survey summary*. Retrieved March 28, 2012, from http://icma.org/en/icma/knowledge_network/documents/kn/Document/302947/EGovernment_2011_Survey_Summary

Leighninger, M. (2011). Citizenship and governance in a wild, wired world: How should citizens and public managers use online tools to improve democracy? *National Civic Review*, *100*(2), 20–29. doi:10.1002/ncr.20056

Madigan, L. (2004). *Guide to the Illinois Open Meetings Act: 5 ILCS 120*. State of Illinois.

Madigan, L. (March 2011). *Public access counselor annual report 2010*. Illinois Attorney General. Retrieved January 12, 2012, from http://foia.ilattorneygeneral.net/pdf/Public_Access_Counselor_Annual_Report_2010.pdf

Madigan, L. (2011). *Public Access Opinion No. 11-006. State of Illinois*. Office of the Attorney General.

PEW Research Center. (2011). *Demographics of internet users*. PEW Internet and American Life Project. Retrieved March 28, 2012, from http://pewinternet.org/Static-Pages/Trend-Data/Whos-Online.aspx

R.A.V. v. St. Paul. (June 22, 1992).505 U.S. 377 U.S. Supreme Court.

Simpson, D., & Moll, L. (1994). *The crazy quilt of government: Units of government in Cook County, 1993*. Office of Publications Services of the University of Illinois.

Smith, A. (2010). *Government online*. Pew Research Center. Retrieved March 20, 2012, from http://www.pewinternet.org/Reports/2010/Government-Online.aspx

Southeastern Promotions, Ltd. V. Conrad et al. (October 17, 1974).420 U.S. 546; 95 S. Ct. 1239; 43 L. Ed. 2d 448; 1975 U.S. LEXIS 3; 1 Media L. Rep. 1140 U.S. Supreme Court.

State of Illinois. (2010). *Census 2010*. State of Illinois. Retrieved April 29, 2012, from http://www2.illinois.gov/census/Pages/default.aspx

State of Illinois. (2010). *2000 census population compared to 1990: Illinois municipalities.* State of Illinois. Retrieved November 22, 2011, from http://illinoisgis.ito.state.il.us/census2000/dplace_census.asp?theSelCnty=001

State of Illinois. (2010). Freedom of Information Act (5 ILCS 140).

Streib, G. D., & Willoughby, K. G. (2005). Local governments as e-governments: Meeting the implementation challenge. *Public Administration Quarterly, 29*(1), 78–110.

Tolbert, C. J., & Mossberger, K. (2006). The effects of e-government on trust and confidence in government. *Public Administration Review, 66*(3), 354–369. doi:10.1111/j.1540-6210.2006.00594.x

Village of Downers Grove. (2012). *Village of Downers Grove info.* Retrieved February 1, 2012, from http://www.facebook.com/pages/Village-of-Downers-Grove-Illinois/156234227805840?sk=wall#!/pages/Village-of-Downers-Grove-Illinois/156234227805840?sk=info

Village of Grayslake. (2012). *Village of Grayslake info.* Retrieved February 2, 2012, from http://www.facebook.com/pages/Village-of-Grayslake-IL/209863695730984?sk=info

KEY TERMS AND DEFINITIONS

Best Practice: A standard method that has constantly shown exceptional results to achieve a certain goal.

E-Governance: Use of technology, electronic communications, and the internet to deliver government services in a way that improves the governance process for both the government and citizens.

E-Government: Use of information and communication technologies for the production and provision of public information and services.

Limited Public Forum: A place created by government for public speech or discussion that is subject to content and user restrictions by the government.

Social Media: Electronic communications that integrate social interaction and information sharing through online applications.

Social Media Policy: Rules and procedures that guide the use of social media.

Web 2.0: The network of online applications that facilitate user participation through interaction, collaboration, and information sharing.

Chapter 3
Citizen–Centric Access to E–Government Information Through Dynamic Taxonomies

Giovanni M. Sacco
Università di Torino, Italy

ABSTRACT

This chapter focuses on dynamic taxonomies, a semantic model for the transparent, guided, user-centric exploration of complex information bases. Although this model has an extremely wide application range, it is especially interesting in the context of e-government because it provides a single framework for the access and exploration of all e-government information and, differently from mainstream research, is citizen-centric, i.e., intended for the direct use of end-users rather than for programmatic or agent-mediated access. This chapter provides an example of interaction and discusses the application of the model to many diverse e-government areas, going from e-services to disaster planning and risk mitigation.

INTRODUCTION

A large quantity of information is currently available from public government sources. E-government is usually associated with prescriptive information such as laws and regulations, whose knowledge and dissemination is obviously fundamental for democracy and order. However, the competencies of governments have considerably grown and they encompass many facets of citizen's everyday life. For instance, online government e-services are available to citizens through the Web in order to simplify and make more efficient

DOI: 10.4018/978-1-4666-3640-8.ch003

common services such as the issuance of ID cards and permits. E-services represent one of the most frequent and critical points of contact between public administrations and citizens. They also represent the only practical way of providing incentives and support to specific classes of citizens, such as those with disabilities, senior citizens, etc. From the similar perspective of improving the quality of life of citizens, governments, especially local ones, use the Web to provide a number of services that are mainly informative and promote the local community, for example job placement services, tourist information, and so forth.

Easy and effective user-centric access to complex information is one of the most critical requirements of e-government. Without timely and accurate information, the participation of citizens in the government is likely to be an illusion: in short, no democracy without knowledge.

Traditional access paradigms are not suited to most search tasks, which are exploratory and imprecise in essence. In most cases, the user does not know exactly what he is looking for and needs to explore the information base, find relationships among concepts and thin out alternatives in a guided way. Unobtrusive but effective guidance in finding relevant information is especially important.

New access paradigms supporting exploration are needed. Since the goal is end-user interactive access, a holistic approach, in which modelling, interface and interaction issues are considered together, must be used. Dynamic taxonomies (Sacco, 2000, Sacco & Tzitzikas, 2009, aka faceted search systems), which provide powerful and simple exploration capabilities, are reviewed in the following. In addition to providing an example, the discussion focuses on e-government areas where the application of dynamic taxonomies can be beneficial: these include e-services, job placement, tourist information, but also areas such as disaster planning and risk mitigation, and medical diagnostic systems.

BACKGROUND

Public information is usually managed by four retrieval techniques, which are frequently used at the same time for different subsets of the information base: a) information retrieval (IR) techniques (van Rijsbergen, 1979) recently called search engines; b) queries on structured databases; c) hypertext/hypermedia links and d) traditional taxonomies.

IR techniques are the obvious choice for laws and regulations, since they are essentially textual in nature. However, their limitations, especially in the legal domain, are well known: Blair and Maron (1985) reported that only 20% of relevant documents in a legal database were actually retrieved. Such a significant loss of information is due to the extremely wide semantic gap between the user model (concepts) and the model used by commercial retrieval systems (words). Other problems include poor user interaction because the user has to formulate his query with no or very little assistance, and no exploration capabilities since results are presented as a flat list with no systematic organization. These latter limitations have been addressed recently. Google and other search engines suggest additional query terms while the user is typing as well as the autocompletion and spelling checking of query terms. Clustering techniques are used to support some sort of exploration, by clustering the documents retrieved by an IR query according to "significant" terms or phrases that occur in them. This approach provides a summary for query results and has been used for instance in the US government portal, firstgov.gov. Cluster summaries do not address the semantic problems inherent in IR and do not increase the recall, which is the critical performance indicator in this context. Rather, they increase the precision of the result because they allow users to quickly skip clusters that are not relevant. In addition, the exploratory capabilities offered by text clustering are quite limited (Sacco, 2000; Hearst, 2006).

Database queries require structured data and are not easily applicable to situations in which most information is textual and not structured or loosely structured. They are extensively used for informative, promotional services (e.g. job placement services, tourist information, etc.) because in most cases they rely on structured information stored in database systems. Like IR, database queries do not support exploration.

Hypermedia (see Groenbaek & Trigg, 1994) is quite flexible, but it gives no systematic picture of relationships among documents; exploration is performed one-document-at-a-time, which is quite time consuming; and building and maintaining complex hypermedia networks is very expensive. They are currently extensively used in public information portals to manage e-services, because the number of e-services is reasonably small.

Traditional taxonomies are based on a hierarchy of concepts that can be used to select areas of interest and restrict the portion of the infobase to be retrieved. Taxonomies support abstraction and are easily understood by end-users. However, they are not scalable for large information bases (Sacco, 2006a), and the average number of documents retrieved becomes rapidly too large for manual inspection. Finally, solutions based on semantic networks, ontologies and the Semantic Web (Berners-Lee et al. 2001) have been proposed. These solutions are more powerful than plain taxonomies but general semantic schemata are intended for programmatic access and are known to be difficult to understand and manipulate by the casual user. User interaction must be mediated by specialized agents, which increases costs, time to deployment and decreases the transparence and flexibility of user access.

DYNAMIC TAXONOMIES

Dynamic taxonomies (Sacco, 2000, Sacco & Tzitzikas, 2009, also called *faceted search systems*) are a general knowledge management model based on a multidimensional classification of heterogeneous data items and are used to explore/ browse complex information bases in a guided yet unconstrained way through a visual interface.

The intensional part of a dynamic taxonomy is a taxonomy, that is a concept hierarchy going from the most general to the most specific concepts. Directed acyclic graph taxonomies modeling multiple inheritance are supported but rarely required. A dynamic taxonomy does not require any other relationships in addition to *subsumptions* (e.g., IS-A and PART-OF relationships).

In the extension, items can be freely classified under n (n>1) topics at any level of abstraction (i.e. at any level in the conceptual tree). This multidimensional classification is a generalization of the monodimensional classification scheme used in conventional taxonomies and models common real-life situations. First, items are very often about different concepts. Second, items to be classified usually have different features, "perspectives" or facets (e.g. Time, Location, etc.), each of which can be described by an independent taxonomy.

In dynamic taxonomies, a concept C is just a label that identifies all the items classified under C. Because of the subsumption relationship between a concept and its descendants, the items classified under C (items(C)) are all those items in the *deep extension* of C, i.e. the set of items identified by C includes the *shallow extension* of C (i.e. all the items directly classified under C) union the deep extension of C's sons. By construction, the shallow and the deep extension for a terminal concept are the same.

There are two important immediate consequences of this approach. First, since concepts identify sets of items, logical operations on concepts can be performed by the corresponding set operations on their extension. This means that the user is able to restrict the information base (and to create derived concepts) by combining concepts through the normal logical operations (and, or, not). Second, dynamic taxonomies can find all the concepts related to a given concept C: these concepts represent the conceptual summary of C. Concept relationships other than subsumptions are inferred through the extension only, according to the following *extensional inference rule*: two concepts A and B are related iff there is at least one item *d* in the knowledge base which is classified at the same time under A or under one of A's descendants and under B or under one of B's descendants. For example, we can infer an unnamed relationship between *Michelangelo* and *Rome*, if an item classified under *Michelangelo* and *Rome* exists. At the same time, since *Rome* is a descendant of *Italy*, also a relationship between *Michelangelo* and *Italy* can be inferred. The extensional inference rule can be seen as a device to infer relationships on the basis of empirical evidence.

The extensional inference rule can be extended to cover the relationship between a given concept C and a concept expressed by an arbitrary subset S of the universe: C is related to S iff there is at least one item d in S which is also in items(C). Hence, the extensional inference rule can produce conceptual summaries not only for base concepts, but also for any logical combination of concepts. Since it is immaterial how S is produced, dynamic taxonomies can produce summaries for sets of items produced by other retrieval methods such as IR or database queries, shape retrieval, etc. and therefore access through dynamic taxonomies can be easily combined with any other retrieval method.

Dynamic taxonomies work on conceptual descriptions of items, so that heterogeneous items of any type and format can be managed in a single, coherent framework. Finally, since concept C is just a label that identifies the set of the items classified under C, concepts are language-invariant, and multilingual access can be easily supported by maintaining different language directories, holding language-specific labels for each concept in the taxonomy. If the metadata descriptors used to describe an item use concepts from the taxonomy, then also the actual description of an item can be translated on the fly to different languages. This feature is extremely important for multilingual governments such as the European Union.

Exploration

The user is initially presented with a tree representation of the initial taxonomy for the entire knowledge base. Each concept label has also a count of all the items classified under it, i.e. the cardinality of items(C) for all C's. The initial user focus F is the universe, i.e. all the items in the information base.

In the simplest case, the user selects a concept C in the taxonomy and *zooms* over it. The zoom operation changes the current state in two ways. First, concept C is used to refine the current *user focus* F, which becomes F∩items(C). Items not in the focus are discarded. Second, the tree representation of the taxonomy is modified in order to summarize the new focus. All and only the concepts related to F are retained and the count for each retained concept C' is updated to reflect the number of items in the focus F that are classified under C'. The *reduced taxonomy* is derived from the initial taxonomy by pruning all the concepts not related to F, and it is a conceptual summary of the set of documents identified by F, exactly in the same way as the original taxonomy was a conceptual summary of the universe. In fact, the term *dynamic taxonomy* indicates that the taxonomy can dynamically adapt to the subset of the universe on which the user is focusing, whereas traditional, static taxonomies can only describe the entire universe.

The retrieval process can be seen as an iterative thinning of the information base: the user selects a focus, which restricts the information base by discarding all the items not in the current focus. Only the concepts used to classify the items in the focus and their ancestors are retained. These concepts, which summarize the current focus, are those and only those concepts that can be used for further refinements. From the human computer interaction point of view, the user is effectively guided to reach his goal by a clear and consistent listing of all possible alternatives, and, in fact, this type of interaction is often called *guided thinning* or *guided navigation*.

Figures 1 to 5 show how the zoom operation works. Figure 1 shows a dynamic taxonomy: the upper half represents the intension with circles representing concepts; the lower half is the extension, and documents are represented by rectangles. Arcs going down represent subsumptions; arcs going up represent classifications. In order to compute all the concepts related to H, we first find, in Figure 2, all the documents classified under H (that is, the deep extension of H, items(H)) by following all the arcs incident to H (and, in general, its descendants): items(H)={ b, c, d }. All the items not in the deep extension of H (see Figure 3) are removed from the extension. In Figure 4, the set of all the concepts under which the documents in items(H) are classified, B(H), is found by following all the arcs leaving each element in the set: B(H)={ F, G, H, I }. The inclusion constraint implied by subsumption states that if items(C) denotes the set of documents classified under C and C' is a descendant of C in the taxonomy, items(C')⊆items(C) (Sacco, 2000). This is equivalent to say that a document classified under C' is also classified under C. Hence, the set of concepts related to H is given by B(H) union all the ancestors of all the concepts in B(H), i.e. the set of all concepts related to H is {F, G, H, I, B, C, A}. Finally, in Figure 5, all the concepts not related to H are removed from the intension, thus producing a reduced taxonomy that fully describes all and only the items in the current focus.

A sample interaction on e-services is reported in Figures 6 and 7. Figure 6 shows the initial dynamic taxonomy, in which every topic is followed by a number indicating how many items are classified under it. As the *Type* facet shows, items include online and offline services, but also guides. It must be stressed that the final goal in

using dynamic taxonomies in e-government portals is to make all information available and findable in an integrated way.

A zoom on *Senior citizens* produces the reduced taxonomy in Figure 7, that shows all and only the topics classified under *Senior citizens*, i.e. those services/guides etc. that specifically apply to them. Drill-down can be iterated as required: for instance, the user might focus on *Housing*, thereby reducing both the number of items and the number of topics that apply.

Advantages

The single, largest advantage of dynamic taxonomies over competing techniques is that they can support exploratory access to any type of

Figure 1. A dynamic taxonomy: the intension is above, the extension below. Arrows going down denote subsumptions, going up classification.

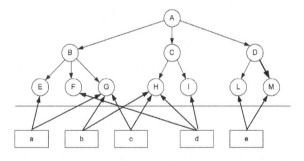

Figure 2. Focusing on concept H: finding all the items classified under H

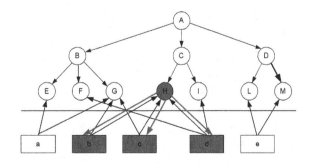

Figure 3. All the items not classified under H are removed

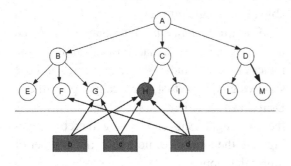

Figure 4. All the concepts under which the items in the focus are classified (and, because of subsumptions) their ancestors are related to H

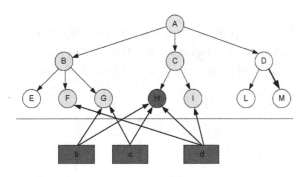

Figure 5. The reduced taxonomy: all concepts not related to the current focus are pruned

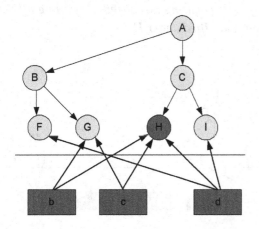

information and therefore provide a single, uniform, coherent and easily understood interface to end users.

Dynamic taxonomies require a very light theoretical background: namely, the concept of a taxonomic organization and the zoom operation, which seems to be very quickly understood by end-users. Hearst et al. (2002) and Yee et al. (2003) conducted usability tests on a corpus of art images. Despite slow response times due to an inadequate implementation, access through a dynamic taxonomy was shown to produce a faster overall interaction and a significantly better recall than access through text retrieval. It is perhaps more important the feeling that one has actually considered all the alternatives in reaching a result, whereas all other non-exploratory search approaches give the user the feeling that more relevant information might be available but he was not able to find it. Although few usability studies exist, the widespread adoption by e-commerce portals (Sacco, 2003) empirically supports this evidence.

An important feature of dynamic taxonomies is the extremely quick convergence of exploratory patterns. Sacco (2006a) shows that 3 zoom operations on terminal concepts are sufficient to reduce information bases of up to ten million items and described by a compact taxonomy with one thousand concepts to an average 10 items. Experimental data on a real news corpus of over 110,000 articles, classified through a taxonomy of 1100 concepts, reports an average 1246 documents to be inspected by the user of a static taxonomy vs. an average 27 documents after a single zoom on a dynamic taxonomy.

The derivation of concept relationships through the extensional inference rule has important implications on conceptual modeling. First, it simplifies taxonomy creation and maintenance. In traditional approaches, only the relationships among concepts explicitly described in the conceptual schema are available to the user for browsing and retrieval. Therefore, all possible relationships must be

Figure 6. E-government portal with a dynamic taxonomy on seven facets (© Knowledge Processors, 2006)

SERVICES	LIFE EVENTS	TYPE	WHERE ARE YOU	CITIZENSHIP	SPECIAL RIGHTS	PROFILE
Income taxes (2)	Working (2)	offline service (5)	Italy (67)	Italian (67)	Women (5)	Sex (67)
Social security contributions (10)	Transportation (13)	online service (62)		EU (4)	**Senior citizens (67)**	Age (67)
Declaration to the police (15)	Housing (7)	guide (19)			Handicapped (16)	Education (67)
Public libraries (9)	Family (5)					
Health related services (20)	Health (20)					
	Police (35)					
	Leisure and culture (14)					
	Retiring (16)					

Figure 7. Reduced taxonomy after a zoom on Senior Citizens: characterization of the 67 items specifically targeted to senior citizens (© Knowledge Processors, 2006)

SERVICES	LIFE EVENTS	TYPE	WHERE ARE YOU	CITIZENSHIP	SPECIAL RIGHTS	PROFILE
Income taxes (2)	Having a child (0)	offline service (5)	abroad (0)	Italian (67)	Women (5)	Sex (67)
Job search services (0)	Studying (0)	online service (62)	Italy (67)	EU (4)	**Senior citizens (67)**	Age (67)
Social security contributions (10)	Working (2)	guide (19)		extra-EU (0)	Handicapped (16)	Education (67)
Personal documents (0)	Transportation (13)				Relationships (0)	
Car registration (40	Housing (7)					
Application for building permission (0)	Family (5)					
Declaration to the police (15)	Paying taxes (0)					
Public libraries (9)	Going abroad (0)					
Certificates (0)	Health (20)					
Enrollment in higher education / university (0)	Sport (0)					
Change of address (0)	Police (35)					
Health related services (20)	Leisure and culture (14)					
	Helping others (0)					
	Retiring (16)					

anticipated and described: a very difficult if not hopeless task. In dynamic taxonomies, no relationships in addition to subsumptions are required, because concepts relationships are automatically and dynamically derived from the actual classification. For this reason, dynamic taxonomies easily adapt to new relationships and are able to discover new, unexpected ones.

Second, compound concepts, which represent the main cause of the combinatorial growth of traditional taxonomies, need usually not be represented explicitly because dynamic taxonomies are able to synthesize them on-the-fly. Sacco (2000) developed guidelines that produce taxonomies that are compact and easily understood by users. Some are similar to basic faceted classification (Ranganathan, 1965; Hearst, 2002), at least in their basic form: the taxonomy is organized as a set of independent, "orthogonal" subtaxonomies (facets or perspectives) to be used to describe data.

As an example of faceted design guidelines, consider a compound concept such as "housing grants for senior citizens in London". It can be split into its *facets*: a *Location* taxonomy (of which *London* is a descendant), a *Citizen* taxonomy (of which *senior citizen* is a descendant), a *Needs* taxonomy (including *housing*) and an *Action* taxonomy (of which *grant* is a descendant). The items to be classified under the compound concept will be classified under Location>London, Citizen>senior citizen, Needs>housing and Action>grant, rather than under the compound concept. The extensional inference rule establishes a relationship among these concepts and the compound concept can be recovered by zooming on any permutation of them. In a conventional classification scheme, in which every item is classified under a single concept, a number of different concepts equal to the cartesian product of the terminals in the four taxonomies has to be defined. Such a combinatorial growth either results in extremely large conceptual taxonomies, which are difficult to understand and navigate, or in a gross conceptual granularity (Sacco, 2000).

Most importantly, faceted design coupled with dynamic taxonomies allows a more effective exploration of the information base makes it simple to focus on a concept, e.g. *senior citizen,* and immediately see all related concepts such as grants, locations, etc., which are recovered through the extensional inference rule. In the compound concept approach, these correlations are unavailable because they are hidden inside the concept label. This feature is extremely important in assisting users to discover all the information relevant to them: for instance, by zooming on *senior citizen* and then on *grants*, the user may discover that, in addition to grants for housing on which she is currently interested into, there might be also grants for education, etc.

Additional advantages include the uniform management of heterogeneous items of any type and format, easy multilingual access which is especially important for transnational e-government applications and easy integration with other retrieval methods, such as IR or database queries.

Dynamic taxonomies do not support reasoning beyond the extensional inference rule, and are therefore less powerful than general ontologies. However, they can be directly manipulated by users without the mediation of specialized agents and represent a quicker, less costly and more user-transparent alternative.

Applications in E-Government

Dynamic taxonomies have an extremely wide application range and a growing body of literature indicates that their adoption benefits most "search" applications, although the main industrial application area is currently e-commerce where dynamic taxonomies represent the *de facto* standard for product selection. Such a widespread use in a key internet application such as e-commerce is quite positive in the present context. First, it should make the adoption of this technology in the public sector considerably easier because major internet players use it and its significant benefits are evident to everyone. Second, e-commerce is extremely popular and this means that the large majority of potential users are already familiar with the paradigm of dynamic taxonomies.

The major application areas in e-government are:

- **Laws and regulations (Sacco, 2005a):** Although the law does not excuse ignorance, citizens have never had complete and up-to-date information on all the laws and regulations that concern them. Traditional information-publishing techniques, usually Official Gazettes on paper, make effective knowledge and awareness virtually impossible. One of the greatest opportunities of e-government is to overcome this information gap and to supply timely and complete information to everybody. This is important not only for the observance of the law, but also for the participation of the citizen in the political decision-making process: no knowledge, no democracy. The on-line, electronic availability of information is but a minor aspect of this problem. Rather, information must be disseminated in effective, timely and accurate ways, in such a way that finding what concerns the citizen is as simple as possible. Dynamic taxonomies can provide guided browsing and personalized exploration for large and complex collections of laws and regulations. Most importantly, dynamic taxonomies can be used to implement push strategies, in which users set a profile of relevant topics, and the system sends them relevant information as soon as it is available.

- **Informative and promotional services (Sacco, 2005c):** Governments, especially local ones, use the web to provide a number of services that are mainly informative and aim at improving the quality of life of citizens and at promoting the local community "abroad". These services include, among others, job placement services, tourist information (hotels, restaurants, etc.), yellow pages to promote local industries and activities, and are supplied in addition to institutional services such as law, regulations and opportunities information bases. A variety of different access techniques, including information retrieval, database queries and hypermedia, are currently used to deploy these services. This approach obviously results in dishomogeneity in access and user interactions that are more complex than required, since different access paradigms are to be mastered. Most importantly, none of these traditional techniques offer the exploration capabilities that are fundamental in this context.

- **Human resources and job placement (Berio, Harzallah & Sacco, 2006):** The current economic situation has considerably increased unemployment, so that one of the task of e-government is to provide job placement services that improve the chances of employment by matching competencies and expectations of citizens. At the same time, limitations on public expenditure requires that public administrations use their human capital in a much better and efficient way than in the past. The only way of doing this is by finding needed competencies inside the public administrations rather than outsourcing specific tasks. Both scenarios are essentially based on exploratory search patterns, in which the user

ranks required competencies by decreasing importance, and the search iterates until no additional competence can be filled or remains. The final set of persons is the set of candidates to be evalutated. This is exactly the type of exploration supported by dynamic taxonomies. By converse, other frequently used techniques, such as text retrieval or structured database queries, result in frustrating hit-and-miss interactions. It is worth mentioning that the combination of dynamic taxonomies with text retrieval allows the use of simpler taxonomies that do not account for all the competencies. Retrieval for rare or less useful competencies can be done by text retrieval. This can considerably simplify the taxonomy and its management, at the cost of not being able to summarize results for these competencies, and of having to standardize them.

- **E-services for citizens (Sacco, 2006b):** Government e-services available to citizens represent one of the most frequent and critical points of contact between public administrations and citizens. In addition to common services such as id cards, permits, e-services represent the only practical way of providing incentives and support to specific classes of citizens. For this reason, discovery of e-services, rather than plain retrieval, is a critical functionality in e-services systems. A solution based on dynamic taxonomies allows the transparent, guided, user-centric exploration of complex information bases. It provides a single framework for the access and exploration of all e-government information and, differently from mainstream research in semantic web, it is intended for direct use by end-users, rather than for programmatic or agent-mediated access.

- **Healthcare, including medical guideline portals (Wollersheim and Rahayu, 2002) and medical diagnosis (Sacco,**

2005b, 2012b): Governments are involved in healthcare either directly, as in most European countries, or indirectly, through national health organizations. In both cases, computerized support for medical diagnosis and medical protocols plays an important role in the effectiveness of the medical system. In traditional knowledge-based diagnostic systems, the system is an oracle and the physician the slave in a master-slave relationship with the system. Not surprisingly, this approach has not proved to be successful. There are a number of reasons for failure, but the most obvious is that highly-skilled personnel resent being treated as a mere input device for a computerized system, and an executor of its decisions. On the other hand, clinical diagnosis through dynamic taxonomies is performed by exploring and thinning out candidate pathologies on the basis of clinical signs and other observable features in a guided yet practitioner-centered way. Dynamic taxonomy systems for medical diagnosis, are immediately perceived as a friendly assistant, that helps the physician to consider all the possible symptoms (consequently guaranteeing a high quality level in diagnosis) while discarding irrelevant data. The physician is always in full control. Early experience with this approach is quite encouraging.

- **Cultural heritage and museum portals (Yee et al., 2003, Hyvönen et al., 2004, Pérez de Celis Herrero et al., 2011):** The preservation of a nation's cultural heritage is certainly one of the tasks of governments. Beyond physical preservation, this entails the creation and maintenance of complex catalogs of art, architectural and cultural works. Dynamic taxonomies allow to classify such complex items according to different perspectives, and provide exploration capabilities that bring to the surface the relationships among items that are essential for the understanding of culture and art.

- **National civil protection, disaster planning and risk mitigation:** These are becoming more and more relevant in many countries, and especially in those which are prone to hydro-geological, seismic or volcanic risks. In this context, the exploration capabilities of dynamic taxonomies and especially the taxonomic summaries that show all related concepts can play a fundamental role (Sacco et al. 2012a) both before and, most importantly, during emergencies. A quick focus on selected concepts can show a complete picture of all relevant information, taxonomically organized in a compact form. This is an invaluable help for situations in which critical and informed decisions must be taken within minutes.

CONCLUSION

Following the quick and widespread adoption by e-commerce portals that has made this new search paradigm quite well known in the Internet, we expect dynamic taxonomies to become pervasive in e-government applications as well, and to replace or integrate traditional techniques. Exploratory browsing applies to most practical situations and search tasks in e-government. In this context, dynamic taxonomies represent a dramatic improvement over other search and browsing methods, both in terms of convergence and in terms of full feedback on alternatives and complete guidance to reach the user goal.

REFERENCES

Berio, G., Harzallah, M., & Sacco, G. M. (2006). Portals for integrated competence management. In Tatnall, A. (Ed.), *Encyclopedia of portal technology and applications*. Idea Group Inc.

Berners-Lee, T., Hendler, J., & Lassila, O. (2001, May 17). The Semantic Web. *Scientific American*, 35–43.

Blair, D. C., & Maron, M. E. (1985). An evaluation of retrieval effectiveness for a full-text document-retrieval system. *Communications of the ACM*, *28*(3), 289–299. doi:10.1145/3166.3197

Groenbaek, K., & Trigg, R. (Eds.). (1994). Hypermedia. *Communications of the ACM*, *37*(2).

Hearst, M. (2006). Clustering versus faceted categories for information exploration. *Communications of the ACM*, *49*(4). doi:10.1145/1121949.1121983

Hearst, M. (2002). Finding the flow in web site search. *Communications of the ACM*, *45*(9), 42–49. doi:10.1145/567498.567525

Hyvönen, E., Saarela, S., & Viljanen, K. (2004). Application of ontology techniques to view-based semantic search and browsing. *Proceedings of the First European Semantic Web Symposium (ESWS 2004), LNCS 3053*, (pp. 92-106).

Pérez de Celis Herrero, C., Lara Alvarez, J., Cossio Aguilar, G., & Somodevilla García, M. J. (2011). An approach to art collections management and content-based recovery. *Journal of Information Processing Systems*, *7*(3), 447–458. doi:10.3745/JIPS.2011.7.3.447

Ranganathan, S. R. (1965). The colon classification. In Artandi, S. (Ed.), *Rutgers series on systems for the intellectual organization of information* (*Vol. 4*). Rutgers, NJ: Rutgers University Press.

Sacco, G. M. (2000). Dynamic taxonomies: A model for large information bases. *IEEE Transactions on Knowledge and Data Engineering*, *12*(2), 468–479. doi:10.1109/69.846296

Sacco, G. M. (2003). The intelligent e-sales clerk: The basic ideas. *Proceedings of INTERACT'03 -- Ninth IFIP TC13 International Conference on Human-Computer Interaction*, (pp. 876-879).

Sacco, G. M. (2005a). No (e-)democracy without (e-)knowledge. In *E-Government: Towards Electronic Democracy, International Conference IFIP TCGOV 2005, Lecture Notes in Computer Science 3416*, Bolzano, (pp. 147-156). Springer

Sacco, G. M. (2005b). Guided interactive diagnostic systems. *18th IEEE International Symposium on Computer-Based Medical Systems (CBMS'05)*, (pp. 117-122).

Sacco, G. M. (2005c). Guided interactive information access for e-citizens. In *EGOV05 – International Conference on E-Government, within the Dexa Conference Framework, Springer Lecture Notes in Computer Science 3591*, (pp. 261-268).

Sacco, G. M. (2006a). Analysis and validation of information access through mono, multidimensional and dynamic taxonomies. *FQAS 2006, 7th International Conference on Flexible Query Answering Systems, Lecture Notes in Artificial Intelligence*. Springer.

Sacco, G. M. (2006b). *User-centric access to e-government information: e-citizen discovery of e-services*. 2006 AAAI Spring Symposium Series, Stanford University.

Sacco, G. M., Nigrelli, G., Bosio, A., Chiarle, M., & Luino, F. (2012a). Dynamic taxonomies applied to a web-based relational database for geo-hydrological risk mitigation. *Computers & Geosciences*, *39*, 182–187. doi:10.1016/j.cageo.2011.07.005

Sacco, G. M. (2012b). Global guided interactive diagnosis through dynamic taxonomies. *18th IEEE International Symposium on Computer-Based Medical Systems* (CBMS'12), Rome.

Sacco, G. M., & Tzitzikas, Y. (Eds.). (2009). *Dynamic taxonomies and faceted search – Theory, practice, and experience. The Information Retrieval Series* (*Vol. 25*). Springer.

van Rijsbergen, C.J. (1979). *Information retrieval*. London, UK: Butterworths.

Wollersheim, D., & Rahayu, W. (2002). Methodology for creating a sample subset of dynamic taxonomy to use in navigating medical text databases. *Proceedings of IDEAS 2002 Conference*, (pp. 276-284).

Yee, K.-P., et al. (2003). Faceted metadata for image search and browsing. Proceedings of ACM CHI 2003, (pp. 401-408).

KEY TERMS AND DEFINITIONS

Extension, Deep: Of a concept C, denotes the shallow extension of C union the deep extension of C's sons.

Extension, Shallow: Of a concept C, denotes the set of documents classified directly under C.

Extensional Inference Rule: Two concepts A and B are related if there is at least one item *d* in the knowledge base which is classified at the same time under A (or under one of A's descendants) and under B (or under one of B's descendants).

Facet: One of several top level (most general) concepts in a multidimensional taxonomy. In general, facets are independent and define a set of "orthogonal" conceptual coordinates.

Subsumption: A subsumes B if the set denoted by B is a subset of the set denoted by A ($B \subseteq A$).

Taxonomy: A hierarchical organization of concepts going from the most general (topmost) to the most specific concepts. A taxonomy supports abstraction and models subsumption (IS-A and/or PART-OF) relations between a concept and its father. Tree taxonomies can be extended to support multiple inheritance (i.e., a concept having several fathers).

Taxonomy, Monodimensional: A taxonomy where an item can be classified under a single concept only.

Taxonomy, Multidimensional: A taxonomy where an item can be classified under several concepts.

Taxonomy, Reduced: In a dynamic taxonomy, a taxonomy, describing the current user focus set F, which is derived from the original taxonomy by pruning from it all the concepts not related to F.

User Focus: The set of documents corresponding to a user-defined composition of concepts; initially, the entire knowledge base.

Zoom: A user interface operation, that defines a new user focus by OR'ing user-selected concepts and AND'ing them with the previous focus; a reduced taxonomy is then computed and shown to the user.

Chapter 4
Technology Design for E–Governance in Nonprofit Organizations

Saqib Saeed
Bahria University Islamabad, Pakistan

Markus Rohde
University of Siegen, Germany

ABSTRACT

Nonprofit organizations are an important sector of society working to support underprivileged citizens. The operations of nonprofit organizations differ from their organizational size, scope, and application domain. Modern computer systems are quite effective in managing organizational tasks, but the nonprofit sector lacks in technological systems concerning organizational settings. In order to foster a successful use of electronic services, it is vital that computer systems are appropriate according to user needs. The diversity of users and their work practices in nonprofit organizations make it difficult for standardized infrastructure to work optimally in diverse organizational settings. In this chapter, the authors discuss the issues and complexities associated with system design for nonprofit organizations. They analyze important open issues that need to be explored for appropriated technology design in this domain.

1 INTRODUCTION

Effective governance methodologies are required to improve the work in every sector of economy. Modern technological artifacts have huge potential to improve the governance in organizational settings. As a result governmental organizations

DOI: 10.4018/978-1-4666-3640-8.ch004

focus on providing of services electronically, which has led to the evolution of an electronic government. Nonprofit organizations are becoming quite important due to their support to underprivileged citizens, and not much literature exists on the e-governance initiative in this particular domain. The operations and compositions of nonprofit organizations are quite different from governmental organizations, and they are not run through standard business models (cf. Saeed et al., 2008). Furthermore nonprofit organizations (NPOs) normally lack funding to invest in establishing technological systems. In this chapter we will argue for a specific research program to support nonprofit organizations in technology design. We will highlight the literature and also briefly discuss the results of our own projects. We will briefly outline the program and the aimed results.

The major obstacle in appropriate technology design in the nonprofit sector, are the various organizational settings, the absence of a stable organizational structure and the lack of financial and human resources. As a result it is pertinent to analyze the organizational work practices before the design of technology, so that technological systems will be well perceived by users. Although structure and working methodology of nonprofit organizations show some similarities with business and governmental organizations, there are considerable differences too. Therefore, the organizational structure needs to be investigated and the application area affecting the IT requirements of nonprofit orgnizations. Figure 1 highlights salient features of nonprofit organisations, which also affect system design. Every application domain

in which nonprofit organizations work, has different work practices, e.g. an organization dealing with human traffickers has different working practices than an organization working to help the victims of natural disasters. So this difference in the application domain is the major factor in setting up system design. Another factor is the nature of nonprofits organizations themselves, whether they work for advocacy only or work in the field. The communication and collaboration needs of both organizations differ, which also affects technological support requirements. In the case of nonprofit organizations working in field settings, office, field, donor, government and general public are major collaborators. As a result communication needs are focused on office-field-government communication, office-public communication, field-public communication and office to donor communication. In the case of advocacy companies, field settings do not play a major role, so their communication needs are mostly office-government communication, office-public communication, and office-donor communication. Another important factor affecting technological needs in nonprofit organization is

Figure 1. Parameters highlighting diversity among nonprofit organizations

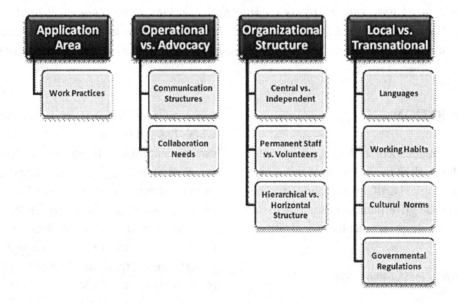

their organizational structure. Technological needs for the governance in nonprofit organizations with central control is different from nonprofits, which are decentralized and work independently at their respective locations. In centrally controlled nonprofit organizations the timely delivery of information regarding strategies, plans of actions is of utmost importance. Similarly, nonprofits relying on volunteers need a motivation and mobilization system to attract and manage volunteers, whereas nonprofits employing permanent staff will focus on managing the staff by employing a payroll system etc. Similarly the hierarchical/horizontal structure of organizations will require different technological needs. The decision-making in horizontal organizations will require quite complex software systems but they may not be required in a hierarchical structure where decision-making is similar to that in business organizations. The transnational nonprofit organizations impose extra requirements for technology design. In such a setting, language differences, cultural norms, government regulations and employees working habits may impose additional technological constraints.

The remaining parts of this chapter are structured as follows: The next section highlights the technology design efforts in nonprofit settings. Section 3 briefly discusses the results of our own research projects and is followed by conclusion where we discuss the research questions which need to be further evaluated by the scientific community.

2 TECHNOLOGY DESIGN IN NONPROFT SETTINGS

The notion of human-centered computing has highlighted this aspect and advocates for gaining in-depth understanding of human work practices before technology design. As a result the technology facilitates existing work practices instead of changing work practices by introduction of computing systems. A successful technology

intervention in an application setting requires that a large majority of users benefit from the technological artifact in practice. Different user-centered approaches have emerged to accommodate the user perspective within system design (cf. Galer et al, 1992; Vredenburg et al., 2001). Despite an enhanced focus on users, system design and system usage timelines were mutually exclusive. Later the "designing for change" notion emerged that supported flexibility during system design in multiple organizational and user contexts, which is also referred to as tailorability (cf. Trigg et al, 1987; Henderson & Kyng, 1991). Designing for change further led to the concept of technology appropriation, which Dourish (2003) described in the following words:

Appropriation is the way in which technologies are adopted, adapted and incorporated into working practice (p. 466).

Many researchers in computer supported cooperative work (CSCW) literature have highlighted how users adopt technology into their work practices (cf. Silverstone & Haddon 1996; Dourish 2003; Pipek, 2005; Balka & Wagner 2006; Stevens 2009). However, Pipek et al. (2006) have described that supporting users in appropriating technology in a less professional settings (such as home/volunteer settings) is also quite beneficial to foster successful usage. Nonprofit organizations are one such domain where supporting efforts for technology appropriation are quite few.

While reflecting on the technology appropriation in nonprofit networks, we categorized the relevant literature into the following three classes: one body of knowledge looks at technology design initiatives, whereas a second body of knowledge mainly focuses on the IT usage by such communities, the last body of knowledge highlights the efforts, made by NPOs themselves, to improve IT appropriation in their operational settings.

The advantages of employing participatory design methods for system design in non-profit organizations have been discussed by Benston (1990). McPhail et al. (1998) applied participatory design methodology to a Canadian non-profit organization in their information system design. A similar initiative was taken by Trigg (2000) to involve a non-profit organization in a database design project. Another project called "Civic Nexus" was carried out at Penn State University aiming at involving regional volunteer organizations in design processes geared towards achieving technological sustainability, participatory design and end user development concepts were applied (cf. Merkel et al., 2004; Farooq et al., 2005; Farooq, 2005; Farooq et al., 2006). Rohde (2004) used participatory design methods to electronically network an Iranian NGO community, so that the NGOs were able to benefit from increased efforts to build social capital. McIver worked on transnational, multi lingual and collaborative legislative work among NGOs on the basis of his involvement in drafting legislation for a civil society's agenda at the World Summit on the Information Society (WSIS) (cf. McIver, 2004; McIver, 2004a). Pilemalm (2002) involved Swedish trade unions in participatory design processes with regard to exploring ICT needs, establishing technological solutions and analyzing their impact on Swedish trade unions.

There have been empirical studies on the adaptation and use of ICT by different voluntary organizations: O'Donnell (2001) analyzed the role of mailing lists in connecting different women's organizations in Northern Ireland. Cammaerts and Van Audenhove (2003) investigated how transnational social movement organizations use the internet in their organizing process. Pini et al. (2004) studied the use of discussion lists by an Australian women's farm group (AWiA). O'Donnell and Ramaioli (2004) analyzed an online information network for the non-profit sector in Ireland. Cheta (2004) investigated the usage of the internet by the social movement organization *Portuguese Accessibility Special Interest Group* (GUIA). Edwards (2004) presented

a case study on the role of internet applications for the Dutch women's movement. Cordoso and Neto (2004) investigated the role of ICTs in the pro-East Timor movement in Portugal. Aelst and Walgrave (2004) analyzed the use of the internet in organizing protests within the anti-globalization movement. O'Donnell et al. (2007) focused on how two community-based organizations used video communication to support economic and social development in remote areas of Canada. Kavada (2005) investigated the use of the internet by three non-governmental organizations in the UK and analyzed how email lists helped the organizing process of the ESF 2004 (Kavada, 2007).

In order to improve ICT use and the appropriation, different projects were voluntarily carried out by the organizations themselves as well. In 2003 Interagency Working Group (IWG) on Emergency Capacity was formed by a consortium of seven NGOs to analyze collaborative capacity-building efforts in the areas of staff capacity, accountability and impact measurement, risk reduction and ICT requirements. The participating NGOs were Oxfam-GB, Save the Children-US, World Vision International, Catholic Relief Services, the International Rescue Committee, CARE International, and Mercy Corps. Currion (2006), based on his visits of the headquarters of the participating NGOs and field visits in Pakistan and Sudan, presented a comprehensive report on the use of information technology by the IWG agencies in their global response to emergencies. This forms a first detailed study of use of ICTs by NGOs in a time of emergency. NetHope is another nonprofit corporation which helps member NGOs by sharing information technology knowledge. The goal of the program is to help NGOs improve their emergency response and enhance organizational effectiveness by strengthening crucial information technology skills (NetHope, 2008). A group named LINGOS offers its services, to improve performance and effectiveness of non-governmental organizations. They offer a Learning Management System (LMS) that enables NGOs to host, develop, track and implement learning content for their staff. They have implemented the "communities of practice" idea to provide a venue for learning professionals from international NGOs to share information on course availability, the best learning practices, sources for learning products, links to sites on learning and development issues and other topics of common interest (LINGOs, 2008).

3 CASE STUDIES

In order to further enrich the body of knowledge, we have worked further in this domain. In this section we will briefly discuss three different projects which we have carried out. These projects were done in different geographical areas to gain an understanding of NPO practices in both developed and developing regions.

3.1 Technology Usage at European Social Forum

In this project we conducted a long-term ethnographical field study of the European Social Forum (ESF) for three years. ESF is a network of heterogeneous political activist organizations who organize a regular biannual event. During our data collection phase, the 5th and 6th European Social Forum, was held in Malmo (2008) and Istanbul (2010), in which 13,000 (Malmo) and 3,000 (Istanbul) activists participated, respectively. Empirical data was gathered from January 2008 to October 2010 using an ethnographic action research approach (Hughes et al., 1994; Randall et al., 2007). The reason for adopting an ethnographic approach was to understand the complexity of work practices as they are performed in real world settings (cf. Garfinkel, 1967; Garfinkel, 1974; Crabtree et al, 2000). The data was collected, by using triangulation of research methods that included semi-structured interviews, participant observation, and a content analysis of relevant documents and web sites. The semi-structured interviews were conducted with thirty-one activists participating in the ESF; overall, we ended up with twenty hours of audio-recordings of these interviews. Moreover, for the participant observation, we carried out eight different field visits, lasting 26 days in total, from 2008 to 2010. We particularly focused on the usage of information and communication technologies (ICTs) when preparing and conducting ESF events, and knowledge sharing practices during the transition phase. We have used the term *fragmented meta-coordination* to highlight coordination practice in this type of action (cf. Saeed et al., 2009; Saeed et al., 2011a). Mundane ICT applications, such as a mailing list (Saeed et al., 2011) and a content management system (Saeed & Rohde, 2010), played a central role in enabling different aspects of fragmented meta-coordination. We were also presented with a specific type of knowledge, termed as *nomadic knowledge,* which is required periodically by different actors and travels along foreseeable paths between groups or communities of actors (cf. Saeed et al., 2010). This knowledge differs from generally held assumptions about the way knowledge is enacted. We also took a historic perspective on the evolution of ICT artifacts within the organizational boundaries. The analysis highlights central organizational and technological challenges related to ICT appropriation in transnational networks of social activists.

3.2 ICT Support in Organizing WSF 2006

World Social Forum (WSF) is the most popular event of civil society organizations and attracts a large number of activists from all over the world (Kavada, 2007). In 2006 WSF was organized as a poly-centric event, two other locations were Bamako, Mali and Caracas, Venezuela. We investigated the organizational practices in the organizing process of the WSF event in Karachi, Pakistan to understand the ICT usage. It was a kind of post analysis so we pursued a combination of research methods to acquire empirical data.

Initial information was gathered using open-ended email questionnaires, which were sent to the people involved in the organizing process. This information was then further refined by review of documents, websites of the event and follow-up interviews over the telephone. In order to understand the communication activities, the contents of emails were analyzed by joining email list (Yahoo Groups). In our findings we discovered information management weaknesses, which hampered the successful organization of the tasks. It was also found that the usage of social media websites was not well conceived. Main modes of communication were telephone calls and physical meetings. Further details of the work can be found in (Saeed et al., 2012a).

3.3 ICT Support for Pakistani NGOs

In this project we used a questionnaire to understand the technology usage among Pakistani NGOs. We specifically selected Pakistan, due to its lower rank at the digital opportunity index. The exact number of civil society organizations there is not known. Some estimates place the presence of nonprofit organizations around 60,000, but the number of registered organizations is only around 12,000 (Sattar & Baig, 2001). The questionnaire was distributed through email to different organizations and it was also accessible online. There were 15 respondents and the responding organizations were located in three different provinces of Pakistan. The survey results highlight the lack of resources as the major obstacle in setting up technological systems. Only six of the fifteen organizations have a dedicated organizational website and only one organization was able to continuously keep it up to date. Email communication was the most widely used ICT service. Further details of the work could be found in (Saeed et al., 2012).

4 CONCLUSION

The objective of this chapter is to highlight the importance of ICT support in nonprofit organizations and stress the need for further research to empower this important sector with modern computing technologies. While we closely analyzed the three case studies, the lack of funding to support technology infrastructure appeared to be the major challenge for technology adoption. Although there are many free and open access software applications available, we have found out that these artifacts were not sustainable enough and are actively used only as long as a certain volunteer had interest in it. Web 2.0 applications are very promising for communication in the nonprofit sector, but in our case studies we have only seen a sporadic use. The results of our case studies brought insights into the technological needs and how diverse they are among nonprofit organizations. It became obvious that HCI/CSCW fields still need to investigate ICT appropriation in nonprofit networks to better understand their technology appropriation needs. The focus of such research efforts could be on a main question of how technology artifacts can help nonprofit networks in their organizational settings and how knowledge management paradigms can help overcome inherent weaknesses due to rapid volunteer turnover.

According to our findings, some appropriations and need-oriented designed systems could improve the governance aspect among nonprofit organizations, but the inherent weaknesses of nonprofit organizations hinder them in adopting such technology infrastructures. Researchers and universities need to contribute to this aspect by fulfilling their social responsibility by establishing joint projects with such organizations by setting up long-term plans and goals (cf. Gurstein, 2011). Such human-centered design initiatives could help the development of sustainable technological systems (cf. Farooq et al., 2005). Although transferability of results from one setting to another is a complex task (cf. Chi et al., 2011), nonprofit organizations working at different geographical levels can benefit from the findings of such projects, when working on their ICT strategy. As results of such initiatives, new technology artifacts could be introduced to this sector, facilitated with HCI/CSCW community knowledge. Furthermore, these technology artifacts need to be evaluated in practice to further refine them. The output of such research initiatives could result in a set of tools and repertoires for the organizational needs of nonprofit organizations.

REFERENCES

Aelst, P. V., & Walgrave, S. (2004). New media, new movements? The role of the internet in shaping the anti-globalization movement. In W. van de Donk, et al. (Ed.), *Cyber protest, new media, citizens and social movements,* (pp. 97-122). London, UK: Routledge.

Balka, E., & Wagner, I. (2006). Making things work: Dimensions of configurability as appropriation work. [ACM.]. *Proceedings of CSCW, 2006*, 229–238. doi:10.1145/1180875.1180912

Benston, M. (1990). Participatory designs by non-profit groups. In *Proceedings of the Participatory Design Conference,* Palo Alto CA, 1990 (pp. 107-113).

Cammaerts, B., & Van Audenhove, L. (2003). *ICT-usage among transnational social movements in the networked society: To organize, to mediate & to influence.* Amsterdam, The Netherlands: ASCoR, Amsterdam Free University.

Cardoso, G., & Neto, P. P. (2004). Mass media driven mobilization and online protest ICTs and the pro-East Timor movement in Portugal. In Van de Donk, W. (Eds.), *Cyber protest, new media, citizens and social movements* (pp. 147–163). London, UK: Routledge.

Cheta, R. (2004). Dis@bled people, ICTs and a new age of activism: A Portuguese accessibility special interest group study. In van de Donk, W. (Eds.), *Cyber protest, new media, citizens and social movements* (pp. 207–232). London, UK: Routledge.

Chi, E. H., Czerwinski, M., Millen, D., Randall, D., Stevens, G., Wulf, V., & Zimmermann, J. (2011). Transferability of research findings: context-dependent or model-driven. In *Proceedings of the 2011 Annual Conference: Extended Abstracts on Human Factors in Computing Systems* (CHI 'EA 11), (pp. 651-654). New York, NY: ACM.

Crabtree, A., O'Brien, J., Nichols, D., Rouncefield, M., & Twidale, M. (2000). Ethnomethodologically informed ethnography and information system design. *Journal of the American Society for Information Science American Society for Information Science, 51*(7), 666–682. doi:10.1002/(SICI)1097-4571(2000)51:7<666::AID-ASI8>3.0.CO;2-5

Currion, P. (2006). *NGO information technology and requirements assessment report.* Emergency Capacity Building Project Obtained through the Internet. Retrieved from http://www.ecbproject.org/publications_4.htm

Dourish, P. (2003). The appropriation of interactive technologies: Some lessons from placeless documents. *International Journal of Computer Supported Cooperative Work, 12*(4), 465–490. doi:10.1023/A:1026149119426

Edwards, A. (2004). The Dutch women's movement online Internet and the organizational infrastructure of a social movement. In van de Donk, W. (Eds.), *Cyber protest, new media, citizens and social movements* (pp. 183–206). London, UK: Routledge.

Farooq, U. (2005). Conceptual and Technical Scaffolds For End User Development: Using scenarios and wikis in community computing. In *Proceedings of the IEEE Symposium on Visual Languages and Human-Centric Computing* Los Alamitos, California, (pp. 329-330).

Farooq, U., Merkel, C. B., Nash, H., Rosson, M. B., Carroll, J. M., & Xiao, L. (2005). Participatory design as apprenticeship: Sustainable watershed management as a community computing application. *Proceedings of the 38th Annual Hawaii International Conference on System Sciences*, Hawaii, January 3-6, 2005.

Farooq, U., Merkel, C. B., Xiao, L., Nash, H., Rosson, M. B., & Carroll, J. M. (2006). Participatory design as a learning process: Enhancing community-based watershed management through technology. In Depoe, S. P. (Ed.), *The environmental communication yearbook* (*Vol. 3*, pp. 243–267). doi:10.1207/s15567362ecy0301_12

Galer, M., Harker, S., & Ziegler, J. (1992). *Methods and tools in user-centered design for information technology (Human factors in information technology)*. Elsevier Science Ltd, North-Holland.

Garfinkel, H. (1967). *Studies in ethnomethodology*. Englewood Cliffs, NJ: Prentice-Hall.

Garfinkel, H. (1974). On the origins of the term ethnomethodology. In Turner, R. (Ed.), *Ethnomethodology* (pp. 15–18). Harmondsworth, UK: Penguin.

Gurstein, M. (2011). Evolving relationships: Universities, researchers and communities. *Journal of Community Informatics*, 7(3).

Henderson, A., & Kyng, M. (1991). There's no place like home: Continuing design in use. In Greenbaum, J., & Kyng, M. (Eds.), *Design at work: Cooperative design of computer systems* (pp. 219–240). Hillsdale, NJ: Lawrence Erlbaum Association.

Hughes, J. A., King, V., Rodden, T., & Andersen, H. (1994). Moving out from the control room: Ethnography in system design. [ACM.]. *Proceedings of CSCW*, 94, 429–439.

Kavada, A. (2005). Civil society organizations and the internet: The case of Amnesty International, Oxfam and the world development movement. In de Jong, W. (Eds.), *Global activism, global media* (pp. 208–222). London, UK: Pluto Press.

Kavada, A. (2007). *The European social forum and the internet: A case study of communication networks and collective action*. Ph.D Thesis, University of Westminster, UK.

LINGOs. (2008). *Obtained through the Internet*. Retrieved from http://www.lingos.org

McIver, W. (2004). Tools for collaboration between tans national NGOs: Multilingual, legislative drafting. In *Proceedings of International Colloquium on Communication and Democracy: Technology and Citizen Engagement*, Fredericton, New Brunswick, Canada, 2004.

McIver, W. (2004). Software support for multilingual legislative drafting. In *Proceedings of the Community Informatics Research Network Conference and Colloquium* Tuscany, Italy, 2004.

McPhail, B., Costantino, T., Bruckmann, D., Barclay, R., & Clement, A. (1998). CAVEAT exemplar: Participatory design in a non-profit volunteer organisation. *Computer Supported Cooperative Work*, 7(3), 223–241.

Merkel, C. B., Xiao, L., Farooq, U., Ganoe, C. H., Lee, R., Carroll, J. M., & Rosson, M. B. (2004). Participatory design in community computing contexts: Tales from the field. In *Proceedings of the Participatory Design Conference*, Toronto, Canada, (pp. 1-10).

NetHope. (2008). *Obtained through the Internet*. Retrieved from http://www.nethope.org

O'Donnell, S. (2001). Analysing the internet and the public sphere: The case of Womenslink. *Javnost (Ljubljana), 8*(1), 39–58.

O'Donnell, S., Perley, S., Walmark, B., Burton, K., Beaton, B., & Sark, A. (2007). Community-based broadband organizations and video communications for remote and rural First Nations in Canada. In *Proceedings of Community Informatics Research Network Conference,* Prato, Italy, 2007.

O'Donnell, S., & Ramaioli, G. (2004). Sustaining an online information network for non-profit organizations: The case of community exchange. In *Proceedings of the Community Informatics Research (CIRN) Network Conference. Sustainability and Community Technology: What Does this Mean for Community Informatics?* Prato, Italy, 2004.

Pilemalm, S. (2002). *Information technology for non-profit organizations extended participatory design of an information system for trade union shop stewards.* PhD Thesis, 2002, Linköping University, Sweden.

Pini, B., Brown, K., & Previte, J. (2004). Politics and identity in cyberspace: A case study of australian women in agriculture online. In de Donk, W. (Eds.), *Cyber protest, new media, citizens and social movements* (pp. 259–275). London, UK: Routledge.

Pipek, V. (2005). *From tailoring to appropriation support: Negotiating groupware usage.* PhD Thesis, University of Oulu, Finland.

Pipek, V., Rosson, M. B., Stevens, G., & Wulf, V. (2006). Supporting the appropriation of ICT: End-user development in civil societies. *Journal of Community Informatics, 2*(2).

PNAC. (2006). *Report of PNAC participation in world social forum 2006 Karachi.* Retrieved from http://www.pnac.net.pk/Reports/WSF-Report.pdf

Randall, D., Harper, R., & Rouncefield, M. (2007). *Fieldwork for design: Theory and practice. Computer Supported Cooperative Work.* New York, NY: Springer.

Rohde, M. (2004). Find what binds: Building social capital in an Iranian NGO community system. In Huysman, M., & Wulf, V. (Eds.), *Social capital and information technology* (pp. 75–112). Cambridge, MA: MIT Press.

Saeed, S., Pipek, V., Rohde, M., & Wulf, V. (2010). Managing nomadic knowledge: A case study of the European Social Forum. In *28th International Conference on Human Factors in Computing Systems,* (pp. 537-546). New York, NY: ACM.

Saeed, S., & Rohde, M. (2010). Computer enabled social movements? Usage of a collaborative web platform within the European Social Forum. In *9th International conference on the Design of Cooperative Systems,* (pp. 245-264). Springer.

Saeed, S., Rohde, M., & Wulf, V. (2008). A framework towards IT appropriation in voulantary organizations. *International Journal of Knowledge and Learning, 4*(5), 438–451. doi:10.1504/IJKL.2008.022062

Saeed, S., Rohde, M., & Wulf, V. (2009). Technologies within transnational social activist communities: An ethnographic study of the European Social Forum. In *Fourth international Conference on Communities and Technologies (C&T '09),* (pp. 85-94). New York, NY: ACM.

Saeed, S., Rohde, M., & Wulf, V. (2011). *Communicating in a transnational network of social activists: The crucial importance of mailing list usage.* In 17th CRIWG Conference on Collaboration and Technology. Springer.

Saeed, S., Rohde, M., & Wulf, V. (2011a). Analyzing political activists' organization practices: Findings from a long term case study of the European Social Forum. *Journal of Computer Supported Collaborative Work, 20*(4-5), 265–304. doi:10.1007/s10606-011-9144-0

Saeed, S., Rohde, M., & Wulf, V. (2012). Civil society organizations in knowledge society: A roadmap for ICT support in Pakistani NGOs. *International Journal of Asian Business and Information Management, 3*(2), 23–35. doi:10.4018/jabim.2012040103

Saeed, S., Rohde, M., & Wulf, V. (2012a). IT for social activists: A study of World Social Forum 2006 organizing process. *International Journal of Asian Business and Information Management, 3*(2), 62–73. doi:10.4018/jabim.2012040106

Sattar, A., & Baig, R. (2001). Civil society in Pakistan: A preliminary report. *CIVICUS Index on Civil Society Occasional Paper Series, 1*(11).

Silverstone, R., & Haddon, L. (1996). Design and the domestication of information and communication technologies: Technical change and everyday life. In Silverstone, R., & Mansell, R. (Eds.), *Communication by design: The politics of information and communication technologies* (pp. 44–74). Oxford, UK: Oxford University Press.

Stevens, G. (2009). *Understanding and designing appropriation infrastructures*. PhD Thesis University of Siegen, Germany.

Trigg, R. H. (2000). From sand box to "fund box": Weaving participatory design into the fabric of a busy non-profit. In *Proceedings of the Participatory Design Conference,* Palo Alto CA, 2000, (pp. 174-183).

Trigg, R. H., Moran, T. P., & Halasz, F. G. (1987). Adaptability and tailorability in NoteCards. [North-Holland.]. *Proceedings of IFIP INTERACT, 87,* 723–728.

Vredenburg, K., Isensee, S., & Righi, C. (2001). *User-centered design: An integrated approach.* USA: Prentice Hall.

Chapter 5
User–Centered Designs for Electronic Commerce Web Portals

Robert Jeyakumar Nathan
Multimedia University, Malaysia

Norazah Mohd Suki
University of Malaysia Sabah, Malaysia

ABSTRACT

Websites connect businesses with customers. They are an important medium that facilitates online transactions, a necessity for businesses. The design and usability of an Electronic Commerce (EC) website play an important role in achieving its objectives (Kumar, Smith, & Bannerjee, 2004; Marcus, 2005; Nielsen, 2003; 2005; Krug, 2006; Cappel & Huang, 2007). Recognizing their importance, design and usability aspects of EC websites have been widely researched in both applied and academic research (Lecerof & Paterno, 1998; Lohse & Spiller, 1999; Nielsen, 2000; Cao, Zhang, & Seydel, 2005; Flavian & Guinaliu, 2006; Nathan, Yeow, & Murugesan, 2008; Nathan & Yeow, 2009; Robins & Holmes, 2008). This chapter discusses the recent work with web design and electronic commerce. The importance of usability and user-centered web designs are highlighted. Usability to specific target groups and industries, such as airlines, government, and services portals, are also discussed. Altogether, design guidelines are given for web industries, and recommendations are made for better usability in designing websites.

WEB DESIGN AND USABILITY

The design of Websites, particularly the usability aspects of Websites, assumes great significance. The *International Organization for Standardization* (IOS) 9241 defines usability as the extent to which a product can be used by specified users to achieve specified goals with effectiveness, efficiency, and satisfaction in a specified context of use (Karat, 1997). Driven by the importance of usability in Web applications, there are several studies on Web usability. For example, Nielsen (2000) and Lecerof and Paterno (1998) studied how Website usability standards reduce errors,

DOI: 10.4018/978-1-4666-3640-8.ch005

enhances accuracy, increases usage, and improve the look of a Website. An empirical study by Lohse and Spiller (1999) found that Website usability explains a high percentage of variance (61%) in online sales.

Moraga *et al.* (2006) compared the different quality models for Web portals and found that researchers paid special attention to visual aspects of interface design. Calero *et al.* (2005) reviewed 60 papers on Web quality (including Web design) from 1992 to 2004 and classified them to the Web Quality Model. Allen (2002) studied the interface design of the University of South Florida's virtual library using usability testing method (University of South Florida Virtual Library, 2012). Chowdhury *et al.* (2006) reviewed many usability studies (including research on interface design) and studied their impact on digital libraries. They concluded digital libraries should be evaluated with respect to their target users. Xie and Cool (2000) compared the search experiences with Web and non-Web interfaces to online databases and found that some of the designs of Web interfaces outperform non-Web interfaces. In essence, all authors agree on the importance of interface design in making an online system usable by users.

Lecerof and Paterno (1998) highlighted the importance of specified users in determining the crucial usability aspects of a system. For example, ease of use may be the crucial usability aspect for Websites targeting children, whereas efficiency may be the crucial usability aspect for business-to-business e-procurement Websites. Ginige and Murugesan (2001) recommended ten key steps for the successful development of a Website. Among them is the need for developers to clearly identify the system's main users. This is in agreement with the ergonomics rule of thumb which states that "one size does not fit all". However, most Website design guidelines are catered to general users, for example, Ergonomic Guidelines for User-Interface Design (Hix and Hartson, 1993) and Web Analysis and Measurement Inventory factors (Kirakowski *et al.*, 1998). Users vary in many ways, thus different groups of Internet users may have different needs for Web interface design. A user-centric design is a vital consideration in designing an interface for Web-based systems (Tilson *et al.*, 1998; Ginige and Murugesan, 2001). Several Web usability factors are introduced based on past empirical studies in the areas of Internet Marketing, Ergonomics and Human Computer Interactions.

The *Overall Web Usability* (OWU) is vital to the success of EC Websites (Agarwal and Venkatesh, 2002; Flavian and Guinaliu, 2006; Tilson *et al.*, 1998). It plays a vital role in the successful design of a Website (Kumar *et al.*, 2004; Liu and Arnett, 2000). In this chapter, seven factors affecting OWU are discussed. These factors are known as Web Interface Usability Factors (WIUFs). The WIUFs have several similarities with the factors in Microsoft Usability Guidelines (MUG) as shown in Table 1.

Graphic and Color Designs in Websites

Kirakowski *et al.* (1998) conducted a survey on the various factors that influenced customers' decision-making in using Websites. He found attractiveness (e.g. UCF and UGM) to be most important compared with factors such as ease in navigation, ease in finding items, and clearly labeled items. Media use refers to the aesthetic appeal (attractiveness) of a Website, which is determined by the extent of the UCF and UGM. Several Web usability studies have also confirmed that appeal does significantly affect the usability of Websites (Lindgaard, 1999; Tractinsky *et al.*, 2000; Brady and Phillips, 2003; Phillips and Chaparro, 2009).

Shenkman and Jonsson (2000) found the most important determinant of the overall judgment of a Website was aesthetic appeal (or beauty). Aesthetic appeal, linked to human emotions, is attractive and will trigger a positive emotion resulting in a positive outcome such as a transaction (Isen, 1993). Another study found that attractiveness will make users feel better when using a product

Table 1. Description of web interface usability factors (WIUFs) and equivalent factor in Microsoft usability guidelines (MUG)

Web Interface Usability Factor (WIUF)	Description	Equivalent factor in MUG
Use of Color and Font (UCF)	The extent of use of color and text font in a Website. This factor is related to the aesthetic aspect of Website.	Media use
Use of Graphics and Multimedia (UGM)	The extent of use of graphics and multimedia objects such as video, audio, and animations in a Website. This factor is related to the aesthetic aspect of Website.	Media use
Clarity of Goals in Website (CGW)	The extent to which a Website offers understandable goals. For example, the goals may be community building, product sales, and research and development.	Goals
Trustworthiness of Website (TOW)	The credibility of the Website which is related to many factors such as the security technology used and the user's familiarity of the Website/brand/company.	Character Strength
Interactivity of Website (IOW)	The extent to which a Website offers interaction. This factor is related to the provision of feedback regarding user action and progress in using the Website. It is also related to made-for-the-medium factors such as personalization, community, and refinement.	Feedback and Made-For-The-Medium
Ease of Web Navigation (EWN)	The extent to which a Website is easily navigated. This factor is related to Website structure, navigation bar/column, site map, site search, directory, and hyperlinks.	Structure
Downloading Speed of Website (DSOW)	The speed of downloading a Webpage. This is related to the pace at which a user receives information.	Pace

MUG – Microsoft Usability Guidelines (Keeker, 1997; Agarwal and Venkatesh, 2002)

Source: Nathan, R.J., Yeow, P.H.P., & Murugesan, S. (2008). Key Usability Factors of Service-Oriented Websites for Students: An Empirical Study. Online Information Review, Vol. 32 (3) pp. 307

and gives them a sense of perceived satisfaction even though the product may not inherently possess qualities that satisfy users (Norman, 2002).

Web developers are also aware of the importance of design in their Website as it has the power to persuade visitors to linger longer in their sites, increasing their likelihood to patronize or purchase.

Goal-Setting in Commercial Websites

Clarity of Goals in Websites (CGW) are the extent to which the Website is understandable to the users (Turban et al., 2006, pp. 662) Websites can have several goals including; cost savings, increased processing efficiency, improved customer service, provided convenient marketing channels, etc. These goals are dependent on the business models of each company. Turban *et al.* (2006, pp. 92) presented *Business to Consumer* models such as transaction brokers, information portals, and community portals. The example given concerned

a Website from a company in New Zealand (www.obo.co.nz). The Website had clearly stated its goals in their "About OBO" link, i.e. for research and development, community building and product sales. The company's Website supports its goals by providing online discussions, sponsored players and an image gallery to achieve and support its stated goals (New Zealand MED, 2000, p.12).

Google at www.google.com clearly states its Website goal as "… to have users leave the Website as quickly as possible" (Google, 2012). As an effective search engine; they promise quick answers and solutions to users' search needs. This information is stated very clearly in their corporate information page.

The business model should be clear in facilitating the full use of the Website for both businesses and customers. Iterated by Rivero (1999), clear business goals in Websites lead to various benefits such as better met customer needs, customers' time saved and company resources optimized because they understand how the Website business can help them.

Consumer Perception of Safety and Security of Websites

Website security is found to be of fundamental importance in various studies. In Tilson *et al.* (1998), Web security for credit card transactions was the most important factor that affects EC usage, among others included: ease of product return, items extensive description, price of items, personal information privacy and pictures of merchandise. Similarly, Brynjolfsson and Smith (2000) found that *Trustworthiness of Website* (TOW) was the most important factor in EC.

Pennington *et al.* (2003) discovered TOW decreases with users' uncertainty towards technology, lack of initial face-to-face interactions and lack of enthusiasm among online buyers and sellers. Turban *et al.* (2006) observed that TOW is dependent on many factors such as competency, benevolence of the seller, reliability, understandability, security of the Website, consumer protection, effective law when using the Website and success stories.

Nathan *et al.* (2009) found that users' risk perception and TOW affect their EC adoption. This relationship is moderated by their knowledge of information technology systems such as *Secure Order Transaction* using *Secure Socket Layer*. Similarly, Turban *et al.* (2006 p. 478) found these technologies increase the TOW, particularly Websites that have online monetary transactions.

User Interactions in Websites

An interactive Website encourages users to participate and interact with it. This feature is commonly explained as the *Interactivity of Website* (IOW). The presence of various media such as text, graphics, animation, audio and video allow users to control Websites (Dix *et al.*, 1993). Watson *et al.* (2000) found that attractive features such as online games, interactive puzzles, prizes, contests, etc. increase user interaction with the Website.

An interactive Website aids users to know and understand what is offered, an example of what could be available include Websites with online newsletters, product demonstrations and customer forums. It provides a way for users to submit their feedback and comments to the administrator through e-mail, comment forms, online surveys and the virtual community (Teo, *et al.*, 2003; Turban *et al.*, 2006 p. 671).

In Agarwal and Venkatesh (2002) Microsoft Usability Guidelines factors, IOW is related to a feedback factor and made-for-the-medium factors which include community, personalization, and refinement. All these factors allow a user to control his/her interaction with a Website.

Getting Around and Navigating Websites

The *Ease of Web Navigation* (EWN) refers to how fast and easy a person can find the information he/she needs. It is determined with navigation bar, navigation column, site map, site search, directory, hyperlinks and Website structure (Turban *et al.*, 2006 pp. 686-689; Chen *et al.*, 1997; Turban *et al.*, 2006 pp. 684-686).

Nielsen (2000) found that users often find it difficult to locate what they are looking for in a Website. EWN is among the important factors that affects users' perception of Web usability and ultimately affects customers' decision making (Gehrke and Turban, 1999; Kirakowski *et al.*, 1998; Nielsen, 2000).

Navigating a Website has to be simple as users want predictable, consistent and intuitive surfing without requiring much thinking (Cheung and Lee, 2005; Iwaarden *et al.*, 2003; 2004). To avoid surfing confusion, Turban et al. (2006, p. 688) proposed consistency in the use of navigational aids, e.g. standardized format and style of navigation bars and standardized placement of navigation aids (i.e. in approximately the same location on every Web page).

Speed of Accessing and Downloading

The *Download Speed of Website* (DSOW) is important to Web surfers. People are normally turned away when they have to wait too long for a Webpage to download (Tilson *et al.*, 1998). It is a vital factor that determines whether a user revisits a Website (Helander, 2000). Large graphics such as animation and video take a long time to download, thus they increase the DSOW.

Turban *et al.* (2006 p. 688) suggested Webpages should not take more than 12 seconds to download lest visitors would be turned away. Nielsen and Tahir (2002) evaluated the 50 most popular Websites and found the average downloading time for their homepage was 26 seconds which exceeds Turban's *et al.* (2006) recommendation. They also found only 28% of the Websites had acceptable DSOW of less than 10 seconds.

USABILITY MEASUREMENT MODELS

Various Web evaluation methods are available for assessing usability of a Website. The most common methods are explained in this section. The 7 popular methods are explained and summarized in Table 2.

Website Industry Effect on Users' Usability Perception

Agarwal and Venkatesh (2002) asserted most Website usability studies were not focused towards specific users and industry, but instead were mostly conducted using a small sample size. They conducted a large scale study (with 1,475 Web evaluations), taking into consideration four industries (airline, bookstore, auto manufacture, and car rental) and two user roles (customer and investor).

Table 2. Web evaluation methods

No.	Method	Description	Used in Other Studies
1	Cognitive Walk-Though	Respondents identify and evaluate each step of their surfing thought process.	Olson and Olson (1990); Lewis, Polson, Wharton and Rieman (1990); Jørgensen (1990).
2	Focus Group	A form of qualitative research approach. Important variables or constructs for usability are determined first through focus group of 8 to 12 people and commonly followed by User Testing.	Kelly, Wacholder, Rittman, Sun, Kantor, Small and Strzalkowski (2007); Rigby (2004); Albers (2009); Ferney and Marshall (2006).
3	Goals, Operator, Methods and Selection (GOMS)	Goals, Operators, Methods, and Selection Rules techniques. Operators focus on the users of the system and their unique requirements while focusing on the goals of using the system.	Card, Moran and Newell (1983); John and Kieras (1994).
4	Prototyping	System is designed in small scale and tested. Potential users are involved in the development of prototype and testing; aspects of interactivity and design are tested for usability.	Kinzie, Cohn, Julian and Knaus (2002)
5	Task-Analysis	Also based on a qualitative research approach like a focus group. Evaluates how people actually use a system and its interface to do a task and accomplish things.	McMullen (2001)
6	Usability Inspection	Group of experts well versed with usability evaluate the usability of Websites based on a set of guidelines.	Frøkjær and Hornbæk (2008)
7	User Testing	Real users of the Website test the Website usability in trying to achieve their own goals of surfing the Website.	Rigby (2004); *Albers* (2009)

Source: Nathan, R.J. & Yeow, P.H.P. (2011) Crucial Web Usability Factors of 36 Industries for Students – A Large Empirical Study. Journal of Electronic Commerce Research.Vol.12 (2) pp. 153

This section presents a simple method of categorizing Web industries into three major categories in order to recommend a more industry-centered design approach for Websites. Based on a study by Nathan, Yeow and Murugesan (2008) a large number of student-related industries (36 industries) were examined to derive a clear understanding of users' usability preference for the various industries.

Based on the nature of the Websites from the 36 industries, three broader categories of industries were noticed. Website industries offering information and services were categorized as "Personal Services". Website industries that were mainly interested in online sales were categorized as "Purchase Services", whereas Website industries offering educational contents were categorized as "Study-Related". Table 3 shows the industries examined and their respective categories.

Design Guidelines and Recommendations

This section presents design guidelines for usable EC Websites, taking into consideration the various WIUFs discussed earlier in the chapter, as well as adapting these elements to 3 major categories of Website industries. The empirical analysis from Nathan et al (2008) pp. 315, reveals the following recommendations adapted in this section:

- Regardless of industry categories, UCF is still the most important factor. The finding confirms the need for aesthetic appeal among EC Website users.
- CGW is important for all industry categories as it is found among the top 3 important factors for all categories.
- TOW is important for Personal and Purchase services as compared to Study-related Websites. Personal Services Websites generally feed users with a wide range of information critical for decision-making, for example, banking and finance

Table 3. Web industries: 36 industries in 3 major categories

Personal Services (Services)	Study Related (Educational)	Purchase Services (Products)
1. Application Service Provider 2. Business to Business 3. Counseling 4. Dating/Match-making 5. Domain Name Registry 6. Employment 7. Entertainment 8. Financial Information 9. Health and Medicine 10. Horoscope 11. Legal Aid 12. Management Consultancy 13. Mass Marketing 14. Publicity 15. Real Estate 16. Stockbrokers 17. Web Design 18. Website Hosting	1. Books 2. Children Education 3. Education 4. Research/Online Library	1. Automotive 2. Clothes 3. Computer Hardware 4. Computer Software 5. Consumer Electronics 6. Cosmetics 7. Flowers 8. Food and Beverage 9. Furniture 10. Music 11. Office Supplies 12. Pets' Products 13. Sports Equipment 14. Travel Packages

Source: Nathan, R.J., Yeow, P.H.P., & Murugesan, S. (2008). Key Usability Factors of Service-Oriented Websites for Students: An Empirical Study. Online Information Review, Vol. 32 (3) pp. 324

Websites facilitates decisions with regards to finances; dating and matchmaking Websites provides crucial information for choosing dating or life partners.

- UGM is important for Personal and Purchase Services Websites compared to Study-Related Websites.
- DSOW is the least important factor for both Personal and Purchase services, but ironically it is the 3rd most important factor for SIUs in Study-Related Websites. This is perhaps due to the study of related Websites where users place higher importance on quick download of information when it comes to education and studies.

- IOW and EWN are found to be somewhat equally important for all categories of industries in the lower ranking of importance. Ranking wise, however, these factors are not the more important WIUFs, but they do significantly contribute to the Overall Web Usability (OWU).

Table 4 summarizes the crucial WIUFs for various Website industries. This helps Web developers to prioritize a specific set of WIUF in designing industry-specific highly usable Websites.

With a better understanding of WIUFs and their relative importance to industry-specific Websites, more specific Web design guidelines can be derived. Table 5 presents specific recommendations for the 3 EC Websites categories.

CONCLUSION

This chapter began with emphasizing the importance of company Websites in connecting what it offered to its intended users. The Website has become a necessity in the economy where intangible elements of marketing dominate purchase decisions of consumers. The case for rising service-dominant logic of marketing could be extended here to cyber marketing (Vargo and Lusch, 2004). In the cyberspace, users' first impression of a company is built by their acceptance of their Website. It connects company and consumers; hence it becomes the face of the organization. Companies are not merely designing Websites, they are building Web impressions.

Web usability was examined from various angles and this chapter concisely presents important factors affecting the overall usability. The factors discussed include the Web use of color, fonts, use of graphics, multimedia, clarity of goals, trustworthiness, interactivity, ease of Navigation and the Download Speed.

These usability factors are then discussed in perspective of online marketing industries, also known Web electronic commerce industries (see Table 3). Thirty Six industries are presented and to enable a more systematic and focused Web design guidelines; these industries are categories into Personal-Services, Study-Related and Purchase-Services Websites.

Table 4. The most crucial Web usability factors for website industries

No.	Most Crucial WIUF	Website Industries	Number of Industries
1	Use of Color and Font (UCF)	Financial information, Web design, employment, flowers, health/medicine, sports equipment, clothes, education, mass marketing, publicity, furniture, music, food and beverage, management consultancy, consumer electronics, children's education and legal aid.	17
2	Use of Graphics and Multimedia (UGM)	Entertainment, automotive, pets products, computer hardware, horoscope, business to business and stockbrokers.	7
3	Clarity of Goals in Website (CGW)	Research/online library, office supplies, dating/matchmaking and cosmetics.	4
4	Trustworthiness of Website (TOW)	Application service provider, computer software, books, travel and real estate.	5
5	Interactivity of Website (IOW)	Counseling, Website hosting and domain name registry.	3
6	Ease of Web Navigation (EWN)	-None-	0
7	Downloading Speed of Website (DSOW)	-None-	0

Source: Nathan, R.J. & Yeow, P.H.P. (2011) Crucial Web Usability Factors of 36 Industries for Students – A Large Empirical Study. Journal of Electronic Commerce Research.Vol.12 (2) pp. 165

Table 5. Web usability guidelines for 3 major Web categories

No.	Web Industry Category	Usability Guidelines based on the Research Findings
1	Personal Services	• Design the Website with emphasis on colors and fonts. • Provide highly trustworthy information taken from reliable sources. • Be clear in defining the goal and main objectives of the Website. • Use of graphics and multimedia are encouraged. • Interactivity and navigation are essential in this category however must not be over-emphasized.
2	Study – Related	• Design the Website with emphasis on colors and fonts. • Be clear in defining the goal and main objectives of the Website. • Download speed is very important given the probability of having heavy information content in these types of Websites. • Provide information in Websites with file formats that are smaller and easily downloadable. • Interactive tools that would involve SIUs in the Website such as, real time chat, online forum, shout-box and blogging are highly encouraged. • Avoid using heavy graphics and images that might compromise the download speed of Website. • Navigational aspects in the Website are important however must not be overemphasized.
3	Purchase Services	• Design the Website with emphasis on colors and fonts. • Be clear in defining the goal and main objectives of the Website. • Use of graphics and images and highly encouraged. • Websites should include information from reliable sources to appear trustworthy at the same time incorporate secure online transaction technologies to facilitate safer online purchases. • Navigational aspects of the Website must not be neglected to facilitate smooth online purchase processes.

Source: Nathan, R.J., Yeow, P.H.P., & Murugesan, S. (2008). Key Usability Factors of Service-Oriented Websites for Students: An Empirical Study. Online Information Review, Vol. 32 (3) pp. 319

Use of Color and Fonts (UCF) and Use of Graphics and Multimedia (UGM) were the highlights of the research findings. Most evaluators emphasized the importance of proper use of these aesthetics elements in Websites. This is found to be true for all three Web industry categories discussed in this chapter. Web evaluators strongly identified aesthetics in Websites to the overall usability of Websites. This finding is in line with Tractinsky (2000) who concluded in his work, what is beautiful is usable.

Clarity of Goals in Websites (CGW) is the first non-aesthetic element that is highlighted as important in ensuring a greater Web usability. Users are perceived to be selective in their process of browsing Websites and highly goal-oriented in their surfing. With the overwhelming number of Websites and portals available for Internet users, Web sites with poor direction, goals and objectives will quickly lose the interest of the online community. The bargaining-power of surfers among Internet users is at its peak with the introduction of more Web sites and increasing number of Web services to users daily. Websites hence has to be clear in communicating their objectives of existence and operations to their target online users.

Interactivity of Websites (IOW) is also highlighted as an important determinant of Web usability. With the merging of various Internet technologies and media including personal computers, mobile phones and mobile entertainment devices; users are now able to surf mobile-rich contents via live streaming. This is a highly connected technological environment that emphasizes connectivity and interactivity with and among users. Future design of Websites has to take into consideration various mediums that users could use to access and retrieve information from Websites. With the possibility of connecting users to the Internet with high speed broadband access, Web designers could further enhance interactivity features in Websites that will draw more users to stay connected with the Website and at the same time with other online users. Interactivity features of Websites will truly enable the Website to delivers to users a seamless online social experience.

The chapter concludes with specific Web design guidelines for the three Electronic Commerce Website categories presented in the chapter (see Table 5). This gist of the chapter is about designing user-centered Websites that will humanize the Web experience to online users. Web designers need to constantly learn from the changing business environment and understand online consumer wants and preferences in making the online experience targeted, relevant and useful for their users.

REFERENCES

Agarwal, R., & Karahanna, E. (2000). Time flies when you're having fun: Cognitive absorption and beliefs about information technology usage. *Management Information Systems Quarterly, 24,* 665–694. doi:10.2307/3250951

Agarwal, R., & Venkatesh, V. (2002). Assessing a firm's web presence: A heuristic evaluation procedure for the measurement of usability. *Information Systems Research, 13,* 168–186. doi:10.1287/isre.13.2.168.84

Albers, J. (2009). When marketing merges with learning, customers profit. *T + D, 63*(6), 72–75.

Allen, M. (2002). A case study of the usability testing of the University of South Florida's virtual library interface design. *Online Information Review, 26*(1), 40–53. doi:10.1108/14684520210418374

Brady, L., & Phillips, C. (2003). Aesthetics and usability: A look at color and balance. *Usability News,* Vol. 5.1. Retrieved from http://psychology.wichita.edu/surl/usabilitynews/51/aesthetics.htm

Brynjolfsson, E., & Smith, M. (2000). Frictionless commerce? A comparison of Internet conventional retailers. *Management Science, 46,* 563–585. doi:10.1287/mnsc.46.4.563.12061

Calero, C., Ruiz, J., & Piattini, M. (2005). Classifying Web metrics using the Web quality model. *Online Information Review, 29*(3), 227–248. doi:10.1108/14684520510607560

Cao, M., Zhang, Q., & Seydel, J. (2005). B2C e-commerce Web site quality: An empirical examination. *Industrial Management & Data Systems, 105*(5), 645–661. doi:10.1108/02635570510600000

Cappel, J. J., & Huang, Z. (2007). A usability analysis of company websites. *Journal of Computer Information Systems, 48*(1), 117.

Card, S., Moran, T., & Newell, A. (1983). *The psychology of human-computer interaction.* Hillsdale, NJ: Lawrence Erlbaum Associates.

Chen, B., Wang, H., Proctor, R. W., & Salvendy, G. (1997). A human-centered approach for designed world-wide-Web browsers. *Behavior Research Methods, Instruments, & Computers, 29,* 172–179. doi:10.3758/BF03204806

Cheung, C. M. K., & Lee, M. K. O. (2005). The asymmetric impact of Web site attribute performance on user satisfaction: An empirical study. *Proceedings of the 38th Annual Hawaii International Conference on System Sciences* (HICSS-38), Big Island, HI, 3-6 January (CD Rom).

Chowdhury, S., Landoni, M., & Gibb, F. (2006). Usability and impact of digital libraries: A review. *Online Information Review, 30*(6), 656–680. doi:10.1108/14684520610716153

Ferney, S. L., & Marshall, A. L. (2006). Website physical activity interventions: preferences of potential users. *Health Education Research, 21*(4), 560–566. doi:10.1093/her/cyl013

Flavian, C., & Guinaliu, M. (2006). Consumer trust, perceived security and privacy policy: Three basic elements of loyalty to a Web site. *Industrial Management & Data Systems, 106*(5), 601–620. doi:10.1108/02635570610666403

Frøkjær, E., & Hornbæk, K. (2008). Metaphors of human thinking for usability inspection and design. *ACM Transactions on Computer-Human Interaction, 14*(4). doi:10.1145/1314683.1314688

Gehrke, D., & Turban, E. (1999). Determinants of successful Website design: relative importance and recommendations for effectiveness. *Proceedings of the 32nd Hawaii International Conference of Information* Systems, Maui, HI. Retrieved from http://ieeexplore.ieee.org/iel5/6293/16785/00772943.pdf

Ginige, A., & Murugesan, S. (2001). The essence of Web engineering – Managing the diversity and complexity of Web application development. *IEEE MultiMedia, 8*, 22–25. doi:10.1109/MMUL.2001.917968

Google. com. (2012). Ten things we know to be true. Retrieved from http://www.google.com/about/company/philosophy

Helander, M. G. (2000). Theories and models of electronic commerce. *Proceedings of the IEA 2000/HFES 2000 Congress*, Vol. 2, (pp. 770-3).

Hix, D., & Hartson, H. R. (1993). *Developing user interfaces: Ensuring usability through product and process*. New York, NY: Wiley.

Isen, A. M. (1993). Positive affect and decision-making. In Lewis, M., & Haviland, J. M. (Eds.), *Handbook of emotions* (pp. 261–277). New York, NY: Guilford.

Iwaarden, J., Wiele, T., Ball, L., & Millen, R. (2003). Applying SERVQUAL to Web sites: An exploratory study. *International Journal of Quality & Reliability Management, 20*, 919–935. doi:10.1108/02656710310493634

Iwaarden, J., Wiele, T., Ball, L., & Millen, R. (2004). Perceptions about the quality of websites: A survey amongst students at Northeastern University and Erasmus University. *Information & Management, 41*, 947–959. doi:10.1016/j.im.2003.10.002

John, B., & Kieras, D. E. (1996). Using GOMS for user interface design and evaluation: Which technique? *ACM Transactions on Computer-Human Interaction, 4*, 287–319. doi:10.1145/235833.236050

Jørgensen, A. H. (1990). Thinking-aloud in user interface design: A method promoting cognitive ergonomic. *Ergonomics, 33*(4), 501–507. doi:10.1080/00140139008927157

Karat, J. (1997). Evolving the scope of user-centered design. *Communications of the ACM, 40*, 33–38. doi:10.1145/256175.256181

Keeker, K. (1997). *Improving Web-site usability and appeal: guidelines compiled by MSN usability research*. Retrieved from http://msdn.microsoft.com/library/default.asp

Kelly, D., Wacholder, N., Rittman, R., Sun, Y., Kantor, P., Small, S., & Strzalkowski, T. (2007). Using interview data to identify evaluation criteria for interactive, analytical question-answering system. *Journal of the American Society for Information Science and Technology, 58*(7), 1032. doi:10.1002/asi.20575

Kinzie, M. B., Cohn, W. F., Julian, M. F., & Knaus, W. A. (2002). A user-centered model for Web site design. *Journal of the American Medical Informatics Association, 9*, 320–330. doi:10.1197/jamia.M0822

Kirakowski, J., Claridge, N., & Whitehand, R. (1998). Human centered measures of success in Web site design. *Proceedings of the 4th Conference on Human Factors & the Web*, Basking Ridge, NJ, 5 June.

Krug, S. (2006). *Don't make me think, second edition: A common sense approach to Web usability*. New York, NY: Pearson Education Inc.

Kumar, R. L., Smith, M. A., & Bannerjee, S. (2004). User interface features influencing overall ease of use and personalization. *Information & Management, 41*, 289–302. doi:10.1016/S0378-7206(03)00075-2

Lecerof, A., & Paterno, F. (1998). Automatic support for usability evaluation. *IEEE Transactions on Software Engineering, 24*, 863–887. doi:10.1109/32.729686

Lewis, C., Polson, P., Wharton, C., & Rieman, J. (1990). Testing a walkthrough methodology for theory-based design of walk-up-and-use interfaces. In *Proceedings of the SIGCHI conference on human factors in computing systems* (pp. 235–242), Seattle, WA, USA.

Lindgaard, G. (1999). Does emotional appeal determine perceived usability of Websites? In L. Straker & C. Pollock (Eds.), *Proceedings of CybErg: The Second International Cyberspace Conference on Ergonomics,* (pp. 202-11). The International Ergonomics Association Press, Curtin University of Technology, Perth, Western Australia.

Liu, C., & Arnett, K. P. (2000). Exploring the factors associated with Website success in the context of electronic commerce. *Information & Management, 38*, 23–33. doi:10.1016/S0378-7206(00)00049-5

Lohse, G., & Spiller, P. (1999). Internet retail store design: How the user interface influences traffic and sales. *Journal of Computer-Mediated Communication, 5*(2).

Marcus, A. (2005). User interface design's return on investment: examples and statistics. In Bias, R. G., & Mayhew, D. J. (Eds.), *Cost-justifying usability*. San Francisco, CA: Morgan Kaufman. doi:10.1016/B978-012095811-5/50002-X

McMullen, S. (2001). Usability testing in a library Web site redesign project. *RSR. Reference Services Review, 29*, 7–22. doi:10.1108/00907320110366732

Moraga, A., Calero, C., & Piattini, M. (2006). Comparing different quality models for portals. *Online Information Review, 30*(5), 555–568. doi:10.1108/14684520610706424

Nathan, R. J., & Yeow, P. H. P. (2009). An empirical study of factors affecting the perceived usability of websites for student internet users. *Universal Access in the Information Society, 8*(3).

Nathan, R. J., & Yeow, P. H. P. (2011). Crucial web usability factors of 36 industries for students – A large empirical study. *Journal of Electronic Commerce Research, 12*(2), 150–180.

Nathan, R. J., Yeow, P. H. P., & Murugesan, S. (2008). Key usability factors of service-oriented websites for students: An empirical study. *Online Information Review, 32*(3). doi:10.1108/14684520810889646

New Zealand Ministry of Economic Development (MED). (2000). *E-commerce: A guide for New Zealand business*. Wellington, New Zealand: Author.

Nielsen, J. (1994). Heuristic evaluation. In Nielsen, J., & Mack, R. L. (Eds.), *Usability inspection methods*. New York, NY: John Wiley and Sons.

Nielsen, J. (2000). *Designing web usability: The practice of simplicity*. Indianapolis, IN: New Riders Publishing.

Nielsen, J. (2003). *Usability 101: Introduction to usability*. Jakob Nielsen's alertbox. Retrieved from http://www.useit.com/alertbox/20030825.html

Nielsen, J. (2005). *Useit.com: Jakob Nielsen's website*. Retrieved from http://www.useit.com

Nielsen, J., & Tahir, M. (2002). *Homepage usability: 50 Web sites deconstructed*. Indianapolis, IN: New Riders Publishing.

Norman, D. A. (2002). Emotion and design: Attractive things work better. *Interaction Magazine, 9*(4), 36–42.

Olson, J. R., & Olson, G. M. (1990). The growth of cognitive modeling in human-computer interaction since GOMS. *Human-Computer Interaction, 5*(2), 221–265. doi:10.1207/s15327051hci0502&3_4

Pennington, R., Wilcox, H. D., & Grover, V. (2003). The role of system trust in business-to-consumer transactions. *Journal of Management Information Systems, 20,* 197–226.

Phillips, A., & Chaparro, A. (2009). Visual appeal vs. usability: Which one influences user perceptions of a Website more? *Usability News,* Vol. 11.2. Retrieved from http://www.surl.org/usabilitynews/112/aesthetic.asp

Rigby, E. (2004). Usability can work for all online marketers. *Revolution (Staten Island, N.Y.),* 56.

Rivero, M. (1999). Web objectives save time, money. *New Hampshire Business Review, 21*(27), 6.

Robins, D., & Holmes, J. (2008). Aesthetics and credibility in Web site design. *Information Processing & Management, 44*(1), 386. doi:10.1016/j.ipm.2007.02.003

Shenkman, B. O., & Jonsson, F. (2000). Aesthetics and preferences of Web pages. *Behaviour & Information Technology, 19,* 367–377. doi:10.1080/014492900750000063

Teo, H. H., Oh, L. B., Lui, C., & Wei, K. K. (2003). An empirical study of the effects of interactivity on Web user attitude. *International Journal of Human-Computer Studies, 58,* 281–305. doi:10.1016/S1071-5819(03)00008-9

Tilson, R., Dong, J., Martin, S., & Kieche, E. (1998). *Factors and principles affecting the usability of four e-commerce sites. Proceedings of Human Factors and the Web, June 5, 1998.* NJ, US: Basking Ridge.

Tractinsky, N. (1997). Aesthetics and apparent usability: Empirically assessing cultural and methodological issues. *Proceedings of CHI 97 Conference: Looking to the Future,* March 22-27, 1997, Atlanta, Georgia, US.

Tractinsky, N., Katz, A. S., & Ikar, D. (2000). What is beautiful is usable. *Interacting with Computers, 13*(2), 127–145. doi:10.1016/S0953-5438(00)00031-X

Turban, E., King, D., Viehland, D., & Lee, J. (2006). *Electronic commerce – A managerial perspective.* New Jersey: Pearson Education Inc.

University of South Florida Virtual Library. (2012). Retrieved from www.lib.usf.edu

Usability First. (2005). Usability in website and software design. *Usability First.* Retrieved from http://www.usabilityfirst.com/methods/index.txl

Vargo, S. L., & Lusch, R. L. (2004). Evolving to a new dominant logic for marketing. *Journal of Marketing, 68,* 1–17. doi:10.1509/jmkg.68.1.1.24036

Watson, R. T., Berthon, P., Pitt, L. F., & Zinkhan, G. M. (2000). *Electronic commerce: The strategic perspective.* Fort Worth, TX: Dryden Press.

Xie, H. I., & Cool, C. (2000). Ease of use versus user control: An evaluation of Web and non-Web interfaces of online databases. *Online Information Review, 24*(2), 102–115. doi:10.1108/14684520010330265

Chapter 6
Authentication Mechanisms for E–Voting

Emad Abu-Shanab
Yarmouk University, Jordan

Rawan Khasawneh
Yarmouk University, Jordan

Izzat Alsmadi
Yarmouk University, Jordan

ABSTRACT

The e-government paradigm became an essential path for governments to reach citizens and businesses and to improve service and public performance. One of the important tools used in political and administrative venues is e-voting, where ICT tools are used to facilitate the process of voting for electing representatives and making decisions. The integrity and image of such applications won't be maintained unless strict measures on security and authenticity are applied. This chapter explores the e-voting process, reviews the authentication techniques and methods that are used in this process and proposed in the literature, and demonstrates few cases of applying e-voting systems from different countries in the world. Conclusions and proposed future work are stated at the end of the chapter.

INTRODUCTION

E-government is more than a phenomenon; it is a paradigm that guides the new governance process. Most countries in the world are embracing e-government initiatives based on confirmed benefits and essential requirements of the new millennium. As one of the main elements in the free democratic world, voting is a key enabler in e-government initiate, not only for the primary reason of citizens' election of their representatives, but to enable quick and reliable feedback from citizens to the government, and evaluating the value of the offered services. E-voting can be an excellent supporting tool for election, decision making, consultation, and participation initiatives.

This chapter reviews e-voting concepts, methods and authentication techniques and tools used to ensure performing a creditable voting process.

DOI: 10.4018/978-1-4666-3640-8.ch006

Once governments are willing to adopt electronic channels for their governance activities, they are entering a new era with new tools that range from simple information systems to more complicated Web 2.0 tools that open doors for all stakeholders to benefit from the power of reaching citizens and fostering a decision making process that activates a more participatory picture.

E-voting is becoming an integral part of any modern election system, where it became the most popular application in this area. E-voting depends on technology in more than one layer to create a very interactive and effective election process from the time of adding and verifying participants, to the last stage of calculating results and winners. All this should be implemented in reliable, correct, secure and fast manner. E-voting can also utilize new hardware, software and network technologies and add convenience to the process by allowing citizens to cast their votes from their homes and using the Internet.

This wide continuum of applications adds more complications to the voting system, where the crucial issues are the following: make sure that voters are from legitimate citizens; can vote and vote once and can do this in a flawless quick pace. Currently, this authentication process utilizes several tools, methods and even technologies to guarantee the legitimacy of elections. This chapter reviews e-voting in general and focuses on authentication methods that are applicable to the e-voting system. The chapter is divided into three sections. Following is a brief description of each section.

The following section reviews e-government concepts and focuses on e-voting domain. It reviews e-voting process in general, its definition, and the general related concepts. Next, authentication methods and techniques are reviewed and described based on a specific typology. Finally, cases from the world related to e-voting systems are explored and analyzed. Conclusions and research recommendations are stated at the end.

E-VOTING SYSTEMS

Electronic government is defined as using Information and Communication Technology (ICT) to enhance government's operations, provide suitable services to citizens, and improve citizens' participation (World Bank, 2007; Yanqing, 2010; Abu-Shanab, 2012). Mason (2011) emphasized the notion of citizen's participation in democratic life, while others related e-government to the provision of information and knowledge to make suitable decisions in political life (Lee, Chang & Berry, 2011). In their pursuit toward improved accountability to citizens (Carter & Belanger, 2004), and improved public service quality (Irani, Al-Sebie & Elliman, 2006), governments try to implement new technologies in all aspects of their operations. E-voting is one of the essential applications especially in the election process.

E-voting is the most researched topic in the area of e-government, although it is not devoted to political cause only. Voting is a needed process in administrative areas, where a decision is needed based on some alternatives. Examples of such applications are used in group decision support systems (GDSS) and policy/agenda setting initiatives. E-voting is defined as using ICT to conduct voting (Buchsbaum, 2005). E-voting, as a democratic activity, includes the voter, the registration authority and a tallying authority, where an electronic system is used to cast votes (Kumar & Walia, 2011a). Research indicated that even with traditional voting systems, using information systems to count votes is considered an e-voting application (Remmert, 2004).

It is important to realize the benefits gained from using electronic systems in the voting process; e-voting offers convenience to the election process, accuracy and accountability of results (Buchsbaum, 2005), time and cost savings (Saveourvotes.org, 2008), increased public participation through open systems and Web 2.0 tools (Bouras, Katris & Triantafillou, 2003), and a flexible process for editing and updating voters' information (Weldemariam, Villafiorita & Mattioli, 2007).

E-voting includes four stages: registering voters, issuing needed documentations, casting votes using electronic systems, where authentication process takes place, and finally, counting and announcing results (Cetinkaya & Cetinkaya, 2007). Some researchers considered the first two stages as one (Gallegos-Garcia, Gomez-Cardenas & Duchen-Sanchez, 2010). E-voting is considered as an integral dimension of e-democracy (Bozinis & Lakovou, 2005).

Types of E-Voting Systems

In the literature, authors tried to list all e-voting system types and channels to make it easier to understand this important application. E-voting can be conducted through two major channels: the first one by using electronic systems installed in public centers for elections while the second channel is conducted using the Internet (Okediran, Omidiora, Olabiyisi, Ganiyu & Alo, 2011). The literature emphasized the web application of e-voting to be more flexible and convenient to citizens' needs. On the other hand, when looking into the history of voting systems, the major e-voting system types are the following:

- **Mechanical punched cards:** Voting is conducted mechanically using paper cards, and an electronic system is attached for the count and analysis of results (Chaum, 1988). Such system is considered the simplest version of electronic systems.
- **Optical-based systems:** The major feature of such systems is the use of an optical reader, with the same information system attached for the count and analysis (Masuku, 2006). Such systems are widely used in standardized online exams where multiple choice questions are commonly used.

- **Using phones in e-voting systems:** Such option is practical as it can extend the election reach to rural areas and utilize the wide spread of mobile networks. The citizens' choices can be casted through predefined lists and choices. Extra authentication and security methods are needed in the case of phone voting (p-voting).
- **Fax-voting:** F-voting is an important type especially for remote areas' cases where citizens' cannot vote inside public centers (e.g. outside cities and villages in rural areas) and hence can vote with predefined conditions and setups. One of the aspects of fax voting is that the voters' choices are revealed and over sighted by recipient of the fax (Puiggali & Rocha, 2007). Similar systems utilize regular mail, which might be considered as a traditional voting rather than an electronic voting case.
- **E-mail voting:** Similar to mail, and fax voting, it is possible to utilize e-mail applications as a communication channel for voting purposes. Tallying authority can print the e-mail to keep it as a record and verification for the voting process, where security and privacy are important (Puiggali & Rocha, 2007).
- **Direct E-voting:** This is accomplished in public voting centers using electronic systems. The process is performed under the supervision of voting committees.
- **Internet voting:** Through the Internet, voting can be conducted and counted on the voting systems directly. Voters log to the e-voting system remotely (e.g. from their homes or offices) where they identify their credentials, get authenticated, and then cast their votes. Votes are counted in the same way through an electronic system installed on a central server (Masuku, 2006).

E-voting systems' legitimacy is guaranteed through few measures that are summarized from many resources (Smith, 2002; Buchsbaum, 2005; Kahani, 2005; Sandikkaya & Orencik, 2006; Anane, Freeland & Theodoropoulos, 2007; Maier, 2010; Patil, 2010; Jafari, Karimpour & Bagheri, 2011; Yumeng, Liye, Fanbao & Chong, 2011); Following is a description of each one:

- **Democracy:** E-voting system should guarantee to all legitimate voters the chance to cast their votes on the election's day. Such condition is important for the integrity of the democratic process, where it is important to ensure the convenience of election procedures for voters, and especially people with special needs. It is also important to accommodate people with low computer literacy. It is important to guarantee the wide use of electronic systems before dropping the traditional system. As seen in the case of Philippines, the government provided all types of backup methods, and especially paper and traditional forms for people who would not or cannot use e-voting machines. It is also important in the first stages to deploy both methods (i.e. traditional and electronic) to gain the critical mass and guarantee the success of e-voting (Bhatnagar, 2004).

- **Eligibility:** Only legitimate voters who are registered and satisfy election conditions can vote (Anane *et al.*, 2007). This rule is also important to prevent repetitive voting.

- **Accuracy:** Accuracy is not a luxurious condition; it is an essential condition for auditing the total votes against the voters cards used (i.e. in comparison with traditional voting). It is one the major characteristics of e-voting systems when compared with traditional paper forms. The system should provide rules for checking the total number of casted votes and the total number of voters.

- **Verifiability:** The system should be verifiable against the total number of casted

votes and the totals of each nominee. All analysis should also be verifiable to guarantee the integrity and image of elections.

- **Integrity:** Votes are not to be changed, forged, or omitted. The system should also provide needed means to audit and discover any attempts to tamper with the system. Further, Langer *et al.* (2010) emphasized the fact that voters need to be assured that their votes will not be changed and omitted. Such issue is important so citizens will not lose faith in e-government and electronic voting systems.

- **Transparency:** The voting procedures should be transparent and clear to all parties. All rules and procedures need to be announced and explained to all stakeholders. Also, it is important to train people and conduct pilot elections on voting procedures for two major reasons: First, to guarantee and verify the presence of the needed skills and resources, and second to raise awareness and gain trust.

- **Privacy:** It is essential not to be able to relate voters to their choices (i.e. votes). In addition, whether it is registered or not, it is important not to reveal such choices to any candidate party. Another aspect of privacy is to prevent other voters from overseeing voters' choices in the election process.

- **Uniqueness:** Some election systems assign one vote for each voter. Other systems assign more. Based on that, it is important to control the number of choices allowed and the totals resulting from the process when implementing the guarding controls for accuracy and verifiability.

- **Receipt-free voting:** This means that voters are not supposed to know the results before the end of elections so that voter's choices will not be affected by results.

- **Flexibility:** The e-voting system should be designed to provide flexibility for voters, with alternative options. Examples of those alternative options include: Electronic vs. paper; compatible systems to all computers and browsers, especially for Internet voting, electrical vs. battery operated; using different languages; utilizing graphical interfaces for illiterate people; and provide tools for people with special needs or impairment.

- **Scalability:** This refers to the system's ability to accommodate the possible size change in the number of voters, their characteristics and related information.

Gerlach (2009) indicated that in order to keep e-voting secure and confidential, compared to paper-based voting, seven design principles should be incorporated: First, proven security: all protocols and techniques must be mathematically proven secure. Second, trustworthy design responsibility: Government security agencies should be responsible for creating a secure voting system. Third, source code must be published and made publicly accessible. Fourth, vote verification: It should be possible to verify that all votes have been correctly accounted for in the final election tally. Fifth, voters' accessibility: System should be accessible to all members, and it should be easy to use. Sixth, ensure anonymization; techniques such as onion routing must be used to ensure anonymization. Finally, expert oversight; to ensure the trust in the e-voting system, a team of experts selected and approved by all major parties taking part in election should be given the options to see and verify all stages of the e-voting process.

E-voting integrity should be maintained, where many conditions need to be satisfied as mentioned in the literature (e.g. Chevallier, 2003; Abu-Shanab, Knight & Refai, 2010). The following are the major rules, where some of them are implied from previous characteristics mentioned before: First, votes cannot be intercepted nor modified. Second, votes cannot be known before the official ballot reading. Third, only registered voters can vote, and also Each voter will have one and only one vote. Fourth, vote secrecy is guaranteed. Fifth, it should never be possible to link a voter and his/her vote. Sixth, the voting web site should be able to resist any denial of service attack, or at least track users who were denied and give them a second chance. Seventh, voters will be protected against identity theft. Eighth, the number of cast votes will be equal to the number of received ballots. Ninth, it will be possible to prove that a given citizen has voted. Tenth, the system will not accept votes outside the ballot opening period. And finally, the system will be auditable.

AUTHENTICATION METHODS

Implementing electronic voting systems is faced by several issues which can be the main reasons for its success or failure; the most important among all focus on security and privacy especially on how to make sure that the one who votes is the legitimate voter and not anyone else. There are several security requirements that electronic voting system must satisfy such as: eligibility, authentication, privacy, robustness, and fairness. Eligibility guarantees that only eligible voters can participate and cast their votes during the election period. Authentication makes sure that the one who votes is the right one and no one else. Voter's privacy enable's voters to vote in a highly private way in which their personal information and voting process information are protected and can't be exposed by others. Robustness means that electronic voting system should be able to handle exceptional situations. For example the system should be protected against any attacks, fraud and disruption. Finally, the system should be fair to all candidates in announcing voting results only at the end of the allowable voting period (Patil, 2010; Yumeng, Liye, Fanbao & Chong, 2011; Jafari, Karimpour & Bagheri, 2011).

The authentication process makes sure that a user has access to the e-voting system and can participate in this e-service, or not, based on his/her original identity. Several authentication schemas and methods are available and can be classified into three main types; knowledge-based methods, token-based methods and biometrics-based methods (Rao & Patil, 2011; Deep, 2011).

Knowledge-Based Methods

Knowledge-based methods, also known as password-based methods, are based on the person's knowledge of something. An example of what a person knows is a password or a PIN number (Rao & Patil, 2011; Deep, 2011). In this chapter, several proposed electronic voting systems that use knowledge-based methods for authentication purposes are presented. The first type proposed a framework for developing a general prototype for electronic voting system, which is trusted by people and has a high level of security through using user ID and password as an authentication scheme. Those user names and passwords are generated when the person registers to use this system (Haziemeh, Khazaaleh & Al-Talafhah, 2011). This proposed system is developed using PHP programming language and MySQL database. The system prevents any unregistered person to access and use the system through its easy to use interface. The system can distinguish unauthorized (unregistered) users who enter incorrect IDs or credentials and notify them that the ID or the password used are wrong and he/she is not an authorized or registered user for the e-voting system. Only authorized users can access this system. The database can also include rules to verify that users can vote one and only one time. However, the simple ID and password authentication system will weaken the security of e-voting systems against hackers especially that IDs and passwords can be easily forgotten, lost, attacked, known and used by others. Using additional methods, which will increase cost, will increase the security of ID and password and prevent unauthorized users from using the system (Rao & Patil, 2011). The complexity or simplicity of passwords can weaken or strengthen the security of the e-voting system; using simple passwords, which will be easily remembered, will weaken the security of the system because these passwords can be guessed by others, easily attacked and easily shared. Although using complex passwords will increase the system security, it makes remembering such passwords more difficult (Kumar *et al.*, 2011). Using complex IDs and passwords will increase the chances of typing errors and as a result rejection of the system, which makes system usage more irritating and inconvenient.

To overcome the security weaknesses associated with using IDs and passwords alone in authentication, Sodiya *et al.* (2011) presented an enhanced electronic voting architecture that uses usernames and PIN numbers that are generated randomly by a computer program during the registration phase in which users are asked to provide several pieces of personal information that are used for authentication purposes during the voting process. Using a system that depends only on a user name and PIN number is not enough to guarantee that the one who votes is the same legitimate person who is supposed to vote. On the other hand, more expensive methods combined with username and PIN number can make the e-voting system more effective. Such methods can depend on using biometrics such as fingerprint, iris or face recognition or other token-based methods.

Another proposed enhanced and secured e-voting system enables citizens to vote from their personal computers, where their computers are connected to the Internet. In the identification phase of this system, the citizen will request a voting certificate (PIN) that makes such citizen authorized to access the voting system after saving his/her personal information in the government election server. This request is issued by using the citizen national ID as a base for such process. This PIN number will be used for authentication purposes in the future and in the e-voting system for which citizens can use it to have a public key that enables them to vote. The public key is received in a message from the e-voting authority using mobile phones after checking personal information and making sure that they comply with what is saved on the system database. This system enables citizens to vote from their personal computers without any additional cost or effort (Al-Anie *et al.*, 2011).

There are several authentication schemes that can be suitable to be used for enhancing the security of remote voting which is one of the main types of electronic voting that enables citizens to vote from their personal computers connected to the Internet, from their mobile phones through sending short messages, or from any mobile terminal with Internet connection (Okediran, Omidiora, Olabiyisi, Ganiyu & Alo, 2011). Sahu and Choudhray (2011) introduced an electronic voting scheme that integrates an electronic voting system with GSM infrastructure which enables people to vote through their mobile terminals anywhere within the election period. Anyone who wants to vote using mobile terminals should have a unique mobile ID given by the election committee based on the phone information. This E-voting system uses 89S52 microcontroller, which have a dual serial communication facility, 16X2 dual line LCD for massage display, and a line converter MAX 232, GSM modem, (Sahu & Choudhray, 2011, p.5644). This microcontroller will receive a message from GSM modem that contains voter's unique mobile and candidate ID, then it will perform the authentication process based on the received voter's mobile ID and the information stored on its database and then send a notification message to voter's mobile whether the voting process is successful or not. This E-voting system, armed with this feature (i.e. voting through mobile terminals using the unique mobile ID in the authentication process), will make the voting process more secured and will increase the percentage of people participating in the e-voting activity. On the other hand, remote voting systems still need to deal with several security or connectivity problems such as the denial of service attacks which will weaken people's trust and usage of this remote electronic voting system (Okediran, Omidiora, Olabiyisi & Ganiyu, 2011). The following table (see Table 1) presents a summary of the advantages and disadvantages of using knowledge-based methods.

Token-Based Methods

Using token-based methods can be more effective than using knowledge-based methods (Khan, 2010) since these methods of authentication are based on what a person possesses (Rao & Patil, 2011; Deep, 2011; Khan, 2010). Even though using token-based methods have several advantages that come from its ability to achieve more secure process than using passwords only; however, there is a need to know the user's password if the token was stolen, which results in having two levels of security. On the other hand, thinking of token attacks issues and the high cost required for purchasing high quality token readers needs to be considered more carefully (Rao & Patil, 2011).

Smart cards can be used as a storage media on which voter's personal information is stored in addition to public key information. It can also store the person's unique physiological characteristics such as fingerprint and iris patterns (Kalaichelvi & Chandrasekaran, 2011a).

Table 1. Advantages and disadvantages of knowledge-based methods

Knowledge-Based Methods (Password-Based)	Advantages	Disadvantages
	Simpler system requirements. Passwords can be easily replaced with new one if the database is attacked or stolen. Easily remembered if it simple passwords. Difficult to attacks if it complex.	Easily forgotten or lost. Easily attacked, known and used by others especially if it's simple. Difficult to remember if it complex passwords.

In this authentication method, several electronic voting systems that used token-based methods are proposed. Rexha, Dervishi and Neziri (2011) introduced an electronic voting system in Kosovo that aims to increase people's trust in using electronic channels for voting compared to traditional methods. The system structure uses smart cards and digital certificates for authentication purposes. This smart card stores people's personal information, digital certificates information and its related private key which is protected by using a personal identification number (PIN) and finger print data.

Kalaichelvi and Chandrasekaran (2011b) introduced an electronic voting system that uses smart cards (i.e. smart tokens) that store media related to voter's information with user biometric data (e.g. iris template for each smart card holder). This data is stored in the system database and when the smart card holder inserts the card into the machine, his\her iris will be captured automatically by a camera and will be compared with the related iris template stored in the database.

This used schema (i.e. smart card with iris data) meets the electronic voting system security requirement in the authentication process especially that this smart card is generated temporarily in the voting preparation stage for voting purposes and the information stored on it can't be reset after it is personalized so each smart card holder can vote only one time. This process of authentication can increase the level of trust in e-voting systems. The system proposed is a robust system. However, there are still some security concerns such as e-voting cards' theft.

Even though using smart tokens with iris in the authentication process has several benefits, there are some disadvantages inherited in such systems especially that using such methods costs more in the designing process and the training afterwards. These token-based systems need to be designed with a high level of security, built in extra training efforts for people using and implementing them, and need some extra cost to purchase high quality cameras and scanners for iris scanning purposes

(Rao & Patil, 2011). The following table (see Table 2) presents a summary of the advantages and disadvantages of using token-based methods.

Biometrics-Based Methods

Biometric-based methods are based on the physical characteristics of the person that are different from one person to another (Rao & Patil, 2011; Deep, 2011). Using biometrics such as: face recognition, fingerprint, iris and vein has several advantages based on its accuracy, uniqueness and complexity especially that it can't be stolen, altered or used by someone else other than the owner or the holder. As a result, systems that use biometrics in authentication process will increase the system's level of security and encourage people to use it in applications where high security is required (Sarkar, Alisherov, Kim & Bhattacharyya, 2010). The performance of biometric-based authentication methods can be evaluated based on two main types of errors: matching errors, which occur during the voting period; and acquisition errors, which occur during the preparation and registration stages (Deep, 2011).

Another proposed scheme for biometric e-voting systems used citizen's fingerprint that is stored in the system database, when a fingerprint scanner scans the voter's fingerprint, the voter ID and other personal information appear on the screen if the scanned fingerprint matched the fingerprint template stored in the database. The

Table 2. Advantages and disadvantages of token-based methods

Token-Based Methods	Advantages	Disadvantages
	Token has more security level than passwords; especially if this token is stolen; still there is a need to know the user passwords. More data can be utilized in the security system.	Tokens easily attacked. Needs more hardware and system requirements Needs high cost for purchasing high quality token readers.

uniqueness feature of a person's fingerprint will increase the level of people's trust in this system (Altun & Bilgin, 2011). Using fingerprint in the authentication process has several drawbacks especially if the fingerprint templates that exist in the database are attacked or stolen. To avoid fingerprint templates attack, an urgent need for enhancing the security and authentication system that uses fingerprint is required. This enhancement can be done through the use of fragile image watermarking techniques that embed additional information to fingerprint image such as an identification number that will be used in addition to the fingerprint in the authentication process (Gothwal, Yadav & Singh, 2011).

On the other hand, skin status, especially dry or wet skin and skin injuries, has a significant impact on the quality of scanning and matching process. This may increase the percentage of matching errors. Such issue will result in a system that may treat eligible voters as ineligible and prevent them from casting their votes and participating in the e-voting stage (Kumar & Walia, 2011b).

In general, using biometrics such as fingerprint, iris and face recognition in the authentication process has several pros and cons based on the extent to which people accept them. The uniqueness and consistency of biometrics will result in a robust authentication process that can leverage the security levels of e-voting systems. Additionally, it will compensate for the additional costs caused by the need to purchase special scanners. Further,

such scanners may need special maintenance and extra technical support (Rao & Patil, 2011).

Even though biometrics can be used as a unique identifier for any person, several risks are associated with using such techniques in the authentication process. Risks come from many resources such as: The ability to scan or capture anyone's iris using hidden cameras; people fingerprints are left on any object that a person touches; if biometrics templates are attacked or stolen, a disaster will happen especially as these templates cannot be replaced by another new one. A person cannot change his/her fingerprint similar to changing or updating his password; iris or other biometric identifier once it is hacked it sets a big challenge to security administration of the election (Deep, 2011; Zahed & Sakhi, 2011).

Roa and Patil (2011) introduced an enhanced authentication method utilizing a three dimensional environment that can be used in several systems for authentication purposes. The security and reliability of such 3-D environment are increased by giving users the option to choose an authentication method according to their preferences that blends a knowledge-based method, token-based method and/or a biometrics-based method. If users prefer not to carry cards, they can choose an authentication method that they prefer such as ID\password, fingerprint or iris. The following table (see Table 3) presents a summary of the advantages and disadvantages of using biometrics-based methods.

Table 3. Advantages and disadvantages of biometrics-based methods

Biometrics-Based Methods	**Advantages**	**Disadvantages**
	High level of security. Unique identifier of the person. Robust. Can't be lost or used by others. Reliable. Accurate especially when high quality scanners are used. Stability over time.	If stolen or hacked, can't be replaced (part of physical nature of humans). May cause health problem as a result of direct scanning. Can be captured or easily scanned using hidden cameras. Fingerprints are picked easily from objects that people touch. Incur high cost on the system. Needs complex processing. Skin status or injury may cause difficulties in scanning.

Discussion of Authentication Methods

This chapter appraised several proposed electronic voting systems that use different methods and techniques in the authentication process to guarantee that the one who votes is the legitimate voter and not anyone else. Different methods and techniques are explored like ID and password, username and randomly generated PIN numbers, PIN numbers and public key (when voting done using personal computers), smart card with biometric (Iris), unique mobile ID (when voting done through mobile terminals) and Biometric (fingerprint).

Biometrics-based authentication (i.e. using iris, fingerprint and face shape) will increase the security level of electronic voting systems more than using knowledge based or token based authentication especially that biometrics have personal characteristics different from one another, can't be stolen or used by someone else unlike knowledge-based or token-based authentication that can be easily stolen, altered, known and used by others. Also, using biometrics-based authentication has several drawbacks especially if the database is attacked and the templates are stolen, so using electronic voting system that uses multi authentication methods (i.e. using a smart card with biometric features) can strengthen the security of the system and overcome the limitations of using each of them alone. On the other hand, such arrangement will be costly and will need more time and effort.

Although there are several proposed electronic voting systems that use different authentication methods and schemas, security problems related to authentication process in electronic voting still exists and developing an enhanced and secure electronic voting system with special specifications is needed to overcome the system security limitations and increase the level of people trust and acceptance of using electronic voting system instead of paper-based voting (traditional way of voting).

E-VOTING EXPERIENCES AROUND THE WORLD

In this section, different experiences from the world are investigated to demonstrate diverse methods applied and the characteristics of each method. Additionally, it is important to emphasize the lessons learned and the recommendations proposed for each case. This chapter explored the literature and detected most reported cases (up to the authors' knowledge) that are related to e-voting application in the election process which focused on security issues and authentication schemes. The cases selected were the ones that reported the authentication schemes used, and thus are chosen from diverse countries and regions of the world.

- **E-voting system in Estonia:** Estonia is one of many countries that provide several electronic services through its e-government portal. One of these provided services is voting through the Internet (I-voting) (Maaten & Hall, 2008; Madise & Vinkel, 2011).

  A small country like Estonia with a population of approximately 1.3 million, with a good infrastructure in communication, wide use of national ID cards, and a fair level of people's acceptance of e-services; all that play a central role in the success of Internet voting (I-voting) in Estonia. The Estonian e-voting experience occurred two times: the local governmental elections in October 2005, and the parliamentary elections in March 2007. Using different methods for the authentication process in e-voting (e.g. public key infrastructure and the digital signature) has a significant impact on Estonia's I-voting process. On the other hand, Estonian people have a high level of trust in their government and in the Estonian I-voting experience, which comes from the political support given to the I-voting system through several laws such as: the digital signature act in 2000 that gives Estonian people the ability to authenticate themselves by using identity cards with two digital certificates (PIN1 and PIN2); European Parliamentary Election Act; the Local Communities Election Act and many others (Maaten & Hall, 2008). These certificates contain the cardholder name and personal information linked with two private keys embedded in the card (Madise & Vinkel, 2011).

- **E-voting in India:** The Indian electronic voting system is developed by the Electronics Corporation of India (ECIL) and Bharat Electronics Limited (BEL) which are government owned companies. Even though this system is characterized by its simple design, ease of use and reliability, several probable attacks and security issues need to be considered especially if any attacker can intercept the voting process and change the casted votes or vote more than one time (Prasad, Halderman & Gonggrijp, 2010).

  Several solutions to enhance the Indian voting system and overcome security issues were proposed. One of these solutions was using biometrics-based methods for authentication purposes based on storing people's fingerprint in a permanent database. This can be done in the voting preparation stage, where the matching process is done during the voting period to ensure that only registered and authenticated people can vote and for only one time (Reddy, 2011).

- **E-voting in Pakistan:** Pakistan is one of the countries that have severe problems in the manual methods of voting in terms of trust, security and buying and selling votes. These problems call for a significant need to use an electronic way of voting that will guarantee voting accuracy and security with high level of people's trust in all stages of the voting process starting from the initial or preparation stage to

counting votes and announcing the results. So an e-voting system is proposed to be used in Pakistan with an easy to use interface, minimum cost and a high level of security through enabling voters to use ID and password for login purposes and their fingerprints for authentication purpose before casting their votes. This proposed system is built using ASP.Net 02.0, Adobe Photoshop 07, Macromedia Flash Player, MS SQL Server 2005 and Veryfinger 2.6 Extended for checking and authentication of the voters and staff (Solehria & Jadoon, 2011).

- **E-voting in Saudi Arabia:** In the case of Saudi Arabia, the government succeeded in applying an authentication system with minimal levels of fraud and errors which opened up several opportunities for the government to increase the number of electronic services provided through its e-government portal. Saudi Arabia adopted a biometric-based identity management system that embeds all of the electronic services provided by electronic government initiative especially since Saudi people have many electronic cards and each one is used for only one purpose. Examples of such cards include: E-gate card which uses the thumbprint and iris in the authentication process, and health insurance card. These cards were used and designed with a high level of security to guarantee that only the cardholder can use the card by using biometric measures such as thumbprint and iris. Developing a biometric-based identity management system opens up several opportunities for the e-government project to enable people to vote via the Internet especially since the authentication process can be guaranteed through already used and existed electronic cards with minimal effort and time (Khan, Khan & Alghathbar, 2010).

- **E-voting in Philippines:** The Philippines case in e-elections is considered a miracle case, where 76300 casting machines were distributed, 700 election offices were opened, 48000 technicians and support officers were hired, and 904 administrators were working in shifts to help make such experience a success. The election administration tried to build all possible contingency plans in the process and provide the logistics of the elections and the topography of the country asserts the need to plan well for the elections.

Focusing on the voting process and to authenticate the voters IDs, the government used a digital signature method, where thousands of machines were distributed to scan signatures. Also, 600 battery operated machines were used as a backup for the process. Such precautions were necessary to guarantee the success of elections. Finally, electrical generators were used in areas that don't have public electricity networks. In all cases, and if such arrangements failed, paper copies of election forms were distributed to all centers to cover any unplanned incidents. The election body spent a hectic 3 days programming the cards, preparing the systems on sites, and downloading needed information. It is stated by public media that the Election Day (11-5-2010) was a phenomenal day (Source: Futuregov.asia, 2011).

- **E-voting in Brazil:** Brazil is one of the world leaders in electronic elections especially that it has adapted such scheme since 1990 (Solehria & Jadoon, 2011). The use of electronic voting machines in Brazil is recommended as the best solution for decreasing fraud and reducing the rate of invalid votes. The first time using such machines was in 1996 and it was expanded with a nationwide implementation in the year 2000 (Kumar & Walia, 2011a). The electronic voting machines that were used in the Brazilian elections consisted of two terminals; one of them is used for casting votes and the other is used for authentication purposes. The system was mainly based on the use of a unique ID for each voter. After entering this unique ID by the voter, a matching process carried out and the casting terminal is activated\deactivated based on this stage results (Esteve, Goldsmith & Turner, 2012).

- **E-voting in USA, the State of Maryland case:** Maryland State used electronic voting machines in the election conducted in the year 2002. In 2006, the electronic poll books were used for authentication purpose and they were similar to those books used in traditional voting. Once the voter was authenticated on the poll books, a one-time activation card is provided to be used by authenticated voter and inserted into the voting machine to cast the vote. Once the vote is casted, the card is erased and returned. Furthermore, it can't be used anymore unless it was activated again for another authenticated voter (Esteve, Goldsmith & Turner, 2012).

CONCLUSION AND FUTURE WORK

This chapter tried to investigate the e-voting systems and some of its important requirements. It is noticeable that an e-government initiative is not an application that supports government's function, but became a way of life, where some researchers considered it as an umbrella for public work or as the new way of providing e-government services to the public. Applications related to e-voting are vital to the success of e-elections and public decision making through the active participation of citizens and the acceptance of such applications. Based on that, this chapter came to focus on the different techniques and methods used to guarantee a higher level of security and tighter authentication measures on the process.

This chapter reviewed the e-voting process and experiences, and tried to describe its different proposed models and the benefits gained from using such system. It is also essential to look into the different characteristics and measures that make the e-voting process a legitimate tool for political usage. It is essential to maintain the democratic reflection on e-voting process through the concepts of eligibility, accuracy, verifiability, and integrity. In addition, many institutions are calling for transparency and privacy of voters. Other measures that are important in the twenty first century include: receipt free voting, uniqueness, scalability and flexibility.

The second section of this chapter reviewed different methods and techniques of authentication. Such methods were reviewed in a typology related to the characteristics of methods in relation to the voter. Three major types of authentication methods were reviewed: knowledge-based, token-based and biometrics-based methods. In the area of knowledge-based methods, a user ID and password are the major technique used. On the other hand, a token-based technique depends on an extra hardware tool (e.g. card or dongle) that verifies the voters ID. Finally, biometric-based methods were explored, where voter's fingerprint, iris or face-shape, where demonstrated in different methods and cases.

In the last section, few e-voting system cases from different countries of the world were explored to demonstrate the used methods and their implications to the e-voting process. Cases showed that many countries started to realize the importance of security and authenticity of the process. Many countries started using multiple methods to improve the security of the system and to avoid any attacks to the system and possibilities of fraud.

This chapter calls for more research in some areas as some methods were not utilized yet because of certain issues such as cost and applicability. It is important to fully utilize biometric methods to increase the robustness of the e-voting systems, and protect them against different possible attacks. It is also important to mix methods to improve the system and keep voters and their votes protected.

REFERENCES

Abu-Shanab, E. (2012). Digital government adoption in Jordan: An environmental model. *The International Arab Journal of e-Technology, 2*(3), 129-135.

Abu-Shanab, E., Knight, M., & Refai, H. (2010). E-voting systems: A tool for e-democracy. *Management Research and Practice, 2*(3), 264–274.

Al-Anie, H., Alia, M., & Hnaif, A. (2011). E-voting protocol based on public-key cryptography. *International Journal of Network Security & Its Applications, 3*(4), 87–98. doi:10.5121/ijnsa.2011.3408

Altun, A., & Bilgin, M. (2011). Web based secure e-voting system with fingerprint authentication. *Scientific Research and Essays, 6*(12), 2494–2500.

Anane, R., Freeland, R., & Theodoropoulos, G. (2007). E-voting requirements and implementation. *Proceedings of the 9th IEEE Conference on E-Commerce Technology, CEC '07* Tokyo, Japan, (pp. 382–392).

Bhatnager, S. (2004). *E-government from vision to implementation*. New Delhi, India: Sage Publications.

Bouras, C., Katris, N., & Triantafillou, V. (2003). An electronic voting service to support decision-making in local government. *Telematics and Informatics, 20*, 255–274. doi:10.1016/S0736-5853(03)00017-0

Bozinis, A., & Lakovou, E. (2005). Electronic democratic governance: Problem, challenge and best practice. *Journal of Information Technology Impact, 5*(2), 73–80.

Buchsbaum, T. (2005). E-voting: Lessons learnt from recent pilots. *International Conference on Electronic Voting and Electronic Democracy: Present and the Future*, Seoul, Korea, March 2005, (pp. 1-22).

Carter, L., & Belanger, F. (2004). Citizen adoption of electronic government initiatives. *Proceedings of the 37th Hawaii International Conference on System Sciences, IEEE Conference,* Chicago, USA, (pp. 1-10).

Cetinkaya, O., & Cetinkaya, D. (2007). Verification and validation issues in electronic voting. *Electronic Journal of E-Government, 5*(2), 117–126.

Chaum, D. (1988). Elections with unconditionally secret ballots and disruption equivalent to breaking RSA. In Guenther, C. G. (Ed.), *Advances in Cryptology - EUROCRYPT '88, LNCS 330* (pp. 177–182). doi:10.1007/3-540-45961-8_15

Chevallier, M. (2003). *Internet voting: Status, perspectives and issues.* Geneva State, Internet Voting Project. Retrieved from www.itu.int/itudoc/itu-t/workshop/e.../e-gov010.pdf

Deep, K. (2011). Various authentication techniques for security enhancement. *International Journal of Computer Science & Communication Networks, 1*(2), 176–185.

Esteve, J., Goldsmith, B., & Turner, J. (2012). *International experience with e-voting.* National Foundation for Electoral Systems (IFES). Retrieved from www.IFES.org

Futuregov.asia. (2011). *Philippines e-elections miracle.* Retrieved from http://www.futuregov.asia/articles/2010/jul/30/philippines-e-election-miracle/

Gallegos-Garcia, G., Gomez-Cardenas, R., & Duchen-Sanchez, G. (2010). Identity based threshold cryptography and blind signatures for electronic voting. *WSEAS Transactions on Computers, 9*(1), 62–71.

Gerlach, F. (2009). Seven principles for secure e-voting. *Communications of the ACM, 52*(2), 1–8.

Gothwal, J., Yadav, S., & Singh, R. (2011). Enhancing fingerprint authentication system using fragile image watermarking technique. *International Journal of Computer Science and Communication, 2*(2), 459–463.

Haziemeh, F., Khazaaleh, M., & Al-Talafhah, K. (2011). New applied e-voting system. *Journal of Theoretical and Applied Information Technology, 25*(2), 88–97.

Irani, Z., Al-Sebie, M., & Elliman, T. (2006). Transaction stage of e-government systems: Identification of its location & importance. *Proceedings of the 39th Hawaii International Conference on System Sciences,* (pp. 1-9).

Jafari, S., Karimpour, J., & Bagheri, N. (2011). A new secure and practical electronic voting protocol without revealing voters identity. *International Journal on Computer Science and Engineering, 3*(6), 2191–2199.

Kahani, M. (2005). Experiencing small-scale e-democracy in Iran. *Electronic Journal on Information System in Developing Country, 22*(5), 1–9.

Kalaichelvi, V., & Chandrasekaran, R. (2011a). Design and analysis of secured electronic voting protocol. *Journal of Theoretical and Applied Information Technology, 34*(2), 151–157.

Kalaichelvi, V., & Chandrasekaran, R. (2011b). Secured single transaction e-voting protocol: Design and implementation. *European Journal of Scientific Research, 51*(2), 276–284.

Khan, B., Khan, M., & Alghathbar, K. (2010). Biometrics and identity management for homeland security applications in Saudi Arabia. *African Journal of Business Management, 4*(15), 3296–3306.

Khan, H. (2010). Comparative study of authentication techniques. *International Journal of Video & Image Processing and Network Security*, *10*(4), 9–15.

Kumar, K., Kumar, N., Md, A., & Sandeep, M. (2011). PassText user authentication using smartcards. *International Journal of Computer Science and Information Technologies*, *2*(4), 1802–1807.

Kumar, S., & Walia, E. (2011a). Analysis of electronic voting system in various countries. *International Journal on Computer Science and Engineering*, *3*(5), 1825–1830.

Kumar, S., & Walia, E. (2011b). Analysis of various biometric techniques. *International Journal of Computer Science and Information Technologies*, *2*(4), 1595–1597.

Langer, L., Schmidt, A., Buchmann, J., Volkamer, M., & Stolfik, A. (2010). Towards a framework on the security requirements for electronic voting protocols. *Proceedings of First International Workshop on Requirements Engineering for e-Voting Systems (RE-VOTE)*, Atlanta, USA, (pp. 61-68).

Lee, C., Chang, K., & Berry, F. (2011). Testing the development and diffusion of e-government and e-democracy: A global perspective. *Public Administration Review*, (May-June): 444–454. doi:10.1111/j.1540-6210.2011.02228.x

Maaten, E., & Hall, T. (2008). Improving the transparency of remote e-voting: The Estonian experience. *Proceedings of 3rd International Conference on Electronic Voting*, Austria, August 6-9, (pp. 31-43).

Madise, U., & Vinkel, P. (2011). Constitutionality of remote internet voting: The Estonian perspective. *Juridica International*, *18*(1), 4–16.

Maier, M. (2010). *Youth Parliament Esslingen – Living e-democracy first binding election to public office over the internet worldwide*. Retrieved from http://www.jgrwahl.esslingen.de/paper.pdf

Mason, D. (2011). *E-government takes policy-making social*. Retrieved from http://www.publicserviceeurope.com/article/667/e-government-makes-policy-making-social

Masuku, W. (2006). *An exploratory study on the planning and design of a future e-voting system for South Africa*, (pp. 1-159). A thesis published in 2006, in the University of the Western Cape, South Africa.

Okediran, O., Omidiora, E., Olabiyisi, S., & Ganiyu, R. (2011). A survey of remote internet voting vulnerabilities. *World of Computer Science and Information Technology Journal*, *1*(7), 297–301.

Okediran, O., Omidiora, E., Olabiyisi, S., Ganiyu, R., & Alo, O. (2011). A framework for a multifaceted electronic voting system. *International Journal of Applied Science and Technology*, *1*(4), 135–142.

Patil, V. (2010). Secure EVS by using blind signature and cryptography for voter's privacy & authentication. *Journal of Signal and Image Processing*, *1*(1), 1–6.

Prasad, H., Halderman, J., & Gonggrijp, R. (2010). Security analysis of India's electronic voting machines. *Proceeding of 17th ACM Conference on Computer and Communications Security (CCS '10)*, October, (pp. 1-24).

Puiggali, J., & Morales-Rocha, V. (2007). Remote voting schemes: A comparative analysis. *Lecture Notes in Computer Science*, *4896*, 16–28. doi:10.1007/978-3-540-77493-8_2

Rao, G., & Patil, S. (2011). Three dimensional virtual environment for secured and reliable authentication. *Journal of Engineering Research and Studies*, *2*(2), 68–75.

Reddy, A. (2011). A case study on Indian E.V.M.S using biometrics. *International Journal of Engineering Science & Advanced Technology*, *1*(1), 40–42.

Remmert, M. (2004). Towards European standards on electronic voting. *Proceedings of the Electronic Voting in Europe Technology, Law, Politics and Society, A Workshop of the ESF TED Programme together with GI and OCG,* July 7th–9th, 2004 in Schloß Hofen/Bregenz, Lake of Constance, Austria. Retrieved from http://www.gi-ev.de/LNI

Rexha, B., Dervishi, R., & Neziri, V. (2011). Increasing the trustworthiness of e-voting systems using smart cards and digital certificates – Kosovo case. *Proceedings of the 10th WSEAS International Conference on E-Activities (E-ACTIVITIES '11),* Jakarta, Island of Java, Indonesia, December 1-3, 2011, (pp. 208-212).

Sahu, H., & Choudhray, A. (2011). Intelligent polling system using GSM technology. *International Journal of Engineering Science and Technology, 3*(7), 5641–5645.

Sandikkaya, M., & Orencik, B. (2006). Agent based offline electronic voting. *International Journal Of Social Sciences, 1*(4), 259–263.

Sarkar, I., Alisherov, F., Kim, T., & Bhattacharyya, D. (2010). Palm vein authentication system: A review. *International Journal of Control and Automation, 3*(1), 27–33.

Saveourvotes.org. (2008). *Cost analysis of Maryland's electronic voting system, 2008.* Retrieved from www.saveourvotes.org

Smith, R. (2002). Electronic voting: Benefit and risks. *Australian Institute of Criminology, 224,* 1–6.

Sodiya, A., Onashoga, S., & Adelani, D. (2011). A secure e-voting architecture. *Proceeding of 8th International Conference on Information Technology: New Generations,* Las Vegas, Nevada, USA, April 11-13, 2011, (pp. 342-347).

Solehria, S., & Jadoon, S. (2011). Cost effective online voting system for Pakistan. *International Journal of Electrical & Computer Sciences, 11*(3), 39–47.

Weldemariam, K., Villafiorita, A., & Mattioli, A. (2007). Assessing procedural risks and threats in e-voting: Challenges and an approach. *Proceeding VOTE-ID'07: 1st International Conference on E-Voting and Identity,* (pp. 1-12).

World Bank. (2007). *The World Bank website: Report from 2007.* Retrieved May 26, 2011, from http://web.worldbank.org

Yanqing, G. (2010). E-government: Definition, goals, benefits and risks. *International Conference on Management and Service Science (MASS),* 24-26 August, 2010, Wuhan, China, (pp. 1-4).

Yumeng, F., Liye, T., Fanbao, L., & Chong, G. (2011). Electronic voting: A review and taxonomy. *American Journal of Engineering and Technology Research, 11*(9), 1937–1946.

Zahed, A., & Sakhi, M. (2011). A novel technique for enhancing security in biometric based authentication systems. *International Journal of Computer and Electrical Engineering, 3*(4), 520–523.

Section 2
Human–Centered E–Government:
Effectiveness and Organizations

Chapter 7
Designing and Implementing E-Government Projects:
Actors, Influences, and Fields of Play

Shefali Virkar
University of Oxford, UK

ABSTRACT

In modern times, people and their governments have struggled to find easy, cheap, and effective ways to run countries. The use of Information and Communication Technologies is gaining ground as a means of streamlining public service provision by shifting tasks from the government to its citizens, resulting in reduced government costs, increased public revenues, and greater government transparency and accountability. The new buzzword is e-Government: the use of ICTs by government, civil society, and political institutions to engage citizens through dialogue to promote greater participation of citizens in the process of institutional governance. However, the implementation of such projects is complicated by the reality that while developmental problems in these countries are many, the resources available to tackle them are scarce. In attempting to investigate the interaction between new technologies, information flows, and the complexities of public administration reform in the developing world, this chapter examines not only the interplay of local contingencies and external influences acting upon the project's implementation but also aims to offer an insight into disjunctions in these relationships that inhibit the effective exploitation of ICTs in the given context.

INTRODUCTION

A popular discourse in international development policy, and one that has been fast gaining ground in India in recent years, is the use of Information and Communication Technology (ICT) platforms

and applications by the public sector as means of reforming government administration and providing citizens with a range of improved services. The new buzzword is e-Governance: "the use of ICTs by government, civil society, and political institutions to engage citizens through dialogue to promote greater participation of citizens in the

DOI: 10.4018/978-1-4666-3640-8.ch007

process of institutional governance" (Bhatnagar, 2003: p.1). This may be achieved through the use of ICTs to improve information and service delivery, and to encourage citizen participation in the decision-making process; thereby making government more transparent, accountable, and efficient, and involving the governing or management of a system using electronic tools and techniques wherever the government offers services or information (Misra, 2005). The essential aims of e-governance are:

- To initiate a process of reform in the way governments work, share information, and deliver services to external and internal clients;
- To produce greater transparency in the functioning of government machinery;
- To help achieve greater efficiency in the public sector;
- To deliver services to citizens and businesses online, targeting tangible benefits such as convenient and universal access (time and place) to such services, and lowering transaction times and costs (Bhatnagar, 2005).

Conceptually, e-Governance may be divided into *e-Democracy*, defined by an express intent to increase the participation of citizens in decision-making through the use of digital media, and *e-Government*, the use of Information and Communication Technologies by government departments and agencies to improve internal functioning and public service provision (Virkar, 2011). e-Government is hence not just about the Internet and the use of Internet- and web-based systems with government and citizen interfaces (Heeks, 2006). Instead it includes office automation, internal management, together with the management of information systems and expert systems (Margetts, 2006); and is, in short, a process of reform in the way governments work, share information and deliver services to internal and external clients through the harnessing of digital Information and Communication Technologies – primarily computers and networks – in the public sector to deliver information and services to citizens and businesses (Bhatnagar, 2003a). Broadly speaking, e-government may be divided into 2 distinct areas: (1) *e-Administration*, which refers to the improvement of government processes and to the streamlining of the internal working the public sector using ICT-based information systems; and (2) *e-Services*, which refers to the improved delivery of public services to citizens through ICT-based platforms.

THE ECOLOGY OF GAMES METAPHOR

From the turn of the century to the present, there has been a progressive movement away from the view that governance is the outcome of rational calculation to achieve specific goals by a unitary governmental actor (Firestone, 1989), and in that context metaphors based on political games have been extremely useful in developing new ways to think about the policy process. The concept of games has been also used as analogy to explain certain features of political behaviour. The best-known and most popular use of the games metaphor is that of Game Theory, the mathematical treatment of how rational individuals will act in conflict situations to achieve their preferred objectives – from the irrationality of life in schools to how coalitions formulate and pass bills in legislatures. The game metaphor, as developed in the social sciences, assumes that during their interactions actors develop strategies for negotiating with others and for maximizing their needs (Fine, 2000). Being in control is central, with actors acquiring the power to direct action and define situations to the extent that they can persuade others that their image of reality should be taken as the primary framework or the model by which the world should be interpreted.

The use of Game Theory and most other game metaphors (although differing widely in their orientation) have had, according to scholars, one major limitation for clarifying policy processes: they focus squarely on a single arena or field of action; be it a school, a county, a legislature, etc. Yet, by their very nature, policy making and implementation cut across these separate arenas, in both their development and impact (Firestone, 1989). In addition, actors at different levels of the policy system encounter divergent problems posed by the system in question and their actions are influenced by varied motives. What is needed, therefore, is a framework that goes beyond single games in order to focus on how games 'mesh or miss' each other to influence governance and policy decisions.

One of the few efforts to look at this interaction and interdependence in a more holistic fashion was proposed by Norton Long (1958) in his seminal discussion of *The Local Community as an Ecology of Games*. The Ecology of Games framework, as first laid out in the late 1950s offers a New Institutionalist perspective on organisational and institutional analysis. As with most theories of New Institutionalism, it recognises that political institutions are not simple echoes of social forces; and that routines, rules, and forms within organisations and institutions evolve through historically interdependent processes that do not reliably and quickly reach equilibrium (March & Olsen, 1989).

Long developed the idea of the Ecology of Games as he believed existing debates about who governed local communities had significant flaws. Contemporary theories on governance, be they élite perspectives (whose idea is of a unified, rational, goal-driven system where a community is governed by the self-interested politics of highly networked economic élites) or pluralistic theories (which propose that economic rationality and goal maximization takes place at the individual level and that a community is governed as a result of coalitions between several élite stakeholder groups), viewed governing as one isolated game in which all players sought to shape policy within the rules defined by the political and economic system (Dutton, 1992). In other words, the Ecology of Games metaphor became an effort to reconcile these two images and sought to provide a more holistic picture of governance based on the presence of multiple fields of play.

Long contended that the structured group activities that coexist in a particular territorial system can be looked at as 'games' (Long, 1958). An "ecology of games" is thus a larger system of action composed of two or more separate but interdependent games; underlining not only the degree to which not all players in any given territory are involved in the same game, but also the fact that different players within that territory are likely to be involved in a variety of interactions (Dutton & Guthrie, 1991). Games can thus be interrelated in several ways. Some actors ('players') might be simultaneously participating in different games, and some players might transfer from one game to another (Long, 1958). Plays (i.e., moves or actions) made in one game can affect the play of others. Also, the outcome of one game might affect the rules or play of another. However, although individuals may play a number of games, their major preoccupation for the most part is with one, central game (Crozier & Friedberg, 1980). A researcher might be able to anticipate a range of strategies open to individuals or organizations if we know what role they play in the game(s) most central to them. Conversely, when the actions of players appear irrational to an observer, it is likely that the observer does not know the games in which players are most centrally involved; the players' moves in one game might be constrained by their moves within other games. The following sections of this chapter examine the issues and challenges that have bearing on the development and implementation of e-government systems from the perspective of the Ecology of Games framework.

ICTS AND THE REFORM OF PUBLIC ADMINISTRATION

Since the early 1990s, a wealth of online applications have emerged which have transformed the original, purely text-based read-only medium of the Internet into one that supports dynamic and modifiable rich-media content useful for the streamlining of government processes. Public sector reform, in its broadest sense, may be generally defined *as change within public sector organizations that seeks to improve their performance* (Heeks, 1998b). e-Governance thus does not merely involve the installation of computers and computer operators in an organisation, instead it involves the *creation* of systems wherein electronic technologies are integrated with administrative processes, human resources, and the desire of public sector employees to dispense services and information to people fast and accurately. The concept thus consists of two distinct but intertwined dimensions – the political and technical – relating to the improvement of public sector management capacity and citizen participation (Bhatnagar, 2003c).

ICT-led reforms are often complex as they involve reforming both organisations and human behaviour, and cannot be made through legislation alone. Such reforms require a change in the way users think, act, how they view their work, and how they share information; together with a simultaneous reengineering of the working of government – its business processes within individual agencies, departments, and across different levels of governments. They work best when part of a broader reform agenda in which the *status quo* is broken down through delegation, decentralisation, and citizen empowerment (Misra, 2005). The development and adoption of e-Government technologies has thus become a large game within the 'meta-game' of a country's development, and brings with it not only an array of benefits, but also numerous challenges and obstacles – all of which shape and are shaped by the perceptions and motivations of a multitude of actors.

E-GOVERNMENT: DEFINITION AND SCOPE

Over the last 10 years, a number of scholars and international organisations have defined e-government in an attempt to capture its true nature and scope. A selection of key definitions is highlighted in *Box A*. Almost all definitions of e-government indicate three critical transformational areas in which ICTs have an impact (Ndou, 2004), illustrating that e-government is not just about the Internet and the use of Internet- and web-based systems with government and citizen interfaces (Heeks, 2006); instead it includes office automation, internal management, the management of information and expert systems, and the design, and adoption of such technologies into the workplace (Margetts, 2006):

- **The Internal Arena:** Where Information and Communication Technologies are used to enhance the efficiency and effectiveness of internal government functions and processes by intermediating between employees, public managers, departments, and agencies. The use of ICTs is thought to improve internal efficiency by enabling reductions in both the time and cost of information handling, as well as improving the speed and accuracy of task processing. In other words, technology is felt to significantly reduce processing times, eliminate inefficient bureaucratic procedures, and skirt manual bottlenecks; allowing information to flow faster and more freely between different public sector entities.
- **The External Arena:** Where ICTs open up new possibilities for governments to be more transparent to citizens and businesses by providing multiple channels that allow

Box A. Definitions of e-Government

Tapscott (1996): "eGovernment is an Internet-worked government which links new technology with legal systems internally and in turn links such government information infrastructure externally with everything digital and with everybody – the tax payer, suppliers, business customers, voters and every other institution in the society."

Fraga (2002): "Government is the transformation of public sector internal and external relationships through net-enabled operations, IT and communications, in order to improve: Government service delivery; Constituency participation; Society."

Commonwealth Centre for E-Governance (2002): "E-Government constitutes the way public sector institutions use technology to apply public administration principles and conduct the business of government. This is government using new tools to enhance the delivery of existing services."

Commonwealth Centre for E-Governance (2002): "E-Government constitutes the way public sector institutions use technology to apply public administration principles and conduct the business of government. This is government using new tools to enhance the delivery of existing services."

World Bank (2010): "eGovernment refers to the use by government agencies of information technologies (such as Wide Area Networks, the Internet, and mobile computing) that have the ability to transform relations with citizens, businesses, and other arms of government. These technologies can serve a variety of different ends: better delivery of government services to citizens, improved interactions with business and industry, citizen empowerment through access to information, or more efficient government management. The resulting benefits can be less corruption, increased transparency, greater convenience, revenue growth, and/or cost reductions."

(Source: Commonwealth Centre for E-Governance (2002), Ndou (2004), World Bank (2010))

them improved access to a greater range of government information. ICTs also facilitate partnerships and collaborations between different government institutions at different levels of a federal structure and between the government and other non-governmental actors.

- **The Relational Sphere:** Where ICT adoption has the potential to bring about fundamental changes in the relationships between government employees and their managers, citizens and the state, and between nation states; with implications for the democratic process and the structures of government.

Thus, although the term e-government is primarily used to refer to the usage of ICTs to improve administrative efficiency, it arguably produces other effects that would give rise to increased transparency and accountability, reflect on the relationship between government and citizens, and help build new spaces for citizens to participate in their overall development (Gascó, 2003). Broadly speaking, e-government may be divided into two distinct areas: (1) *e-Administration*, which refers to the improvement of government processes and

to the streamlining of the internal workings of the public sector using ICT-based information systems, and (2) *e-Services*, which refers to the improved delivery of public services to citizens through ICT-based platforms. The adoption of e-government often involves interactions to reform the way governments, their agencies, and individual political actors work, share information, and deliver services to internal and external clients by harnessing the power of digital Information and Communication Technologies – primarily computers and networks – for use in the public sector to deliver information and services to citizens and businesses (Bhatnagar, 2003).

In order to further the analysis of issues affecting the impact of ICTs on administrative reform, this paper sets out a four-fold categorisation of existing e-governance projects based on this author's research. Case studies may be discussed along different axes depending on the level of the participating government agency, the geographic focus of the project (rural or urban), the nature of the initiating agency, and the central relationship impacted by the project. The four categories that may be derived from this author's research are explored briefly below:

- **Level of Government:** Case studies may be classified according to the level of government at which they are implemented; more specifically as projects implemented by *local government* agencies, at the level of the *state government*, or at the *national government* level:
  - **Local Government:** Includes those E-governance projects of note, which are initiated at the level of local government.
  - **State Government:** Covers those e-governance projects initiated by state government departments and agencies.
  - **National Government:** Those E-governance initiatives begun by or within national government ministries and other national-level agencies and institutions.
- **Geographic Focus:** Projects may also be categorised and discussed according to the location of their target audience or in terms of the section of the population from whom feedback is sought – namely *rural* or *urban* populations:
  - **Rural:** Those projects whose target population or target audience is primarily based in rural areas.
  - **Urban:** Those projects that impact people living primarily in urban areas.
- **Nature of Collaborative Process:** E-governance projects may also be classified according to nature of the initiating agency or according to the context of the political dynamic between the public and private sectors within which the project was conceived and implemented. More specifically, they may be discussed as *government-led initiatives, civil society-led projects,* or *collaborative ventures*:
  - **Government-led Initiatives:** Are projects initiated either wholly by government departments and agencies or those in which the government take a leading role.
  - **Civil Society-led Projects:** Include those projects initiated within the broader sphere of governance, involving efforts initiated wholly or primarily by civil society bodies and NGOs.
  - **Collaborative Ventures:** Cover those projects initiated across sectors, generally conceived as a joint venture between a government agency and a private sector/ civil society entity, and having a variety of different stakeholders.
- **Central Relationship Impacted:** The final axis against which case studies may be classified is based on the central relationship impacted by the project under study. Existing projects deal with improving *government-to-government* functioning, *government-to-citizen* interactions or *government-to-business* dealings:
  - **Government to Government:** Electronic service delivery can result in productivity gains within government organisations. Data may be easily shared across government agencies electronically, resulting in a tighter monitoring of employee productivity, the identification of bottle-necks in service delivery, and the accumulation of historical data which may be mined for policymaking purposes.
  - **Government to Citizen:** A number of States across India have developed online systems for the delivery of municipal services to their citizens. Citizens benefit from shorter processing times, the availability of a plethora of services in one place, fewer visits to government departments, greater government accountability and reduced corruption through

the elimination of intermediaries. In some parts of South Asia, a number of these applications have resulted in a limited empowerment of local communities that could previously not acquire government information easily either due to physical distance or corruption.

- ○ **Government to Business:** These projects involve the online delivery of public services to businesses and industry and include systems such as tax collection and e-procurement, thus providing businesses with an easier channel through which they may interact with government.

DIFFERENT E-GOVERNMENT DELIVERY MODELS

e-Government applications tend to develop in two stages (Bhatnagar, 2003b). Initially, a back-office system is set up within the adopting agency to handle online processes and information about services provided by the agency is published on a website. The second step involves the setting up of the 'front-office': the use of ICTs in the actual delivery of a service, where citizens can interact with the site to download application forms and information sheets for a variety of services such as filing a tax return or renewing a license, with more sophisticated applications being able to process online payments.

A key three-stage strategy used by actors in games related to the design and development of e-government systems and technology policy, particularly those in developing countries who wish to radically transform public administration by moving government services from manual processes to online systems, is to adopt different models of service delivery at different stages of the development process. The first move generally involves the automation of basic work pro-

cesses and the online provision of information and services by government departments from computers based within the departmental premises (Bhatnagar, 2003e).

Citizens interact with a designated government employee or private computer operator who accesses data and processes transactions on their behalf. Locating online terminals within agency premises tends to result in greater ownership of the system by government staff, reducing resistance to technology and facilitating easier acceptance of change. However, the downside of this mode of delivery is that citizens are still required to visit different government departments to avail of different public services, all within their fixed hours of work. In addition, the dependence of an entire agency office on a single person (or small group of people) to operate the system may cause friction (see Box B).

The second stage in the evolution of service delivery is the use of conveniently located citizen kiosks or service centres in public places, again manned by public or privately hired operators (Basu, 2004). This mode of delivery scores over the previous one as multiple services – municipal, state, or federal – may be offered at each location. Kiosks also generally stay open longer than government offices, both before and after regular office hours, maximising system coverage by allowing working individuals to access services at times more convenient to them. In recent years, citizen service centres have become popular, particularly in countries where Internet penetration is low.

The final platform of e-government service delivery, popular in countries where Internet penetration and skills are high, is the one-stop shop online portal from where citizens with a computer and an Internet connection may, at any time of day, access a whole range of public information and services themselves without having to visit a kiosk or depend on a computer operator (West, 2004). However, for such a mode of delivery to become ubiquitous, a number of conditions need

Box B. Innovative Solutions to e-Government Service Delivery

In developed countries, services are generally offered through self-service Internet portals that become a single point of interaction for the citizen to receive services from a large number of departments. In developing countries, owing to the different sets of socio-economic and political rules and constraints that circumscribe interactions related to technology design and adoption, new models of service delivery different to those found in industrialised nations must be explored.

Unlike self-service models, where citizens interact directly with a one-stop portal from their homes or offices, most applications in the developing world are department-specific and are usually accessed by citizens at online service counters or public kiosks, where government or contracted private sector employees interact with citizens and mediate between them and computer screens to process transactions. More recently, Citizen Service Centres are created at convenient locations where citizens may access online services provided by several departments, though again citizens do not interact directly with web interfaces and the collection of processing charges and other dues is usually handled through conventional means.

In the context of an ecology of games, the employment of external computer operators in service delivery projects results in creation of a new group of actors whose primary role is to act as intermediaries between both sets of technology users (government servants and the public). Whilst the roles and actions of these players are often generally greatly curtailed by decisions and policies taken by their bosses in government administration, their mere presence in an office may have a significant impact on other players' perceptions of a given system – either speeding up the acceptance and adoption of technology by creating a positive impression in the minds of government employees or hindering it by stirring up feelings of jealousy and resentment.

(Bhatnagar, 2003a)

to be in place – citizens must have the technological hardware and skills to access the system, the back-end of the government agency must be fully computerised, government staff must be trained on the new technology, security and privacy loopholes must be closed, and trust in online transactions must be built up.

The step-by-step strategy outlined above is generally adopted by key political and administrative actors involved with the implementation of e-government projects, and if followed may reduce political tensions and controversies that might arise as the result of change by not only ensuring maximum citizen access to services, but also an increased acceptance of the technology by agency staff.

GAINS FROM E-GOVERMENT

e-Government has become an influential concept for the reform of public administration, and is increasingly being seen as the answer to a plethora of problems that country governments at all levels face in serving their citizens effectively (Heeks, 2000). A number of gains from e-government applications that accrue to both government and citizens may be found in current literature, many of which are often cited by project designers and

champions as important reasons for pursuing ICT project design and adoption games within their respective organisations.

Cost Reductions and Efficiency Gains

Efficiency gains and expenditure reductions in the private sector have long been associated with the introduction of ICTs, where studies have linked their deployment to increased competitiveness resulting from reductions in the cost of setting up and running an enterprise. It is now widely believed in policy circles that these benefits may be realised by the public sector as well, and that putting services online and automating processes can substantially decrease working costs, increase operational efficiency, and enhance the transparency of many activities relative to the manual handling of tasks – all of which would have a significant impact on government finances (Edmiston, 2003). A frequently cited example in developing countries is that of public procurement (the letting of contracts for major public works and the sale of public assets) which has traditionally been thought of as an expensive, time consuming process prone to so-called 'grand corruption' that often results in huge revenue losses for the government. The use of electronic procurement (or e-procurement) systems, according to supporters of e-government, can not only reduce the amount of revenue lost to corruption by increasing the transparency of the bidding process, but also save money by cutting down on paperwork, doing away with advertising, and lowering the price of procurement through increased competition (Bhatnagar, 2004).

The use of ICTs in government is also thought to reduce the number of inefficiencies in processing by allowing information sharing across employees and departments, thereby contributing to the elimination of mistakes from manual processes and reducing the time required for transactions; to the benefit of both the transacting agency and the citizen (Ndou, 2004). Scholars such as Gramlich (1990) note that the savings accruing to individual consumers of public services in the form of time and travel may in some cases be much larger than any pecuniary savings accruing to the providers, as often, saving time is the single most important benefit derived by a citizen from an e-government project (Gramlich, 1990). The same can be said of time and convenience gains made by government employees who use a digitised system. Efficiency, according to Edmiston (2003), is therefore not about simply minimising the government's cost of providing a given level of public services, but also about minimising social cost – a large part of which is the cost to those employees and constituents using or receiving public services. This may be attained through the streamlining of internal processes by enabling faster, more informed decision-making and by speeding up transaction processes.

Quality of Service Delivery

Under the paper-based model of public service delivery, evidence from the field suggests that processes are often long, time-consuming affairs that lack transparency and generally result in poor service quality and high levels of citizen/business dissatisfaction (Fang, 2002). e-Government initiatives that put services online and simplify procedures through automation are believed to have the potential to enhance service delivery in terms of time, content, and openness by reducing red tape, offering round-the-clock accessibility, and enabling fast and convenient transactions (Ndou, 2004). To start with, proponents of e-government argue that easily accessible information means that citizens need to spend less effort in finding out how a good or service might be obtained and are required to make fewer visits to government departments to avail of the required service. By automating routine clerical work, staff time can be freed up for more substantial tasks.

Quick processing times reduce the total time spent on transactions and reduce waiting periods, whilst automating processes ultimately results in more efficient services through the introduction of competition between departments (Bhatnagar, 2004). ICTs thus also potentially allow for the replacement of what has traditionally been labelled "street-level bureaucracies", substituting these with what Reddick (2005) calls "system-level" or "server-based" public organisations that permit citizens to access public services online using ICTs, entirely side-stepping any face-to-face contact with the actual providers and facilitators.

In cases such as these, it is important to note that the expectation that ICTs will somehow improve the trustworthiness of government agencies (thereby increasing citizen's trust in government and raising the image of public service providers) entails the premise that ICTs themselves are seen as trusted mediators or actors involved in the performance of an agency's task (Avgerou, Ganzaroli, Poulymenakou & Reinhard). However, the validity of such an assumption – as demonstrated by the vast literature on the subject – cannot be taken for granted. Instead, scholars such as Carter & Bélanger (2005) and Wankentin, Gefen & Pavlou (2002) have recognised that even if successfully implemented, e-government services themselves may not be trusted enough by the population at large, and their adoption by government agencies requires the implementation of transparent mechanisms that inspire trust.

Transparency, Anticorruption, and Accountability

Another strong argument in favour of e-government is the impact that ICTs are thought to have on corruption and accountability. Corruption is seen to flourish when there is no transparency in government functioning, particularly under manual or paper-based systems where citizens have to visit government departments in person, hand over application forms, and pay fees to designated officials in order to obtain a service (Bhatnagar, 2003c). Complex and ambiguous rules and cumbersome procedures, coupled with extensive face-to-face contact give officials the opportunity to extract bribes and use their monopoly powers to commit corrupt acts, particularly by delaying or denying services to citizens.

Information and Communication Technologies offer benefits not found in conventional information systems that make anti-corruption reforms possible by introducing transparency in the data, decisions, actions, rules, procedures, and performance of government agencies; thereby simplifying processes and rules, taking away discretion by automating processes, building accountability, introducing competition between delivery channels, standardising documentation for effective supervision, and centralising and integrating data for better audit and analysis (Bhatnagar, 2003c). As scholars and practitioners such as Colby (2001), Budhiraja (2003), and Chaurasia (2003) have noted; ICTs allow for greater accessibility and instant communication; facilitating automatic record keeping, the systematic classification and recovery of data, better knowledge management, and the improved sharing of information.

The government and its officials are thus made accountable for their actions as, unlike in manual or paper-based environments, those responsible for particular decisions or activities are readily available and administrative actions are easily traceable; thereby transforming public administration and improving government-citizen interactions (Bhatnagar, 2003d). e-Government projects also eliminate face-to-face contact, bringing services to citizens' doorsteps and, in doing so, improving the quality of information, striking at the root of corruption by enhancing transparency, reducing opportunities to commit corrupt acts, and increasing the likelihood of exposure for those who indulge in them (Bhatnagar, 2004). The availability of a variety of official documents and publications regarding the activities of public agencies together with the economic and legislative aspects of government helps raise transparency as well (Ndou, 2004).

Improvements in the Quality of Decision Making

Community creation and the continuous interaction between the government and its citizens made possible through the use of ICTs is felt by scholars to have a positive impact on the quality of decision-making by allowing policy élites to tap into wider sources of information, perspectives and solutions (OECD, 2001). It is often argued that the speed and immediacy of ICT networks allow citizens to communicate, give feedback, ask questions, complain, exchange information effectively and build relationships with their representatives; whilst at the same time allowing policymakers the opportunity to interact directly with the users of public services and solicit their opinions (Virkar, 2007).

Citizens benefit from being able to contribute their own ideas and share knowledge and information through active participation in political and governmental discussions, and political elites are at the same time able to take better decisions after listening to and understanding the needs and requirements of their constituents (Ndou, 2004). This two-way relationship can help build trust in government and improve the relationship between the government and the public, as well as between various departments and public agencies. However, it is often forgotten that improvements in the speed and quality of decisions also depend on a number of factors; including the willingness of the government to empower its citizens with new information, the ability of government staff to work on new systems and process large amounts of data, the prevailing organisational values within the public sector, and the ability of the adopting public agency to make the transition from a hierarchical model of public administration to a more flexible, decentralised one (Ndou, 2004).

Increases in the Capacity of Government

The use of ICTs is also thought to offer opportunities to increase government capacity by making necessary the reorganisation of internal administration, transactions, communication, and easy information exchange (Ndou, 2004). ICTs allow for databases to be shared between departments and skills and capacities to be pooled; resulting in faster information exchanges, quicker and cheaper provision of goods and services, and overall faster decision making processes – benefitting both government and businesses/citizens. In this way, such technologies have an impact on the structure of government and various aspects of public management including bureaucratic and political arrangements within departments, decision-making processes, and the means of ensuring transparency and accountability (Radoki, 2003).

Further, the introduction of technology in government has transformed the way in which roles and responsibilities are allocated within and across public sector agencies at different levels (Jae Moon, 2002). The creation of an information infrastructure is therefore central to the transition of a public agency from pre-existing information systems, structures, and procedures to a fully digitised organisation that would support the process of government reform (Navarra & Cornford, 2005). However, Navarra & Cornford (2005) note that this change requires more than just simply introducing electronic versions of existing services: instead the focus should be on designing and implementing systems that might be exploited across and within new service channels, essentially supporting both the 'invention' and 're-invention' of government.

E-government initiatives have also been used by authoritarian regimes to not only provide services to citizens and make the business environment more competitive, but also to consolidate the power of the central government over its own local representatives and to contribute towards the depoliticisation of society through tighter grassroots control. In his seminal paper on the use of ICTs by authorities in China, Kluver (2005: p.76) notes how the use of such initiatives has been to add "…stability and order to a chaotic governing process and social change, and to re-establish the control of governing authorities, including improving the quality of surveillance and data gathering, and hence policy making, the elimination of corruption, and ultimately the re-legitimation of the Communist Party in China." However, although the significance of these goals is worth noting as they raise important questions regarding the role of ICTs in the transformation of governance, it is important to remember that the powers exercised by the central authority in China are often greatly different from those held by either developed or developing democratic governments, and the wider applicability of the lessons drawn might thus be altogether fairly limited.

In sum, from an administrative point of view, the new Information and Communication Technologies – and more specifically the Internet – may change the way governmental actors pursue their goals. Firstly, new opportunities to improve efficiency arise when the government and its agencies use ICTs to create and maintain networks. To quote Mechling (2002: p.155) "…public organizations are rapidly becoming networked [both within government departments and across them] and they are using these networks to produce and deliver services. This will ultimately lead to efficiency improvements, much as has happened with the private sector".

Secondly, digitization has cut transaction costs, in some cases all the way to zero. This holds particularly true for cost savings generated by the payment of bills online and document downloads where, according to Fountain (2001: p.5), the "movement from paper-based to web based processing of documents and payments typically generates administrative costs savings of roughly 50 per cent". The possibility of lowering costs gives rise to further efficiency gains as governments and their agencies have the chance to achieve their objectives using fewer resources such as time, money, and physical inputs. Finally, the introduction of ICTs brings about important transformations at the organisational level as they affect the chief characteristics of the Weberian bureaucracy; reshaping the processes of production, coordination, control, and direction that take place within the public sector (Fountain, 2002). Many simple clerical tasks prevalent in paper-based bureaucracies have already been replaced by computerised databases and digital documents: desktop computing capacity and the availability of multiple databases and analytical tools have collapsed the work of different positions into a few or one position that deals with many tasks, and the use of digital tools has allowed relatively unskilled employees to make sophisticated evaluations.

From a citizen perspective, services can be delivered more rapidly, with shorter processing and information retrieval times increasing the quality and efficiency of service delivery. Waiting times may be reduced, as routine cases are dispensed with quickly and access to different databases allows civil servants to cut down processing times. e-Government helps in the creation of "transaction-capacity governance", where citizens are empowered, are no longer information-poor, no longer required to wait in long queues for services and are no longer exposed to the socio-economic consequences of corruption (Prahalad, 2005). The result is greater civic engagement and less corruption, leading to better governance. Citizens have easier access to service agencies through, for example, information kiosks and have access to public information at the click of a button. In turn, increased access to services can stimulate the openness of government by taking away discretion, curbing opportunities for arbitrary action, and increasing chances for disclosure (Bhatnagar, 2003c). Indeed, e-government pilots in some developing countries have already demonstrated a marked positive impact on corruption, transparency, and quality of service provision (Bhatnagar, 2003a).

PUBLIC SECTOR REFORM GAMES: PROCESS RE-ENGINEERING AND E-GOVERNMENT

Information and communication technologies bring about rapid changes in management patterns, such as the breakdown of traditional administration hierarchies and the streamlining of decision-making within and across agencies. Steps to adopt and use ICTs in government are thus generally taken as part of a broader reform or change-management agenda driven by actors from different levels, where new technology is introduced to solve existing administrative problems. The re-engineering of administrative processes is possibly the most important step in implementing an application, as it requires that an agency undertakes substantial reform of its organisational structure (Bhatnagar, 2004). This is particularly true as using ICTs with out-dated or inappropriate processes can increase corruption and other forms of poor governance by providing opportunities for officials to perform dishonest activities faster and still avoid detection (Pathak & Prasad, 2005).

Re-engineering processes often involves playing games to change the mind-sets and culture of an organisation's workforce, including using strategies that recognise the need to train employees, improve skill sets, and deploy appropriate supporting infrastructure to enable online processes that are useful to both the user and the implementing organisation. A common strategy in a successful implementation game is to map existing methods and procedures, usually followed by the simplification of these procedures in such a way that the overall task can be completed in as few steps as possible (Misra, 2005).

The looked-for outcome of such an exercise is that of mutual cooperation; where all the players in the game accept the modification of processes that result in fewer steps, any eventual reduction in the number of people needed to perform tasks, and the automation of certain operations that result in eventual back-end computerisation. However, this is not always the case, and re-engineering games may get stymied in conflicting moves made by different key players. This thesis argues, therefore, that the use of ICTs alone will not guarantee the success of a project in achieving its objectives and reaching its full potential. Successful e-government systems and re-engineered processes standardise rules and procedures, but it is the well thought-out games played by project designers and implementers which ultimately bring down resistance and fear, reduce opportunities for exercising discretion, and create an environment conducive to the adoption of the new technology.

Related to this, project managers, whilst implementing a project, have to decide whether they will adopt a top-down approach to decision-making or whether they will select a more participatory style. Whilst a top-down approach to project management does yield a number of benefits – including the speeding up of decisions that might otherwise be difficult to make (particularly true for cases like the one under study, where employees might attempt to resist the introduction of technology when faced with dramatic changes) – such an approach means that during the planning of a system, priority goes to those features and aspects which are seen as important by a select, centralised group of planners. There is a danger that some of the priorities of the main users, the staff on the ground, may be overlooked and any mismatch between design and user needs may result in employees rejecting the system. To add to this, most of the literature on organisational change in the private sector stresses the importance of employee participation in the planning of change-inducing projects (Lefebvre & Lefebvre, 1996), particularly to enhance staff morale, an idea which is catching on in public sector management and e-government circles (Heeks, 2006).

GAME CHANGERS: CHALLENGES TO E-GOVERNMENT IMPLEMENTATION

Whilst e-government in particular and ICTs in general have the potential to be powerful drivers of wealth creation and growth, the multidimensionality and complexity of such initiatives implies the existence of a wide variety of factors that hinder constructive interactions and the smooth acceptance and adoption of ICTs. Employee resistance is still the biggest barrier to successful change, and many e-government projects face substantial internal resistance from government staff as computerisation changes workloads, work profiles, and work content (Bhatnagar, 2004). Government employees often fear both change and ICT applications, as they see computerisation as leading to a loss of power and responsibility, causing redundancy and job losses, and view easily and widely accessible information as a loss of control. In addition, politics and individual interests, considered by some to be the strongest determinants of e-government success or failure (Heeks, 2006), also create resistance to change, internal conflicts, and turf issues within public organisations (Barki, Rivard & Talbot, 1993). Fear of the unknown – an uncertainty of the benefits that may accrue from the new system or a perception whose disadvantages outweigh advantages – can also lead to problems (Bhatnagar, 2004). Addressing resistance successfully, therefore, requires the existence of incentives for employees to learn and change, and the establishment of well-structured plans that embrace employee participation throughout all stages of the implementation process (Ndou, 2004).

From the point of view of the analytical framework used to explain actor behaviour within such projects, issues that create discord and resistance may be seen as game-changers as their presence (or absence) impacts and changes the actor perceptions and motivations which underlie game objectives, moves, and strategies. Although there is no single list of challenges to e-government initiatives available, a look through the literature reveals the existence of several issues consistent across different disciplines (Gill-Garcia & Pardo, 2005).

Challenges Relating to ICT Infrastructure

Poor infrastructure is recognised by many scholars as one of the main challenges to the successful implementation of e-government projects (Ndou, 2004), for inhospitable working environments and badly designed systems are bound to have a negative impact on employee mind-sets. For instance, while the use of ICTs in government offers the potential for substantially improving public sector employee performance, these gains are often not realised owing to difficulties arising from flaws in system design, issues of compatibility, access, and lack of basic computer literacy.

Organisational and Managerial Challenges

e-Government systems affect the civil service in many ways. The computerisation and automation of work processes alters accountability, reduces discretion and flexibility, and makes performance visible and easy to monitor. It requires that staff retrain and retool. By making information available to all, computerisation alters the power and authority vested in different levels, flattening the hierarchy within the organisation (Bhatnagar,

2004). Change management issues must therefore be addressed as new work practices, new processes, and new ways of performing tasks are introduced so that uncertainties and fears are addressed during the early stages of implementation.

Human Capital Challenges

A major barrier to the successful implementation of e-government projects in the public sector is the lack of skills required to deal with new technologies and new ways of working; as e-government projects require hybrid human capital capacities, together with technological skills for system design, installation, and operation; and commercial and managerial skills for handling online processes, functions, and customers (Silcock, 2001). This is a particular problem in developing countries where there is often a chronic lack of qualified staff and inadequate human resource training, generally resulting in extreme frustration and resentment amongst employees if not dealt with properly (UNPAN, 2002).

Legal, Institutional, and Environmental Challenges

Additional challenges to the successful implementation of ICTs in government relate to a more general institutional framework and policy environment within which the adopting government organisation operates (Bajjaly, 1999). As e-government systems increase in sophistication, they often require the introduction of a new range of rules, policies, laws, and legislative changes to address electronic activities; such as freedom of information, electronic archiving, data protection, computer crime, intellectual property rights, and copyright issues (Ndou, 2004). There is also need to reform complementary (though un-automated) processes and systems through legislation to ensure that they do not conflict with laws and instead give support to the new system. External pressures such as policy agendas and politics may also pose challenges to the successful outcome of an ICT-for-government initiative as, in making any kind of decision, public managers have to take into account a large number of (often) restrictive laws and regulations, leading to an innate desire to resist change and preserve the *status quo* (Bellamy, 2000).

Information and Data Issues

e-Government is about the capture, storage, use, and dissemination of information by the private sector, and thus a number of challenges to its successful implementation relate to the quality of data and data structures. First and foremost, data quality problems that could hamper the uptake of an e-government system, the quality of decisions, and processes flowing from it include inaccuracies, inconsistencies, and incompleteness of data (Redman, 1998). Scholars such as Ballou & Tayi (1999) further identify the lack of proper appropriate data as a challenge to the implementation of ICT initiatives, and one that might cause users of a system to reject it in the long-term.

Strategy Challenges

Another reason that some e-government projects result in user frustration and can degenerate into negative interactions is the failure on the part of project planners to develop and establish a feasible, context-tailored strategy (Heeks, 2006). Many public institutions go in for e-government projects by simply transferring information, processes, and services online without considering the organisational changes needed to take advantage of their full benefits (Ndou, 2004). Devising a good strategy is a difficult task, as it requires planners to take a holistic view of a project, its different aspects, and current processes; as well as think about its long-term focus and objectives. A clear strategy needs to engage in a rigorous assessment of the reality on the ground: the laying out of an inventory of project details and a statement

of costs and benefits, followed by a continuous monitoring and evaluation of the process of project implementation and use. Only then will actors be completely aware of the situation and the role that they are required to play.

Leadership Challenges

A final challenge to effective e-government implementation is the development and maintenance of committed leadership (GAO, 2001). As e-government is a complex process accompanied by large investments, high risk, and radical change; there is a need for senior public managers who can understand the real costs and benefits associated with projects and be able to lead, motivate, influence, and support employees under their charge. Every project needs its 'e-champions': leaders who have the vision, commitment, and skills to oversee the implementation of the project, empower workers in the new digital office environment, and defend experimental action (Allen, Julliet, Paquet & Roy, 2001). Strong leadership and clear lines of accountability are thus vital to overcoming employees' resistance to change, marshalling the resources needed to improve management, and building loyalty to the project within an organisation.

DISCUSSION AND CONCLUSION

Prompted by declining computing costs, a number of developing countries have begun to direct administrative reform towards achieving decentralised development planning through the diffusion of technology to relatively small units of administration (Madon, 1993). Those behind this programme are often strong leaders motivated, not only by the spirit of public service, but by the power, influence, and reputation that associating themselves with such a trendy, cutting-edge agenda can bring.

The use of ICTs in public service provision is also encouraged by actors originating from civil society bodies, multinational companies, and the public at large; people who believe that technology can, over time, shift the balance of power between different players in the governance game by not only making government agencies more citizen-centric and responsive, but also enhancing their own status as serious players in policy and decision-making circles. However, despite the potential of ICTs to support reform, there remain substantial problems for (a) the public sector of a developing country to enter the ICT-reform era at all, and (b) for a country to move on within a short time period from a paper-based environment to an integrated approach that effectively uses ICTs to enable the delivery of reform objectives.

Introducing e-government initiatives into public bodies is therefore a tricky game to play, as computerisation alters the work-load, work profile, and work content of the average public sector employee; impacting accountability, reducing the opportunities for exercising discretion, making performance more visible, flattening the hierarchy, often forcing the need for retraining and retooling, and sometimes creating redundancy (Bhatnagar, 2006). Many projects tend to face internal resistance from staff – particularly from the middle to lower levels of the civil service – with moves made to re-engineer processes and effect back-end computerisation having a profound effect on the way civil servants perform their duties and perceive their jobs. Very often in developing countries, it is the fear of the unknown that drives this resistance, particularly if the introduction of new technology results in a change of procedures and the need for new skills. Further, in corrupt service delivery departments, there may be pressure to slow down or delay the introduction of technology-led reforms due to the impending loss of additional income.

The implementation of e-government projects in the developing world is also often complicated by the reality that, whilst developmental problems in these countries are many, the resources available to tackle them are scarce. At the national level, the players of games to develop technology policy compete with those who seek to advance other development objectives. Low rates of Internet penetration in these countries, for instance, mean that in order for any project to be successful, governments would have to first invest heavily in the infrastructure and skills needed to generate an atmosphere conducive to increased public sector employee use and citizen access (Roy, 2005). Large investments in both the initial groundwork and the project itself would inevitably divert resources away from other high-priority areas such the provision of basic infrastructure, primary education, health services, and water and sanitation requirements. It is thus doubly imperative that projects are successful: not only must they improve standards of living, but in doing so must justify the investment made in them. Success, however, cannot be guaranteed, and failures risk compounding the problems of the developing world.

In modern times, people and their governments have struggled to find easy, cheap, and effective ways to run countries. In the long term, ICT-based applications have the potential to revolutionise patterns of communication between authority and citizenry, radically restructuring politics and governance at all levels by making systems more integrated, transparent, and efficient. The use of ICTs and web-based software would, as discussed in previous sections of this chapter, streamline public service provision by shifting tasks from the government to its citizens; resulting in reduced government costs, increased public revenues, and greater government transparency and accountability. Additionally, such an outcome would encourage citizen interest and participation, and decrease instances of deliberate attempts (by both citizens and those in positions of authority) to flout the law.

However, the broader debate surrounding the prioritisation of issues in the setting of a development agenda still rages in scholarly and policy circles. Critics of e-government, and particularly of its introduction in a developing country context, contend that administrative reform is not an important enough issue to justify exposing cash-strapped governments to the risks and opportunity costs associated with ICT projects. It comes as no surprise, therefore, that whilst e-government initiatives have been pursued with vigour in the developed world over the past few decades, governments in the developing world have been less enthusiastic about investing heavily in the necessary infrastructure (Kluver, 2005). It is only recently that government leaders, particularly in Asia, have made efforts to implement technology-driven reforms within government bodies across the region (Parks, 2005).

The four-fold classification laid out in Section III, and the issues discussed in previous sections of this chapter illustrate that the development and implementation of ICT-for-development projects carry deep political implications: the politics of power and influence often drive the design of the project, political reputations may be staked on the outcome, opinions generally vary on whether certain aspects of the system are economically and politically viable or desirable, politics circumscribes what can and cannot be implemented, and reactions to reform the system always have deep political implications. This chapter thus aimed to give the reader a thorough understanding of key concepts and challenges involved in the study of e-government initiatives and the larger issues that underlie their conception and development.

Whilst it is widely recognised that ICTs are strategically important to a country, and the need for investment in e-government is generally well-accepted, questions related to the balancing of investment in ICTs with the need to give priority to other basic infrastructural requirements still need to be answered, and there is apprehension in some quarters that money used for e-government will absorb scarce developmental resources whilst not delivering on potential benefits. Only time and further research will be able to tell.

REFERENCES

Agarwal, A. (Ed.). (2007). *eGovernance case studies*. Hyderabad, India: Universities Press.

Aluko, B. T. (2005). Building urban local governance fiscal autonomy through property taxation financing option. *International Journal of Strategic Property Management*, 9(1), 201–214.

Asquith, A. (1998). Non-elite employees' perceptions of organizational change in English local government. *International Journal of Public Sector Management*, 11(4), 262–280. doi:10.1108/09513559810225825

Avgerou, C., & Walsham, G. (2000). Introduction: IT in developing countries. In C. Avgerou & G. Walsham (Eds.), *Information technology in context: Studies from the perspective of developing countries* (pp. 1-8). Ashgate, UK: Aldershot.

Bahl, R. W., & Linn, J. F. (1992). *Urban public finance in developing countries*. New York, NY: Oxford University Press.

Bangalore Mahanagara Palike. (2000). *Property tax self-assessment scheme handbook: Golden Jubilee Year 2000*. Bangalore, India: BBMP.

Bangalore Mahanagara Palike. (2007). *Assessment and calculation of property tax under the capital value system (New SAS): 2007-2008*. Unpublished Handbook.

Bardhan, P. (1997). Corruption and development: A review of issues. *Journal of Economic Literature*, 35, 1320–1346.

Bhagat, R. B. (2005). Rural-urban classification and municipal governance in India. *Singapore Journal of Tropical Geography*, 26(1), 61–73. doi:10.1111/j.0129-7619.2005.00204.x

Bhagwan, J. (1983). *Municipal finance in the metropolitan cities of India: A case study of Delhi Municipal Corporation*. New Delhi, India: Concept Publishing.

Bhatia, D., Bhatnagar, S., & Tominaga, J. (2009). *How do manual and e-government services compare? Experiences from India. Information and Communications for Development* (pp. 67–82). World Bank Publications.

Bhatnagar, S. (2003a). E-government: Building a SMART administration for India's states. In Howes, S., Lahiri, A., & Stern, N. (Eds.), *State-level reform in India: Towards more effective government* (pp. 257–267). New Delhi, India: Macmillan India Ltd.

Bhatnagar, S. (2003b). Public service delivery: Does e-government help? In S. Ahmed & S. Bery (Eds.), *The Annual Bank Conference on Development Economics 2003* (pp. 11-20). New Delhi, India: The World Bank and National Conference of Applied Economic Research.

Bhatnagar, S. (2003c). *Transparency and corruption: Does e-government help?* Draft paper for the compilation of the Commonwealth Human Rights Initiative 2003 Report 'Open Sesame: Looking for the Right to Information in the Commonwealth.

Bhatnagar, S. (2003d). *The economic and social impact of e-government.* Background technical paper for E-government, the Citizen and the State: Debating Governance in the Information Age, the proposed UNDESA publication (World Public Sector Report for 2003).

Bhatnagar, S. (2003e). Role of government: As an enabler, regulator, and provider of ICT based services. Asian Forum on ICT Policies and e-Strategies, Asia-Pacific Development Information Programme, United Nations Development Programme.

Bhatnagar, S. (2004). *E-government: From vision to implementation.* New Delhi, India: Sage Publications.

Bhatnagar, S. (2005). *E-government: Opportunities and challenges.* World Bank Presentation. Retrieved from http://siteresources.worldbank.org/INTEDEVELOPMENT/Resources/559323-1114798035525/1055531-1114798256329/10555556-1114798371392/Bhatnagar1.ppt

Bresciani, P., Donzelli, P., & Forte, A. (2003). Requirements engineering for knowledge management in egovernment. *Lecture Notes in Artificial Intelligence, 2645*, 48–59.

Budhiraja, R. (2003). *Electronic governance: A key issue in the 21st century.* Electronic Governance Division, Ministry of Information Technology, Government of India. Retrieved from http://www.mit.gov.in/eg/article2.htm

Centre for Policy Research. (2001). *The future of urbanisation: Spread and shape in selected states.* New Delhi, India: Centre for Policy Research.

Cheema, G. S. (2005). *Building democratic institutions: Governance Reform in developing countries.* Bloomfield, CT: Kumarian Press, Inc.

Colby, S.-S. (2001). *Anti-corruption and ICT for good governance.* Deputy Secretary-General, OECD in Anti-Corruption Symposium 2001: The Role of Online Procedures in Promoting and Good Governance.

Commonwealth Centre for E-Governance. (2002). *E-government, e-governance and e-democracy: A background discussion paper.* International Tracking Survey Report, no. 1.

Cornwell, B., Curry, T. J., & Schwirian, K. P. (2003). Revisiting Norton long's ecology of games: A network approach. *City & Community, 2*(2), 121–142. doi:10.1111/1540-6040.00044

Crozier, M., & Friedberg, E. (1980). *Actors and systems.* Chicago, IL: University of Chicago Press.

Curtin, G., Sommer, M., & Vis-Sommer, V. (Eds.). (2003). *The world of e-government.* New York, NY: Hayworth Press.

Dada, D. (2006). The failure of e-government in developing countries: A literature review. *The Electronic Journal on Information Systems in Developing Countries, 26*(7), 1–10.

Datta, A. (1984). *Municipal finances in India.* New Delhi, India: Indian Institute of Public Administration.

Datta, A. (1999). Institutional aspects of urban governance in India. In Jha, S. N., & Mathur, P. C. (Eds.), *Decentralization and local politics* (pp. 191–211). New Delhi, India: Sage Publications.

De, R. (2007). *Antecedents of corruption and the role of e-government systems in developing countries.* Paper presented at the Electronic Government 6th International Conference, EGOV 2007, Ongoing Research, Regensburg, Germany, September 3-7, 2007.

Dunleavy, P., & Margetts, H. (2000). *The advent of digital government: Public bureaucracies and the state in the information age.* Paper to the Annual Conference of the American Political Science Association, September 2000.

Dutton, W. H. (1992). The ecology of games shaping telecommunications policy. *Communication Theory, 2*(4), 303–324. doi:10.1111/j.1468-2885.1992.tb00046.x

Dutton, W. H. (1999). *Society on the line: Information politics in the digital age.* Oxford, UK

Dutton, W. H., & Guthrie, K. (1991). An ecology of games: The political construction of Santa Monica's public electronic network. *Informatization and the Public Sector, 1*(4), 279–301.

Edmiston, K. D. (2003). State and local e-government: Prospects and challenges. *American Review of Public Administration, 33*(1), 20–45. doi:10.1177/0275074002250255

Fine, G. A. (2000). Games and truths: Learning to construct social problems in high school debate. *The Sociological Quarterly, 41*(1), 103–123. doi:10.1111/j.1533-8525.2000.tb02368.x

Firestone, W. A. (1989). Educational policy as an ecology of games. *Educational Researcher, 18*(7), 18–24.

Flatters, F., & MacLeod, W. B. (1995). Administrative corruption and taxation. *International Tax and Public Finance, 2*, 397–417. doi:10.1007/BF00872774

Fountain, J. E. (2001). *Building the virtual state: Information technology and institutional change.* Washington, DC: Brookings Institution.

Fountain, J. E. (2002). A theory of federal bureaucracy. In Kamarck, E., & Nye, J. S. Jr., (Eds.), *Governance.com: Democracy in the information age* (pp. 117–140). Washington, DC: Brookings Institution.

Gascó, M. (2003). New technologies and institutional change in public administration. *Social Science Computer Review, 21*(1), 6–14. doi:10.1177/0894439302238967

General Accounting Office. (2001). *Electronic government: Challenges must be addressed with effective leadership and management.* (GAO-01-959T). Retrieved from http://www.gao.gov/new.items/d01959t.pdf

Guhan, S., & Paul, S. (Eds.). (1997). *Corruption in India: Agenda for action.* New Delhi, India: Vision Books.

Gupta, M. P., & Jana, D. (2003). E-government evaluation: A framework and case study. *Government Information Quarterly, 20*, 365–387. doi:10.1016/j.giq.2003.08.002

Gupta, P., & Bagga, R. K. (Eds.). (2008). *Compendium of egovernance initiatives in India.* Hyderabad, India: Universities Press.

Hammond, A. L. (2001). Digitally empowered development. *Foreign Affairs, 80*(2), 96–106. doi:10.2307/20050067

Haque, S. M. (2002). E-governance in India: Its impacts on relations amongst citizens, politicians and public servants. *International Review of Administrative Sciences*, *68*, 231–250. doi:10.1177/0020852302682005

Heeks, R. (1998a). *Information technology and public sector corruption* (Working Paper 4). Institute for Development Policy Management, University of Manchester.

Heeks, R. (1998b). *Information age reform of the public sector: The potential and problems of IT for India.* (Information Systems for Public Sector Management Working Paper Series Paper No. 6). IDPM, University of Manchester.

Heeks, R. (2000). The approach of senior public officials to information technology related reform: Lessons from India. *Public Administration and Development*, *20*(3), 197–205. doi:10.1002/1099-162X(200008)20:3<197::AID-PAD109>3.0.CO;2-6

Heeks, R. (2001). *Building e-governance for development: A framework for national and donor action.*

Heeks, R. (2002a). i-Development not e-development: Special issue on ICTs and development. *Journal of International Development*, *14*(1), 1–11. doi:10.1002/jid.861

Heeks, R. (2002b). Information systems and developing countries: Failure, success and local improvisations. *The Information Society*, *18*, 101–112. doi:10.1080/01972240290075039

Heeks, R. (2003). *Most egovernment-for-development projects fail: How can the risks be reduced?* (iGovernment Working Paper Series – Paper No. 14), University of Manchester.

Heeks, R. (2005). eGovernment as a carrier of context. *Journal of Public Policy*, *25*(1), 51–74. doi:10.1017/S0143814X05000206

Heeks, R. (2006). *Implementing and managing egovernment – An international text.* New Delhi, India: Vistar Publications.

Hindriks, J., Keen, M., & Muthoo, A. (1999). Corruption, extortion and evasion. *Journal of Public Economics*, *74*, 395–430. doi:10.1016/S0047-2727(99)00030-4

Huque, A. S. (1994). Public administration in India: Evolution, change and reform. *Asian Journal of Public Administration*, *16*(2), 249–259.

Jalal, J. (2005). Good practices in public sector reform: A few examples from two Indian cities. In Singh, A. (Ed.), *Administrative reforms: Towards sustainable practices* (pp. 96–116). New Delhi, India: Sage Publications.

Kluver, R. (2005). The architecture of control: A Chinese strategy for egovernance. *Journal of Public Policy*, *25*(1), 75–97. doi:10.1017/S0143814X05000218

Kumar, R., & Best, M. L. (2006). Impact and sustainability of e-government services in developing countries: Lessons learned from Tamil Nadu, India. *The Information Society*, *22*, 1–12. doi:10.1080/01972240500388149

Layne, K., & Lee, J. (2001). Developing fully functional e-government: A four stage model. *Government Information Quarterly*, *18*(2), 122–136. doi:10.1016/S0740-624X(01)00066-1

Lefebvre, E., & Lefebvre, L. (1996). *Information and telecommunication technologies: The impact of their adoption on small and medium-sized enterprises.* International Development Research Centre. Retrieved from http://www.idrc.ca/en/ev-9303-201-1-DO_TOPIC.html

Legrain, P. (2002). *Open world: The truth about globalisation.* London, UK: Abacus.

Lewis, A. (1982). *The psychology of taxation.* Oxford, UK: Martin Robertson & Company.

Long, N. E. (1958). The local community as an ecology of games. *American Journal of Sociology, 64*(3), 251–261. doi:10.1086/222468

Madon, S. (1993). Introducing administrative reform through the application of computer-based information systems: A case study in India. *Public Administration and Development, 13*, 37–48. doi:10.1002/pad.4230130104

Madon, S. (2004). Evaluating the developmental impact of e-governance initiatives: An exploratory framework. *Electronic Journal of Information Systems in Developing Countries, 20*(5), 1–13.

Madon, S., & Bhatnagar, B. (2000). Institutional decentralised information systems for local level planning: Comparing approaches across two states in India. *Journal of Global Information Technology Management, 3*(4), 45–59.

Madon, S., Sahay, S., & Sahay, J. (2004). Implementing property tax reforms in Bangalore: An actor-network perspective. *Information and Organization, 14*, 269–295. doi:10.1016/j.infoandorg.2004.07.002

Maheswari, S. R. (1993). *Administrative reform in India.* New Delhi, India: Jawahar Publishers and Distributors.

March, J. G., & Olsen, J. P. (1989). *Rediscovering institutions: The organisational basis of politics.* New York, NY: The Free Press.

Margetts, H. (1998). *Information technology in government: Britain and America.* London, UK: Routledge.

Margetts, H. (2006). Transparency and digital government. In Hood, C., & Heald, D. (Eds.), *Transparency: The key to better governance?* (pp. 197–210). London, UK: The British Academy. doi:10.5871/bacad/9780197263839.003.0012

Mathew, G. (2006). A new deal for municipalities. In *Proceedings of the National Seminar on Urban Governance in the Context of the Jawaharlal Nehru National Urban Renewal Mission,* India Habitat Centre, New Delhi 24th – 25th November 2006, (pp. 102–116).

Mechling, J. (2002). Information age governance. In Kamarck, E., & Nye, J. S. Jr., (Eds.), *Governance.com: Democracy in the information age* (pp. 171–189). New York, NY: Brookings Institution.

Meijer, A. (2002). Geographical information systems and public accountability. *Information Policy, 7*, 39–47.

Minogue, M. (2002). Power to the people? Good governance and the reshaping of the state. In Kothari, U., & Minogue, M. (Eds.), *Development theory and practice* (pp. 117–135). Basingstoke, UK: Palgrave.

Misra, S. (2005). eGovernance: Responsive and transparent service delivery mechanism. In A. Singh (Ed.), *Administrative reforms: Towards sustainable practices* (pp. 283–302). New Delhi, India: Sage Publications.

Mitra, R. (2000). Emerging state-level ICT development strategies. In Bhatnagar, S., & Schware, R. (Eds.), *Information and communication technology in development: Cases from India* (pp. 195–205). New Delhi, India: Sage Publications.

National Institute of Urban Affairs (NIUA). (2004). *Reforming the property tax system. Research Study Series No. 94.* New Delhi, India: NIUA Press.

Ndou, V. (2004). E-government for developing countries: Opportunities and challenges. *The Electronic Journal on Information Systems in Developing Countries, 18*(1), 1–24.

Newman, J. (Ed.). (2005). *Remaking governance: Peoples politics and the public sphere.* Bristol, UK: The Policy Press.

Nilekani, N. (2004, 25th October). Redemption in this world, this land. *The Economic Times*, Editorial Page. Retrieved from http://economictimes.indiatimes.com/articleshow/897648.cms

Norris, P. (2001). *Digital divide: Civic engagement, information poverty and the internet worldwide*. Cambridge, UK: Cambridge University Press. doi:10.1017/CBO9781139164887

Odendaal, N. (2002). ICTs in development – Who benefits? Use of geographic information systems on the Cato Manor Development Project, South Africa. *Journal of International Development*, *14*, 89–100. doi:10.1002/jid.867

Olowu, D. (2004). *Property taxation and democratic decentralisation in developing countries* (Working Paper Series No. 401). The Hague, The Netherlands: Institute of Social Studies.

Parks, T. (2005). *A few misconceptions about egovernment*. Retrieved from http://www.asiafoundation.org/pdf/ICT_eGov.pdf

Pathak, R. D., & Prasad, R. S. (2005). The role of egovernment in tackling corruption: The Indian experience. In R. Ahmad (Ed.), *The Role of Public Administration in Building a Harmonious Society, Selected Proceedings from the Annual Conference of the Network of Asia-Pacific Schools and Institutes of Public Administration and Governance (NAPSIPAG)*, December 5-7, 2005, (pp. 343 – 463).

Paul, S., & Shah, M. (1997). Corruption in public service delivery. In Guhan, S., & Paul, S. (Eds.), *Corruption in India: Agenda for action*. New Delhi, India: Vision Books.

Pratchett, L. (1999). New technologies and the modernization of local government: An analysis of biases and constraints. *Public Administration*, *77*(4), 731–750. doi:10.1111/1467-9299.00177

Rao, N. R. (1986). *Municipal finances in India (theory and practice)*. New Delhi, India: Inter-India Publications.

Rao, V. (2003). *Property tax reforms in Bangalore*. Paper presented to the Innovations in Local Revenue Mobilisation Seminar. Retrieved from http://www1.worldbank.org/publicsector/decentralization/June2003SeminarPresentations/VasanthRao.ppt

Rhodes, R. A. W. (1996). The new governance: Governing without government. *Political Studies*, *44*, 652–667. doi:10.1111/j.1467-9248.1996.tb01747.x

Ribeiro, E. F. N. (2006). Urban growth and transformations in India: Issues and challenges. In *Proceedings of the National Seminar on Urban Governance in the Context of the Jawaharlal Nehru National Urban Renewal Mission*, India Habitat Centre, New Delhi 24th – 25th November 2006, (pp. 1 – 11).

Ronaghan, S. A. (2002). *Benchmarking e-government: A global perspective*. The United Nations Division for Public Economics and Public Administration (DPEPA) Report.

Rosengard, J. K. (1998). *Property tax reform in developing countries*. Boston, MA: Kluwer Academic Publications. doi:10.1007/978-1-4615-5667-1

Roy, S. (2005). *Globalisation, ICT and developing nations: Challenges in the information age*. New Delhi, India: Sage Publications.

Sachdeva, P. (1993). *Urban local government and administration in India*. Allahabad, India: Kitab Mahal.

Schware, R. (2000). Useful starting points for future projects. In Bhatnagar, S., & Schware, R. (Eds.), *Information and communication technology in development: Cases from India* (pp. 206–213). Delhi, India: Sage Publications.

Singh, A. (1990). Computerisation of the Indian income tax department. *Information Technology for Development*, *5*(3), 235–251. doi:10.1080/02681102.1990.9627198

Singh, N. (1996). *Governance and reform in India*. Paper presented at Indian National Economic Policy in an Era of Global Reform: An Assessment, Cornell University, March 29-30 1996.

Singh, S. S., & Misra, S. (1993). *Legislative framework of Panchayati Raj in India*. New Delhi, India: Intellectual Publishing House.

Stoker, G. (1998). Governance as theory: Five propositions. *International Social Science Journal*, *50*(155), 17–28. doi:10.1111/1468-2451.00106

Taylor, J., & Williams, H. (1988). *Information and communication technologies and the transformation of local government* (Working Paper 9). Centre for Urban and Regional Development Studies (Newcastle).

The Economic Times. (2008, June 8). Urban India gets under the digital mapping radar. *The Economic Times*, p. 14.

The eGovernments Foundation. (2003). *Street naming and property numbering guide*. Bangalore, India: Author.

The eGovernments Foundation. (2004). *The property tax information system with GIS*. Presentation document. Bangalore, India: Author.

The Government of India. (2003). *Electronic governance – A concept paper*. Retrieved from http://egov.mit.gov.in

The Government of Karnataka. (2005). *A note on the process of implementation of computerisation etc – Guidance* notes. Unpublished.

The Times of India. (2006, July 22). E-governance, GIS: New face of BMP. *The Times of India*, p.1

The Times of India. (2009a, January 8). Popular debut for online tax calculator: Applicable for residential properties, citizens rue increase in net amount. *The Times of India*, p. 2

The Times of India. (2009b, January 10). E-calculator spreads its wings. *The Times of India*, p. 2

The World Bank. (2004). *Building blocks of egovernment: Lessons from developing countries* (PREM Notes No. 91), August 2004.

United Nations Development Programme. (1997). *Corruption and good governance: Discussion paper 3*. New York, India: UNDP.

Vijayadev, V. (2008). *Private communication*. (SAS 02-03 to 06-07 Excel spreadsheet), State Nodal Officer, Municipal Reforms Cell, Directorate of Municipal Administration.

Vincent, S. (2004). A new property map for Karnataka. *IndiaTogether.org*. Retrieved from http://www.indiatogether.org/2004/mar/gov-karmapgis.htm

Virkar, S. (2011). Exploring property tax administration reform through the use of information and communication technologies: A study of e-government in Karnataka, India. In J. Steyn & S. Fahey (Eds.), *ICTs and sustainable solutions for global development: Theory, practice and the digital divide, Vol. 2: ICTs for development in Asia and the Pacific*. Hershey, PA: IGI Global.

Wade, R. H. (1985). The market for public office: Why the Indian state is not better at development. *World Development*, *13*(4), 467–497. doi:10.1016/0305-750X(85)90052-X

World Bank. (2010). *Definition of e-government*. Retrieved from http://web.worldbank.org

Chapter 8

E-Government System Design and Port Authorities:
A Survey of Approaches and a Case Study Combining Internet and E-Learning Technologies

Jim Prentzas
Democritus University of Thrace, Greece & University of Patras, Greece

Gregory Derekenaris
University of Patras, Greece

Athanasios Tsakalidis
University of Patras, Greece

ABSTRACT

Port authorities constitute very active organizations that frequently interact with citizens as well as public and private organizations. The employees and administration of port authorities require effective e-government services in order to implement their tasks. The required services should provide effective information flow and collaboration to improve decision making, governance, and integration of all sectors. In this chapter, the authors briefly outline issues concerning the usefulness of intranets in organizations and corresponding services provided to organization employees. They briefly present key aspects of certain recent approaches concerning e-governance and intranets in ports. The authors also present a case study involving the e-government services implemented for Patras's Port Authority in Greece. The specific port authority has a lot of workload because the corresponding port is the third largest in Greece and a main gate to countries abroad. The case study combined Internet-based technologies with e-learning technologies. E-learning services assist employees in acquainting themselves with newly introduced e-government services. Therefore, e-learning may contribute in the successful realization of e-government projects.

DOI: 10.4018/978-1-4666-3640-8.ch008

INTRODUCTION

Port authorities are governmental or quasi-governmental organizations operating ports. They own property as well as vehicles and machinery (e.g. cranes, trailing trucks) useful for performing various tasks in ports. Port authorities collect fees for services provided such as ship anchorage and use of machinery. Fees are also collected from passenger and vehicle tickets as well as customs. Several ports also include marinas attracting visitors owning crafts. Port authorities charge fees for services provided in marinas.

Port authorities are usually governed by boards or councils. Port authorities consist of several divisions and departments. Important tasks performed by a port authority's departments involve utilization of resources, planning and construction of works, maintenance of land and marine installations, maintenance of machinery, administration, economics, marketing and public relations. Corresponding data is recorded in databases. We highlight a few of the tasks performed. Departments responsible for utilization of resources schedule the use of resources (e.g. machinery, marina docking spots), compute/collect appropriate charges and coordinate trade activities related with marine traffic. Departments responsible for planning and construction of works design programs for development works involving the port authority, develop studies for future work concerning infrastructure and supervise all corresponding work. For machinery owned by the Port Authority, a department is responsible for repairs and maintenance, to draft reports with technical specifications of machinery and spare parts that need to be acquired and to record data concerning operation/maintenance of available machinery. In total, port authorities play a key role in commerce, transport, logistics, navigation and tourism.

Port authorities very frequently interact with shipping companies and shipping agencies. Shipping agencies are responsible for handling routine tasks at ports regarding ships and cargo after agreements with corresponding ship owners. There are also other actors closely interacting with port authorities' internal departments in several activities that also require efficient data exchange. Such actors are shipping agents, customs agencies, towage companies, tourism offices, shipping companies, craft owners, environmental agencies, coast guard, accident and rescue services, banks, ministries, prefectures, municipalities, police and security agencies.

Port authorities constitute organizations which frequently interact with citizens as well as public and private sectors. E-government projects involving ports require careful design due to the several services provided (Pallis & Lambrou, 2007). Automation and efficient data exchange are among the functionalities required in a variety of activities and functions that take place in a port. Coordination of administrative organs and public-private organizational models are also required (Pallis & Lambrou, 2007). The variety of port activities and participants entails a careful design of e-government infrastructures especially taking into consideration human factors and the need for collaboration.

Nowadays, the Internet plays a key role in almost all, if not all, business activities. An increasing number of organizations are adopting the use of Internet and Intranet technologies in their working environment in order to facilitate and automate their working tasks (Robertson, 2009, 2010; Blackmore, 2010; Casselberry, R. et al., 1996). With the effective dissemination of information within organizations, employees save time and paperwork is reduced. The advantages are multiplied in case of an organization that frequently interacts with the public. A significant amount of information becomes available to the public through the Web reducing the employees' workload and facilitating the citizens' contact with the organization. Just like other organizations, port authorities may exploit Internet and Intranet technologies. The provided services may promote information sharing, communication and collaboration among its internal departments, its administration and interesting parties interacting with port authorities.

In this chapter, we briefly outline the usefulness of intranets in e-government projects and survey approaches concerning design and implementation of e-services for the management of port activities. We highlight key points of such approaches and factors affecting their successful realization. We also present a case study involving Patras' Port Authority in Greece.

Patras' Port Authority has a lot of workload because the corresponding port is the third largest one in Greece and a main gate to countries abroad. The Port is connected via ferry-boat lines with ports of neighboring countries (e.g. Italy) as well as the nearby islands. A large part of merchandise exported to and imported from EU countries passes through Italy and Patras' Port. A large number of travelers also move through the Port due to the nearby touristic attractions (i.e. islands and archaeological sites).

Intranet-based services have been implemented for Patras' Port Authority aiming to modernize working practices. The main goals were to facilitate exchange of information among the Port Authority's various departments, to provide useful information to the public regarding activities of the Port, to facilitate the exchange of statistical data between shipping agents and Port Authority's corresponding employees. Among the implemented services were an application for the Port's Marina and an application for the acquisition and management of shipping agents' statistical data. E-learning applications were also developed to acquaint employees with the new e-services.

It should be mentioned that there are other public or quasi-public organizations exhibiting similar characteristics as port authorities. E-government services such as the ones discussed in this chapter could thus be applied to other organizations. The discussion presented in this chapter could be applied to other organizations exhibiting one or more of the following characteristics: (a) they consist of several departments that frequently interact with citizens as well as public and private sectors, (b) they are responsible for the planning and construction of works or implementation of services, (c) they possess electromechanical equipment used to perform various works and (d) they are responsible for coordination of transport and trade activities. Examples of public organizations with certain of the aforementioned characteristics (i.e. characteristics (a)-(c)) are prefectures and municipalities.

This chapter is organized as follows. The following section briefly presents issues concerning the usefulness of Intranets in organizations and outlines popular Intranet-based services. Further on, approaches concerning the design and implementation of e-services for the management of port activities are briefly discussed. Afterwards, a case study involving Patras' Port Authority is presented. Finally, the conclusions of the chapter are presented.

INTRANETS IN ORGANIZATIONS

During the last decades, Internet-based technologies have been employed in the working environments of many organizations to facilitate their working tasks. Internet-based technologies provide the means for communication and distribution of information among organization employees as well as among organization departments and citizens (or other organizations). Internet-based technologies have proven useful to organizations consisting of several departments that frequently interact with each other and require valid and on-time information.

Intranets are networks based on Internet technologies that are confined to an organization. The main purposes of an intranet are information and service sharing, improved productivity, communication, collaboration and teamwork within an organization. To facilitate the transactions of an organization with third parties (e.g. organization partners), extranets were introduced. An extranet is a network extending an organization's intranet by allowing controlled access to specific users outside the organization. Network security is an important aspect in intranets in order to prevent unauthorized external access to vital organization information and services. Intranets and extranets appeared in the 1990s and since then have expanded in many organizations.

The success of Intranets is due to the fact that they 'inherit' advantages offered by Internet technologies. Such advantages, among others, are the following (Robertson, 2009, 2010; Blackmore, 2010; Casselberry, R. et al., 1996):

- **Open standards:** Internet-based technologies are available for all Operating Systems and also devices besides computers. Internet-based content is accessible by any device having a proper client installed.
- Development of Web-based applications is facilitated with the availability of tools and the reuse of implemented code.
- **Extendibility:** Intranets are extendable to handle increasing needs of an organization.
- **User-friendly interface:** The interface of Intranet-based services is similar to the interface of Internet-based services. Most Intranet-based services are accessed through a Web browser.

The exploitation of intranets in organizations substantially improves working conditions, enhances administrative work and assists in reaching documented decisions as effective information dissemination is provided in all administrative levels. Intranets may assist in automation and reorganization of existing processes. In brief, intranets provide advantages to organizations such as the following:

- **Increase of productivity:** Productivity of employees is increased as organization applications, databases and all types of files are available in the intranet and may be conveniently accessed and managed with organization computers. This means that information is distributed effectively within the organization and employees have access to and may submit relevant information whenever it is necessary.
- **Time savings:** It has been reported that due to the introduction of intranets, employees save time as they complete their tasks faster.
- **Reduction of paperwork and printings:** Paperwork and printings are reduced as all tasks are performed with computer-based processing. Organizations thus may save money.
- **Effective data maintenance:** With the availability of data through a server, data duplication among organizations is avoided and the overhead of maintaining data is reduced.
- **Data views based on department/employee privileges:** It is easy to provide different views to the same data according to department and employee privileges.
- **Data visualization:** Executives may have fast access to on-line data visualization facilities (e.g. organization statistics) useful in reaching strategic decisions.
- **Ability for offline and online communication among employees:** Communication and collaboration may be also enhanced with the availability of offline and online tools.
- Cultivation of relations among employees and also among employees and the administration.

Popular Intranet-Based Services

Intranets constitute the backbone for providing various services. The exact services provided within an Intranet depend on the needs of an organization. Popular Intranet-based services are the following:

- **Group management:** Available tools enable administrators to manage groups of employees and assign relevant access privileges to corresponding services.
- **Shared file repositories:** The files of an organization that need to be accessed by various employees (with proper privileges) are stored to a Web server organized in folders. Each organization department (or specific department groups) may be assigned a shared workspace containing all files accessed by corresponding department employees. Furthermore, there are certain files that need to be accessed by employees of every department and so a workspace with such files is accessible by every employee. Such a scheme facilitates file retrieval and update, eliminates data duplication and reduces printings. Examples of files that may be accessible through shared workspaces include, among others, telephone number catalogues, catalogues of email addresses, templates, logos, CVs, application forms, technical reports, deliverables, meeting proceedings, papers, manuals, tutorials, e-books, images, audio and video files.
- **Bulletin board service:** Announcements concerning an organization are stored in a Web server and are accessible through a user-friendly interface. The bulletin board service replaces conventional bulletin boards on which printed announcements are posted. It is easier to maintain e-announcements and to provide search and filtering facilities. Multimedia announcements may also be provided. Each organi-

zation department (or a specific group of departments) may have its own bulletin board service.
- **E-mail service:** An e-mail server may be used to assign an e-mail address to each employee of the organization. Mailing lists of employees' e-mail addresses may also be formed.
- **Discussion forum services:** Discussion forums enable asynchronous communication among employees by submitting offline messages and replies to messages. Each organization department (or a specific group of organizations) may have its own discussion forum.
- **Chat and videoconference services:** Chat and videoconference services are useful as they facilitate the synchronous communication and collaboration among employees as well as among organization departments. Such services are also useful to organizations with geographically dispersed departments.
- **Providing access to organization databases and applications:** Intranet-based applications may be developed by exploiting databases maintained within an organization. Such applications enable employees to access and manage data stored in the databases by using a Web browser. Web forms provide convenient means of data management. Stored data may be also used to provide on-line statistics. By exploiting such applications, distribution of information and decision-making are facilitated.
- **Shared bookmark services:** Employees may share bookmarks that is, URLs of websites relevant to their work. Bookmarks may be organized into folders.
- **Shared calendar services:** Shared calendar services are necessary as organization employees perform various tasks in collaboration. Calendar services enable management and scheduling of events (e.g.

meetings) as well as scheduling of the allocation of resources such as meeting/conference rooms and equipment.

Exploitation of intranets and extranets in port authorities may offer advantages. More specifically, they may promote collaboration and information sharing within the port authorities and also among port authorities and cooperating actors. These issues will be made clearer after the discussion in the following section.

E-GOVERNANCE IN PORT SERVICES

In this section, we briefly outline some of the most recent approaches concerning e-governance in ports. We also highlight the advantages that intranets may provide.

In (Alavi, 2008) lessons and experiences concerning the use of Information and Communication Technologies (ICT) to facilitate port trade in Tunisia are discussed. The specific approach involves sectors beyond port authorities such as customs brokers, traders, shipping agents, customs agency, freight forwarders, cargo handlers, banks and Ministry of Commerce. Prior to the integration of the corresponding ICT services, Tunisia's international competitiveness was hindered. Trade transactions were costly and inefficient till the 1990s. This was so due to time-consuming processes resulting from customs clearance requirements, port logistics, customs port and technical control procedures and inefficient document exchange and processing among involved parties. The introduced services concerned adoption of international standards for trade documentation as well as close and rapid coordination among all the involved parties. Meanwhile, the necessary administrative and political commitment was present to facilitate introduction of the new e-services. Cooperation among private and government stakeholders was a key factor to the project's

success. Corresponding committees involving public and private stakeholders were created. The role of these committees was multifold. Prior to the design and implementation of the e-services, the committees worked to simplify, rationalize and eliminate procedures and documents. The committees played an important role in surpassing problems and delays during the design and implementation of the services. A phased approach was adopted to reduce complexity, acquire early user feedback and evaluation. Information concerning all stakeholders was standardized to facilitate information exchange and processing. Furthermore, the electronic processing framework of all stakeholders was extended. Therefore the design, implementation and application of the corresponding services involved several steps and prerequisite success factors. With the introduced services, trade in Tunisia is facilitated significantly reducing import and export processing times.

In (Amato et al., 2008) a web-based multi-agent approach is presented. The approach promotes electronic data interchange among seaport operators and provides interoperability among ports. More specifically, the implemented platform coordinates information flow characterizing port operations. The approach includes several actors besides port authorities. A survey was carried out concerning both public and private seaport actors. Based on the results of the survey, the workflow was divided into different cycles (e.g. export terminal cycle). The agents consider these cycles and their corresponding processes. The overall architecture is able to provide replies to (user or software) queries by acquiring information from distributed information sources such as web pages. The overall architecture employs an ontology-based approach to integrate different information sources. Both queries and replies are characterized by a high semantic level. Examples of information regarding queries and replies may concern the estimated time of arrival or departure of ships. The agent architecture consists of the following types of agents: interface, broker, host and mobile agents. Interface agents receive queries and provide replies. They act as active and personalized collaborators by performing user profiling and semantic knowledge representation. A host agent associated with each information source, acquires required information and passes it to the requesting mobile agent. Broker agents instantiate as many mobile agents as the corresponding information sources to a query. It is mentioned that the system can be generalized to any port although it focuses on the Italian port of Taranto which is the second largest in the country for load handling.

In (Swift, 2011) the need to provide port-wide access to Wi-Fi and WiMax is stressed out and the results of a corresponding survey are presented. The availability of corresponding technologies for access and use by seafarers may enhance their welfare, recruitment and retention. Seafarers need cheap and easy ways to connect to the Internet, communicate with their families and friends and access further e-services (e.g. e-banking, e-commerce). By providing port-wide access to the Internet, seafaring may become more appealing to younger generations and also the issue involving the global shortage of officers may be dealt with. It is argued that these services should be available on board docked ships besides seafarer centers. Survey results showed that few ports provide such services. Among the ports that provide these services are the Port of Antwerp and the Port of Singapore. Ports lack such services for various reasons such as argued lack of demand, the cost required to install and maintain the services and the questionable revenue in return, security concerns and potential threat to port welfare organizations. The cost to install and operate networking technologies depends on the port's area and topography as well as the existence (or inexistence) of relevant infrastructure. However, there is potential for generating revenue. The services may act as a marketing device, seafarers and other citizens may be charged for the provided services and the fees ports pay to port agents may be reduced since captains may connect to port authority informa-

tion systems without the intervention of agents. Security concerns mainly involve potential threats to the port and ship information systems, downloading of illegal material and use of the services by citizens living nearby the port area. In certain ports, such security issues have been handled with virtual private networks, separate Internet connection and independent service provider for the port departments, special security protocols and software, user registration and accounts to obtain access to Internet services. Port administrations need to be kept informed about the existing and emerging networking technologies. A general conclusion is that ports around the world could share knowledge and experiences under the guidance of the organizations such as the International Association of Ports and Harbors.

In (Li & Zhang, 2010) the significance and advantages of China E-port are analyzed. Certain of the application systems are also presented. China E-port is an integrated data exchanging platform improving the efficiency of port logistics and involving all types of nationwide administrative organizations. China's port logistics is developing rapidly due to the growing volume of import and export trade. The combination of logistics, business flow and information flow in import and export trade enhances economic competitiveness. China E-port runs steadily providing real-time online services to several hundreds of thousands enterprise users. It consists of a data center and several application systems. The project's advantages, among others, are information sharing, access to the services any time and at any place, low cost, easy learning and operating as well as strong safety precautions. Main application systems of China E-port are the following: (a) manifest declaration system, a sub-system of customs agency, (b) export foreign exchange collection system, a sub-system of port law enforcement system, (c) import and export express clearance, (d) processing trade, (e) online payment and (f) identity authentication.

In (Toh, Welsh & Hassall, 2010) a collaboration model for a global port cluster is presented. The model facilitates collaboration among interdependent organizations and business entities to create port communities. The provided services must be flexible, achieve seamless integration of applications and extend the boundaries of existing service catalogues. The ICT architecture of a port cluster needs to include collaborative service catalogue to enhance business synergies. Portal platforms with dynamic service alignment are required to provide better utilization of the infrastructure, knowledge sharing and collaboration.

In (Norzaidi et al., 2011) the indirect effects of intranet functionalities on port industry managers' performance are studied based on questionnaire results. The paper investigates if intranet functionalities predict perceived usefulness, if perceived usefulness influences intranet usage and if intranet usage affects managers' performance. The results gave positive answers to the aforementioned questions. The research involved managers in various port services besides port authorities such as terminal operators, customs and marine departments. Most of them work in non-IT departments. The most important technology functionalities of intranets identified in a descending order of importance are immediacy of communication, physical interface and concurrency. Findings showed that frequently used communication tools must be made available to facilitate communication with superiors, colleagues and partners. High-speed line is considered an important aspect of physical interface that assists in performance of tasks. Concurrency is required to perform several tasks simultaneously. These findings could be exploited in the design, implementation and maintenance of intranet services in organizations. It should be mentioned that in previous work (Norzaidi et al. 2007, 2009), the task-technology-fit model was used to investigate the impact of intranet usage on managers' in the port industry.

Table 1 summarizes key points for each one of the approaches. The presented approaches involve certain key aspects. More specifically, they put emphasis on issues such as the following:

- Communication, collaboration and information sharing among employees (Alavi, 2008; Amato et al., 2008; Li & Zhang, 2010; Toh, Welsh & Hassall, 2010; Norzaidi et al., 2011), among port services and private/public sectors (Alavi, 2008; Amato et al., 2008; Li & Zhang, 2010; Toh, Welsh & Hassall, 2010) and among different ports (Amato et al., 2008; Toh, Welsh & Hassall, 2010).

- Operations characterized by a high semantic level (Amato et al., 2008).

Table 1. Key points for outlined approaches

Approach	Key Points
(Alavi, 2008)	Lessons and experiences concerning use of ICT to facilitate port trade in Tunisia. Cooperation among private and government stakeholders a key factor to the project's success.
(Amato et al., 2008)	Multi-agent approach, electronic data interchange among seaport operators, interoperability among ports. Information acquisition from distributed information sources and provision of replies to queries. Queries and replies characterized by a high semantic level.
(Swift, 2011)	Arguments and survey results for the need to provide port-wide access to Wi-Fi and WiMax. Such services may enhance seafarers' welfare, recruitment and retention.
(Li & Zhang, 2010)	Outlines China E-port, an integrated data exchanging platform improving the efficiency of port logistics and nationwide administrative organizations.
(Toh, Welsh & Hassall, 2010)	Collaboration model for interdependent organizations and business entities in a global port cluster.
(Norzaidi et al., 2011)	Questionnaire to record indirect effects of intranet functionalities on port industry managers' performance.

- Need to provide infrastructure enhancing seafarers' welfare, recruitment and retention (Swift, 2011).

From the aforementioned approaches, it is evident that port services should put emphasis on issues such as competiveness, cost reduction, time-efficiency, facilitation of trade and transport across borders and automation of information exchange with cooperating actors. Exploitation of intranets and extranets may have a substantial contribution in the aforementioned issues. More specifically, they could provide the platform for the aforementioned services. Furthermore, services involving seafarers could also be provided by port governance systems.

CASE STUDY: IMPLEMENTED SERVICES FOR PATRAS' PORT AUTHORITY

Patras' Port is the third largest port in Greece. Patras was found in the 11th century B.C. Since then, Patras's Port has uninterruptedly been a busy port as a result of the port's geographical location. Patras is located in the eastern Mediterranean and more specifically in southwestern Greece quite close to the Italian peninsula and the regions of the Adriatic Sea. Commercial relations among Greece and the corresponding countries of the region have always been very tight. Patras' Port has always played an important role in the financial growth of the city and the surrounding region as it has always constituted an important export and import center of the country.

Throughout the centuries, there have been a few periods during which the Port declined. However, since the 1960s the Port's workload reached a peak due to commerce and tourism. The role of the Port has been upgraded since the early 1990s when former Yugoslavia dissolved. Since then, Patras' Port has become a main gate of Greece to countries abroad. A large part of merchandise exported to and imported from EU countries passes through Italy and Patras' Port. Most of merchandise is carried on trucks. Consequently, a lot of merchandise (imports and exports) moves through the Port.

Patras's Port is connected via ferry-boat lines with ports of neighboring countries (mainly Italy) and nearby Ionian Islands (i.e. Corfu, Cephalonia, Ithaca and Zakynthos). More than forty (40) ferryboats connect Patras' Port with Italian ports. Many travelers move through the Port since nearby there are two of the most significant archaeological sites in the world. These sites gather many visitors from abroad. Very close to the southwest of Patras is Olympia, the birthplace of the Olympic Games. The archaeological site of the Delphi Oracle is close to the northeast of Patras. Therefore, Patras can be a good starting point or stopover for visitors traveling by sea.

In total, it is estimated that the Port concentrates about half of the total foreign passengers traveling by sea all over the country. The number of travelers and vehicle traffic volume reach a peak during summertime, which constitutes Greece's main tourist period. According to Patras' Port Authority website, the average traffic data for Patras's Port for the period 2001-2011 is the following:

- To/From foreign destinations, approximately 1,200,000 passengers, 270,000 trucks and 200,000 vehicles.
- To/From local destinations, approximately 500,000 passengers, 18,000 trucks and 100,000 vehicles.
- To/From foreign/local destinations, approximately 1,700,000 passengers, 300,000 trucks and 300,000 vehicles.

Patras' Port Authority owns mechanical equipment (e.g. cranes) useful for tasks such as the work of heavers. Patras' Port Authority charges shipping companies and cargo truck owners for the use of corresponding machinery. Patras' Port includes a Marina that can serve up to 450 boats. The Port Authority charges boat owners for services provided in the Port's Marina. It should be mentioned that free Wi-Fi Internet access and electronic information signs are provided in Patras' Port area.

Patras's Port has similarities to several other ports around the world. This is especially the case for ports in the Mediterranean Sea such as Italian ports with which relations are close.

Intranet-Based Services: Introductory Aspects

Intranet-based services have been implemented for Patras' Port Authority aiming to facilitate the dissemination of information among the Port Authority's departments and among the Port Authority and shipping agents.

The user gains access to the Intranet-based services through a Web browser. The provided services are briefly outlined below:

- Distribution of information among the Port Authority's departments and access to databases adapted in the Intranet.
- A bulletin board service supporting the posting and viewing of announcements involving the Port Authority's personnel and the public.
- A discussion forum for the Port Authority's personnel.
- An application for the Port's Marina enabling craft owners to apply on-line for docking space and the Marina's personnel to manage the corresponding data through a Web-based user interface.
- An application enabling shipping agents to submit statistical data to the Port Authority and the Port Authority's corresponding personnel to manage it.
- E-learning applications for the training of Port Authority's employees to the use of the newly introduced technologies.

The implementation of the services is based on the "shared workspace" notion. The workspace comprises information and tools stimulating cooperation and communication. The use of the workspaces is restricted (password-controlled). They may be accessed via Web browsers using the login/password scheme. After a successful log on, only the workspaces that the user has access privileges are presented to him/her. Within the boundaries of a workspace, the user may manage and share the different kinds of information with other members of the workspace. The following sections present a more detailed description of the system's functionality.

For the development of the system, the Microsoft Internet Information Service for Windows was used. Active Server Pages were used to manipulate the information stored in the databases and to dynamically produce the contents of applications presented to users. Further tools were used to implement the e-learning applications (see corresponding section).

Bulletin Board Service

The bulletin board is a shared location for posting and viewing electronic announcements. The goal of this service is to unify the Port Authority's various paper bulletin boards thus facilitating the management of announcements as well as the distribution of information among employees and the public. Employees save time through this e-service as they may manage or view e-announcements from their own computer without having to go to the position of the conventional paper bulletin board. A further advantage of this service is that printing of announcements and of their updated versions is generally avoided. It is needless to say that a paper bulletin board has limited space for posting announcements. Due to the limited space in conventional bulletin boards, printings of recent announcements usually covered previous announcements. The e-service also facilitates searching and sorting of announcements. The purpose of this service is twofold. On the one hand, it supports announcements exclusively involving the Port Authority's personnel. On the other hand, the service may be used to publish announcements involving the public to the Port Authority's Web site.

Each announcement contains its subject, its submission date and its body. On the first page of a bulletin board, the user views a list of the announcements excluding their body. By clicking on an announcement's subject, the user can view its body. Sorting of announcements is performed according to submission date. Searching is performed based on subject and submission date. An icon to the left of an announcement's subject denotes whether the specific announcement has or has not been read by the specific user. This feature enables the user to focus on new announcements. The user may also filter the displayed announcements according to the date of their posting. For instance, the user may choose to view today's announcements, announcements of the last two days, announcements of the last month, etc.

Different types of announcements are supported. More specifically, there are two main types of announcements provided in two distinct bulletin boards:

- Internal announcements concerning internal affairs of the Port Authority and accessed only by the Port Authority's employees. The user may manage only the announcements he/she posts whereas the system administrator manages all internal announcements.
- Announcements concerning the public. Additional options are provided in the corresponding bulletin board.

The purpose of the announcements concerning the public is to inform the public for various issues involving the Port Authority. A typical port authority employee may submit such announcements for posting. However, the system administra-

tor decides which of the announcements will be published to the Web site of the Port Authority. Whenever an employee submits an announcement for posting, the system administrator is notified with an email message. The interface provided for the management of the announcements is similar to the interface of the bulletin board concerning internal announcements. The status of an announcement may be either 'published' if it has been published to the Web site or 'pending' otherwise. The status of an announcement submitted by an employee for posting is 'pending' until the system administrator decides that it should be published to the Web site. The status of an announcement is displayed beside its subject. The system administrator may update or delete pending or published announcements.

An interesting aspect in the specific bulletin board service would be to provide categorized announcements addressed to specific actors the port authority interacts with such as shipping agents, towage companies, tourism offices, shipping companies and craft owners. Similarly, internal announcements could be categorized. The current service may be easily customized to provide such categorized announcements.

Discussion Forum

The discussion forum is a shared location for posting and viewing text messages. It is an 'offline' way of communication provided to Port Authority personnel as it enables asynchronous discussions. The service supports messages in a simple text format. Each message contains its subject, its submission date and its body. The interface is similar to the interface of the bulletin board service. However, the user may submit replies to messages and replies to replies creating thus a chain of messages. For each message in the first page of the discussion forum, the author name and the number of replies are displayed besides its subject and submission date.

Currently the discussion forum involves only the Port Authority's departments. However, it would be useful to extend the service in order to include other actors with which the Port Authority frequently interacts (e.g. shipping agents).

Port's Marina Service

The Marina service involves the employees of the Marina Department and craft owners. It enables the management of data deriving from applications submitted by craft owners for docking space in the Port's Marina during specific time periods. Corresponding data is stored in a database. Craft owner applications are submitted on-line and are subsequently processed by the employees of the Marina Department. The employees may respond positively or negatively to the applicants via email or fax. The response depends on the availability of docking spots during the time period in which the applicant desires to dock the Marina. The number of docking spots is limited because not all of them are suitable for every type of craft. The length and width of a craft define the possible docking spots it may occupy.

The service is accessed through a Web-based interface. The database contains data concerning craft owners, crafts, pending and accepted applications. Craft attributes among others, concern craft name, code number, nationality, length and width. Among the application attributes are the docking time period, the docking spot assigned to the corresponding craft (set only for accepted applications) and whether the required amount of money has been paid by the craft owner or not. The interface provides facilities to insert new records as well as to search for and update existing records. The employees may search the database according to a variety of criteria and manage corresponding data. Employees may search for crafts and applicants based on criteria such as name and code number. Search for applications is one of the most usual operations performed and so this type of search is based on several criteria. Search

criteria for applications involve the name or code number of corresponding craft, the name or code number of corresponding applicant, application date, the docking time period, whether the applications are accepted or pending, the assigned docking space (in case of accepted applications), whether (for accepted applications) the required amount of money has been paid by craft owners or not. The user may also choose how the retrieved applications will be sorted (e.g. according to craft name, applicant name, application date, docking time period, etc.).

Furthermore, employees may obtain various statistics in the form of bar charts. The current version of the service provides statistics involving a specific year chosen by the user from an available list. For a specific year, the bar charts depict either the fluctuations of applications per month or the fluctuations of applications per craft nationality. The current service may be easily customized to provide further statistics if desired by the Marina Department personnel. Such statistics could involve longer time periods (e.g. previous five years or previous decade). Therefore, further statistics could involve the fluctuations of applications per craft nationality for a long time period, the craft nationalities with the greatest number of applications, the years with the greatest number of applications, etc. The employees of the Marina Department may exploit the generated statistics when preparing reports for the Port Authority's administration. On-line access to statistics pertaining to marina applications may be useful to other Port Authority departments involved with administration and economics.

To facilitate processing of pending applications and selection of spots in which crafts may dock, a map of the Marina depicting the vacant and occupied docking spots for a given time period is made available on-line to the employees. Three colors are used to display the status of the docking spots for the specific time period. More specifically:

- Green is assigned to vacant docking spots,
- Red is assigned to docking spots occupied throughout the given time period,
- Orange is assigned to docking spots that are vacant for some days of the given time period and occupied the remaining days.

This functionality enables the effective management of pending applications and docking spots.

The current version of the service is quite simple. However, it may become more sophisticated by providing further assistance in decision making concerning pending applications. The Port Authority may benefit by rejecting specific applications and accepting other ones instead. The decision of accepting or rejecting an application may depend on a combination of factors and not only on the number of vacant docking spots during the period that an applicant desires to dock the Marina. It may depend on the number and type of other applications made for the same period. The applicant's interaction with the Marina Department in previous time periods should also be taken into account. An applicant using the Marina's services several times in the past may receive top priority compared to other applicants. The combination of such factors could affect the decisions taken by the employees of the Marina Department. Therefore, a future version of this service could incorporate decision making policies to assist in application handling to the benefit of the Port Authority.

Acquisition and Management of Shipping Agents' Data

The Port Authority and more specifically the Shipping Agents Department records statistical information regarding the amount of cargo, the number of passengers and vehicles carried by ships and ferries arriving to and departing from the Port. A significant part of this information is obtained from the corresponding shipping agents. Up till now, the Shipping Agents Department obtained this data in hard copy format and then had to update manually the database used to store this data. The specific application automates this procedure facilitating the employees' working task.

On the one hand, the shipping agents can log on a password-protected location and submit the statistical data required by the Port Authority through a user-friendly interface mainly consisting of forms. This data is stored in a database.

On the other hand, the employees of the Shipping Agents Department log on a shared workspace and obtain access to the data stored in the database. This data pertains to information regarding shipping agents located in the city, the shipping lines connecting the Port with other ports as well as the amount and type of cargo, the number of passengers and vehicles carried by ships arriving and departing from the Port. The employees of the Shipping Agents Department may insert new records, update or delete existing records. In addition, various search facilities are provided to the employees. Search criteria involve, among others, name of a shipping agent, name of an agent's ship and name of a port. Moreover, employees may obtain various statistics in the form of bar charts. Such statistics may be useful to other Port Authority departments as well as the administration.

E-Learning Applications

The introduction of the new technologies raised the issue of acquainting the Port Authority's employees and the shipping agents with their use. Training, however, employees with varying backgrounds in Internet and Intranet technologies may prove to be a rather difficult, time-consuming and expensive task. Some employees were not accustomed with the use of Internet tools such as a Web browser or an email client and Web-based applications such as discussion forums and bulletin boards. Others however were more proficient. To facilitate the training process, two alternative e-learning applications were developed: a stand-alone and an intranet-based application. The stand-alone application was developed with the Adobe Authorware multimedia authoring tool and can be run in a conventional personal computer having a DVD drive. The intranet-based application consists of Web pages and Adobe Flash animations and may be accessed whenever the employees log on the Intranet.

Stand-alone and network-based e-learning technologies may generally prove useful during the introduction stage of new e-government services. E-learning applications may be also used to train an organization's employees under the supervision of experienced tutors prior to the introduction of new services. Employees may also exploit e-learning facilities at home. The standalone application proved useful to employees that were not very accustomed to Internet-based technologies. Such employees used the stand-alone application at work and at home to acquaint themselves with the introduced ICT technologies. Video-based assistance available on the application was extremely useful. Employees that did not have an Internet connection at home also found the standalone application useful. It seems that incorporation of e-learning facilities in the services' help could be beneficial. The Intranet-based e-learning application is useful in eliminating problems deriving from the employees' interaction with the e-government services mostly during the initial operation period. Generally speaking, the main purpose of the e-learning applications is to achieve a smooth adoption of the new services. Furthermore, certain issues presented (e.g. popular Intranet services developed for organizations) may provide motives for the development of further Port Authority e-services in the future.

The e-learning applications consist of courses in Internet and Intranet technologies. The courses start from introductory concepts appropriate for beginners in these technologies and scale up to concepts for more advanced users. The courses are discerned into two main sections: (i) Internet technologies and (ii) Intranet technologies. Each main section is organized into subsections.

The first main section provides an introduction to popular Internet technologies and tools. At the end of the section, the learner acquires the knowledge skills to use popular Internet tools. The section first includes subsections introducing main concepts regarding the Internet and the World Wide Web which are mainly addressed to learners with minimal Internet experience. Further subsections present issues involving, among others, the use of a Web browser and an email client, how to search for information in the World Wide Web and main functionalities of Web-based bulletin boards and discussion forums.

The second main section first presents introductory concepts concerning Intranets, their general usefulness within an organization and popular Intranet services addressed to organization employees. Further on, specific Intranet Services addressed to the Port Authority's employees are presented with relevant examples (e.g. how to perform search, updates and insertions of records in the various applications, how to retrieve and manage announcements, how to retrieve statistics, how the functionality of the on-line marina map works, etc.). The purpose of the second section is to acquaint learners with the notion of an Intranet, typical Intranet services and how working process is changed with the introduction of the implemented services.

Furthermore, the user may consult a glossary of primary terms in order to enrich his/her knowledge. The glossary may be activated either by clicking the appropriate button at the bottom of the screen or by clicking on various entries lying within the text of the sections in the form of hyperlinks.

Emphasis was put on the design and implementation of the e-learning user interface, since interactivity and functionality are factors that totally determine acceptance. The primary objective was to design a user interface that could be usable by users with diverse abilities, needs, requirements and preferences. The main issue was to develop training facilities with rich interactions that would support user functionality efficiently and effectively. Functionality concerns both training activities, like interactions during a simulation demonstrating the use of an Internet tool (e.g. email client), and navigational ones.

The information provided by the e-learning applications is presented in a variety of ways, such as interactive simulations demonstrating key aspects of the implemented system, hypertext, appropriate images, animations and videos. Every screen (or page) follows one of some predefined presentation patterns. On the top area of the screen lies the title of the specific section. Moreover, at the bottom of the screen, there are several buttons facilitating interaction among the user and the application. Such buttons provide learners the option to navigate to the first, last, next or previous screen of the current subsection, to navigate to the first screen of the corresponding main section, to activate the corresponding audio file, to activate the glossary and to exit the application. The main content is displayed at the central area of the screen, the active screen.

User Views

The user interface is unified offering different view to different users. We may distinguish the following basic views as far as users are concerned:

1. **Typical Port Authority Employee View:** In this view, the user may read and post announcements to the bulletin board. Moreover, the user may modify announcements he/she has posted. Finally, access to the intranet-based e-learning application is possible.
2. **Shipping Agent Department View:** In this view, the user may perform all actions permitted to a typical Port Authority employee and additionally manage data acquired from shipping agents.

3. **Marina Department View:** In this view, the user may perform all actions permitted to a typical Port Authority employee and additionally access the Port's Marina service.
4. **Shipping Agent View:** This view enables shipping agents to submit statistical data to the Shipping Agents Department.
5. **Port Authority Statistics View:** This view is addressed to the Port Authority's administration and departments involved in economics and marketing. The user may perform all actions permitted to a typical Port Authority employee. Additionally he/she may view generated statistics concerning shipping agents data and applications for marina docking.
6. **System Administrator View:** In this view, the user may access all available information.

Discussion Concerning Implemented Services and Future Extensions

The implemented services were an effort to introduce Intranet-based services to Patras' Port Authority. The purpose was to enhance collaboration, team work and information sharing among Port Authority departments and also among Port Authority departments and partners (e.g shipping agents). Reduction of paperwork was also a goal of the services. Employees may also save time when performing their tasks. Specific applications (i.e. shipping agent data, marina service) were made available in the intranet. Data visualization through generated bar charts provides a brief summary of stored data enabling employees and the administration to reach conclusions.

Compared to the previous status, the implemented services offer a number of advantages. Further advantages could be offered by implementing extensions to existing services. More specifically:

- The bulletin board service facilitates the posting and management of announcements reducing printings. Employees save time and dissemination of information among the different Port Authority departments and among the public and the Port Authority is made easier. Future extensions to this service were also highlighted. One extension involves categorization of posted internal announcements. Another one involves the implementation of similar services for other actors with which the port authority frequently interacts such as shipping agents and shipping companies. This last extension will enhance information sharing among the Port Authority and collaborating parties.

- The discussion forum provides an opportunity to enhance offline communication and teamwork among employees and also among employees and the administration. The employees may also provide useful suggestions for improvements required in the services the Port Authority offers to the public. A dedicated forum for suggestions may be employed. The trend in organization intranets is to promote socialization and a discussion forum may contribute towards this direction. The advantages would be multiplied in case the discussion forum service is extended to include other parties cooperating with the Port Authority such as shipping agents.

- The marina service provides advantages to craft owners and employees of the corresponding department. In the previous status, craft owners had to fax their applications for docking space whereas the implemented service enables them to submit applications online. The task of employees is facilitated as electronic processing of record searching, insertion and updates is available. Selection of docking spots for pending applications is facilitated with the on-line map depicting the status of the marina's docking spots. As

mentioned, a future version of the marina service could provide further assistance in the handling of pending applications. Furthermore, the generated statistics provide a visual representation of summary information concerning craft applications. Such statistics could also be useful to the Port Authority's administration and to other Port Authority departments (e.g. Economics, Public Relations and Marketing Departments). Useful decisions could be reached with the availability of corresponding information.

- The service pertaining to the acquisition and management of shipping agents' data provides advantages to shipping agents and employees of the Port Authority reducing paperwork. Shipping agents may submit the corresponding data on-line without having to submit it to the Port Authority in hard copy. The task of the Port Authority's employees is simplified as they do not have to update manually the corresponding database. This means that employees save time. Available search facilities and generated statistics are additional bonuses for the employees and the administration.

In corresponding sections, directions for future extensions were highlighted. More specifically, the existing services may be extended to provide further functionalities. Additional services may be also incorporated into the intranet. Further applications concerning specific departments besides the Marina and Shipping Agents Department may become available in the intranet. Other types of offline communication tools could also be incorporated into the intranet. Online communication tools such as chat could also be exploited. The intranet-based e-learning application could also be extended to include further learning sections that may interest the Port Authority's employees.

CONCLUSION

This chapter discussed issues concerning Intranet-based services in the design of e-government systems. The discussion focused in the context of port authorities. Port authorities constitute very active organizations that require carefully designed e-government services to facilitate distribution of information, communication, collaboration and decision making. The chapter presented a case study involving Patras' Port Authority in Greece. The case study combined Internet-based technologies with e-learning technologies. E-learning facilities were deemed necessary to acquaint employees with the newly introduced services. Introduction of e-government services changes the way working tasks are performed. E-learning may contribute in the acceptance of newly introduced e-government services among organization employees and thus in the ultimate success of the e-government services.

We believe that other port authorities could benefit from a system having the features discussed in the case study. Most of the port authorities face similar issues such as need for: (a) effective dissemination and management of announcements, (b) asynchronous collaboration among employees by using discussion forums, (c) submission and management of shipping agent data, (d) management of applications for marina docking spots, (e) data visualization through statistical graphs to highlight relevant information, (f) e-learning activities teaching employees IT issues concerning their work.

Other governmental organizations could benefit from a similar system or similar features as well. As mentioned in the introduction of the chapter, prefectures and municipalities have certain similar characteristics to port authorities. Efficient distribution and management of information is required among the internal departments of prefectures and municipalities, among the departments of different types of local government as well as among citizens, private sectors and local government departments (Bouras et al., 1999).

It should be mentioned that ports attract the interest of investors due to prospects for financial gains and promotion of trade. The geographical location of Greece has attracted such investors. During the last years, Greek ports also started to attract cruise tourists. Cosco, a large Chinese company, has invested in Piraeus, the largest port in Greece. A main aim of the investment, among others, was to tighten trade relations between Europe and China. The investment was successful and it is very likely that similar investments from the same and/or other companies will be made in Greek ports in the near future.

One could point out certain future directions regarding intranets in port authorities. In the following, we discuss relevant issues involving available applications, decision-making, socialization and lifelong learning.

As mentioned in (Scaplehorn, 2012), there is no limit to the functionality that may be provided on an organization's intranet. In a port authority, there are several applications used by its departments to perform their corresponding tasks. Most of these applications could become available through the intranet to facilitate dissemination of information and decision-making. For instance in (Yu et al., 2012) a GIS application is made available through the port authority's intranet. Obviously, access should be restricted to users with appropriate privileges. Furthermore, information overloading should be avoided when sharing information among the different departments. Applications for port authority partners should also be available through an extranet-based approach. The last issue is of great importance due to the increasing role of cooperation among port authorities and other sectors.

Administration and decision-making are significant tasks in a port authority. A drawback of intranets and IT systems in general is the availability of enormous amounts of information to their users. Assistance is required for exploiting the available information in order to reach decisions. The corresponding contribution of a port authority's intranet may prove important as long as appropriate functionalities are provided. Such functionalities should route critical information to port authority staff, provide information filtering to cope with information overload and provide data visualization (Denton & Richardson, 2012). The contribution of intranets in management decision-making could be enhanced enabling executives to carve successful strategies that could add value to port authorities. As ports are attracting the interest of investors, effective management and planning will be required by port authorities.

Socializing intranets is also an aspect of interest (Scaplehorn, 2012). The increasing popularity of social media has led to thoughts about incorporating relevant aspects in intranets. Obviously, such aspects could be pursued in port authority intranets. Besides discussion forums, blogs could also be incorporated within intranets. Short, concise and informal blog announcements from executives could be appreciable from port authority employees (Scaplehorn, 2012). Furthermore, commenting and rating functionalities could also be provided so that intranet users may provide feedback on the usefulness of provided services.

Port authorities could cultivate a lifelong learning culture among the employees and the administration in order to be kept up-to-date with evolving and developing technologies. Lifelong learning ICT infrastructures (e.g. Learning Management Systems) could be exploited for this purpose (Kats, 2013). Learning Management Systems have become popular in supporting conventional classroom instruction and conducting pure distance learning courses (Antonis, Lampsas and Prentzas, 2008). They provide functionalities such as learning content management and sharing, synchronous and asynchronous collaboration, convenient and flexible control to tutors. In this context, e-learning systems employing Artificial Intelligence methods to effectively adapt to learner characteristics (Prentzas and Hatzilyger-

oudis, 2011; Hatzilygeroudis and Prentzas, 2006; Hatzilygeroudis, Koutsojannis, Papavlasopoulos, & Prentzas, 2006) may be useful. Cooperation with nearby universities to periodically teach port authority staff issues involving ICT may also prove fruitful. Finally, collaboration among port authorities around the world could also assist in sharing useful experiences (Swift, 2011).

REFERENCES

Alavi, H. (2008). Trading up: How Tunisia used ICT to facilitate trade. *IFC SmartLessons*, June 2008.

Amato, A., Calabrese, M., Di Lecce, V., & Quarto, A. (2008). Multi agent system to promote electronic data interchange in port systems. In *Proceedings of the 21st IEEE Canadian Conference on Electrical and Computer Engineering* (pp. 729-734). IEEE.

Antonis, K., Lampsas, P., & Prentzas, J. (2008). In Leung, H., Li, F., Lau, R., & Li, Q. (Eds.), *Adult distance learning using a Web-based learning management system: Methodology and results* (*Vol. 4823*, pp. 508–519). Lecture Notes in Computer Science Heidelberg, Germany: Springer.

Blackmore, P. (2010). *Intranets: A guide to their design, implementation and management*. New York, NY: Routledge.

Bouras, C., Destounis, P., Garofalakis, J., Triantafillou, V., Tzimas, G., & Zarafidis, P. (1999). A co-operative environment for local government: an Internet-Intranet approach. *Telematics and Informatics*, *16*, 75–89. doi:10.1016/S0736-5853(99)00020-9

Casselberry, R. (1996). *Running a perfect intranet*. Que Publications.

Denton, K., & Richardson, P. (2012). Using intranets to reduce information overload. *Journal of Strategic Innovation and Sustainability*, *7*(3), 84–94.

Hatzilygeroudis, I., Koutsojannis, C., Papavlaso-poulos, C., Prentzas, J. (2006). Knowledge-based adaptive assessment in a Web-based intelligent educational system. In *Proceedings of the Sixth IEEE International Conference on Advanced Learning Technologies* (pp. 651-655). IEEE.

Hatzilygeroudis, I., & Prentzas, J. (2006). Knowledge representation in intelligent educational systems. In Ma, Z. (Ed.), *Web-based intelligent e-learning systems: Technologies and applications* (pp. 175–192). Hershey, PA: Information Science Publishing.

Kats, Y. (Ed.). (2013, in press). *Learning Management Systems and Instructional Design: Best Practices in Online Education*. Hershey, PA: Information Science Reference.

Li, Y., & Zhang, X. (2010). Study on development and system application of China E-port. In *Proceedings of the 3rd International Conference on Information Management, Innovation Management and Industrial Engineering* (pp. 430-433). IEEE.

Norzaidi, M. D., Chong, S. C., Murali, R., & Salwani, M. I. (2007). Intranet usage and managers' performance in the port. *Industrial Management & Data Systems, 107*(8), 1227–1250. doi:10.1108/02635570710822831

Norzaidi, M. D., Chong, S. C., Murali, R., & Salwani, M. I. (2009). Towards a holistic model in investigating the effects of Intranet usage on managerial performance: A study on Malaysian port industry. *Maritime Policy & Management, 36*(3), 269–289. doi:10.1080/03088830902861235

Norzaidi, M. D., Chong, S. C., Salwani, M. I., & Lin, B. (2011). The indirect effects of Intranet functionalities on middle managers' performance – Evidence from the maritime industry. *Kybernetes, 40*(1-2), 166–181. doi:10.1108/03684921111117988

Pallis, A. A., & Lambrou, M. (2007). Electronic markets business models to integrate ports in supply chains. *Journal of Marine Research, 4*(3), 67–85.

Prentzas, J., & Hatzilygeroudis, I. (2011). Techniques, technologies and patents related to intelligent educational systems. In Magoulas, G. D. (Ed.), *E-infrastructures and technologies for lifelong learning: Next generation environments* (pp. 1–28). Hershey, PA: Information Science Reference.

Robertson, J. (2009). *What every intranet team should know*. Broadway, Australia: Step Two Designs.

Robertson, J. (2010). *Designing intranets - Creating sites that work*. Broadway, Australia: Step Two Designs.

Scaplehorn, G. (2012). *Bringing the internet indoors: Socializing your intranet*. Retrieved July 30, 2012, from http://www.contentformula.com/articles/2010/bringing-the-internet-indoors-socializing-your-intranet

Swift, O. (2011). *Developments in new technology & implications for seafarers' welfare – Seafarers' access to WiFi and WiMax in ports*. Technical Report, International Committee on Seafarers' Welfare.

Toh, K. K. T., Welsh, K., & Hassall, K. (2010). A collaboration service model for a global port cluster. *International Journal of Engineering Business Management, 2*(1), 29–34.

Yu, J. J., Qin, X. S., Larsen, L. C., Larsen, O., Jayasooriya, A., & Shen, X. L. (2012). A GIS-based management and publication framework for data handling of numerical model results. *Advances in Engineering Software, 45,* 360–369. doi:10.1016/j.advengsoft.2011.10.010

KEY TERMS AND DEFINITIONS

Authoring Tool/System: A software tool enabling non-experts to create content that can be delivered to end users. The term is most often used in an e-learning context. The created e-learning content usually conforms to standards such as SCORM (Shareable Content Object Reference Model).

Bulletin Board Service: An e-service that replaces conventional bulletin boards on which printed announcements are posted. Announcements are accessed and maintained through a user-friendly interface providing search and filtering facilities.

Discussion Forum Service: A Web-based e-service enabling asynchronous (offline) communication among its users. Communication is performed by submitting offline messages and replies to messages.

E-Learning: The term refers to all types of electronically supported learning and teaching. The employed technologies are constantly expanding based on the evolution of Information and Communication Technologies. Many types of e-learning services have been developed. E-learning services may be standalone or network-based, may be utilized within classrooms or out of classrooms, may promote individualized or collaborative learning, may be self-paced or led by instructors. E-learning services may be accessed with devices besides computers such as mobile phones and televisions. Distance learning and lifelong learning are usually based on e-learning services.

Extranet: A network employing Internet technologies that extends an organization's intranet by allowing controlled access to specific users outside the organization (e.g. organization partners). The users from the outside may not be granted access to all services available to organization users but to specific services. Network security mechanisms are required.

Intranet: A network within an organization that employs Internet technologies. The main purposes of an intranet are information and service sharing, improved productivity, communication and collaboration. Network security mechanisms are employed to protect an intranet from unauthorized external access. An intranet facilitates the working process and information flow within an organization.

Port Authority: A port authority is a governmental or quasi-governmental organization operating ports. Port authorities collect fees for services provided (e.g. ship anchorage, use of cranes and other machinery, port marina service, passenger/vehicle tickets, customs). Port authorities usually regulate investment and pricing in ports.

Shipping Agency: A company responsible for handling routine tasks at ports regarding ships and cargo after agreements with corresponding ship owners. There are different types of shipping agencies providing specific services. Examples of services provided are arrangements for docking spots, ship provisions, ship repairs, cargo collection, preparation of required documents and provision of information to shipping companies and ship personnel. A shipping agency receives fees for the provided services.

Web-Based Learning: A type of e-learning employing Web-based technologies. Web-based learning may support conventional classroom instruction or pure distance learning. Web-based learning may be employed individually or may support the creation of learning communities with the use of synchronous and asynchronous communication and collaboration tools. Platforms such as Learning Management Systems are frequently used to support Web-based learning.

Chapter 9
An Evaluation Framework to Assess E-Government Systems

Jaffar Alalwan
Institute of Public Administration, Saudi Arabia

Manoj Thomas
Virginia Commonwealth University, USA

ABSTRACT [1]

Evaluating e-government systems is a difficult task involving multi-faceted perspectives. Although a review of the literature discovers several e-government evaluation frameworks, numerous shortcomings still exist. The objective of this chapter is to propose a formative and holistic framework to remedy the current research gaps. The formative position of the evaluation framework ensures the evaluation objective achievement, and the holistic approach ensures completeness and continuity of the evaluation process. The framework can be used as a template for researchers and practitioners to assess e-government projects. The authors demonstrate the applicability and practicability of the framework by applying it to the Korean Government-for-Citizen (G4C) project.

INTRODUCTION

Many countries adopt e-government systems in order to establish government reforms and raise efficiency of government transactions. In developing and developed countries, investment in e-government systems is estimated to be greater than 1% of the gross domestic products (Petricek et al., 2006). However, current empirical validation is not enough to determine the effects of e-government systems on governmental performance (Lim and Tang, 2008). Research shows that evaluation of information systems (IS) in general is a difficult undertaking (Jones and Hughes, 2001; Serafeimidis and Smithson, 2000). In addition, the evaluation process has to address multiple perspectives that complicate enumerating the benefits of the IS (Symons and Walsham, 1988). Evaluation of an e-government system is

DOI: 10.4018/978-1-4666-3640-8.ch009

no exception since determining the benefits of an e-government system is complicated and multi-faceted involving myriad perspectives (i.e. social, technical, political) (Beynon-Davies, 2005; Liu et al., 2004; Khalifa et al., 2004). Evaluation also entails the exploration of the diverse needs of the different citizen groups (e.g., students, lawyers, architects) (Jansen, 2005).

Farbey et al. (1993) claim that IS evaluation is a critical factor to IS success and the choice of an IS evaluation approach should reflect the right organizational context. Funilkul et al. (2006) defined the evaluation framework for e-Government services as "the comprehensive guidance for a government organization which can be used to develop the quality and efficiency of the objectives and strategies of its services and for conforming to citizens' requirements". Furthermore, there are many approaches that are designed to evaluate e-government. While some approaches are called "hard" approaches (e.g., return on investment, payback period, etc.) others may be postulated as "soft" (e.g., satisfaction of employees and citizen, degree of customization). Hard approaches address tangible benefits and risks while soft approaches are proposed to assess intangible benefits and risks. Evaluating e-government systems (and IS, in general) based on hard approaches that depend on tangible measures is the more commonly adopted evaluation basis in many countries. Hard approaches are not without drawbacks. Some of these drawbacks are - the limited view of stakeholders, the complete dependence on accounting and financial instruments (Farbey et al., 1995), the ignorance of human and organizational aspects of the users (Serafeimidis and Smithson, 2000), and the failure to include the intangible benefits and costs associated with its use (Hochstrasser, 1992).

There is no IS evaluation approach that is suitable for every firm (Khalifa et al., 2004). Furthermore, evaluation approaches that combine both hard and soft facets are limited (Orange et al., 2006). Borrowing from the body of IS literature may be pragmatic, but challenging, as IS research-

ers still debate actively about the approach most suitable for IS domain (Alshawy and Alalwany, 2009). Many studies acknowledge that evaluation of e-government is an important research area that needs more investigation (Fountain, 2003; Jones et al., 2006; Remenyi et al., 2000). A holistic evaluation approach is necessary to determine the needs of citizens and businesses, and to help government and private firms measure the return on investment of e-government (Sakowicz, 2006).

Funilkul et al. (2006) summarize the purposes of the evaluation of e-government services. The first and foremost is to ensure that e-government services meet the institution's institutional goals and objectives. This type of a formative evaluation (i.e. evaluation by achieving systems objectives), although widely accepted, is rarely deployed in e-government studies (Hamilton and Chervany, 1981; Bertot et al., 2008). Formative evaluation is continuously monitoring for the systems activities and the objectives. Bertot et al. (2008) define formative evaluation as the "ongoing evaluation that monitors program activities with the goal of modifying and improving the program on a regular basis". An incessant evaluation process is crucial to enhance e-government services. We propose formative evaluation as one necessary pillar in the framework suggested in this paper.

Although the literature identifies several e-government evaluation frameworks, numerous shortcomings exist in the prior work. First, some frameworks focus on a few selected dimensions of e-government (e.g., citizen services, awareness initiatives, IT collaboration) and pay less attention to other dimensions (e.g., mobilization, standard setting). These studies design the evaluation framework based on the technical perspective and focus less on the social perspective. Second, many frameworks are designed to evaluate specific e-government systems in specific countries. These frameworks are usually unique to the county context and may not be applicable in a different setting. Third, the continuous achievement of e-government objectives, or formative evaluation, is not considered in most of the current frameworks. Thus, the objective of this paper is to propose a formative and holistic framework to remedy the aforementioned drawbacks. To include the social and technical aspects of the e-government evaluation framework, we lean heavily on Bostrom and Heinen's (1977) Social-Technical-System (STS) model. We adopt Sakowicz's (2006) four dimensions of e-government for a holistic view that includes all e-government dimensions.

The goal of the proposed framework is not only to increase the knowledge in the field of e-government evaluation but also to provide a template for researchers and practitioners to assess e-government projects. The conceptual framework contributes to research by integrating the formative evaluation approach, STS model, and other relevant theories from the e-government evaluation field. The formative side of the evaluation framework ensures the achievement of objectives and continuity of the evaluation process. Furthermore, by classifying the e-government system into four interactive quadrants, the framework facilitates determining the objectives achievement gaps.

The rest of this manuscript is arranged as follows. We first review the related e-government evaluation literature. The conceptual framework and the theoretical background will be discussed in the following section. We then demonstrate the applicability of the framework by using it to assess the Korean Citizen-for-Government (C4G) project. Finally, we conclude the paper by highlighting future research directions.

RELATED LITERATURE REVIEW

One of the challenges that face practitioners in public and private sectors is how to assess the effectiveness of information technology investments (Kaisara and Pather, 2011). E-government projects are no exception since they are expensive, risky, and difficult to accomplish. Therefore agencies seeking to implement e-government solutions need to carefully analyze the feasibility of the initiatives (Shan et al., 2011). Post-implementation evaluation of e-government projects is essential to ensure that the government agencies are capable of effectively delivering the required e-services (Gupta and Jana, 2003). Research shows a significant slowdown in the development of e-government projects due to the lack of assessment and monitoring methods (Li et al., 2007; Kunstelj and Vintar, 2004; Lessen et al., 2009).

A major part of e-government evaluation research focuses on website assessment and environmental evaluation. For instance, Torres et al. (2005) propose a method to evaluate the usage and the quality of 23 European Union cities' websites. Gupta and Jana (2003) suggest a comprehensive evaluation method to assess the development of the Indian e-government projects. Accenture (2003) develops a maturity model to assess e-government based on a study of e-government websites in 24 countries. Kaisara and Pather (2011) evaluate e-government by adopting a service quality approach. They develop a multi-item instrument to evaluate e-government website in South Africa. Signore (2005) identifies three perspectives in his proposed model for assessing e-government websites: managers' view, developers' view, and users' view. The proposed model evaluates website quality by incorporating internal and external quality based on ISO standards. Xin et al. (2010) design an evaluation index system based on balanced scorecard. They adopt information entropy and rough set theory to process the dataset of their research.

Another direction of e-government evaluation literature focuses on measuring e-readiness that is related to human and technological infrastructure (Accenture, 2003, 2006). In addition, the literature review also identifies many studies that attempt to design a framework for e-government evaluation. Some of these studies position the citizen as the foci of the evaluation model. Wang et al. (2005) propose a citizen-centric approach that consists of three parts: information users, information problem, and information pool. In addition to the evaluation of e-government services, the framework has the ability to address why citizens fail to find the needed information. Eschenfelder and Miller (2005) propose a socio-technical toolkit that focuses on the value of social and political context for citizens. Carter and Belanger (2004) discuss the adoption of e-government services by citizens. The study identifies several factors that influence the citizen's perspective. They are usefulness, relative advantage, compatibility, perceived ease of use, image, and trust in the Internet and in governments. Zhang et al. (2007) propose a user-centric evaluation model for e-government in China based on IS diffusion. Alshawy and Alalwany (2009) propose a citizen evaluation model for use in developing countries; the proposed model is based on social, technical, and economic factors. Based on the Delone and McLean model, Van Der Westhuizen and Edmond (2005) develop an assessment framework that evaluates product and project dimensions. Wang and Liao (2008) propose an evaluation framework (also based on the DeLone and McLean model) where the evaluation data are collected from direct surveys. Victor et al. (2007) focus on the significance of post-implementation evaluation using the process maturity framework. The researchers assert that conclusions gained from post-implementation assessment can be useful for the development of future projects. Liu et al. (2008) provide a framework that focuses on e-government stakeholders and their key performance indicators. E-government strategy is one dimension of this framework, and the success of e-government is assessed based on the strategic requirements. Batini et al. (2009) propose the GovQual framework as an evaluation framework for e-government project selection. GovQual has several layers to evaluate project quality in different organizational environments.

In addition, many general evaluation frameworks have also been proposed. Funilkul et al. (2006) propose a generic evaluation framework based on Control Objectives for Information and related Technology (COBIT), ISO 9000, and Technology Acceptance Model (TAM). Griffin and Haplin (2005) present an evaluation model for UK government based on local accountability. It consists of scrutiny processes, principal stakeholders, joined-up accountability, sanctions and the political dimension. The study showed that a scrutiny committee has more influence than executives, but the executive participation affects that influence. Esteves and Joseph (2008) present a comprehensive assessment framework that examines three dimensions: e-government maturity level, stakeholders, and assessment levels. Gupta and Jana (2003) suggest a flexible evaluation framework based on hard measures, soft measures, and hierarchy of measures to assess the tangible and intangible benefits of e-government. They implement the framework in an Indian case study and find that e-government projects should be in a mature stage in order to conduct a proper evaluation. Sorrentino et al. (2009) address e-government evaluations from a cognitive level based on organization theory and policy studies. The research suggests the cognitive resource role to e-government evaluation. Based on analytic hierarchy process, Ji (2009) proposes a quantitative evaluation method that determines a weight of evaluation index and a gray correlation analysis. The Canadian report of Evaluation Framework for Government On-line Initiative (PWGSC 2005) includes a measurement instrument that is used to evaluate the performance of technical, cultural, and learning initiatives of Canadian e-government. Irani et al. (2005) propose a framework to evaluate public sectors information systems.

Last but not least, literature discusses the assessment model of e-government projects at a country level. Examples include the Indian e-government assessment model, the European e-government economics project, the Indian impact assessment model, and the British public value framework (Gupta, 2007). World Bank suggests evaluating five dimensions of e-government: (1)

services for citizens, (2) services for businesses, (3) empowerment through information, (4) government purchasing, and (5) anti-corruption and transparency (www.worldbank.org). United Nations also proposes an evaluation model to assess a countries' quality of public services and the efficiency of public administration (United Nations, 2008).

In short, the importance of e-government cannot be overlooked. E-government offers advantages such as information sharing and security, reduced operation costs, citizen participation and empowerment, and high transparency (United Nations, 2008). The literature review in this section highlights the importance of an e-government evaluation process. The evaluation is essential for many reasons. It serves to detect shortcomings in the existing e-government system. It helps specify ways to improve the current system. It also helps in justifying the Information, Communication and Technology (ICT) investment, and verifying whether corrective actions are needed. Finally, evaluating e-government is important to ensure that organizations meet government standards and legislative requirements. This paper proposes an interpretive evaluation approach that emphasizes the role of stakeholder involvement and the social context. By using a formative evaluation framework, the evaluation process focuses on the socio-technical nature of interaction between the citizen and the e-government system, a critical construct that is not considered in any of the studies mentioned above.

THE THEORETICAL BACKGROUND AND THE PROPOSED FRAMEWORK

Bostrom and Heinen's (1977) present a Social-Technical-System (STS) model where organizations are divided into two complementary systems: social systems and technical systems. Social systems consist of structure and people, while technical systems consist of technology and tasks.

All four components are interrelated as depicted in Figure 1. STS examines the appropriateness of the design elements of the information systems (Bostrom and Heinen, 1977; Lyytinen et al., 1998). Bostrom and Heinen, (1977) spotlight two important points: First, the social and technical systems are complementary. Second, systems failures can be reduced by focusing on the social components (i.e. user skills) and technical components (i.e. technology and tasks). Because of the interaction between citizens and e-government systems, the STS model is applicable in e-government evaluation.

The framework we propose is shown in Figure 2. For the evaluation framework to be holistic, the main components, objectives, and the objectives measures need to be identified. Similar to the STS model, the framework consists of four quadrants: management of e-government, e-government technology, e-government stakeholders, and e-government processes. The proposed evaluation process consists of two phases. The first phase is determining the objectives that fall within four dimensions (management, technical, stakeholders, and process) and their measures. The objectives and their measures are dynamic factors that may vary contextually from country to country. The lists of objectives and the objective measures are by no means comprehensive. In the proposed framework we include the most common ones

Figure 1. Social-technical-system (STS) model, source: Bostrom and Heinen's (1977)

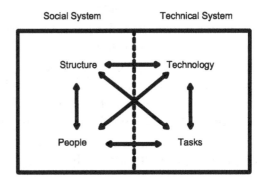

Figure 2. The objectives and their measures for each quadrant

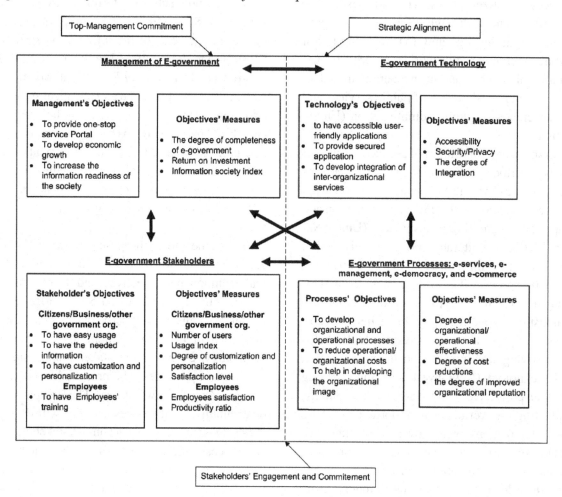

identified from literature. The second phase is evaluating the interaction among the objectives. This is accomplished by following the flowchart suggested in Figure 3.

To ensure a successful evaluation process, several conditions have to be satisfied before identifying the objectives and measures of each quadrant in the framework. First, e-government strategies should be linked to the wider institutional strategies in order to avoid strategy misalignment (Porter, 2001; Alalwan, 2010). Second, top executive commitment is crucial to the evaluation success. To assure an effective evaluation practice, the continuing support of senior executive officers is just as important as having formal

evaluation tools (Jones et al., 2006). The management should address factors such as developing the evaluation committee, determining the duration of the project, and delegating funds required for the completion of the evaluation process. Identifying and provisioning the resources have to be based on the varying situation, needs, and financial capabilities of the organization. Last but not least, stakeholders' engagement and commitment to the use of e-government services are essential to develop, implement, and evaluate e-government projects (Bertot et al., 2008). In addition to encouraging users' participation, openness allows stakeholders to understand what services the government firms are performing,

Figure 3. The formative evaluation

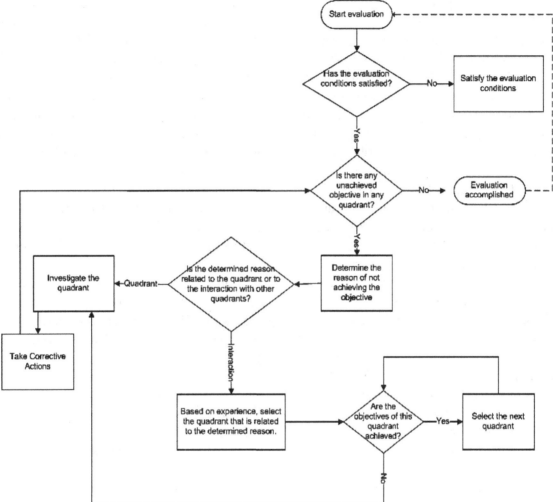

and why certain things are done in a particular way (Eschenfelder and Miller, 2005). The following sub-sections discuss the details of the two phases of the evaluation process.

Phase I: Determining the Objectives and Their Measures for Each Quadrant

Establishing measures provide a means to assess the corresponding objective. In this phase, we determine the objectives of each quadrant (see Figure 2) and how assessment results for each objective

can be measured. Observations and results offer valuable feedback that can be used to implement corrective actions if needed.

Management of E-Government

This part of the framework refers to the managerial aspects of e-government. It addresses what the government managers aim to achieve from the e-government system. Many studies have shown that providing a one-stop portal, or a 'one-stop government' is necessary for e-government projects to eliminate multi-level bureaucratic processes and raise the efficiency of the organization (Kamal et al, 2009). One-stop government is defined as "providing integrated e-government services to the stakeholders through e-government portals" (Hangen and Kubicek, 2000; Dias and Rafael, 2007). Affisco and Soliman (2006) discuss the strategic importance of the one-stop portal to differentiate modern e-government services from traditional ones. Lee et al. (2005) indicate that "e-government initiatives aim to enable government agencies to work together more efficiently and provide one-stop service to citizens and businesses". The achievement of one-stop government can be measured by the degree of completeness of e-government.

Economic growth is also an important objective behind designing e-government systems. Gant et al. (2002) find that adopting web portals to provide e-services at state government level has the potential to yield economic growth. The return on investment metric is widely used to measure the efficiency of e-government system (Gils, 2002). The lack of society readiness for e-services is also noted as an initiator for e-government projects (Goldstein, 2008). Information society index is commonly used as a measure to determine the society's readiness for e-services (Asgarkhani, 2005; Mutula and Brakel, 2006; Bui et al., 2003).

E-Government Technology

Hardware, software, and IT infrastructure are the enablers of the e-government technology quadrant. Literature review identifies many objectives that are attributable to IT infrastructure and the interacting constituents. One objective is to have accessible and user-friendly applications. Wimmer and Holler (2003) suggest that accessible and user-friendly portal interfaces is a crucial e-government requirement since users are heterogeneous and have different levels of expertise. This objective can be measured by accessibility rate, and interface effectiveness that satisfies the users' needs (TerryMa and Zaphiris,

2003). Disability access is becoming a standard requirement for most e-government services. West (2007) notes that there is progress in disability access, with 54% of federal websites offering disability access in 2007 compared to 47% in 2003. Another important objective in this quadrant is secure access to applications and privacy concerns. West (2007) study concludes that security usage in e-government websites have also increased dramatically. The study shows that 73% of U.S government websites have some form of security policy in 2007 while the percentage was only 7% in 2000. These statistics reveal the established need and compulsion to punctuate the importance of securing e-government applications. This objective can be measured by the degree of implemented and planned e-government security initiatives and privacy policies. A third objective is to promote integration of inter-organizational services. Lam (2005) concludes that developing integration of inter-organizational services is one of the critical success factors of e-government. This objective can be measured by the degree of service integration between different organizations.

E-Government Processes

We adopt Sakowicz's (2006) dimensions to define this quadrant. According to Sakowicz (2006), e-government processes have four dimensions. The first is 'e-services' that describe the delivery of government information, programs, and services to the external stakeholders. The second is the 'e-management' process, which refers to the functions that support the public institution internally such as electronic records management and information flow management. The third is 'e-democracy' which refers to the public activities that raise the citizen involvement with e-government system such as e-voting, cyber campaign, and virtual town meetings. The fourth one is the 'e-commerce' processes which involve the monetary exchange over the internet. It relates to the transaction management in the e-government ecosystem such as, paying taxes, utility bills, and vehicle registration. We synthesize three objectives for the aforementioned e-government processes, i.e., to develop organizational and operational processes, to reduce operational and organizational costs, and to enhance organizational image. Kral and Zemlicka (2008) use e-government as an example of service-oriented systems that can develop business processes. In addition, many studies (Gant et al., 2002; Edmiston, 2003; Kim et al., 2007) conclude that e-government technology has the potential to reduce organizational costs and raise the efficiency. Grabow (2003: cited in Amberg et al., 2005) finds that organizational image, as perceived by employees, citizens, and other companies, is improved in response to improved electronic techniques. Many measures can be used in this regard, and we suggest the following - the degree of organizational efficiency, the degree of improved organizational reputation, and the percentage of cost reduction.

E-Government Stakeholders

E-government systems are designed for different stakeholders such as employees, citizens, business, and other government organizations. Gupta (2007) classifies the key stakeholders of e-government into five categories: service users (i.e. end customers), government users, funding agencies, public/private partners, and other stakeholders (e.g., academics and researchers). For the purpose of this paper, stakeholders are classified into two groups according to their shared service delivery objectives. Citizens, businesses, and other government organizations form the first group (external stakeholders) and e-government employees fall in the second group (internal stakeholders). The service delivery for the external users target objectives such as, providing ease of use, ensuring accuracy of information, and enabling customization and personalization features. Bélanger and Carter (2008) suggest that the accuracy of stakeholders' information is critical

for e-government adoption. Chiang (2009) notes that ease of use is an essential construct of an e-voting system. Tat-Kei Ho (2002) argues that e-government paradigm has shifted the service delivery principle from standardization and equity to customization and personalization. The use of e-government portals is positively associated with user satisfaction (Welch, 2005). These objectives can be assessed by measures such as satisfaction level, number of users, degree of customization and personalization, and perceived ease of use. For the internal stakeholder group, the main objective is to establish training on the new e-government systems. Chen et al. (2006) suggest that in order for the employees to accept the system, training and technical staffing should be considered during e-government planning. This objective can be measured by employees' satisfaction level and productivity ratio.

Phase II: The Formative Evaluation Flowchart

The flowchart in Figure 3 depicts the formative evaluation process. As indicated earlier in the introduction, objectives achievement is the main driver that guides the formative evaluation process.

We should keep in mind that, as depicted in Figure 2, each quadrant interacts, affects, and is affected by every other quadrant. First and foremost, the evaluator needs to make sure that the required evaluation conditions have been satisfied. After that, the evaluator needs to detect any objective deviation by using the objective's measures. If there is any objective deviation, the evaluator needs to determine the source of this deviation. Questions that can assist the evaluator in this assessment include – "is the source of deviation limited to one or more objectives in the quadrant?", "is it contained within the quadrant itself?", or "is it related to an interaction with other quadrants?". If the source of the deviation is within the quadrant itself, the evaluator needs to check and investigate the related factors that cause the deviation. After determining the deviation source, the evaluator can suggest corrective actions and start the evaluation process again. If the source of the deviation is linked to another quadrant, the evaluator needs to determine the closest quadrant that may interact with the previous quadrant and identify the cause of the deviation. After the quadrant selection, the evaluator checks whether the objectives of this quadrant are achieved or not. If they are achieved, the evaluator selects another quadrant that has a deviation in the objectives. After selecting the component that has the deviation, the evaluator investigates and checks the related factor that may cause the deviation and take corrective action to solve the problem. The evaluation process is formative in that it can cycle iteratively to identify deviations from the drawn objectives. The dotted line (see Figure 3) captures the ongoing evaluation cycle and the continuous feedback that provisions the formative evaluation process.

THE KOREAN GOVERNMENT-FOR-CITIZEN (G4C) PROJECT CASE STUDY

Figure 4 prepares the ground for demonstrating the applicability of the framework in evaluating e-government system. We use the suggested

Figure 4. The objectives and their measures for the Korean Government-For-Citizen (G4C) Project

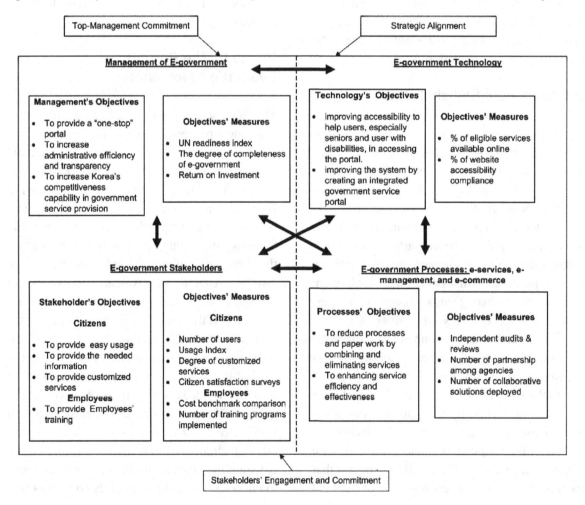

framework to assess the Government-For-Citizen (G4C) project implemented in South Korea. The G4C project, which began in 2002, is a "one-stop" e-government system that is designed to improve the public services provided to the citizens. From its inception, the project was driven by the vision of integrating the latest information technology to actualize government services so that the number of physical visits to the public offices could be minimized and the required paperwork could be reduced.

G4C ranked number one in the United Nations E-Government Survey 2010 (UN, 2010), and it is considered one of the most successful and advanced e-government service systems. Our assessment method is based on the analysis of the project description and evaluation document that is published by the Ministry of Public Administration and Security (2009). We analyze the objectives and their measures from the publicly available documentation and classify them using our framework. The documents (Ministry of Public Administration and Security, 2009) not only allow us to evaluate the G4C but provide an insight into the system development and post-implementation use from 2000 to 2009. Figure 4 summarizes the analysis by mapping the objectives and related measures.

Although the e-government evaluation process should start by determining where the necessary conditions have been satisfied, the information from the public documents does not provide adequate data to determine the extent to which they are fulfilled. Available documentation does provide ample evidence to indicate that there is a strong top-management commitment to develop, implement, and maintain the project since the early phases of G4C system. Evidence gathered from user testimonials, media reports and system description support citizens' engagement and commitment, and government sponsored initiatives to promote online services (via advertisement and commercials). Incentives introduced since 2007

include an arrangement to waive fees for all online services and transactions. No evidence was found to determine the extent of employees' engagement or about linking the IT strategy to the corporate strategy. Considering the success of the G4C project, it is likely that these intentions have been actualized, although not documented publicly.

In the 'Management of E-government' quadrant, the objectives include providing a "one-stop" portal, increasing administrative efficiency and transparency, and increasing Korea's competitiveness capability in government service provision. In addition to UN readiness index, cost saving is measured quantitatively. The economic benefit is estimated to be $47.2 billion between 2003 and 2008. In the 'E-government Technology' quadrant, the objectives include improving the system by creating an integrated government service portal, electronic document verification system, and electronic official document management system for civil petition. It also includes capabilities to help seniors and those with disabilities to access and use the portal. Another objective aims to improve general accessibility, which is represented by offering information about a wide variety of government services regardless of time and place.

In the 'E-government Processes' quadrant, G4C processes include e-services (e.g., national identification registry certificates), e-management (e.g., official electronic document management system for civil petition), and e-commerce (e.g., fees payment). The objectives include reducing processes and paperwork by combining and eliminating services, and enhancing service efficiency and effectiveness. Since the government system is not democratic, e-democracy is irrelevant. In the 'E-government Stakeholders' quadrant, G4C categorizes two types of stakeholders - citizens and employees. G4C promotes numerous citizens' objectives. They include encouraging the use of portal by having a simple verification process, offering customizable online service packages, and providing online tutorials. Employees' objectives include developing and designing training programs to improve the utilization of the system and enhance the processing methods.

Once the interaction dimensions, their objectives, and the measures of the objectives have been determined, the G4C system is ready for evaluation. This is phase II of the evaluation process and the evaluator can follow the flowchart as depicted in Figure 3. The case study shows that the proposed framework can assess e-government systems effectively. The G4C case study also shows the adaptability of the framework to different e-government agencies. By dividing the evaluation process into two phases, the framework offers the flexibility that makes it suitable for varying contexts. In addition, the formative nature of the framework enables measuring the socio-technical objective achievements and ensuring the continuity of the evaluation process.

CONCLUSION

In this paper, we present a framework to evaluate the efficiency of e-government systems. In addition to increasing the knowledge in the field of e-government evaluation, the goal of the proposed framework is to remedy the current gaps in e-government evaluation literature. The formative part of the evaluation framework ensures the objectives achievement and the continuity of the evaluation process. The framework facilitates identifying and filling gaps in the existing e-government evaluation systems by classifying the e-government systems into four interactive quadrants.

The framework contribution is twofold. First, it contributes to research by integrating the formative evaluation approach, STS model, and other relevant theories from the e-government evaluation field. Second, the framework can be used as a template for researchers and practitioners to assess e-government projects. We demonstrated the use of the framework by applying it as a tem-

plate to assess the South Korean G4C project. The case study corroborates the applicability and practicability of the framework.

It must be mentioned that the proposed approach is not without limitations. The framework could benefit from more validation by testing each quadrant in real-world settings. Furthermore, the G4C case study is based solely on publicly available documents, reports and testimonials regarding the South Korean e-government project. Future research agenda will focus on improving the validation of the suggested model by designing a survey and conducting interpretive studies in a real-world public organization.

REFERENCES

Accenture. (2003). *E-government leadership: Engaging the customer*. Retrieved from http://www.accenture.com/xdoc/en/newsroom/epresskit/egovernment/ egov_epress.pdf

Accenture. (2006). *Leadership in customer service: Building the trust*. Retrieved from http://www.accenture.com/xdoc/en/industries/government/acn_2006_govt_report_FINAL2.pdf

Affisco, J. F., Khalid, S., & Soliman, K. (2006). E-government: A strategic operations management framework for service delivery. *Business Process Management Journal*, *12*(1), 13–21. doi:10.1108/14637150610643724

Alalwan, J. (2010). Can IT resources lead to sustainable competitive advantage. *Proceedings of the Southern Association for Information Systems Conference*, Atlanta, GA, USA March 26th-27th, (pp. 231-236).

Alshawi, S., & Alalwany, H. (2009). E-government evaluation: Citizen's perspective in developing countries. *Information Technology for Development*, *15*(3), 193–208. doi:10.1002/itdj.20125

Amberg, M., Markov, R., & Okujava, S. (2005). A framework for valuing the economic profitability of e-government. *Proceedings of International Conference on E-government*, (pp. 31-41).

Asgarkhan, M. (2005). Digital government and its effectiveness in public management reform. *Public Management Review*, *7*(3), 465–487. doi:10.1080/14719030500181227

Batini, C., Viscusi, G., & Cherubini, D. (2009). GovQual: A quality driven methodology for e-government project planning. *Government Information Quarterly*, *26*, 106–117. doi:10.1016/j.giq.2008.03.002

Bélanger, F., & Carter, L. (2008). Trust and risk in e-government adoption. *The Journal of Strategic Information Systems*, *17*(2), 165–176. doi:10.1016/j.jsis.2007.12.002

Bertot, J., Jaeger, P., & McClure, C. (2008). Citizen-centered e-government services: Benefits, costs, and research needs. *Proceedings of the 2008 International Conference on Digital Government Research*, Vol. 289, (pp. 137-142).

Beynon-Davies, P. (2005). Constructing electronic government: The case of the UK inland revenue. *International Journal of Information Management*, *25*(1), 3–20. doi:10.1016/j.ijinfomgt.2004.08.002

Bostrom, R., & Heinen, J. (1977). MIS problems and failures: A socio-technical perspective part I: The causes. *Management Information Systems Quarterly*, *1*(3), 17–32. doi:10.2307/248710

Bui, T., Sankaran, S., & Sebastian, I. (2003). A framework for measuring national e-readiness. *International Journal of Electronic Business*, *1*(1), 3–22. doi:10.1504/IJEB.2003.002162

Carter, L., & Belanger, F. (2004). Citizen adoption of electronic government initiatives. *Proceedings of 37ᵗʰ Annual Hawaii International Conference on System Sciences*, Big Island, Hawaii, (pp. 5–8).

Chen, Y., Chen, H., Huang, W., & Ching, R. (2006). E-government strategies in developed and developing countries: An implementation framework and case study. *Journal of Global Information Management*, *14*(1), 23–46. doi:10.4018/jgim.2006010102

Chiang, L. (2009). Trust and security in the e-voting system. *Electronic Government: An International Journal*, *6*(4), 343–360. doi:10.1504/EG.2009.027782

Dubin, R. (1978). *Theory building*. New York, NY: The Free Press.

Edmiston, K. (2003). State and local e-government prospects and challenges. *American Review of Public Administration*, *33*(1), 20–45. doi:10.1177/0275074002250255

Eschenfelder, K. R., & Miller, C. (2005). *The openness of government websites: Toward a socio-technical government website evaluation toolkit*. MacArthur Foundation/ALA Office of Information Technology Policy Internet Credibility and the User Symposium, Seattle, WA.

Esteves, J., & Joseph, R. (2008). Comprehensive framework for the assessment of eGovernment projects. *Government Information Quarterly*, *25*, 118–132. doi:10.1016/j.giq.2007.04.009

Farbey, B., Land, F., & Targett, D. (1993). *How to assess your IT investment: A study of methods and practice*. Oxford, UK: Butterworth-Heinemann Ltd.

Farbey, B., Land, F., & Targett, D. (1995). A taxonomy of information systems applications: The benefits evaluation ladder. *European Journal of Information Systems*, *4*, 41–50. doi:10.1057/ejis.1995.5

Fountain, J. (2003). Prospects for improving the regulatory process using e-rule making. *Communications of the ACM*, *46*(1), 43–44. doi:10.1145/602421.602445

Funilkul, S., Quirchmayry, G., Chutimaskul, W., & Traunmuller, R. (2006). *An evaluation framework for e-government services based on principles laid out in COBIT, the ISO 9000 standard, and TAM*. 17th Australasian Conference on Information Systems.

Gils, D. (2002). Examples of evaluation practices used by OECD members countries to assess e-government. *Draft Point*, *9*, 1–64.

Goldstein, R. (2008). *Community informatics, electronic government and inclusion: Strategies for the consolidation of a citizens' democracy in Latin America*. Prato CIRN 2008 Community Informatics Conference: ICTs for Social Inclusion: What is the Reality? Refereed Paper.

Griffin, D., & Halpin, E. (2002). Local government: A digital intermediary for the information age? *Information Polity*, *7*(4), 217–231.

Gupta, M., & Jana, D. (2003). E-government evaluation: A framework and case study. *Government Information Quarterly*, *20*, 365–387. doi:10.1016/j.giq.2003.08.002

Gupta, P. (2007). Challenges and issues in e-government project assessment. *Proceedings of the 1st International Conference on Theory and Practice of Electronic Governance ACM*, USA.

Hamilton, S., & Chervany, N. (1981). Evaluating information system effectiveness - Part I: Comparing evaluation approach. *Management Information Systems Quarterly*, *5*(3), 55–69. doi:10.2307/249291

Hangen, M., & Kubicek, H. (2000). *One-stop government in Europe: Results of 11 national surveys*. University of Bremen.

Hochstrasser, B. (1992). Justifying IT investment. *Proceedings of the Advanced Information Systems Conference; The New Technologies in Today's Business Environment*, UK, (pp. 17–28).

Irani, Z., Love, P., Elliman, T., Jones, S., & Themistocleous, M. (2005). Evaluating e-government: Learning from the experiences of two UK local authorities. *Information Systems Journal*, *15*, 61–82. doi:10.1111/j.1365-2575.2005.00186.x

Jansen, A. (2005). *Assessing e-government progress—Why and what*. University of Oslo. Retrieved from http://www.uio.no/studier/emner/jus/afin/FINF4001/h05/undervisningsmateriale/AJJ-nokobit2005.pdf

Ji, Z. (2009). The research on the evaluation of e-government system. *International Conference on Industrial and information Systems*, (pp. 220-223).

Jones, S., & Hughes, J. (2001). Understanding IS evaluation as a complex social process: A case study of a UK local authority. *European Journal of Information Systems*, *10*(1), 189–203. doi:10.1057/palgrave.ejis.3000405

Jones, S., Irani, Z., Sharif, A., & Themistocleous, M. (2006). E-government evaluation: Reflections on two organizational studies. *Proceedings of the 39th Hawaii International Conference on System Sciences*, Kauai, Hawaii. January 4–7.

Kaisara, G., & Pather, S. (2011). The e-government evaluation challenge: A South African Batho Pele-aligned service quality approach. *Government Information Quarterly*, *28*(2), 211–221. doi:10.1016/j.giq.2010.07.008

Kamal, M. M., Weerakkody, V., & Jones, S. (2009). The case of EAI in facilitating-government services in a Welsh authority. *International Journal of Information Management*, *29*(2), 161–165. doi:10.1016/j.ijinfomgt.2008.12.002

Khalifa, G., Irani, Z., Baldwin, L. P., & Jones, S. (2004). Evaluating information technology with you in mind. *Electronic Journal of Information Systems Evaluation*, *4*(5), 246–252.

Kim, H., Pan, G., & Pan, S. (2007). Managing IT-enabled transformation in the public sector: A case study on e-government in South Korea. *Government Information Quarterly*, 24(2), 338–352. doi:10.1016/j.giq.2006.09.007

Kral, J., & Zemlicka, M. (2008). Implementation of business processes in service-oriented systems. *International Journal of Business Process Integration and Management*, 3(3), 208–219. doi:10.1504/IJBPIM.2008.023220

Kunstelj, M., & Vintar, M. (2004). Evaluating the progress of eGovernment development: A critical analysis. *Information Polity*, 9(3–4), 131–148.

Lam, W. (2005). Barriers to e-government integration. *Journal of Enterprise Information Management*, 18(5), 511–530. doi:10.1108/17410390510623981

Lee, S. M., Tan, X., & Trimi, S. (2005). Current practices of leading e-government countries. *Communications of the ACM*, 48(10), 99–104. doi:10.1145/1089107.1089112

Lessen, V. T., Nitzsche, J., & Leymann, F. (2009). Conversational web services: Leveraging BPEL for expressing WSDL 2.0 message exchange patterns. *Enterprise Information Systems*, 3(3), 347–367. doi:10.1080/17517570903046300

Li, W., Zheng, W., & Guan, X. (2007). Application controlled caching for web servers. *Enterprise Information Systems*, 1(2), 161–175. doi:10.1080/17517570701243273

Lim, J. H., & Tang, S.-Y. (2008). Urban e-government initiatives and environmental decision performance in Korea. *Journal of Public Administration: Research and Theory*, 18, 109–138. doi:10.1093/jopart/mum005

Liu, J., Derzs, Z., Raus, M., & Kipp, A. (2008). Lecture Notes in Computer Science: *Vol. 5184. Egovernment project evaluation: An integrated framework* (pp. 85–97). New York, NY: Springer.

Liu, T., Liu, R., & Zhao, P. (2004). Research on e- government system assessment methods. *Journal of Wuhan Automotive Polytechnic University*, 26(3).

Lyytinen, K., Mathiassen, L., & Ropponen, J. (1998). Attention shaping and software risk—A categorical analysis of four classical risk management approaches. *Information Systems Research*, 9(3), 233–255. doi:10.1287/isre.9.3.233

Ministry of Public Administration and Security. (2009). *G4C: Government for Citizens*. Korean Government. Retrieved from http://korea.go.kr/html/files/intro/001.pdf

Mutula, S., & Brakel, P. (2006). An evaluation of e-readiness assessment tools with respect to information access: Towards an integrated information rich tool. *International Journal of Information Management*, 26(3), 212–223. doi:10.1016/j.ijinfomgt.2006.02.004

National Office for the Information Economy. (2003). *E-government benefits study*. Canberra, Australia: NOIE.

Orange, G. Burke, A. Elliman, T., & Kor, A. (2006). CARE: An integrated framework to support continuous, adaptable, reflective evaluation of e-government systems. *European and Mediterranean Conference on Information Systems*, Alicante, Spain, July 6–7.

Petricek, V., Escher, T., Cox, I. J., & Margetts, H. (2006). *The web structure of e-government developing a methodology for quantitative evaluation*. International World Wide Web Conference, Edinburgh, UK.

Porter, M. (2001). Strategy and the Internet. *Harvard Business Review*, (March): 63–78.

Remenyi, D., Money, A., Sherwood-Smith, M., & Irani, Z. (2000). *Effective measurement and management of IT costs and benefits*. Oxford, UK: Butterworth-Heinemann.

Sakowicz, M. (2006). *How to evaluate e-government? Different methodologies and methods.* Warsaw School of Economics, Department of Public Administration. Retrieved from http://unpan1.un.org/intradoc/ groups/public/documents/NISPAcee/UNPAN009486.pdf

Serafeimidis, V., & Smithson, S. (2000). Information systems evaluation in practice: A case study of organizational change. *Journal of Information Technology, 15*(2), 93–105. doi:10.1080/026839600344294

Shan, S., Wang, L., Wang, J., Hao, Y., & Hua, F. (2010). Research on e-government evaluation model based on the principal component analysis. *Information Technology Management, 12,* 173–185. doi:10.1007/s10799-011-0083-8

Sorrentino, M., Naggi, R., & Luca Agostini, P. (2009). E-government implementation evaluation: Opening the black box. *Proceedings of the 8th International Conference on Electronic Government,* Linz, Austria.

Symons, V., & Walsham, G. (1988). The evaluation of information systems: A critique. *Journal of Applied Systems Analysis, 15,* 119–132.

Terry Ma, H., & Zaphiris, P. (2003). *The usability and content accessibility of the e-government in the UK.* London, UK: Centre for Human-Computer Interaction Design, City University. Retrieved from http://www.soi.city.ac.uk/~zaphiri/Papers/

Torres, L., Pina, V., & Acerete, B. (2005). E-government developments on delivering public services among E.U. cities. *Government Information Quarterly, 22*(2), 217. doi:10.1016/j.giq.2005.02.004

United Nations. (2008). *UN e-government survey – From e-government to connected government.* United Nations. Retrieved from http://unpan1.un.org/intradoc/groups/public/documents/UN/UNPAN028607.pdf

United Nations E-Government Survey. (2010). Retrieved from http://www2.unpan.org/egovkb/global_reports/10report.htm

Van Der Westhuizen, D., & Fitzgerald, E. P. (2005). Defining and measuring project success. *Proceedings of the European Conference on IS Management, Leadership and Governance,* Reading, United Kingdom.

Victor, G. J., Panikar, A., & Kanhere, V. K. (2007). E-government projects–Importance of post completion audits. *Proceedings of the 5th International Conference of e-Government.* Retrieved from http://www.iceg.net/2007/books/1/20_308.pdf

Wang, L., Bretschneider, S., & Gant, J. (2005). Evaluating web-based e-government services with a citizen-centric approach. *Proceedings of 38th Annual Hawaii International Conference on Systems Sciences,* Big Island, Hawaii, January 3–6.

Wang, Y. S., & Liao, Y. W. (2008). Assessing eGovernment systems success: A validation of the DeLone and McLean model of information systems success. *Government Information Quarterly, 25,* 717–733. doi:10.1016/j.giq.2007.06.002

Welch, E. (2005). Linking citizen satisfaction with e-government and trust in government. *Journal of Public Administration: Research and Theory, 15*(3), 371–391. doi:10.1093/jopart/mui021

West, D. (2007). *State and federal e-government in the United States.* Brown University. Retrieved from http://www.insidepolitics.org/egovt07us.pdf

Wimmer, M., & Holler, U. (2003). Applying a holistic approach to develop user-friendly, customer-oriented e-government portal interfaces. *Lecture Notes in Computer Science, 2615,* 167–178. doi:10.1007/3-540-36572-9_13

Xin, C., Ding, R., & Xie, W. (2010). Performance evaluation of e-government information service system based on balanced scorecard and rough set. *Proceedings of International Conference on Information Management, Innovation Management and Industrial Engineering.*

Zhang, N., Guo, X., & Chen, G. (2007). Diffusion and evaluation of e-government systems: A field study in China. *Proceedings of the 11th Pacific Asia Conference on Information Systems*, (pp. 271-283).

ENDNOTES

[1]   A preliminary version of this manuscript was presented at the 17th American Conference on Information System (AMCIS), Detroit, Michigan, August 4-7, 2011.

Chapter 10
Measuring Effectiveness of an E-Governance System:
A Human-Centric Approach

Bijaya Krushna Mangaraj
XLRI School of Business & Human Resources, India

Upali Aparajita
Utkal University, India

ABSTRACT

In the era of economic liberalisation, institutions of higher education in the government sector, particularly universities, are facing tremendous challenges in terms of academic, general, and financial administration, which need effective governance. Recently, some of the universities are trying to adopt e-governance as a platform for such a purpose. However, the design of such a system is very much important, as it has to cater to the needs of various stakeholders in the public system. In this context, the effectiveness measurement of such an e-governance system is really necessary either to improve its performance level by re-aligning its organisational culture or by providing inputs for re-designing the system in order to make it more effective. Hence, the performance of such a system can be known if a human-centric approach with multiple criteria of evaluation is considered in the governance environment. This chapter attempts to determine those criteria by multiple factor analyses carried out for the purpose of considering multiple stakeholders. Analytic hierarchical processes as well as fuzzy analytic hierarchical processes have been then employed to measure the effectiveness of e-governance systems along those criteria, taking an Indian university as a case study.

INTRODUCTION

Electronic governance (E-governance) has gained tremendous importance and popularity all over the world. In India, this has found its origin during the seventies with a focus on in-house government applications in the areas of census, defence, economic monitoring, etc., where significant improvements have been realised in terms of organisational performance. The efforts of the National Informatics Centre (NIC) of the government of India in this direction were praiseworthy. However, in the ad-

DOI: 10.4018/978-1-4666-3640-8.ch010

ministration of higher education in the government sector, particularly in university administration, it has limited applications to date. Governance is a concept that involves the interactions among structures, processes and traditions that determine how power is exercised, how decisions are taken and how their say has been utilised by citizens or other stakeholders (Davidson, 2005). E-governance is the public sectors' use of information and communication technologies with the aim of improving information and service delivery encouraging users' participation in the decision-making process and making organisations more accountable, transparent and effective. The issues include function and data requirements within an organisation (i.e., actors, responsibilities), the need of information management strategies (i.e., key elements, actors, the structure and mission/business objectives of an organisation and data requirements), the factors of modern informa-tion–based organisations and the critical factors of success (Davidson, 2005). The last few years have seen the spread of e-governance in India with the help of government incentives. It has enabled the delivery of government services to a large base of people across different segments and geographi-cal locations. Its effective use in the government administration has generally enhanced existing efficiencies, reduced communication costs to a greater extent and increased transparency in the functioning of various departments. In addition to simplifying the process of data collection, analysis and audits making it less tedious, the government has also benefitted from reduced duplication of work. It has helped the government in cutting red-tapism, avoiding corruption and reaching citizens directly. The government has changed its role from "an implementer to facilitator and regulator". It has also given to its citizens' the easy access to tangible benefits be it through simple application such as online form-filling, bill securing and pay-ments, booking tickets or complex application such as distance education and tele-medicine. Hence, e-governance is now no longer looked at

as an option; rather a necessity for organisations aiming for good governance. However, people and organisational culture play a dominant role in making e-governance a success. Successful organisations try to measure the effectiveness of such systems from time, to time for educating their employees on performance measures and uses as they are benefitted from the processes which e-governance systems deliver and, if necessary, bring changes in the design aspects of such sys-tems. But for the measurement of performance, human aspect is most vital as users mostly decide the success of such systems. They use their experi-ences, expertise and have their own expectations which are mostly culture- specific and hence, a human-centric evaluation is quite meaningful in performance measurement of such systems.

In an e-governance approach, the technology involves automation and computerisation of ex-isting paper-based procedures that is prompting new style of leadership, new ways of debating and deciding strategies, new methods of transacting the governance, new techniques of listening to users and new strategies for organising and delivering information. Therefore, e-governance should not be viewed as a new type of government; rather a new tool of governance through the use of technology. In India, the National e-governance plan was initiated in 2003 which reflected the strategic intent of government of India in the right perspective. Many projects were earmarked under the plan with the government trying to address the digital divide. The World Bank, the Asian Development Bank, and the United Nations had been approached and in response they have funded e-governance projects in India. The IT policy of state of Odisha in India has been replaced by ICT policy of Orissa in 2004. This policy was designed to unleash the power of IT for the betterment of common citizens through e-governance and provide easy and comfortable access to informa-tion by public. A number of core ICT projects have been implemented in the state to establish the basic technological infrastructure. Some of them which are in operation are Secretariat LAN, Secretariat Training Facility, GRAMSAT, Odisha State Portal, Unicode Odia Project, Bhulekh (Land Records Computerisation), ORIS (Registration Office Computerisation), e-Shishu, e-Procurement, PRIASOFT, RURALSOFT etc. Fifteen government departments were also identified to provide more than 200 services under the Mission Mode Programme (MMP) of National e-governance programme. At present, the IT department of Odisha is set to implement around 50 e-governance projects in the state in the next two-three years under the state and national e-governance plans.

Government universities in India cater mostly to the vast middle- class segment of the country and their effectiveness matter a lot for the development of human resources of the country. In the higher education sector of Government of Odisha, Utkal University is the first university to adopt e-governance as a mode of practice of good governance due to the deteriorating trend of university administration in the government sector. Its effectiveness will help meritorious students to achieve the fruit of higher education at an affordable price. But, to know the system effectiveness of this project, it needs to be evaluated in a scientific as well as holistic manner based on the evaluation methods available in the literature. This chapter focuses on human- centric evaluation of e-governance project which is strengthened by the widely prevailing view that there is a need to have such an approach in the provisions of public services. A model has been presented for measuring effectiveness of e-governance system of Utkal University in this perspective based on the concepts of "users' experience" and "users' satisfaction". They come from the study of" usability "which is an approach to understand the use of objects that has developed with an emphasis on defining, characterising, analysing and improving well- defined tasks. In the context of technology, it can change the individual when used, and individual can also change the technology in use. So, to understand the technology in operation, one

must understand the users as well as the interaction between the users and the technology. Kline & Pinch (1996) emphasized the way in which technologies are appropriated and influenced by their users in ways that were not conceived of by their creators due to such interaction. This interaction primarily depends upon their experiences with the technology which can be observed looking at their expressions. Turner and Bruner (1986) describe how reality is presented itself to our consciousness through experience, and the expression is how this individual experience is framed and articulated. Hence, studying expressions can give an idea about the users' experience in general which needs to be evaluated for measuring usability as well as effectiveness. The International Standard Organisation defines measuring usability in ISO 9241 as being "the measurement of effectiveness, efficiency and satisfaction of users trying to carry out tasks". Hence, experience-focussed evaluation in contrast to task- focussed one is more meaningful as it is a holistic approach encompassing larger issues of user experience and his relationship with the technology. Unlike task-focussed evaluation, where well -defined tasks can be studied under controlled conditions, the act of observing, recording and evaluating users' experience is fundamental to the experience-focussed evaluation. This may be termed as "thick descriptions" or "situated actions" which may be studied through an ethnographic approach including societal culture, holistic perspective, contextualisation, emic perspectives and multiple realities, etic perspective, non-judgemental orientation, inter and intra-cultural diversity, structure and function, symbol and ritual, micro and macro operationalism etc. Thus, an ethnographic approach helps not only in getting a detailed and thick description of users' experience to understand and appreciate the full complexity of the lived experience, but also to understand the situation within which the technology was used. Therefore, in this chapter, users' experience and satisfaction have been captured in an analytical framework to measure system effectiveness in a quantitative scale which can be comparable to similar type of systems in operation, or in the improvement of the said system for better effectiveness.

EVALUATING E-GOVERNANCE IN A STATE-OWNED UNIVERSITY

In the pre-independence period higher education in India was offered only in government-owned universities. Even after independence, the same trend continued and universities in India apart from imparting higher education became instrumental in conducting advanced research, linking themselves to R&D organisations of industries to provide latest know-how and advising governments in administration as well as long-term perspective planning. But gradually new private owned universities/institutions were developed in order to meet the growing demand in the higher education sector in a populous country like India. However, government-owned universities still remain as the service provider of higher education at a much cheaper price as compared to their private counterparts. As a result of which, a vast majority of students get the opportunity of availing higher education at a very nominal cost in government-owned universities. But, in the recent years, the low output from most of the government--owned universities are due to problem of governance, be it academic or non-academic. This requires administrative reforms for universities in the government sector to face the current reality. Brussell (2012) offers a unique account of new era of administrative reforms characterised by the use of digital technologies to deliver public services. In this context, e-governance can be considered as a major tool in solving this problem to a great extent for the good governance of government-owned universities.

Utkal University, established in 1943 is the first university of the state of Odisha and is the seventeenth oldest university in India. It is one of the largest universities located in Bhubaneswar having about more than 3,000 post-graduate and doctoral students enrolling every year. At present, the university has 27 Post-Graduate departments for study and research in various fields such as, science, humanities, law, and business administration. It also runs 30 self- financing, job-orienting courses, such as 5- Year Integrated MBA, MCA and Law; Mass Communication,, Forensic Science, Bio-Technology, Information Technology, etc.. The organisational chart of Utkal University has been presented in Figure 1 to show the flow of information as well as the power structure amongst various units. The e-governance project of this University was assigned to Teledata Informatics Ltd. (TIL), which aimed at bridging the digital divide and making IT and IT–enabled education and service, a reality in Odisha. In order to offer high quality products and services, TIL has undergone quality screening processes for design, delivery and support systems of the services offered, which were tested and certified by Lloyd's Register Quality Assurance Ltd., London. It worked under the guidance of Ministry of Human Resource Development, Government of India and has implemented number of projects in various states of India. For this purpose, this company has exclusively developed software and e-governance support for the various types' administration. Apart from several projects of Government of India, viz., Sarrva Shiksha Abhiyan (SSA), Computer Aided-Learning (CAL) in the Central and North-Eastern states of India, e-governance project for Utkal University was implemented by TIL with an objective of developing similar type of systems for other universities in the state. When TIL was asked to develop an e-governance system of Utkal university, it carried out some discussions with the university officials for a need assessment of information for various levels as shown in Figure 1. Based on such information, an e-governance system was implemented in Utkal university for its routine and strategic management. However, the effectiveness of such a system needs to be

Figure 1. Organizational chart of Utkal University

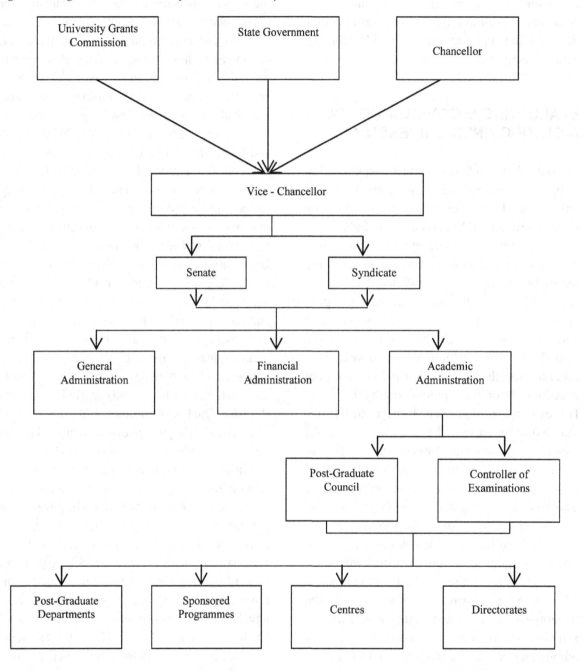

evaluated in terms of its cost as well as associated benefits available to various stakeholders at different levels of this public system. This exercise largely depends on the system design of such a project which TIL has implemented it for developing an effective e-governance system for Utkal university so that similar types of projects can be implemented in other universities of the state as well. As a result, a need was created for measuring the system effectiveness not only for improving the service delivery of the system but also, popularising e-governance in the university administration of the state. As discussed earlier, in order to decide the success and failure factors rather the acceptability of e-governance projects, the service mechanism has to cope with the requirements of individuals which can be done through an experience-focussed evaluation. The present chapter aims at doing the job in a human–centric perspective using analytical tools in a sequential manner.

The literature on evaluation of e-governance offers some hard measures, viz., cost benefit analysis and benchmarking. Cost benefit analysis is a technique that assesses projects through a comparison between their costs and benefits. Generally, cost benefit analyses are comparative; they compare the costs and benefits of the situation with and without the project. On the other hand, benchmarking provides a method of evaluating performance against best practice and provides strategic guidance. Benchmarking e-governance suffers from the vague definition of e-governance. However, Janssen et al (2004) made a study to analyse what was actually measured in the form of indicators used so as to judge the relevance of benchmarking results. Hartson et al. (2001) highlight multiple criteria for user evaluation whereas Law & Hvannberg (2004) describe a heuristic evaluation method for effectiveness. El-Kiki & Lawrence (2007) attempt to measure the effectiveness and efficiency of m-government services based on citizens' and business needs. Looking at all these, a multi-perspective approach seems to be very much suitable for this evaluation considering various stakeholders and their experience and satisfaction as the basis for the evaluation process. Therefore, for effectiveness evaluation of e-governance system in the university, this chapter considers multiple evaluation criteria along which different sectors of governance can be evaluated which has been shown in Figure 2. Hierarchical analyses have been employed to determine the effectiveness of this technology in different sectors of governance.

MEASURING EFFECTIVENESS IN A MULTI-PERSPECTIVE FRAMEWORK

The multi-perspective approach to problem solving was introduced by Mitroff and Linestone (1993) based on the principle of unbounded system thinking. It is more a philosophy than a method and contains little guidance as to how it should be implemented in practice. Courtney (2001) used this approach as a basis for a new paradigm in decision support systems. Courtney's work has attracted significant interest, as can be seen in some of the works, e.g. Cil et al (2005), Chae et al (2005), Hall et al (2005), Hall and Davis (2007) and Petkov et al (2007) etc.. This approach classifies the possible perspectives on a situation into the categories of technical, organizational, personal, ethical and aesthetic. Based on this approach, Linggang and Hitoshi (2005) developed a methodological tool for policy-set design from multi-perspective viewpoints that would be practical and applicable to real world and large-scale problems. Initially, a comprehensive logical system of factors and variables on expressing system design, with considerations on various viewpoints of different stakeholders in the system were wholly constructed. A mathematical model followed by an algorithm was proposed for the optimal set of expressway expansion project in the construction plan taking a case study of Japan. In the similar manner, evaluation based on multi-perspective approach can also consider multi-perspective viewpoints, i.e., different perspectives of various stakeholders.. Therefore, in multi-perspective evaluation, the fundamental aspect of understanding users' experience and satisfaction is to recognize the different stakeholders whose experiences, and hence expressions, along various criteria of evaluation must be included for an effective evaluation. Outranking methods for ranking criteria/alternatives can be very well applied to measure effectiveness either by a way of comparison or based on expectation. Some of the methods in this direction are due to Saaty

Figure 2. Effectiveness measurement model of e-governance project

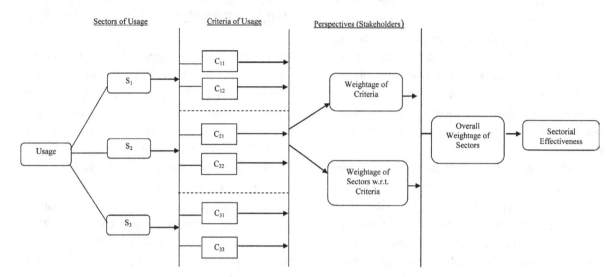

[(1980, 2008)], Buckley [1984, 1985(a), 1985(b)], Chang, (1992,1996), Kahraman et al. (2004), etc.

Nielsen (1993) explained that usability is not a single, uni-dimensional property of the interface and could be defined in terms of learnability, efficiency, memorability, errors and satisfaction. A formal definition of the term is derived from ISO-9244-11(1998) as the extent to which a product can be used by specified users to achieve specified goals with effectiveness, efficiency and satisfaction in the specified context of use. Roy & Bouchard (1999) reviewed existing instruments and concepts related to user satisfaction measurement and discussed their limitations. They also proposed through a case study, how this existing body of knowledge can be used, in spite of its limits, as a starting point for developing one's methods and tools. Sedera, et al. (2001) felt the growing consensus that enterprise systems in many cases failed to provide expected benefits. The increasing role of information technology and the uncertainty of large investment in enterprise systems have created a strong need to monitor and measure enterprise system performance. Their work identified how enterprise system benefits can be usefully measured, with a "balance" between qualitative and quantitative factors. Gupta

and Jana (2003) suggested a flexible framework to choose an appropriate strategy to measure the tangible and intangible benefits of e-governance taking an Indian case study. Chang & King (2005) developed an instrument that might be used as an information system functional scorecard (ISFS). It was based on a theoretical input-output model of the information system's functions role in supporting business process effectiveness and organisational performance. Hosseini & Mazinani (2006) proposed a straightforward approach for measuring the amount of IT effectiveness on business processes by using fuzzy reasoning approach. The amount of this metric would help managers to find out their strengths and weaknesses and improve performance of businesses processes by utilization of IT service and elevate IT to its rightful place in the organisation. De (2006) presented an Indian case for evaluation of e-government projects by using development theory as propounded by Amartya Sen. Griffths et al. (2007) considered user satisfaction as a multi-dimensional and subjective variable which can be affected by many factors other than performance of the system. They focused on information retrieval and information system literature in an attempt to understand what user satisfaction is, how it is measured, what factors affect it and why findings on user satisfaction have been so varied and contradictory.

Huang et. al.(2008) proposed a generalizable methodology, based on survival analysis, to quantify user satisfaction in terms of session times, i.e., the length of time users stay with an application. Unlike subjective human surveys, their methodology was based solely on passive measurement. In the model development process, they identified the most significant performance factors and their impacts on user satisfaction. They also discussed how they could be exploited to improve user experience and optimise resource allocation..Mangaraj and Upali (2008) developed a fuzzy logic based methodology for evaluating development projects based on users' experience in a multi-perspective framework. Wang & Zheng (2010) used models of data envelopment analysis (DEA) for the evaluation of performance of complex information systems. Taking the information systems of 30 sales enterprises, they used DEA models to analyse the information system from angles of both technical efficiency and sales efficiency. Al-Maskari & Sanderson (2010) investigated the factors influencing user satisfaction in information retrieval and found that user satisfaction was a subjective variable, which could be influenced by several factors such as system effectiveness, user effectiveness, user effort and user characteristics and expectations. Therefore information retrieval evaluators should consider all these factors in obtaining user satisfaction and using it as a criterion of system effectiveness. Fitsilis et al. (2010) proposed an approach which resulted in the development of an e-government balanced scorecard to evaluate e-governance projects from various perspectives, interrelated to strategic planning, such as product, process and project management as well as service quality management. Mahalik (2010) discussed the outsourcing process in e-governance through a multi-criteria decision-making (MCDM) approach. He attempted through a case analysis, with the help

of analytic hierarchy process (AHP), to find out an optimum level of outsourcing and suggested a balanced strategy between in-house and external agency, which in turn helps to reduce the failure rates in e-governance projects. Sahu et al. (2010) also used AHP for evaluation of e-governance projects for CSI-Nihilent e-governance awards in India. Huang and Lee (2010) examined Taiwan's e-government performance based on the impact ICT applications in administrative service and democracy improvement have on citizens I terms of cost and benefit. Hermana and Silfianti (2011) evaluated the performances of public services through the websites of the local governments in Indonesia. Manian et al. (2011) constructed an approach based on a fuzzy TOPSIS (Technique for Order Preference by Similarity to Ideal Situations) and balanced score card (BSC) for evaluating an IT department. The BSC concept was applied to define the hierarchy with four major perspectives, viz., financial, customer, internal business processes and learning and growth, and performance indicators were selected for each perspective. A fuzzy TOPSIS information system was constructed to facilitate the solution process regarding strategies for improving department performance. Chander and Kush (2012) have considered metrics like delivery, usability, broken links, and feedback and traffic analysis for web-portal assessment and also made a comparative analysis for two states in India. Rizky et al. (2012) utilized linear regression and independent sample t-tests to analyze whether in every local government website provides transparent information about the activities of the local governments in Java with those outside Java. Jukic et al. (2012) present a comparative review of existent (inter-) national approaches to e-government project evaluation with the stress on intangible/hidden costs and public value dimension which are the two components usually missed out in traditional investment evaluation methods. They developed a model for ex-ante and ex-post evaluation of e-governance projects for Slovenian public administration. Wijaya et al. (2012) presented findings of a study devoted to investigating the factors based on item analysis, which influence successful implementation of e-government at local level in Indonesia. Morgeson (2012) outlined two types of performance measurement-internal and external measurement – and emphasized the importance of external citizen-centric performance measurement in the e-government context. Looking at all these concepts and methods employed in e-governance evaluation, this chapter develops an integrative methodology starting from factor identification to determination of effective evaluation scores. MCDM tools have been used for pairwise comparison values as well as linguistic variables for the purpose.

METHODOLOGY OF EFFECTIVE EVALUATION

This section discusses a four-step methodology for measuring the effectiveness of an e-governance system based on the evaluation of the users. Since technological effectiveness is a complex concept to be understood by users, this construct needs to be dimensionalised into measurable components to be evaluated by users either in quantitative or ordinal scales. Once significant factors are identified, stakeholders are asked to give their opinion about the importance of different sectors, different criteria etc. in a pairwise comparison method and based on users' expectations so far as e-governance is concerned. These opinions are to be pooled based on suitable aggregation operators to compute the evaluation scores. The algorithm of the methodology is stated as follows:

Step 1: Perform Item Analysis for the Selection of Valid Items (Variables)

Identify the items after conducting validity and reliability tests.

Step 2: Perform Exploratory Factor Analysis and Considering Only the Dominant Components

Exploratory factor analysis is a multi-variate statistical technique which can be used to explore the possible underlying factor structure of a set of observed variables. This also helps in reducing data complexity by reducing number of variables being studied. It involves a set of techniques which, by analyzing correlations between variables, reduces their number into fewer factors which explain much of the original data more economically. In this method, one decides on the number of factors by examining output from a principal component analysis (i.e., Eigen values are used). In this method, one does not have any hypothesis about the nature of underlying factor structure. This analysis is exploratory in the sense that it does not impose any structure on the relationships between the observed variables.

Step 3: Perform Saaty's Analytic Hierarchical Process (AHP) to Determine the Weights of Sectors & Criteria Independently and Sectors with Respect to Criteria Jointly

AHP as conceptualized by Saaty (1980, 2008) is a multi-criteria decision-making (MCDM) technique of measurement through pairwise comparison and relies on the judgments of experts to obtain priority weights. It is these weights that measure intangibles in relative terms. However, the method also has a scope to find out inconsistent judgments which can be modified for their inclusion in the evaluation process. The methods structure the decision hierarchy from the top with the goal of the decision, then objectives from a broad perspective, through the intermediate levels, i.e. the criteria. The lowest level of the hierarchy is the alternatives whose priority weights need to be determined. Figure 3 shows a pairwise comparison matrix for k-criteria whose elements N_{mk} are in the set S (see Table 1).

Figure 3. Pairwise comparison matrix for n criteria

where, S is the set of values in a 9-pint ordinal scale, which is explained in table 1.

The numerical values in between, viz., 2, 4, 6 and 8 represent intermediate importance levels and can be interpreted accordingly. Once a pairwise comparison matrix is obtained for the criteria, their weights can be obtained for by employing Saaty's AHP process.

Step 4: Perform Fuzzy AHP Method to Determine the Sectors with Respect to Criteria Jointly

In the context of effectiveness evaluation, certain indicators, viz., awareness, timeliness, content, usability etc., may not be expressed in a quantitative manner by a user. Also, the user's satisfaction for these types of indicators cannot be expressed in discrete manner. In these cases, linguistic variables in terms of fuzzy modelling can be used to represent the degree of such indicators and can be manipulated by fuzzy aggregation operators. For example, the various levels of "importance" for the criterion in terms of linguistic values can be expressed as:

- **P:** Absolute important (100%)
- **VH:** Very highly important
- **H:** Highly important
- **M:** Moderately important
- **L:** Lowly important
- **VL:** Very lowly important
- **φ:** Not important (0%)

In case of use of linguistic variable being used in the expression of opinions, the linguistic values can be structured in an ordinal scale. Also, various types of operators, viz., competitive, compensatory, competitive-cum-compensatory etc. can be defined to aggregate these variables. A fully competitive operator results in a "MIN" value whereas "MAX" value can be obtained through a fully compensatory operator. In that sense, "MEDIAN" becomes the appropriate average giving a value between "Min" and "Max" and can be termed as a competitive-cum-compensatory operator. For example, let $S = \{s_0, s_1, \ldots, s_L\}$ be the ordinal scale of linguistic information to measure the effectiveness of various indicators used by

Table 1. Intensity of importance, definitions and explanations

Intensity of importance	Definition	Explanation
1	Equal importance	Two activities contribute equally
3	Moderate importance	Experience and judgment slightly favour one activity over other
5	Strong importance	Experience and judgment strongly favour one activity over other
7	Very strong	An activity is favoured very strongly over another
9	Extreme	The evidence favouring one activity over another is of the highest possible order of affirmation

stakeholders.. This scale of linguistic variables can be linearly ordered and hence:

$$s_0 < s_1 < s_2 < \ldots < s_L.$$

where, $L = \{1.2, 3, 4, \ldots, l\}$. This means that, s_i, where, I belongs to L are not numbers and no other structure is assumed to exist on S. In that case, s_i could be $\{\phi, VL, L, M, H, VH, P\}$, where $\phi =$ none, VL = Very low, L = Low, M = Medium, H = High, VH = Very high, and P = Perfect.

If the stakeholders assign VL, L, VH, VH, L, VH to some issue, then assigning $\phi = 0$, VL = 1, L = 2, M = 3, H = 4, VH = 5, P = 6, the arithmetic mean for the series becomes $20/6 = 3.22$ which does not represent to any ordinal number in the scale S. But, the competitive (Min) and compensatory (Max) operators will result in VL and VH by aggregating the fuzzy sets. However for this, median is between $S_2 = L$ and $S_5 = VH$. Hence, Med (median roundup) for

$$t = \frac{2+5+1}{2} = 4$$

is $S_4 = H$. Similarly, Med (round down) for

$$t = \frac{2+5-1}{2} = 3$$

is $S_3 = M$. Otherwise,

$$t = \frac{i+j}{2},$$

for I and j, both even or both odd and represent the indices of s's for which the median is to be determined. Hence, median becomes a competitive–cum-compensatory operator which gives a value within these limits averaging qualitative information. Similarly, when a factor gets a value H with respect to stakeholder and at the same time the factor has an overall importance H for a particular situation, then the factor which gets a

rating as H*H cannot be similarly put in the same ordinal scale if H gets a value 4 as mentioned. Hence, pooling information in a hierarchical set-up necessitates fuzzy aggregation procedures involving various operators.

There are various reasons for the evaluator to use an ordinal scale S for the purpose. The evaluation process performed by the stakeholders may be very subjective, and then it seems more appropriate to use an ordinal scale. Also, it is probably easier for stakeholders to assign $s_i \in S$ to the sectors and criteria than to assign numbers or ratios of numbers and some of the criteria may be vaguely or imprecisely understood by the stakeholders. In that case, the linguistic values like "low", "high" are preferable. The stakeholders assign $s_i \in S$ for the satisfaction level of each of the sectors. Based on the facts, each stakeholder S_j of type k assigns a fuzzy set $\mu_j^k(S_i)^l$ in S for the sectors $I_1, I_2, \ldots I_m$. Then $\mu_j^k(S_i)^l$ measures the degree of satisfaction of I_i for stakeholder S_j of type K with respect to l-th criterion. Similarly, ach stakeholder S_j has a fuzzy set $\lambda_j^k(C)^l$ defined over the criteria $C_1, C_2, \ldots C_m$, with values in S. Then $\lambda_j^k(C)^l$ indicates the importance of criterion C_m with respect to the overall effectiveness for expert S_j. In this case, the same scale S has been used both for sectors and criteria. The data collected by the analyst may be displayed in matrices T_l^k and P_k as:

$$T_1^k = \begin{array}{c} S_{1k} \\ S_{2k} \\ \cdot \\ \cdot \\ \cdot \\ S_{ik} \end{array} \begin{bmatrix} & I_1 & & I_2 & \cdots & I_n \\ & & & & & \\ & \mu_j^k(S_i)^1 = a_{ij}^{kl} \in S & & & \\ & & & & & \\ & & & & & \\ & & & & & \end{bmatrix}_{jxn}$$

for the type of stakeholder k, where $1 \leq k \leq K$. Similarly,

$$P^k = \begin{array}{c} S_{1k} \\ S_{2k} \\ \cdot \\ \cdot \\ \cdot \\ \cdot \\ S_{ik} \end{array} \left[\begin{array}{cccccc} C_1 & & C_2 & \cdot & \cdot & \cdot & \cdot & C_l \\ & & & & & & & \\ \lambda_j^k(C)^1 = b^{kl}{}_j \in S & & & & & & \\ & & & & & & & \\ & & & & & & & \\ & & & & & & & \\ & & & & & & & \end{array} \right]_{jxm}$$

Given the data T^k_1 and P_k, the evaluator has to determine the effectiveness measure in the same ordinal scale. If the averaging is done across the group of stakeholders, then the matrices T^k_1 for all criteria l, $1 \le l \le m$ are used to complete matrix M^l, where

$$M^l = \begin{array}{c} S_1 \\ S_2 \\ \cdot \\ \cdot \\ \cdot \\ \cdot \\ S_k \end{array} \left[\begin{array}{cccccc} I_1 & & I_2 & \cdot & \cdot & \cdot & \cdot & I_n \\ & & & & & & & \\ M_{nk} \in S & & & & & & \\ & & & & & & & \\ & & & & & & & \\ & & & & & & & \\ & & & & & & & \end{array} \right]$$

and matrices P_k can be used to produce matrix N, where,

$$N = \begin{array}{c} S_1 \\ S_2 \\ \cdot \\ \cdot \\ \cdot \\ \cdot \\ S_k \end{array} \left[\begin{array}{cccccc} C_1 & & C_2 & \cdot & \cdot & \cdot & \cdot & C_l \\ & & & & & & & \\ N_{mk} \in S & & & & & & \\ & & & & & & & \\ & & & & & & & \\ & & & & & & & \\ & & & & & & & \end{array} \right]$$

In the similar manner, averaging across criteria matrices M^l can be reduced to $[m_{ij}]$ and N can be reduced to column matrix $[n_i]$ based on averaging across stakeholders. These matrices provide ratings of sectors with respect to criteria as well as the criteria. The averaging procedure for getting

M^l and N as well as $[m_{ij}]$ and $[n_i]$ can be accomplished by using functions A,B and F,S, where,

$$m_{ij} = F(m_{n1}, m_{n2}, \ldots, m_{nk}) \text{ and } n_i = G(n_{m1}, n_{m2}, \ldots, n_{mk})$$

Now m_{ij} and n_i need to be combined to obtain the weighted ranking for each sector with respect to a criterion through a function λ, where, λ: S X S $\rightarrow$ S, and define $p_i = \lambda(m_i, n_i)$. Then p_i are the result of combining a criterion's weight n_k and sectorial rating. Next, we need to average, across all the indicators as:

$$Q: \prod_{i=1} S \text{-----------} \rightarrow S$$

and define $W = Q(p_1, p_2, \ldots p_i)$ where W becomes the overall measure of effectiveness being measured in the same ordinal scale. At this point A, B, F, G, λ and Q are defined in terms of median operator and the function λ has to satisfy the following three properties:

1. **Symmetry:** $\lambda(x,y) = \lambda(y,x)$
2. $\lambda(x,x)$ is strictly increasing
3. $\lambda(S_0,S_0) = S_0$ and $\lambda(S_L,S_L) = S_L$

where S_0 and S_L represent the lowest and highest possible ranking by the stakeholders, i.e., if for some criterion C_m, if a sector receives the lowest possible ranking s_0 and that criterion also has the lowest possible weight, then when these are combined, the result p_i is the lowest possible ranking. Similarly when the highest possible ranking S_L and S_L are combined, the result will be the highest possible ranking. But, max. and min. operators do not appear to be appropriate operators for combining sectorial ranking m_i and the criteria weights n_i. A mixed operator can be an appropriate one for the purpose which can be defined as:

$$MM(x,y) = \begin{cases} \max(x,y) & \text{if } x,y \geq S_a \\ \min(x,y) & \text{if } x,y \leq S_a \\ \text{med.}(x,y) & \text{Otherwise} \end{cases}$$

where $S_l < S_a < S_L$. The med. operator may be the round up med. or the round down med. For example, if $S = \{\phi, VL, L, M, H, VH, P\}$, then S_a could be H or VH. Considering $S_a = H$ for the above scale, a λ table using a mixed operator and a median (round up) operator can be given as in Table 2.

E-GOVERNANCE EVALUATION

The proposed algorithm was used to evaluate the implementation of e-governance project in Utkal University based on a multi-perspective framework using users' experience and satisfaction. This methodology has the capability of evaluating multiple sectors, viz., academic, administrative and financial by multiple stakeholders, i.e. teachers, students and administrative staffs, along multiple evaluation criteria. Here, the measurement of the effectiveness has been considered in two aspects, e.g.,

1. Measuring importance of the evaluation criteria with respect to the overall evaluation across multiple stakeholders of different types;
2. Measuring evaluation scores of different sectors along these criteria across the same stakeholders.

Keeping this model in view, data were collected from three groups of stakeholders, viz., students, teachers, administrative staffs. The sample was a stratified random sample consisting of 10% elements of the entire population. Responses were obtained from them based on their experience and satisfaction regarding the ambitious e-governance project whose success would take a lead role for the implementation of similar type of projects in other universities of the state as well as similar institution. Since the general feeling of the university does not speak about the effectiveness of this project, the authors wanted to measure it through various dimensions so that this monitoring can help the administration to take necessary steps to make it more user-oriented.

Initially, 29 items were selected for the purpose. These items were put to validity and reliability tests. Finally, for the items having Cronbach-α were more than 0.8 were considered for the analysis in the next step. Exploratory factor analysis was conducted for 26 selected items from the previous step to identify the factor structure in a hierarchical manner. For example, six factors consisting of 18 items were constructed based on their respective factor loadings. These factors were again grouped into two categories, viz., and value for money and efficiency in transactions. This has resulted in a two level factorisation of necessary criteria for effectiveness evaluation of e-governance project. The hierarchy of criteria as given in Table 3 can be considered for evaluation of three governance sectors, viz., academic, administrative and financial. The hierarchical structure of the process in terms of criteria and alternatives has been presented in Figure 4.

The opinion of the stakeholders regarding the effectiveness of the system has been collected for

Table 2. λ-table

m_i	n_i						
	ϕ S_0	VL S_1	L S_2	M S_3	H S_4	VH S_5	P S_6
$\phi = S_0$	S_0	S_0	S_0	S_0	S_0	S_3	S_3
$VL = S_1$	S_0	S_1	S_1	S_1	S_1	S_3	S_1
$L = S_2$	S_0	S_1	S_2	S_2	S_2	S_4	S_1
$M = Z_3$	S_0	S_1	S_2	S_3	S_3	S_4	S_5
$H = S_4$	S_0	S_1	S_2	S_3	S_4	S_5	S_6
$VH = S_5$	S_3	S_3	S_4	S_4	S_5	S_5	S_6
$P = S_6$	S_3	S_4	S_4	S_5	S_6	S_6	S_6

Table 3. Factorisation of e-governance criteria

	Conceptual Level	Operational Level(I)	Operational Level(II)
Perspectives	**Criteria**	**Indicators**	**Items**
Students/ Teachers/ Employees/ Officials	Value for Money	Pricing	• Initial cost of e-governance project compared to other development projects. • Revenue generation due to this transformation. • Annual maintenance cost of this project.
		Content	• Precision of the information. • Correctness of the information. • Sufficiency of the information.
		Availability	• Availability of the service at anytime. • Availability of the service at anywhere. • Availability of service to anybody
	Efficiency of transaction.	Usability	• Understanding of the working of the system. • Easiness of using the system. • Easiness of learning the system.
		Quality	• Timely information. • Up-to-date information. • Reliability of information
		Communicability	• Ability to communicate with Teachers. • Ability to communicate with the students. • Ability to communicate with the officials

Figure 4. Hierarchical structure of e-governance evaluation process

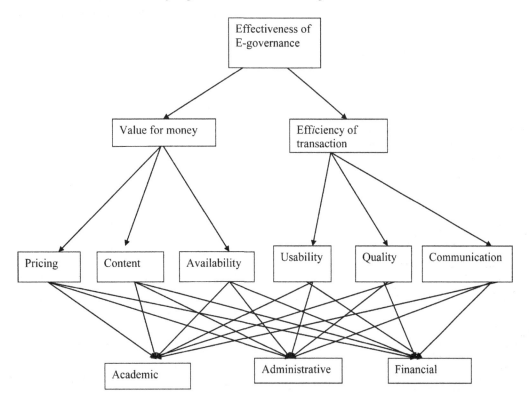

all the criteria and sectors of governance with respect to those criteria as mentioned in Table 3 through pairwise comparison. Based on step-.III, the criteria score across stakeholders for individual factors were obtained which are shown in Table 4. Also, the sectorial evaluation based on the criteria across the stakeholder was carried out for the existing e-governance set-up and is shown in Table 5.

Hence, Tables 4 and 5 give sufficient information about the effectiveness of e-governance to different sectors as has been perceived by different stakeholders. This has been measured with respect to the present set-up and one can observe that administration and financial wings of the university were not benefitted to the desired extent. The same trend has been experienced by all stakeholders irrespective of their types. Since this evaluation has been carried out in a pairwise comparison manner; it does not evaluate a criterion or an alternative as per the expectation of the evaluator. It simply compares one criterion over the other or one alternative over other with respect to a criterion only. Hence, step-IV of the algorithm

was carried out to evaluate the criteria as well as the sectors in terms of effectiveness as per the expectation of the evaluator. Here, each criterion/sector is evaluated by a stakeholder in terms of effectiveness and that to again in the framework of his expectation. Since assigning a quantitative value is really difficult to make, effectiveness measures in terms of linguistic values have been considered. The analysis of the data obtained from the users was subjected to the method stated in step-IV. However, the outputs do not show any encouraging result regarding the overall effectiveness as can be seen from the values depicted in Tables 4 and 5. This clearly highlights the effectiveness of the project in various sectors of governance as per the expectations of the stakeholders. An ethnographic study conducted for the purpose reveals that the various features of present e-governance system as well their advantages are not known to most of the users, as a result of which, most of them are ignorant of using the system and getting the benefit.. At the same time, users express their inability for handling the system. This is mostly due to lack of awareness among

Table 4. Criteria scores across stakeholders

Operational Level(II)	Stakeholders		
Items	Students	Teachers	Officials
• Initial cost of e-governance project compared to other development projects.	0.47	0.42	0.52
• Revenue generation due to this transformation.	0.33	0.36	0.24
• Annual mantainanance cost of this project.	0.2	0.22	0.24
• Precision of the information.	0.32	0,36	0.31
• Correctness of the information.	0.42	0.39	0.36
• Sufficiency of the information.	0.26	0.25	0.33
• Availability of the service at any time.	0.42	0.36	0.34
• Availability of the service at anywhere.	0.34	0.27	0.32
• Availability of service to anybody.	0.24	0.37	0.34
• Understanding of the working of the system.	0.23	0.19	0.27
• Easiness of using the system.	0.36	0.43	0.39
• Easiness of learning the system.	0..41	0.38	0.34
• Timely information.	0.42	0.32	0.35
• Up-to-date information.	0.39	0.37	0.32
• Reliability of information.	0.19	0.31	0.33
• Ability to communicate with Teachers.	0.34	0.37	0.36
• Ability to communicate with the students.	0.39	0.32	0.36
• Ability to communicate with the officials.	0.27	0.31	0.28

Table 5. Sectorial evaluation in the present e-governance set-up

Criteria	Sectors	Stakeholders			
		Students	Teachers	Officials	Overall
Value for Money	Academic	0.5(M)	0.54(M)	0.52(M)	0.52(M)
	Administration	0.3(L)	0.26 (VL)	0.25(M)	0.27(L)
	Financial	0.2(VL)	0.2(L)	0.23(M)	0.21(L)
Efficiency of transaction.	Academic	0.6(M)	0.58(M)	0.56(H)	0.58(M)
	Administration	0.2(L)	0.30(L)	0.25(M)	0.25(L)
	Financial	0.2(M)	0.12(L)	0.19(M)	0.17(M)

the users and also the system not being quite user-friendly. Besides, the present administration is yet to come out for a result-oriented approach involving a total process change even in the same organisational structure, which has created a negative image of e-governance due to a lack of a strong will for educating its own members for a better productivity of university system even in the presence of such a technology. Also, it was realised that the design phase of the system did not include the views of the various stakeholders for the effective operation of the system. In this context, a strong initiative should be undertaken for a change in the organisational culture to make this technology productive. At the same time, the system needs to be redesigned considering the viewpoints of the stakeholders as they constitute the major component of a public system. Therefore, apart from simply altering the platform of governance due to the availability of this technology in recent times, the implementation of e-governance in the state-owned universities needs to take care of the above discussed dimensions.

CONCLUSION

E-governance is the use of information and communication technologies to support good governance. It moves beyond old IT application in government. There are some initiatives in developing countries for such an approach for good governance. In this context, the effectiveness of e-governance projects needs to be measured for monitoring the performance of such projects. This is necessary for the improvement of the design aspect of such systems in order to make them more user-friendly as well as user-supportive. The key dimensions in this approach are the value for money and efficiency in the study of effectiveness which can be explained in terms of sub-dimensions, viz., pricing, content, availability, usability, quality and communicability. These sub-dimensions are then factored into various items along which the opinions of the stakeholders can be obtained either in a scale through pairwise comparison method or in terms of linguistic variables. However, these criteria at the conceptual level need to be operationalized through various indicators so that these can be evaluated in terms of satisfaction by various stakeholders. As the ultimate goal of effective evaluation is to improve the system of interest in the said context, the evaluation tool or method designed can be proved effective only if such a goal can be attained ideally not only at an optimal cost, but also in a participative basis. This needs a sound methodology which aims at getting right information from various stakeholders, and at the same time, depends upon a scientific methodology for analysis of such information. The present chapter provides such a methodology using few analytic techniques and has been applied to a case study in an Indian context to study the effectiveness of e-governance project through users' experience and satisfaction measure. This methodology has the scope of studying effectiveness in two approaches, viz. effectiveness in various sectors in a comparative manner and effectiveness based on the expectations of the users. The methodology can also be applicable to hierarchical situations even with multiple numbers of levels. This will definitely help in identifying the areas where emphasis has to be given to strengthen the process in order to make e-governance as the basis of governance of the present day, so that the objectives of the universities in the government sector can be realised in order to enable meritorious students to acquire higher education at an affordable price.

SCOPE FOR FURTHER RESEARCH

Ultimately e-governance must be about meeting the needs of citizens and improving the quality of service. Taking the lesson from marketing management, e-governance must be citizen-centric and service oriented. This means that, a vision of e-government implies providing greater access to information as well as better service availability. Even when e-governance projects seek to improve internal processes, the ultimate goal of the government should be to serve citizens better and in the university perspective, this means recognising diverse stakeholders such as, students, teachers, employees and investors. To know the effectiveness of such a project, appropriate methodologies are to be to be applied considering the diverse nature of its users.

Traditionally, evaluation methodologies employ hard measures viz., cost benefit analysis and benchmarks in e-government which are difficult for operationalization as most of the times quantification of the costs and the associated benefits with it cannot be made. But the significance of soft approaches is that, it employs multidimensional attribute measures of information value, which is meaningful in the context of e-governance.

However, in a human-centric approach, a good method requires multiple criteria as well as multiple stakeholders, and to evaluate these with the help of the relevant data. Depending upon the type of data whether quantitative or ordinal, one should look for a methodology which can incorporate the multi-dimensional aspect of the evaluation environment. Multiple regression analysis, fuzzy analytic hierarchical process (FAHP), TOPSIS, fuzzy TOPSIS are some of the techniques which can utilised for a multi-perspective, multi-criteria evaluation in a hierarchical framework. Besides, as confirmatory factor analysis is a relevant technique for the validation of scales in the measurement of constructs, this technique using structural equation modelling can also be used for testing the multidimensionality of the human-centric effective evaluation of e-governance projects.

REFERENCES

Al-Maskari, A., & Sanderson, M. (2010). A review of factors influencing user satisfaction in information retrieval. *Journal of the American Society for Information Science and Technology*, *61*(5), 859–868. doi:10.1002/asi.21300

Brussell, J. (2012). *Corruption and reforms in India: Public service ii the digital age*. Cambridge University Press. doi:10.1017/CBO9781139094023

Buckley, J. J. (1984). The multiple judge, multiple criteria ranking problem: A fuzzy set approach. *Fuzzy Sets and Systems*, *13*, 25–37. doi:10.1016/0165-0114(84)90024-1

Buckley, J. J. (1985a). Fuzzy hierarchical analysis. *Fuzzy Sets and Systems*, *17*, 233–247. doi:10.1016/0165-0114(85)90090-9

Buckley, J. J. (1985b). Ranking alternatives using fuzzy numbers. *Fuzzy Sets and Systems*, *15*(1), 21–31. doi:10.1016/0165-0114(85)90013-2

Chae, B., Paradice, D., Courtney, J. F., & Cagle, C. J. (2005). Incorporating an ethical perspective into problem formulation: Implications for decision support system design. *Decision Support Systems, 40*(2), 197–212. doi:10.1016/j.dss.2004.02.002

Chander, S., & Kush, A. (2012). E-governance web portals assessment of two states. *International Journal of Advanced Research in Computer Science and Software Engineering, 2*(2).

Chang, D. Y. (1992). Extent analysis and synthetic decision, Optimisation techniques and applications. *World Scientific, 1,* 352-355.

Chang, D. Y. (1996). Application of extent analysis method to fuzzy AHP. *European Journal of Operational Research, 95*(3), 649–655. doi:10.1016/0377-2217(95)00300-2

Chang, J. C., & King, W. R. (2005). Measuring the performance of the information system: A functional scorecard. *Journal of Management Information Systems, 22*(1), 85–115.

Cil, I., Alpturk, O., & Yazgan, H. R. (2005). A new collaborative system framework based on a multiple perspective approach: Inteli team. *Decision Support Systems, 39*(4), 619–641. doi:10.1016/j.dss.2004.03.007

Courtney, J. F. (2001). Decision-making and knowledge management in inquiring organisations: towards a new decision-making paradigm for DSS. *Decision Support Systems, 31,* 17–38. doi:10.1016/S0167-9236(00)00117-2

Davidson, F. (2005). *Information and organizations/use of indicators-governance syllabus.* Rotterdam, The Netherlands: IHS.

De, R. (2006). Evaluation of e-government systems: Project assessment vs. development assessment. In Wimmer, M. A. (Eds.), *EGOV 2006, LNCS 4084* (pp. 317–328). Berlin, Germany: Springer-Verlag. doi:10.1007/11823100_28

El-Kiki, T., & Lawrence, E. (2007). *Mobile user satisfaction and usage analysis model of m-government services.* Retrieved from http://www.mgovernment.org/resurces/euromgv022006/PDF/11_El-Kiki.pdf

Fitsilis, P., Anthopoulos, L., & Gerogiannis, V. C. (2010). An evaluation framework for e-government projects. In Reddick, C. (Ed.), *Citizens and e-government: Evaluating policy and management* (pp. 69–90). Hershey, PA: Information Science. doi:10.4018/978-1-61520-931-6.ch005

Griffths, J. R., Johnson, F., & Hartley, R. J. (2007). User satisfaction as a measure of system performance. *Journal of Librarianship and Information Science, 39*(3), 142–152. doi:10.1177/0961000607080417

Gupta, M. P., & Jana, D. (2003). E-government evaluation: A framework and case study. *Government Information Quarterly, 20,* 365–387. doi:10.1016/j.giq.2003.08.002

Hall, D., Guo, Y., Davis, R. A., & Cegielski, C. (2005). Extending unbounded system thinking with agent oriented modelling: Conceptualizing a multiple perspective decision-making support systems. *Decision Support Systems, 41*(1), 279–295. doi:10.1016/j.dss.2004.06.009

Hall, D. J., & Davis, R. A. (2007). Engaging multiple perspectives: A value based decision-making model. *Decision Support Systems, 43*(4), 1588–1664. doi:10.1016/j.dss.2006.03.004

Hartson, H. R., Andre, T. S., & Williges, R. C. (2001). Criteria for evaluating usability evaluation methods. *International Journal of Human-Computer Interaction, 13*(4), 373–410. doi:10.1207/S15327590IJHC1304_03

Hermana, B., & Silfianti, W. (2011). Evaluating e-government implementation by local government: Digital divide in internet-based public services in Indonesia. *International Journal of Business and Social Science, 2*(3).

Hosseini, R., & Mazinani, M. (2006). A fuzzy approach for measuring IT effectiveness of business processes. *Proceedings of the 6th WSEAS International Conference on Applied Informatics*, (pp. 1-6). Elounda, Greece.

Huang, T., Chen, K., Huang, P., & Lei, C. (2008). A generalizable methodology for quantifying user satisfaction. *IEICE Transaction Communication. E (Norwalk, Conn.), 91-B*(5), 1260–1268.

Huang, T., & Lee, C. (2010). Evaluating the impact of e-government on citizens: Cost-benefit analysis. In Reddick, C. (Ed.), *Citizens and e-government: Evaluating policy and management* (pp. 37–52). Hershey, PA: Information Science Reference. doi:10.4018/978-1-61520-931-6.ch003

ISO_9241-11. (1998). *Ergonomic requirements for the office works with visual display terminals (VDTs)-part-11: Guidance on usability* (No. ISO 9241-11:1998(E)). Geneva, Switzerland: International Organisation for Standardisation.

Janssen, D., Rotthier, S., & Snijkers, K. (2004). If you measure it they will score: An assessment of international e-government benchmarking. *Information Polity, 9*(3/4), 121–130.

Jukic, T., Bencina, J., & Vintar, M. (2012). Multi-attribute evaluation of e-government projects: Slovenian approach. *International Journal of Information Communication Technologies and Human Development, 4*(1), 82–92. doi:10.4018/jicthd.2012010106

Kahraman, C., Cebeci, U., & Raun, D. (2004). Multi-attribute comparison of catering servicing companies using fuzzy AHP: The case of Turkey. *International Journal of Production Economics, 87*(2), 171–184. doi:10.1016/S0925-5273(03)00099-9

Kline, R., & Pinch, T. (1996). Users as agents of technological change: The social construction of the automobile in the rural United States. *Technology and Culture, 37*(4), 763–795. doi:10.2307/3107097

Law, E. L.-C., & Hvannberg, E. T. (2004). Analysis of strategies for estimating and improving the effectiveness of heuristic evaluation. In *Proceedings of NordiCHI,* 23-27 October, Tampere, Finland.

Linggang, C., & Hitoshi, I. (2005). Expressway policy-set analysis from multi-perspective viewpoints: Model, algorithm and application. *Journal of the Eastern Asia Society for Transportation Studies, 6*, 4144–4159.

Mahalik, D. K. (2010). Outsourcing in e-governance: A multi-criteria decision-making approach. *Journal of Administration and Governance, 5*(1), 21–35.

Mangaraj, B. K., & Upali, A. (2008). Multi-perspective evaluation of community development programmes: A case study for a primitive tribe of Orissa. *Journal of Social and Economic Development, 10*(1), 98–126.

Manian, A., Fathi, M. R., Zarchi, M. K., & Omidian, A. (2011). Performance evaluating of IT department using a modified fuzzy TOPSIS and BSC methodology. *Journal of Management Research, 3*(2:E10), 1-20.

Mitroff, I., & Linestone, H. A. (1993). *The unbounded mind*. New-York, NY: Oxford University Press.

Morgeson, F. V. (2012). E-government performance measurement: A citizen-centric approach in theory and practice. In Chen, Y., & Chu, P. (Eds.), *Electronic governance and cross-boundary collaboration: Innovations and advancing tools* (pp. 150–165). Hershey, PA: Information Science Reference.

Nielsen, J. (1993). *Usability engineering*. San Diego, CA: Academic Press.

Petkov, D., Petkova, O., Andrew, T., & Nepal, T. (2007). Mixing multiple criteria decision making with soft system thinking techniques for decision support in complex situations. *Decision Support Systems, 43*(4), 1615–1629. doi:10.1016/j.dss.2006.03.006

Rizky, A., Medawati, H., & Hermana, B. (2012). Do information's richness of provincial government websites will support regional economies in Indonesia. In *Proceedings of 3rd International Conference on E-education, E-business, E-management and E-learning*, (IPDER, Vol. 27, pp. 200-204). Singapore: IACSIT Press.

Roy, M. C., & Bouchard, L. (1999). Developing and evaluating methods for user satisfaction measurement in practice. *Journal of Information Technology Management, 10*(3-4), 49–58.

Saaty, T. L. (1980). *The analytic hierarchy process*. New York, NY: McGraw-Hill.

Saaty, T. L. (2008). Decision making with the analytic hierarchy process. *International Journal of Services Sciences, 1*(1), 83–98. doi:10.1504/IJSSCI.2008.017590

Sahu, G. P., Panda, P., Gupta, P., Ayaluri, S., Bagga, R. K., & Prabhu, G. S. N. (2010). E-governance project assessment: Using analytical hierarchical process methodology. In Gupta, P., Bagga, R. K., & Ayaluri, S. (Eds.), *Enablers of change: Selected e-governance initiatives in India* (pp. 21–51). India: IUP.

Sedera, D., Gable, G., & Rosemann, M. (2001). A balanced scoreboard approach to enterprise system performance measurement. *Proceedings of the Twelfth Australasian Conference on Information Systems.*

Turner, V. W., & Bruner, E. M. (1986). *The anthropology of experience*. Urbana, IL: University of Illinois Press.

Wang, H., & Zheng, L. (2010). Measuring the performance of the information system using data envelopment analysis. *International Journal of Digital Context Technology and Applications, 4*(8), 92–101. doi:10.4156/jdcta.vol4.issue8.10

Wijaya, S., Dwiatmoko, A., Surendro, K., & Sastramihardja, H. S. (2012). A statistical analysis of priority factors for local e-government in a developing country: Case study of Yogyakarta Local Government, Indonesia. In IRMA (Ed.), *Digital democracy: Concepts, methodologies, tools, and applications* (pp. 559-576). Hershey, PA: Information Science.

ADDITIONAL READING

Allen, A. B., Juillet, L., Paquet, G., & Roy, J. (2001). E-government and government on-line in Canada: Partnership, people and prospects. *Government Information Quarterly, 18*, 93–104. doi:10.1016/S0740-624X(01)00063-6

Bailey, J. E., & Pearson, S. W. (1983). Development of a tool for measuring and analyzing computer user satisfaction. *Management Science, 29*(5), 530–545. doi:10.1287/mnsc.29.5.530

Bhatnagar, S. (2004). *E-government vision to implementation*. New Delhi, India: Sage.

Bhusan, N., & Ria, K. (2004). *Strategic decision-making: Applying the analytic hierarchy process*. London, UK: Springer-Verlag.

Blumenthal, A. L. (1977). *The process of cognition*. Englewood Cliffs, NJ: Prentice-Hall.

Brynjolfsson, E., & Hitt, L. M. (1998). Beyond the productivity paradox. *Communications of the ACM, 41*(8), 49–55. doi:10.1145/280324.280332

Buckley, J. J., & Chanas, S. (1989). A fast method of ranking alternatives using fuzzy numbers. *Fuzzy Sets and Systems, 30*(3), 337–338. doi:10.1016/0165-0114(89)90025-0

Charnes, A., Cooper, W. W., & Rhodes, E. (1978). Measuring the efficiency of decision making units. *European Journal of Operational Research, 2*(4), 429–444. doi:10.1016/0377-2217(78)90138-8

Deng, H. (1999). Multi-criteria analysis with fuzzy pairwise comparison. *International Journal of Approximate Reasoning, 21*, 215–231. doi:10.1016/S0888-613X(99)00025-0

Farbey, B., Land, F., & Targett, D. (1993). *How to assess your IT investment: A study of methods and practices.* Oxford, UK: Butterworth-Heinemann.

Fong, S., & Meng, H. S. (2009). Performance monitoring system for e-government services. In *Proceedings of the 3rd ACM International Conference on Theory and Practice of Electronic Governance*, Vol. 322, (pp. 74-82).

Freeman, R. J. (2009). Goals measurement and evaluation of e-gov projects. In Reddick, C. (Ed.), *Handbook of research on strategies for local e-government adoption and implementation: Comparative studies* (pp. 479–496). Hershey, PA: Information Science Reference. doi:10.4018/978-1-60566-282-4.ch025

Gatian, A. W. (1994). Is user satisfaction a valid measure of system effectiveness? *Journal of International Management, 26*(3), 119–131.

Glazer, A., Kanniainen, V., & Niskanen, E. (2002). Bequests, control rights, and cost–benefit analysis. *European Journal of Political Economy, 19*, 71–82. doi:10.1016/S0176-2680(02)00130-1

GOI. (2000). *Electronic governance: A concept paper.* New Delhi, India: Ministry of Information Technology, Government of India.

GOI. (2003). *E-readiness assessment of states in India.* New Delhi, India: Report prepared by National Council of Applied Economic Research.

GOI. (2010). *Toolkit for monitoring and evaluation for e-governance under JNNURM.* New Delhi, India: Ministry of Urban Development, Government of India.

Guba, E. G., & Lincoln, Y. S. (1981). *Effective evaluation.* San Francisco, CA: Jossey-Bass.

Hair, J., Black, W., Babin, B., & Tatham, R. (2006). *Multivariate data analysis.* New Jersey: Pearson Prentice Hall.

Heeks, R. (2001). *Understanding e-governance for development.* iGovernment, Working Paper Series, Paper No.11, Institute for Development policy & management, University of Manchester.

Heeks, R. (2006). *Implementing and managing e-government: An international text.* New Delhi, India: Vistar.

Horan, T., & Abhichandani, T. (2006). Evaluating user satisfaction in an e-government initiative: Results of structural equation modelling and focus group discussions. *Journal of Information Technology Management, 17*(4), 33–44.

Hufnagel, E. M. (1990). User satisfaction—Are we really measuring system effectiveness? *Proceedings of the Twenty-Third Annual Hawaii International Conference on System Science*, Vol. 4, (pp. 437-446).

Hwang, C. L., & Yoon, K. (1981). *Multiple attributes decision-making methods and applications.* Berlin, Germany: Springer. doi:10.1007/978-3-642-48318-9

Jackson, P. (2001). Public sector added value: Can bureaucracy deliver? *Public Administration, 79*(1), 5–28. doi:10.1111/1467-9299.00243

Kaplan, R. S., & Norton, D. P. (1996). Using the balanced scoreboard as a strategic management system. *Harvard Business Review, 74*(1), 121–129.

Kaylor, C., Deshazo, R., & Eck, V. D. (2001). Gauging e-government: A report on implementing services among American cities. *Government Information Quarterly, 18*, 293–307. doi:10.1016/S0740-624X(01)00089-2

Keller, K. L. (1999). Conceptualizing, measuring and managing customer based brand equity. *Journal of Marketing, 57*(1), 1–22. doi:10.2307/1252054

Kjaer, A. M. (2004). *Governance*. Cambridge, UK: Polity Press.

Kline, P. (1996). *An easy guide to factor analysis*. New York, NY: Routledge.

Kooiman, J. (2003). *Governing as governance*. London, UK: Sage.

Kunstelj, M., & Vintar, M. (2004). Evaluating the progress of e-government development: A critical analysis. *Information Polity*, *9*, 131–148.

Layne, K., & Lee, J. (2001). Developing fully functional e-government; A four-stage model. *Government Information Quarterly*, *18*, 122–136. doi:10.1016/S0740-624X(01)00066-1

Lin, H., Chong, Y. Y., & Salvendy, G. (1997). A proposed index of usability: A method of comparing the relative usability of different software systems. *Behaviour & Information Technology*, *16*(4), 267–278. doi:10.1080/014492997119833

Lindgaard, G., & Dudek, C. (2003). What is the evasive beast we call user satisfaction. *Interacting with Computers*, *15*, 429–452. doi:10.1016/S0953-5438(02)00063-2

Lunnen, K. M., & Ogles, B. M. (1998). A multi-perspective, multi-variable evaluation of reliable change. *Journal of Consulting and Clinical Psychology*, *66*(2), 400–410. doi:10.1037/0022-006X.66.2.400

Lunnen, K. M., Ogles, B. M., & Pappas, L. N. (2008). A multi-perspective comparison of satisfaction, symptomatic change, perceived change and end point functioning. *Professional Psychology, Research and Practice*, *39*(2), 145–152. doi:10.1037/0735-7028.39.2.145

Melone, N. P. (1990). A theoretical assessment of the user-satisfaction construct in information system research. *Management Science*, *26*(1), 76–91. doi:10.1287/mnsc.36.1.76

Michael, R., & Jens, W. (1999). Measuring the performance of ERP software—A balanced scoreboard approach. *Proceedings of the Tenth Australasian Conference on Information Systems*.

Munro, I., & Mingers, J. (2002). The use of multi-methodology in practice—Results of a survey of practitioners. *The Journal of the Operational Research Society*, *53*, 369–378. doi:10.1057/palgrave.jors.2601331

Pappu, R., Quester, P. G., & Cooksey, R. W. (2005). Consumer based brand equity: Improving the measurement-Empirical evidence. *Journal of Product and Brand Management*, *14*(3), 143–154. doi:10.1108/10610420510601012

Pfeilsticker, A. (1981). The systems approach and fuzzy set theory bridging the gap between mathematical and language oriented economist. *Fuzzy Sets and Systems*, *6*, 209–233. doi:10.1016/0165-0114(81)90001-4

Rossa, M. B., Michela, O., & Francesca, P. (2010). A multi-perspective approach to the evaluation of a multi-user system in the field of TEL. *International Journal of Technology Enhances Learning*, *2*(3), 201–214. doi:10.1504/IJTEL.2010.033577

Saaty, T. L. (1994). How to make a decision: The analytic hierarchy process. *Interfaces*, *24*(6), 19–43. doi:10.1287/inte.24.6.19

Shinjo, K., & Zhang, X. (2003). Productivity analysis of IT capital stock: The U.S.A.–Japan comparison. *Journal of the Japanese and International Economies*, *17*(1), 81–100. doi:10.1016/S0889-1583(03)00005-4

Thong, J. Y. L., & Chee-Sing, Y. (1996). Information system effectiveness: A user satisfaction approach. *Information Processing & Management*, *12*(5), 601–610. doi:10.1016/0306-4573(96)00004-0

UN. (2003). *E-governance at crossroads*. World Public Sector Report, UN.

Van Laahoven, P. J. M., & Pedrycz, W. (1983). A fuzzy extension of Saaty's priority theory. *Fuzzy Sets and Systems, 11*(3), 229–241.

Wilson, M., & Howcroft, D. (2000). The politics of IS evaluation: A social shaping perspective. *Proceedings of 21ˢᵗ ICIS*, (pp. 94-103).

Wolstenholme, E. F. (1999). Qualitative versus quantitative modeling: The evolving balance. *The Journal of the Operational Research Society, 50*, 422–428.

Xiu-Tian, Y. (2003). A multiple perspective product modelling and simulation approach to engineering design support. *Concurrent Engineering, 1193*, 221–234.

Yeh, S., & Chu, P. (2010). Evaluation of e-government services: A citizen-centric approach to citizen e-complaint services. In Reddick, C. (Ed.), *Citizens and e-government: Evaluating policy and management* (pp. 400–417). Hershey, PA: Information Science Reference. doi:10.4018/978-1-61520-931-6.ch022

Zadeh, L. A. (1965). Fuzzy sets. *Information and Control, 18*, 338–353. doi:10.1016/S0019-9958(65)90241-X

Zaied, A. N. H. (2012). An e-service success measurement framework. *International Journal of Information Technology and Computer Science, 4*, 18–25. doi:10.5815/ijitcs.2012.04.03

Zimmermann, H. J. (1996). *Fuzzy set theory and applications*. Kluwer, Academic Publisher.

KEY TERMS AND DEFINITIONS

Exploratory Factor Analysis (EFA): In many real life situations, one comes across variables such as "usability", 'leadership" etc., which can't be measured directly. Such variables, called the latent variables can be measured by a set of factors reflecting the underlying variables of interest. EFA is a statistical technique aimed at identifying these common factors based on the principle of correlation technique to determine meaningful cluster of shared variance. It begins with a large number of variables and then trying to reduce the interrelationships amongst the variables to a few numbers of clusters or factors in such a manner that variables are maximally correlated within the factors and minimally correlated with others.

Fuzzy Modelling: This represents how fuzzy logic can be used for modelling real system or processes. To demonstrate fuzzy modelling we use many functions from fuzzy logic along with standard mathematical functions. Fuzzy modelling has been proven effective in dealing with complex systems involving impreciseness that are otherwise difficult to model. Technology based on this methodology has been applied to many real world problems, be in the area of consumer products or even in decision-making. This approach tries to model a complex system as a fuzzy model which in turn uses the concept of fuzzy set theory for a variety of analysis.

Linguistic Variable: It is a variable whose values are words or sentences in a natural or artificial language. Each linguistic variable may be assigned on or more linguistic values, which are in turn connected to a numeric value through a mechanism of membership functions. These variables associate a linguistic condition with a crisp variable. They represent a range of values corresponding to a crisp variable and are central to fuzzy logic manipulations. In a variety of common applications, including product pricing, process control, market condition etc., they can be defined and used to facilitate the expression of facts.

Multiple-Criteria Decision-Making (MCDM): It is a discipline aimed at supporting decision-makers faced with making multiple and sometimes conflicting evaluations. It allows subjective as well as objective factors to be considered in the decision-making process allowing active participation of a decision-maker or a group of decision-makers for a rational foundation to make decision. MCDM aims at highlighting these conflicts and derives a way to come to a compromise in a transparent process. The criteria involved in a decision-making process must be measurable—even if the measurement is performed at the nominal or ordinal scale. Criterion outcomes provide the basis for comparison of choices and consequently facilitate the selection of the satisfactory one.

Organisational Culture: It has been conceived as the root metaphor for conceptualizing organisations and reflects the personality of an organisation. This explains all about how the organisation organises itself, its principles, rules, procedures and above all the belief, which constitute the culture of the organisation. This concept is particularly important when attempting to manage organisational change. It can be identified as an influential factor affecting the success and failure of organisational change efforts.

Performance Measurement: It is an aspect of performance management and is a fundamental building block of a total quality organisation. Traditional performance measures based on financial information provide very little to support organizations in their quality journey, because they do not consider process performance and improvements seen by the customers. However, in a successful total quality organization, performance will be measured by benefits obtained by the customers as well as the results delivered to other stakeholders. Hence, measuring performance helps in monitoring and evaluation for system effectiveness.

System Effectiveness: Measuring the extent to which targets are being met, and identifying the factors that hinder or facilitate their realization. It determines the causal effects of the system and involves trying to know if the system has achieved its intended output. This can involve using various techniques to find out the effects of the system. In a multi-perspective framework, this involves multiple perspectives as well as multiple stakeholders in the measurement process.

User Satisfaction: There has been a growing interest in user experience and its assessment. This is due to the fact that user experience is essential for many technological products and services and consists of some sense of satisfaction relating to the perceived acceptability of the system. Hence, user satisfaction is the product of the interaction between the user and the system and has received considerable attention of researchers since the 1980s as an important surrogate measure of information system success. Several models for measuring user satisfaction were developed. However, this can be done effectively in terms of a complete set of factors that affect the usability of the system.

Chapter 11
E–Voting System Usability:
Lessons for Interface Design, User Studies, and Usability Criteria

M. Maina Olembo
Technische Universität Darmstadt, Germany

Melanie Volkamer
Technische Universität Darmstadt, Germany

ABSTRACT

The authors present a literature review, carried out by searching through conference proceedings, journal articles, and other secondary sources for papers focusing on the usability of electronic voting (e-voting) systems and related aspects such as ballot design and verifiability. They include both user studies and usability reviews carried out by HCI experts and/or researchers, and analyze the literature specifically for lessons on designing e-voting system interfaces, carrying out user studies in e-voting and applying usability criteria. From these lessons learned, the authors deduce recommendations addressing the same three aspects. In addition, they identify for future research open questions that are not answered in the literature. The recommendations hold for e-voting systems in general, but this chapter especially focuses on remote e-voting systems providing cryptographic verifiability, as the authors consider these forms as most promising for the future.

INTRODUCTION

Electronic Voting (e-voting) systems continue to be used in different countries and contexts around the globe, enabling governments to obtain information on citizens' preferences more quickly and efficiently. Systems in use are both e-voting machines and remote e-voting systems. Four reasons can be stated why usability is important in e-voting systems. First, due to the election principle of universal suffrage, anyone who meets the voting age requirement[1] should be able to use these systems to cast his vote. This includes first time voters, elderly persons, and even those who

DOI: 10.4018/978-1-4666-3640-8.ch011

do not frequently interact with technology. Second, a voter should be able to easily express his wishes. The interface design should neither cause him to make mistakes nor influence his decision. Poor interface design can easily cause a voter not to cast a vote for his desired candidate. Third, voters remain novices due to a lack of training and irregular interaction with these systems, since elections are held infrequently in many countries and contexts. As such, the learning that occurs from continuous interaction with systems over a period of time is less likely to occur. Finally, if the usability of e-voting systems is not considered, frustration is likely to occur, reducing acceptance among voters, and thus decreasing voter turnout. Usability issues are especially important in e-voting systems that provide verifiability and in particular when using cryptographic verifiability.

Verifiable voting systems have been discussed since their proposal by Cohen and Fischer (1985). We mainly distinguish e-voting systems which implement voter verifiable paper audit trails (VVPAT) and those using cryptographic means for verifiability. The premise is that individual voters are able to verify that their vote is cast as they intended, and stored as they cast it. In addition, voters and any interested parties are able to verify that all votes are tallied as stored. Voters will then have to carry out certain steps to verify votes, and may encounter unfamiliar terminology, such as encryption. However, verifiable systems are only beneficial if any voter who wants to verify his vote can do so without being a specialist. The German Federal Court Decision (2009) backs this stance, requiring that the correctness of all essential steps in the election are publicly examinable without having specialist knowledge. Verifiable e-voting systems are thus even more challenging from a usability point of view; however, the future of e-voting lies in verifiable e-voting as black box systems (where one has to trust that technical and organizational processes are correct, yet no way of testing this is provided) continue to face criticism (Alvarez & Hall, 2008, p. 31).

There is a lot of literature available on usability studies of e-voting systems. Researchers and developers of future e-voting systems should take this into account, especially if they address the most complex systems, namely verifiable and cryptographically-verifiable remote e-voting systems.

In this book chapter, we review existing literature on usability of e-voting systems. Note that the literature is primarily from a western perspective, specifically from American and European contexts. The literature is available either as user studies or usability reviews carried out by HCI experts and/or researchers. We *summarize lessons learned* for e-voting system interface design, user studies, and usability criteria, from which we *extract relevant recommendations* for the same. We argue for extensions in these three areas to take into account verifiable and cryptographically-verifiable e-voting systems.

The content of this book chapter is especially relevant to designers of e-voting systems for whom the information provided will form basic input for designing future e-voting systems, in particular, verifiable e-voting systems, where user interaction for verifiability is required, and therefore understanding is critical. Researchers who are interested in replicating user studies or carrying out their own studies will obtain invaluable information, as will researchers and practitioners who are interested in the usability of e-voting systems, and the criteria used to determine usability. Further we identify future research beneficial to researchers seeking open questions in this field.

The reader is advised that accessibility issues in e-voting systems will not be discussed, and are left for future work. However the Voluntary Voting Systems Guidelines (VVSG), which we review, addresses the evaluation of accessibility aspects of e-voting systems.

In the next section we give background information for the reader to better understand the content presented. Following that, in the methodology section, we describe the approach used to identify literature included in this book chapter, and the approach by which the lessons learned and recommendations were obtained. We then present lessons learned from the literature regarding interface design, user studies, and usability criteria for e-voting systems. In each of the lessons learned sections, we derive recommendations from the literature as a take-away for the reader. In addition we present future research directions and conclude with a brief discussion on the future of usable e-voting systems.

BACKGROUND INFORMATION

In order for the reader to understand the content presented in this book chapter, we briefly introduce and categorize the *E-Voting Systems* which have been studied in the literature. We introduce terms for, and briefly describe the working of, *Verifiability*, and present the *Voluntary Voting Systems Guidelines (VVSG)* which have been recommended for use in determining the usability of e-voting systems. These guidelines have also been used in the literature surveyed. We give an overview of the *ISO 9241-11 standard (ISO, 1998)* which, in the literature surveyed, has primarily been used as criteria to evaluate the usability of e-voting systems. We will review the *Human Computer Interaction (HCI) Usability Evaluation Techniques* applied in the literature surveyed, summarize the approach used to design many of the user studies, and present terms we will use to refer to some of the experiments carried out. Finally we will touch on *Mental Models*, which determine how people perceive voting and voting aspects. This is a recent yet relevant consideration in the design of e-voting systems, especially those systems providing verifiability.

E-Voting Systems

E-voting systems can be categorized based on their use, whether in person *at the polling station* or *remotely*. We only discuss those e-voting systems that are later presented in the book chapter:

- **Mechanical Voting Machines:** Though the lever voting machine is a type of *mechanical voting machine* we include it in our discussion as its usability has been evaluated in the literature surveyed. Lever voting machines are used to tabulate votes at polling stations. They maintain no record of individual votes, only the tally of votes, which is kept by a mechanical register (Jones, 2001). The lever voting machine was commonly used in polling station elections in the USA.

- **E-Voting Machines:** Nowadays, e-voting machines, also referred to as *direct recording electronic* (DRE) voting machines, are more commonly used in polling stations. These devices are designed to electronically record votes cast by voters, and can be classified as:

  - **Classical or early DRE voting machines:** These mostly use touch buttons. An example is the NEDAP voting machine which was used in the Netherlands until 2007 (Loeber, 2008).

  - **Advanced DRE voting machines:** Provide voters with a touchscreen interface. An example is the Diebold AccuVote TS.

  - **DREs with paper audit trail (PAT):** Also referred to as a voter verifiable paper audit trail (VVPAT) (Mercuri, 2001). These provide a printout of the voter's selection. The VVPAT is a paper record generated during vote casting that can be checked at the point of casting a vote, or audited at a later stage, to verify correctness of the digital tally. With the implementation of VVPATs, it is common to have DRE machines fitted with printers for this purpose[2]. An example is the Avante Vote-Trakker, which prints a paper record of the vote, behind a plastic screen.

    ○ **DRE with cryptography:** These provide verifiability using cryptographic means. In many of these systems, a voter can leave the polling station with a paper receipt which he can later use to verify that his vote was stored correctly. An example is Bingo Voting (Bohli, Müller-Quade & Röhrich, 2007).

- **Paper-based or scan-based e-voting systems:** These are used in polling stations. Punch card voting systems use paper ballots onto which the voter punches holes at designated points to indicate the candidate he has selected. Optical scan systems use paper ballots that are then scanned with scanning devices, either at the polling station (precinct-count), or at a central tallying location (central-count) (Jones, 2003). In Demirel, Frankland and Volkamer (2011), optical scan systems are further classified into three. First, a scanning device can be employed by voters during the vote casting process, for example the Digital Voting Pen (Arzt-Mergemeier, Beiss & Steffens, 2007). In the second category, voters have to scan their paper ballots with a Precinct Optical Scan (PCOS) system after marking their candidates on the ballot. In the third category, the paper ballots are scanned by the election authority during the tallying process, using an optical scanner or a barcode scanner.

- **Remote E-Voting Systems (REV):** These include SMS voting systems (Storer, Little & Duncan, 2006), and Internet or online voting systems. Some Internet-based e-voting systems have been deployed, for example in Estonia (Madise & Martens, 2006) and in Switzerland (Gerlach & Gasser,

2009). These systems are considered black box systems as they provide no verifiability of the internal processes (non-verifiable). We can distinguish between fully-verifiable Internet voting systems, for example Helios (Adida, 2008), and partially-verifiable Internet-based e-voting systems, for example the Norwegian e-voting system (Stenerud & Bull, 2012). We briefly discuss verifiability in the next subsection.

Verifiability

Verifiability has been introduced in e-voting systems to enable voters to check that the systems do not manipulate their votes. There are three aspects of verifiability that we consider. The first is referred to as *cast-as-intended*. Here the assurance is that the vote has been sent out from the system as the voter intended it, without any unauthorized modification. The second aspect is *stored-as-cast* which provides assurance that the vote is stored in the ballot box as it was sent, and has not been manipulated. The availability of stored-as-cast verifiability is dependent on cast-as-intended verifiability being provided. These two aspects constitute *individual verifiability*, which is specifically available to an individual voter. *Tallied-as-stored*, which refers to *universal verifiability*, can be made available to any interested party to ensure that all votes counted in a given election have not been modified since the time that voters cast them. A system that provides both individual and universal verifiability is *end-to-end* or *fully-verifiable*.

Verifiability is available in e-voting systems in polling stations, for example through voter verifiable paper audit trails (VVPAT) (Mercuri, 2001), and in remote electronic voting systems through cryptographic means, for example, the Benaloh Challenge (Benaloh, 2006). According to de Jong, van Hoof and Gosselt (2008), paper audit trails may be incorporated in the voting process in various ways. Voters may be required to verify the printout (voter verified paper audit trail), or they may be enabled to do so (voter verifiable paper audit trail). The paper printouts may also have different functions in the elections: a) They are considered the final election result, therefore all printouts are tallied; b) They are used to verify the results of voting machines and results are compared from a random sample; and c) They are recounted when disputes arise.

The Benaloh Challenge provides the first step of individual verifiability. The voter makes his candidate selection, which the system encrypts and then commits to. This commitment is displayed to the voter, for example in Helios (Adida, 2008), as a hash value. A voter can now choose whether to cast a vote after making his candidate selections, or verify it (that is, challenge the system). If he chooses to verify it, the system has to provide the information used for encrypting the vote, which can be verified independently to prove that it correctly encrypted the vote. Since the system does not know which vote a voter will verify and which one he will cast, it is forced to behave correctly. The voter can then check that the same commitment is posted on a bulletin board. This provides the second step of verifiability. In the third step of verifying, interested parties can download election data and cryptographic material to then tally the votes and verify correct processing and tallying. We do not go into the detail of these steps in this book chapter, but further reading on Electronic Voting is available for the interested reader in the Additional Reading section.

Voluntary Voting Systems Guidelines (VVSG)

The Voluntary Voting Systems Guidelines (VVSG) have been adopted by the United States Election Assistance Commission (EAC) (2005) for the certification of voting systems. A revision (VVSG 1.1.) was proposed in 2009. The VVSG provide requirements against which to test voting systems for compliance regarding functionality, accessibility and security. Volume 1 of the VVSG focuses on voting system performance. Section three of this same volume focuses specifically on usability and accessibility and section nine focuses on VVPAT usability. The recommendations for consideration regarding usability are summarized below:

- The vendor should carry out usability tests on the voting system using participants that are representative of the general population.
- The voting process should provide functional capabilities, such as alert the voter of overvotes[3] or undervotes[4] on the ballot.
- The voting system should present information in any language required by law.
- The voting process should minimize cognitive difficulties for the voter, for example, by providing clear instructions.
- The voting process should minimize perceptual difficulties for the voter, for example, taking into account color-blind voters.
- The voting system should minimize interaction difficulties for the voter, for example, minimizing accidental activation.
- The voting process should maintain privacy of the voter's ballot.

ISO 9241-11 Standard

In the literature surveyed, the ISO 9241-11 standard (ISO, 1998) from the International Organization for Standardization (ISO) has commonly been used to evaluate the usability of e-voting systems. According to the standard, usability is defined as *'the extent to which a product can be used by specified users to achieve specified goals with effectiveness, efficiency and satisfaction in a specified context of use'.*

Effectiveness focuses on accuracy and completeness of user tasks. It is commonly measured using error rate. Other measures prescribed by the standard are percentage of goals achieved, percentage of users successfully completing a task and average accuracy of completed tasks. Efficiency considers resources used to achieve effectiveness in user tasks. It is commonly measured using time taken to perform a task. Other measures include mental or physical effort, materials or financial cost of performing the task.

Satisfaction focuses on user's attitudes towards the system, and is measured subjectively by asking users to report on their opinions of a given system. The standard also gives frequency of discretionary use and frequency of complaints as measures.

One approach in subjectively measuring user satisfaction is to use a standardized instrument, usually in the form of a Likert scale (Laskowski, Autry, Cugini, Killam & Yen, 2004). The instrument most commonly used in the literature reviewed has been the System Usability Scale (SUS), (Brooke, 1996). SUS is a ten-item scale, which is presented to a user immediately after they have interacted with the system under evaluation. A subjective usability (SU) score is then calculated based on the responses given. A high SUS score indicates high voter satisfaction with the e-voting system.

According to Rubin and Chisnell, (2008), the attributes of usability, in addition to efficiency, effectiveness and satisfaction, are usefulness, learnability and accessibility. The authors explain that learnability is a part of effectiveness and relates to a user's ability to operate the system to some defined level of competence after some predetermined amount and period of training, and also considers how a user performs with an interface after long periods of disuse.

Note that these three metrics for usability, namely, effectiveness, efficiency, and satisfaction, have also been recommended by the National Institute of Standards and Technology (NIST) (Laskowski et al., 2004) to assess the usability of voting systems.

Human Computer Interaction (HCI) Usability Evaluation Techniques

We discuss in this section the variety of techniques from the human computer interaction (HCI) field that have been used to evaluate the usability of e-voting systems in the literature surveyed. Readers are directed to Sharp, Rogers and Preece (2007) for more detail on HCI usability evaluation techniques. Further literature on HCI Research Techniques, Usability, and Design is also provided in the Additional Reading section.

The techniques used in the literature surveyed are:

- **Interviews:** There are *structured, unstructured, and semi-structured interviews*, which involve interviewing individuals, as well as *group interviews*, for example, focus groups (Sharp et al. 2007). In interviews, participants answer questions asked by an interviewer and the situation is similar to the interviewer and participant having a discussion about a topic of interest. The interviewer can ask for clarification where responses are unclear, or further explore comments that participants make:
  - **Focus groups:** These fall under group interviews, where a group of three to ten people are interviewed, and the discussion is moderated by a trained facilitator. *Scenarios* or *prototypes* can be used to help guide the discussion in interviews.
    - **Scenarios:** These are narrative descriptions of tasks, for example one could write a narrative describing how a voter would cast his vote using an e-voting system. Participants would then read the scenario and discuss the same with the facilitator and other participants.

- **Prototypes:** These are mock-ups, ranging from paper-based to software-based, that are designed early in the development process, to give stakeholders an idea of a design, for example, of an interface. Participants would then interact with these prototypes during the focus group.

- **Analytical evaluation:** These are carried out by experts who have a background in HCI. The methods in this category include heuristic evaluation and cognitive walkthroughs:
  - **Heuristic evaluation:** This is an inspection technique where experts use usability principles to evaluate the e-voting system. These heuristics can be obtained from literature, for example, Nielsen's Heuristics (Nielsen, 1993).
  - **Cognitive walkthrough:** Experts mimic a user's interaction with the e-voting system for a specific scenario of use. Their thoughts and observations are captured for further analysis and give insight to potential problem areas in the design of the e-voting system.

- **Personas:** These are descriptions (using fictional characters) of typical users that would use a system. These are not real users, but are designed to represent profiles of real users of systems. Designers then focus on these users and design the e-voting system to fit their needs.

- **Lab Studies:** Participants interact with the e-voting system to cast a vote, and the time taken to do so, and any errors made are recorded. The lab may be set up to represent the environment in a real polling station.

- **Field study:** Here participants interact with the e-voting system in a 'normal' uncontrolled environment, as opposed to the controlled environment of a lab setting.

Techniques for data collection in user studies, as seen in the literature surveyed, are:

- **Pre- and post-voting questionnaires:** The pre-voting questionnaire frequently collects demographic data or data that can be used to include or exclude participants from the study. Post-voting questionnaires collect participants' opinions after they have interacted with the system.

- **Exit poll:** Participants are asked a series of questions at the end of a voting session. In many cases, this is after they have cast a vote in a real election.

- **Direct Observation:** A participant may be observed as they interact with the e-voting system. In this type of observation, the participant is aware of the observation, for example, since the researcher is present in the lab, or recording tools are utilized:
  - **Video recording:** A camera can be set up to record the interaction.
  - **Audio recording:** Audio capture tools can be used to capture the participants' thoughts if they are asked to think aloud.
  - **Collecting eye-tracking data:** This data captures the points on the screen that the participant focuses on, which can then inform placement of interface elements for greatest impact.

- **Indirect Observation:** In this type of observation, data can be collected unobtrusively, without the participant being aware of it, unless they are informed:
  - **Log file analysis:** Researchers use software to collect information about participants' interaction with the e-voting system. This data is stored in a log file that is later examined.

The literature surveyed shows that a common approach in the design of user studies for e-voting has been to split participants into several groups. One group of participants receives a slate instructing them which candidates to vote for. This is the '*directed*' group. A second group of participants receives a voter guide with relevant information on the candidates to help guide their selection. Participants in this '*voter guide*' group are then free to vote for candidates of their choice. A third group is '*undirected*', receiving neither a slate nor a guide, and participants are free to vote as they choose.

In some voting experiments, a group of participants are asked to skip some races in order to simulate *roll-off*[5]. In other cases, participants in the undirected group are given an exit interview for researchers to determine voting intent. This is then compared to votes cast to obtain the error rate. Finally, the order of participants' interaction with the systems is usually randomized (using a Latin Square) in order not to introduce bias.

Mental Models

Mental models describe people's perceptions of objects, determining how people then interact with the objects (Staggers & Norgio, 1993). Mental models are an important consideration in design, as designers are then able to use knowledge of the user's mental model in designing an object, thus lessening frustration, and greatly improving satisfaction. It is only recently (Campbell & Byrne, 2009b) that researchers have considered voters' mental models in e-voting systems.

METHODOLOGY

In this section, we discuss and justify how we selected the papers reviewed, how the lessons learned are extracted, and how the recommendations and future work are identified.

How the literature was identified: The literature reviewed in this book chapter was identified by searching scientific literature repositories, such as www.scholar.google.com, and the digital libraries of the Institute of Electrical and Electron-

ics Engineers (IEEE), and the Association for Computing Machinery (ACM). Search criteria included 'usability and e-voting', 'usability and remote e-voting' and 'ballot design'. The search was not limited to a specific time period. From the results of the electronic repository search, we reviewed conference papers and journal articles. Further papers and reports were obtained from the references of the identified papers. The papers reviewed were published between the years 1998 and 2012.

The conference papers were proceedings of HCI, usability, e-voting, security, democracy, and governance conferences and workshops. We reviewed these to identify papers that focused on user studies and usability of e-voting systems. These conferences are the ACM SIGCHI Conference on Human Factors in Computing Systems (CHI), the International Conference on Human Computer Interaction (HCII), the IADIS International Conference Interfaces and Human Computer Interaction, the Symposium on Usable Privacy and Security (SOUPS), the USENIX Security Symposium, the International Conference on E-Voting and Identity (VOTE-ID), the Electronic Voting Technology Workshop/Workshop on Trustworthy Elections (EVT/WOTE), the International Workshop on Requirements Engineering for Electronic Voting Systems (REVOTE), the International Conference on Electronic Governance (EGOV), the International Conference for E-Democracy and Open Government (CEDEM), the International Conference on Theory and Practice of Electronic Governance (ICE-GOV), the Workshop on Socio-Technical Aspects in Security and Trust (STAST), and the International Conference on Information Integration and Web-based Applications and Services (iiWAS). We included a paper (Everett, Byrne, & Greene, 2006) that was published in the proceedings of the Annual Meeting of the Human Factors and Ergonomics Society (HFES). References from the identified papers were reviewed and further relevant papers identified and selected for inclusion.

Relevant journals were from the fields of usability, user studies, computing, electronic governance, information design, and political science. Here we reviewed articles from the Journal of Usability Studies, International Journal of Human-Computer Studies, IEEE Transactions on Information Forensics and Security, International Journal Universal Access in the Information Society, the European Journal of Information Systems, the Social Science Computer Review, the International Journal of Electronic Governance, Information Design Journal, American Politics Research, the Journal for the American Political Science Association, Perspectives on Politics, and Public Opinion Quarterly, to identify relevant papers. Once again, the references from the selected papers were reviewed to identify other relevant articles.

How the lessons learned were obtained: The literature was reviewed for findings that could guide researchers on (a) designing e-voting interfaces, (b) conducting user studies, and (c) determining usability criteria used in e-voting usability studies.

How the recommendations were made: Recommendations are given as a take-away pointing to what researchers should do, in light of the research findings. For example, where studies found that first time and elderly voters need help while voting, we recommend integrating help features in e-voting system interfaces. Recommendations will be numbered R-ID-(subsection)-(number) in the Interface Design section, R-US-(subsection)-(number) in the user studies section, and R-UC-(subsection)-(number) in the Usability Criteria section. For example, recommendation R-ID-BD-1 is the first recommendation focusing on Ballot Design, contained in the Interface Design section.

How the future work is deduced: Where we identified open questions in the literature surveyed, we recommend these as future work. Future work recommendations are identified as FW-ID-(subsection)-(number), FW-US-(subsection)-(number), and FW-UC-(subsection)-(number) for the three sections. For example, FW-ID-BD-1 is

first recommendation for future work on ballot design, in the Interface Design section.

LESSONS LEARNED AND RECOMMENDATIONS FOR INTERFACE DESIGN

In this section, we summarize lessons learned from the literature regarding design of e-voting interfaces. Additionally we provide recommendations as a take-away for e-voting interface designers, and researchers looking for future research and open questions in this field. We first address issues related to the ballot and vote casting, namely *Standardized Ballot/Interface Design, Simple and Clear Ballot Instructions, and Providing of Review/Confirmation Screens. Considering Time, Speed, Voting Tasks and Effort, Providing Help Features*, and *Educating Voters and Poll Workers* are the next subsections, and relate to supporting voters and poll workers. *Identifying Mental Models* is the next subsection and here the lessons learned are applicable for providing voters with help during voting and educating voters and poll workers, in addition to systems providing Voter Verifiable Paper Audit Trails (VVPAT) and cryptographic verifiability. Further individual lessons learned and recommendations for interface design of these two types of e-voting systems are discussed in the next two subsections, that is, *Use of the Voter Verifiable Paper Audit Trail (VVPAT),* and *Understanding in Cryptographically-Verifiable Voting.* We concentrate on these e-voting systems separately as they are especially challenging from a usability point of view. Note that in this section we refer to voters, as opposed to participants.

Standardized Ballot/Interface Design

In the literature surveyed, we find attention is given to screen guides for voters. For example, the e-voting machine in Roth (1998) used flashing lights to indicate to voters races that were not yet completed. The importance of appropriate ballot format is seen in Kimball and Kropf (2005) where they indicate that poor ballot format can result in unrecorded votes. In addition, ballot interfaces that are not standardized can be unfamiliar to voters and cause confusion (Niemi & Herrnson, 2003). Participants in the studies by Greene, Byrne and Everett (2006), Everett, Byrne and Greene (2006), and Byrne, Greene and Everett (2007) were most satisfied with the bubble ballot. The bubble ballot design is similar to a multiple-choice exam question where voters are presented with candidate information in a listing. Next to each candidate name, is an oval or square box on which the voter can mark their selection with an X or by shading the box. Considering large candidate listings, MacNamara, Gibson and Oakley (2012) adapted the Prêt à Voter backend to accompany the Dual Vote frontend, and acknowledge that the paper ballot used was not suitable for a large candidate list. Intuitively, one might think that a full-face ballot, that is, one that displays all offices and candidates on a single screen is a solution. However, Norden et al. (2006) found a high rate of residual votes[6] with full-face DREs. Selker, Hockenberry, Goler and Sullivan (2005) describe a Low Error Voting Interface (LEVI) designed after several iterations, with the aim of reducing the number of errors by voters. This was achieved by providing voters with information on which races they had voted in, which races were pending, and if they had made a selection on a screen.

Recommendations

- **R-ID-BD-1:** Ballot design should be standardized such that the ballots are familiar to voters, for example, imitating the paper ballot design on the e-voting system interface.
- **R-ID-BD-2:** The ballot should indicate to voters when their vote has been successfully cast and should clearly indicate if the vote casting process has been completed.

- **R-ID-BD-3:** The interface should inform voters if their vote is invalid based on the selections they have made.
- **R-ID-BD-4:** Researchers should use the bubble ballot as a standard design where the ballots and candidate listing supports it.

Future Work

- **FW-ID-BD-1:** Future research should investigate the design of large ballots on e-voting systems and their usability.

Simple and Clear Ballot Instructions

Ballot instructions guide voters on how they can successfully mark their desired candidate(s) on the ballot presented. These can be placed at the top of the ballot, or the instructions can precede each race that voters have to fill out, if the ballot contains more than one race. Roth (1998) calls for the use of simple and clear instructions in the ballot to avoid voter confusion. The literature surveyed points to the use of the Flesch-Kincaid readability test (Kincaid, Fishburne, Rogers & Chissom, 1975), to test the comprehension level of instructions in the ballot. The location of the instructions on the ballot has been considered as in Western culture, for example, voters usually look at the upper left-hand corner of the ballot (Kimball & Kropf, 2005). Instructions should appear before the task as voters may not read the instructions. Laskwoski and Redish (2006) recommend four best practices for ballot instruction design. First the consequences of an action, for example, `*You will not be able to make further changes*' should be mentioned before the action to be carried out, for example, `*Click to Confirm*'. This is because voters are likely to carry out the action if it appears first, and then read of the consequences later. Second, at the sentence level, voters should be given the context of use before the action, for example: `*To vote for your desired candidate, mark an x in the box next to the candidate name'*. In this example, the sentence is in two parts, the first is the context, the second is the action to be carried out. Third, is to use words that are familiar to voters, and the fourth best practice is to place instructions in logical order, following the structure of the ballot the voter will encounter, for example, `*To vote for your desired candidate, mark an x in the box next to the candidate name. To proceed to the next race, click the Next button'*.

Recommendations

- **R-ID-BI-1:** Simple and clear instructions should be designed for ballots.
- **R-ID-BI-2:** The consequences of an action in an instruction should precede the call to act.
- **R-ID-BI-3:** Each instruction sentence on the ballot should give the result of the action before stating the action to be carried out.
- **R-ID-BI-4:** The instructions should use words that are familiar to voters, and additionally should match the order of tasks on the ballot, in a logical sequence.
- **R-ID-BI-5:** Instructions should be placed at the upper left-hand corner of the ballot, and should be given before the task to be carried out, for instance, having instructions how to mark a candidate correctly, just above where this action will be carried out.

Providing Review/ Confirmation Screens

In the literature surveyed there are calls for the implementation of review screens in e-voting systems. A review screen presents the voter with all the candidate selections they have made on the ballot, and supports their re-checking the correctness of the selections, before casting the ballot. Herrnson et al. (2006) recommend that

voters should be given an option to review their vote(s) and correct the ballot when errors are noted. Norden, Creelan, Kimball and Quesenbery (2006) found that lower residual vote rates were observed when the voter was given the chance to correct errors before casting his ballot. Voters are also seen to expect this, as MacNamara et al. (2010) report that a system that did not implement a confirmation graphical user interface received a low voter satisfaction rating.

The question arises whether the review screen would help voters notice changes made to their votes. Everett (2007) found only 32% of participants noticed if races were added to or missing from the review screen, and only 37% of participants noticed vote flipping on their review screen. Campbell and Byrne (2009a) did a follow up on these findings, since participants in Everett's study were not explicitly instructed to check the review screen, and the interface may not have simplified the process of checking. They investigated if the rate of anomaly detection could be improved by instructing participants to check the review screen, and re-designing the review screen such that undervotes were highlighted. Participants were later directly asked if they noted any anomalies on the review screen, and 50% reported noticing anomalies. As undervotes were highlighted on the review screen, more participants noticed them (61%) compared to 39% who noticed their votes were changed to another candidate.

Recommendations

- **R-ID-RS-1:** Review screens should be implemented in e-voting systems.
- **R-ID-RS-2:** Voters should be instructed to pay attention to the review screen, in order to ensure that they notice any changes made to their votes.
- **R-ID-RS-3:** Techniques such as additional coloring or highlighting should be used to draw voters' attention to races where they have not yet voted.

Future Work

- **FW-ID-RS-1:** Future research should investigate how to instruct voters to pay more attention to the review screen especially for them to notice when their candidate selections are modified.

Considering Time, Speed, Voting Tasks, and Effort

Note that we consider voting tasks to be the activities that voters would carry out while casting their vote, for example, changing their vote and writing in a candidate's name for a write-in vote. Herrnson et al. (2008) found that more actions meant more time to complete the voting process and lowered voter satisfaction. Kiosk-voting participants in Oostveen and Van den Besselaar (2009) commented that they found it faster to mark a cross (X) next to their candidate's name, fold the ballot or put it in an envelope and then in to a ballot box, compared to inserting a smart card, typing in a PIN code, scrolling through several computer screens before making a selection and confirming their vote.

Herrnson et al. (2006) and Herrnson et al. (2007) found that participants' accuracy declined when they were asked to carry out less straight forward tasks, such as voting for two candidates in one office. Everett et al. (2008) showed that increasing voting speed for voters may decrease accuracy. They also noted a 'fleeing voter' error, where some participants left the DRE before completing a final step. Greene (2008) compared direct access and sequential access interfaces for DREs and found that voters were dissatisfied with the direct access voting interface, which was faster, but also had more undervote errors and premature ballot casting. Conrad et al. (2009) found that the more effort users needed to put into casting their vote, the less satisfied they were. Lower satisfaction also resulted from voting incorrectly and an increased effort to vote.

Recommendations

- **R-ID-TS-1:** Researchers should aim to reduce the amount of time and effort that voters need to take in order to cast their vote, that is, the number of voting tasks should be reduced, in particular regarding poll workers and voters enabling e-voting machines or in checking the voter's right to vote in Internet voting.
- **R-ID-TS-2:** Care should be taken in facilitating voters to vote quickly, as in direct access DREs, since faster voting may lead to more voter errors.

Future Work

- **FW-ID-TS-1:** Further research should investigate how much time voters would be willing to spend in carrying out voting tasks, and how much time they would consider to be too much.

Providing Help Features

Both Herrnson et al. (2006) and Herrnson et al. (2008) found that although voters can cast their votes unassisted, not all of them are able to do so, and help facilities need to be provided. Prosser, Schiessl and Fleischhacker (2007) found in their study that voters hardly took notice of the help information provided, and recommend providing help 'just-in-time' when it is needed.

Recommendations

- **R-ID-HF-1:** It is recommended that designers integrate help facilities to give voters information when they need it and to guide them on what next steps are required. In Internet voting, for example, the help facilities should be placed on every page the voter will access on the e-voting system, as well as next to tasks that are likely to be confusing for voters.

Future Work

- **FW-ID-HF-1:** Future research should investigate appropriate help for voters given the different voting contexts, for example hotlines and email in Internet voting.

Educating Voters and Poll Workers

Participants in the study by Herrnson et al. (2005b) responded more favorably to the voting machines, and the authors credit this to an education campaign carried out to familiarize voters with the machines. Kalchgruber and Weippl (2009) report on a short case study where two groups of students, the first group with an IT security background, and the second group with basic IT security knowledge, were given explanations of how the Scratch and Vote system (Adida & Rivest, 2006) worked. They found that a lot of education is necessary for voters to understand the difference between non-verifiable, partially-verifiable and end-to-end verifiable systems. Three studies in the literature surveyed focused on the poll worker (Chisnell, Becker, Laskowski & Lowry, 2009a; Chisnell, Bachmann, Laskowski & Lowry, 2009b; Goggin, 2008), but none to our knowledge concentrates on poll worker education.

Recommendations

- **R-ID-ED-1:** Voters should be educated before introducing new e-voting technology.
- **R-ID-ED-2:** The techniques used for educating voters, for example, videos or handouts, should take into account the diversity of voters, in terms of age, experience with voting, and education.

Future Work

- **FW-ID-ED-1:** Future research should investigate effective means of voter education, particularly to introduce voters to, and

enable them to use, new e-voting systems and new approaches to voting, such as verifiable e-voting systems.
- **FW-ID-ED-2:** As poll workers also require education for new e-voting systems and approaches, and as this has not yet been addressed comprehensively in the literature, future research should investigate poll worker education.

Identifying Mental Models

Campbell and Byrne (2009b) carried out an online web-based survey in order to identify voters' mental model of straight-party voting (SPV)[7]. The findings show that voters are confused by SPV, which can be due to a gap between how voters think SPV should work and how it actually does work. Karayumak, Kauer, Olembo, Volk, and Volkamer (2011b) found that voters were confused why they needed to verify their vote, and were concerned about compromising secrecy of the vote.

Recommendations

- **R-ID-MM-1:** Voters' mental model should be investigated as voter confusion may be due to differences between how voters think an e-voting system should work, and how it has been designed and implemented.
- **R-ID-MM-2:** Voters should be educated about verifying their vote to deal with the problem of confusion and secrecy concerns.

Future Work

- **FW-ID-MM-1:** Future research should investigate specific approaches to extending voters' mental model to take into account new e-voting systems and voting approaches, as the literature surveyed does not give information on how to close the gap identified.

Use of the Voter Verifiable Paper Audit Trail (VVPAT)

In the study by MacNamara et al. (2011a), 89% of participants thought a paper audit trail was important in e-voting systems. The following literature, however, shows that users have challenges with the VVPAT. Participants in the study by Selker et al. (2005) reported no errors after voting, although errors actually existed. Selker, Rosenzweig and Pandolfo (2006) also report on a study where it was observed that voters, when asked to pay close attention to the VVPAT, did not notice any errors, yet errors did exist. In de Jong et al. (2008), the paper audit trail was observed not to affect the voters' experiences with the voting machine. Herrnson et al. (2006) observed that some voters ignored the VVPAT, and those who did not, seemed to get confused. They also found that voters had less confidence that their vote was accurately recorded on the systems with a VVPAT. In another field study, although participants were instructed to pay attention to the verification systems, it was noted that they did not spend as much time as needed to verify every selection (Center for American Politics and Citizenship [CAPC], 2006).

Recommendations

- **R-ID-VV-1:** Voters should be provided with clear instructions that are easy to understand and follow, guiding them on how to check their vote selection when VVPATs are provided.

Future Work

- **FW-ID-VV-1:** Further research is necessary to identify the best approach for instructing voters, as well as to further investigate voter challenges with VVPATS.
- **FW-ID-VV-2:** Voters' mental model regarding VVPATS should be investigated in future research.

Understanding in Cryptographically-Verifiable Voting

Bär, Henrich, Müller-Quade, Röhrich, and Stüber (2008) discuss the use of Bingo voting (Bohli et al., 2007) in a student election at Karlsruhe University. They report that some voters did not trust the random number generator as they were not familiar with its internal operation, and that a majority of voters did not bother to check any random number in order to verify that their vote was properly recorded. Using a focus group approach, Schneider et al. (2011) assessed the usability of Prêt à Voter version 1.0 (Bismark et al., 2009). Although participants were able to cast their votes, audit them, and check for inclusion on the bulletin board, comments showed that they did not understand some of the design decisions, and in verifying their ballot, expected to see the name of the candidate they had voted for. Karayumak et al. (2011b) identified that although voters were able to verify their vote when instructed to do so using modified interfaces of the Helios voting system (Adida, 2008), they lacked an understanding of the need for verifiability. Carback et al. (2010) observed from comments by voters who had voted with Scantegrity II in a municipal election that voters did not understand about verifying their votes online.

Recommendations

- **R-ID-CV-1:** Voters should be provided with clear instructions that are easy to understand and follow to verify whether their vote is properly encrypted (cast as intended) and stored (stored as cast).
- **R-ID-CV-2:** Appropriate help features should be integrated for cryptographically-verifiable e-voting, taking into account the different types of voters, ranging from first-time voters, to frequent voters.
- **R-ID-CV-3:** Voter education is recommended in cryptographically-verifiable

e-voting as voters may not understand the need for verifiability, and instead expect a secure system that does not require one to carry out verifiability steps.

Future Work

- **FW-ID-CV-1:** It is recommended that voters' mental model and voter education be investigated in future research, with the aim of eliminating confusion and the reluctance to carry out the necessary verifiability steps.
- **FW-ID-CV-2:** Further research should investigate voters' view of the amount of time it takes to verify votes in the case of cryptographically-verifiable e-voting systems, whether they find it too long or acceptable.

LESSONS LEARNED AND RECOMMENDATIONS FOR USER STUDIES

In this section, we summarize lessons learned from the literature surveyed regarding conducting user studies to evaluate the usability of e-voting systems. As such, we use the term participants, rather than voters. The discussion here focuses on *Relevant Methodology*, *Considering Ecological Validity*, and *Maintaining Vote Secrecy*. We notice interaction between these three areas as they relate to the approach used in e-voting studies. *Incentives for Participants, Number of Participants in E-Voting User Studies, Considering Voter Information Processing Capabilities when Designing User Tasks for Studies,* and *Considering Ethical Issues* focus specifically on participants. We wrap up with *Errors due to Technology used in User Studies.* We make recommendations that are relevant for researchers interested in carrying out user studies, and researchers interested in open questions for further research. Contradictions and open research questions will be recommended for further research.

Relevant Methodology

Some researchers in the literature surveyed apply several HCI methodologies on the same system, beginning with an evaluation by experts, then moving on to user studies with actual users. Bederson, Lee, Sherman, Herrnson and Niemi (2003) used an expert review, close-up observation and a field study. Herrnson et al. (2006) used an expert review by computer-human interaction experts, a laboratory experiment, a large field study and natural experiments in two US states. Traugott et al. (2005) describe natural experiments as observing changes in the real world and investigating their consequences. Karayumak, Kauer, Olembo and Volkamer (2011a) started with a cognitive walkthrough on Helios version 3.0 and the improved interfaces were tested in a user study by Karayumak et al. (2011b).

A high number of interface and interaction issues are likely to be identified by experts and these can be dealt with before users are involved. System reviews with experts can take a shorter amount of time, and identify a high number of critical errors, compared to testing with users, a process which may take longer and cost more. Furthermore, an all-rounded view of the systems is obtained using different research methodologies. Pilot studies are critical in identifying issues before users interact with the systems (Sharp et al., 2007).

Recommendations

- **R-US-RM-1:** We recommend the following methodology: A usability study should begin with an evaluation involving experts, after which changes can be made to the e-voting aspect under study based on the feedback received. As a second step, a user study should be carried out. Note that before the user study is carried out, pilot studies are necessary to identify any difficult or unclear issues in the study design. If a lack of understanding of the e-voting system aspect under study is observed, the voters' mental model should be investigated. Based on feedback from participants after the user study, the e-voting should be re-designed. The re-design should be tested in subsequent user studies, and several iterations at this stage may be necessary, switching between re-design and small user studies for user feedback. Finally, field studies should be carried out, where the re-designed e-voting system can be tested in a real election with real voters. Exit polls should accompany the field studies, to obtain voters' feedback on the e-voting system, and related aspects being studied.

Considering Ecological Validity

Ecological validity considers the extent to which the results of an experiment can be applied to real-life conditions (Patrick, 2009). As a result, researchers want their studies to be as realistic as possible. Schneider et al. (2011) designed a ballot similar to that of UK elections in order to give participants a familiar voting experience. The study in Byrne et al. (2007) found that the frequency of errors was not affected by the use of fictional candidates.

Cross II et al. (2007) set up two voting booths in the lobby of a student union building, in order to provide a noisy environment similar to that available during actual elections. In Fuglerud and Røssvoll (2011) participants could decide in which location to conduct the test and were encouraged to use their own PC and equipment. Weber and Hengartner (2009) had each participant used their own laptop and email address.

Field studies, or exit polls, can also be carried out to obtain feedback from voters after they have voted in a real election. Carback et al. (2010), report on an exit poll carried out after voters voted in a real election using Scantegrity II at Takoma Park. Herrnson et al. (2005) carried out an exit poll out after participants voted in an actual election. Van Hoof, Gosselt & de Jong (2007), and de Jong, van Hoof and Gosselt (2008) also recruited participants immediately after they cast a vote in a real, national election. If this is done, however, there may be legal requirements to satisfy, for example, van Hoof et al. (2007) had to have election questions that were unrelated to the real election questions. A second challenge is that participants may cast a vote for the same candidate they voted for in the real election, or try to avoid revealing their true vote, in the case where real ballots are used, by randomly selecting candidates. This then makes error detection difficult as participants cannot recall their vote selection (Selker et al., 2006).

The tasks that participants carry out can be structured as real tasks similar to those in real elections. Herrnson et al. (2008) gave participants tasks to carry out, for example change a vote, cast a write-in vote or omit a vote for one office, but also asked them to make selections on their own in response to information about specific candidates to keep them engaged in the process. How participants interact with the voting processes can also resemble real scenarios. In Chisnell et al. (2009b), participants worked in pairs, to mimic a realistic situation for poll workers.

Rather than setting up a mock election, we find attempts made to give a real election whose results participants can be interested in. De Jong et al. (2008) had participants vote for a charity organization to receive a 1,000 Euro donation. This gave the participants a clear purpose and immediate implications to their voting. Winckler et al. (2009) also asked participants to vote for a charity organization. In MacNamara et al. (2012) participants voted for their favorite country from four available options. Participants in the study by Cross II et al. (2007) voted on the best burger and fries from options given.

Recommendations

- **R-US-EV-1:** Ballots used in user studies should be similar to those used in real elections to minimize confusion and the number of errors from participants.
- **R-US-EV-2:** Researchers should select one of the following approaches to provide ecological validity: a) The ballot used in the study should be similar to a real ballot, for example based either on the candidates listed, the design of the ballot, or the number of races provided. b) The voting environment should be similar to that of a real election, either by the voting machines used, or holding the election in a location where real elections are held, for example in a town hall. c) Giving voters tasks similar to tasks in a real election, for example, picking a ballot paper, marking their ballot in the voting booth, and dropping their ballot in a ballot box. Though this example is for polling station and paper-based elections, the same can be used for Internet voting where voters are made aware that they are in the voting booth (marking candidates) and dropping their ballot in the ballot box (submitting the vote). d) Running an election for which participants are more likely to be interested in the results, for example a charities' election.
- **R-US-EV-3:** This recommendation includes suggestions that are optional for the researcher. Where mock elections are set up for user studies, they should give a realistic feel to the participants, for example, in the design of the ballot, or in the location of the study. Fictitious candidates can be included in ballots for user studies. As argued in the literature surveyed, this increases the life span of the research instruments used, and they do not become obsolete in a short period of time. User studies can also be set up in the participants' natural environment, or in the case of Internet voting, should use the participant's equipment where possible, in order to be realistic.

Maintaining Vote Secrecy

A trade-off between vote secrecy and identifying participants' voting errors in user studies is seen, for example in van Hoof et al. (2007), where participant numbers were marked on the paper ballot to check if participants voted correctly. A camera recorded the voter interaction with the voting machine, and researchers could see the actual voter input. Some participants seemed to vote for the same candidate they voted for in the real election as participants, though instructed which candidates to vote for, instead cast votes for other candidates. In Conrad et al. (2009), the participant first indicated to an experimenter which candidates he intended to vote for. Some researchers seem to be aware of this challenge, for example, MacNamara et al. (2010) report that they did not identify voter intention in their study in order to preserve secrecy of the ballot.

Recommendations

- **R-US-VS-1:** Where possible, researchers should preserve vote secrecy, or inform participants when it will not be preserved. As an example, researchers can direct voters which candidates they should vote for, but should not link the vote to the participants (for example, by randomly giving voter slates in the study). Researchers can then compare expected votes with actual votes to identify error the rate.

Incentives for Participants

In user studies, participants are often given financial or in-kind incentives to motivate them and improve their engagement in the study. However, this might not apply in e-voting studies. Goggin (2008) attempted to mimic voter motivation as in a real election, and offered participants a $5 USD bonus for voting accurately, that is, voting for the same candidates in all ballots in the study. The bonus was found not to increase voter motivation for accuracy.

Recommendations

- **R-US-IP-1:** Researchers should offer financial or in-kind incentives to participants in user studies.

Future Work

- **FW-US-IP-1:** Future research should investigate if incentives will improve participants' engagement in e-voting studies.
- **FW-US-IP-2:** A second question should explore the role of intrinsic motivation in engaging participants in e-voting user studies, for example, by motivating them to participate in studies in order to be part of improving the usability of e-voting systems.

Number of Participants in E-Voting User Studies

The user studies reviewed in the literature surveyed have recruited varying numbers of participants; the smallest we identified was 7, and the largest 1,540. There are guidelines available on the number of users necessary for results of a study to have statistical significance (Cook and Campbell, 1979). With usability studies, the number of participants is determined by the resources available for the study, including time and finances, as well as the study design. Lazar, Feng and Hochheiser (2010) recommend 15 – 20 participants pointing out that smaller studies may miss out on some useful results.

Recommendations

- **R-US-NP-1:** It is recommended that researchers determine the number of participants for their e-voting studies based on the resources available, the study design, and whether statistically significant results are required. For statistically significant results, statistical techniques can be used, based on the desired degree of confidence, to identify the number of participants from the sample population. If statistically significant results are not required, studies should have, as a minimum, 15 – 20 participants, depending on the goals of the study.
- **R-US-NP-2:** Field studies should have a large number of participants, as the results will be representative of a larger population.

Considering Voter Information Processing Capabilities when Designing User Tasks for Studies

In the study by van Hoof et al. (2007), participants were instructed which candidate to vote for, and were expected to memorize this and vote for the same candidate twice, using a voting machine and a paper ballot. The paper reports that participants voted for the wrong candidate as they could not remember the voting task. Research in psychology points to human limits in processing information (Miller, 1956), with proposals being made for grouping information that is presented to aid recall.

Recommendations

- **R-US-VIP-1:** Researchers should not require participants to remember voting tasks, and instead should provide both written and verbal instructions on what tasks participants are expected to carry out.

Considering Ethical Issues

As user studies involve human participants, ethical issues have to be taken into account (Lazar et al., 2010). In the literature surveyed, we see that researchers may be required to obtain approval from a university institutional review board (IRB) (Carback et al., 2010). In some studies, we see participants being informed about the goals and objectives of the study and signing consent forms before taking part in the study (Everett et al. (2006); Greene et al. (2006); Fuglerud & Røssvoll (2011)). Informing participants fully of the objectives of a study however presents a challenge since participants may change their behavior once they are made aware of the goal of the study, or try to act in a manner they think appropriate, for example in a study on verifiable e-voting systems, they may verify their vote in the study, but they may not do so in a real election.

Recommendations

- **R-US-EI-1:** Usability study design as well as tasks for participants should be reviewed to ensure that they do not violate ethical requirements.
- **R-US-EI-2:** Participants should be informed about the goals of the study either before or after the study.
- **R-US-EI-3:** Participants should sign consent forms before participating in e-voting usability studies.
- **R-US-EI-4:** Where a university ethics board or institutional review board is available, these should check that the study design and materials consider ethical issues. Where these bodies are not available, or cannot offer necessary guidance, researchers should take the responsibility of sepa-

rately reporting how they have met standard ethical requirements (see Burmeister, 2000).

Future Work

- **FW-US-EI-1:** Future research in usability of e-voting systems should investigate how to handle ethical issues, besides requiring full disclosure to participants as this may not match with e-voting study objectives.

Errors Due to Technology Used in User Studies

In MacNamara et al. (2010) the e-voting system being tested was found to have misclassified 38 out of 332 votes (an error rate of 11.4%), majority of which were caused by a faulty hybrid pen. When these misclassifications were removed, the error rate was much lower, standing at 3.9%.

Recommendations

- **R-US-ET-1:** Researchers should use equipment whose development has been completed and tested, in order to avoid errors arising in actual user studies.

LESSONS LEARNED AND RECOMMENDATIONS FOR USABILITY CRITERIA

In this section we discuss lessons learned from the literature regarding *Metrics for Usability Evaluation*, which describes the measures used in the literature surveyed to measure usability. In the next subsection, *Usability and Design Guidelines*, we focus on guidelines identified in the literature surveyed. We close by discussing *The Need for Baseline Data* in evaluating usability. Researchers seeking to identify usability criteria for their research will find useful information. We also indicate open research questions for future research.

Metrics for Usability Evaluation

In the literature surveyed, we observe that a large number of studies used the three metrics for usability from the International Organization for Standardization (ISO) namely, effectiveness, efficiency and satisfaction. (Goggin, 2007; Goggin, 2008; Campbell & Byrne, 2009a; Everett et al., 2006; Greene et al., 2006; Byrne et al., 2007; Everett, 2007). These three metrics have also been recommended by the National Institute of Standards and Technology (NIST) as appropriate to evaluate the usability of e-voting systems (Laskowski et al., 2004). In some studies, the criteria used were not identified as those recommended by ISO, yet the data collected concerned voter errors, ballot completion time, and voter satisfaction (MacNamara et al., 2011b; Conrad et al., 2009; MacNamara et al., 2010).

Everett et al. (2008) used errors and the System Usability Scale (SUS) (Brooke, 1996) while van Hoof et al. (2007) only used the error rate to evaluate usability of the e-voting system. The SUS has also been used by MacNamara et al. (2011a), while Winckler et al. (2009) used SUS in addition to the Unified Theory of Acceptance and Use of Technology (UTAUT) model (Venkatesh, Morris, Davis and Davis, 2003).

Finally, a number of other approaches have been used, for example, Fuglerud and Røssvoll (2011) used accessibility guidelines, for example, the Web Content Accessibility Guidelines (WCAG) 2.0 from the World Wide Web Consortium (W3C) (W3C, 2008), expert evaluation using personas, and subjective ranking of prototypes by participants and Cross II et al. (2007) used Likert-scale type questions to evaluate usability. Bär et al. (2008) compare the usability of Bingo Voting to Prêt à Voter (Bismark et al., 2009), Punchscan (Popoveniuc & Hosp, 2010), and a scheme by Moran and Naor (Moran & Naor, 2006), all

verifiable voting schemes. The authors use the effort needed for voting and additional steps for the voter to ensure correctness, as measures to compare the usability of these schemes. CAPC (2006), Oostveen and Van den Besselaar (2009), Karayumak et al., (2011b) and Herrnson et al., (2008) gave voters questionnaires developed by the authors to obtain voters' subjective assessments, and used these to evaluate usability. CAPC (2006) also used heuristics developed by the researchers for the expert review. Chisnell et al. (2009a) and Chisnell et al. (2009b), Herrnson et al. (2006), Laskowski and Redish (2006), Kimball and Kropf (2005) and Roth (1998) used guidelines developed by the researchers themselves.

The challenge with using different approaches to evaluate usability of e-voting systems is that it makes it difficult for researchers to effectively compare the usability of the systems, particularly if these approaches are not replicable by other researchers.

Recommendations

- **R-UC-MUE-1:** It is recommended that a standardized approach to evaluate usability is adopted, for example, using the three ISO measures of effectiveness, efficiency and satisfaction.

Future Work

- **FW-UC-MUE-1:** Where existing usability metrics are insufficient to evaluate usability aspects, for example in verifiable e-voting systems, future research should explore new metrics, for example, number of actions for voters, and learnability of interfaces.

Need for Baseline Data

Baseline data provides a benchmark against which researchers can compare the usability of e-voting systems. Three studies were identified as having been carried out with the specific objective of providing baseline data. Everett et al. (2006) measured the usability of the three traditional paper ballots, namely bubble, arrow and open ballots. Greene et al. (2006) focused on lever machines, arrow and bubble ballots, and Byrne et al. (2007) investigated paper ballots, punch cards and lever machines.

Recommendations

- **R-UC-BD-1:** Usability evaluations of traditional e-voting systems should be carried out with the specific objective of providing baseline data to allow for comparison between traditional and new e-voting systems.

FUTURE RESEARCH DIRECTIONS

There is a need for further research to determine how to design usable e-voting systems, how to evaluate the usability and which usability criteria to apply. This is especially the case for verifiable and cryptographically-verifiable remote electronic voting systems, where users can carry out extra steps to verify their vote before casting it. Mental models have only recently been investigated in e-voting but they are an important aspect to understand so as to improve the usability of e-voting systems. Consequently, more research should be done to identify voters' mental model, particularly applied to cryptographically-verifiable e-voting systems, where voters are seen to lack understanding (see for example, Carback et al., 2010; Schneider et al., 2011; and Karayumak et al., 2011b).

Research has been carried out that compares the usability of DREs to traditional voting methods in order to provide baseline data for comparison with new e-voting systems. This should be extended for different types of elections, ballots, and races.

In many of the e-voting studies surveyed, the criteria that have commonly been used to evaluate usability have been efficiency, effectiveness and satisfaction. There are no studies where learnability of e-voting systems has been tested. We consider learnability to be relevant measures, especially for cryptographically-verifiable e-voting systems, because it can show how easy it is for a voter to re-learn how to use an interface after a period of time where they have not used it (Rubin & Chisnell, 2008). We therefore recommend that this be investigated in future research.

Furthermore, although there is a standard available for criteria with which to assess the usability of e-voting systems (Laskowski et al., 2004), our survey of the literature shows that this is currently not in use across studies of usability in e-voting. Researchers have applied different criteria to determine usability of e-voting systems making it difficult to compare the usability. Some research however has applied the ISO 9241-11 standard (ISO, 1998), which gives metrics for usability. More research is needed to identify criteria that can be applied uniformly for usable e-voting systems. If the existing criteria are insufficient, there is need for further research to expand them to accommodate newer e-voting systems or develop new criteria.

Bederson et al. (2003) refer to research on ballot position having an effect on the candidates that are selected. As such, they point out that ballots designed in user studies can incorporate techniques to randomize candidate order. They also indicate that this can create difficulty for voters who have pre-planned their voting. Given that Prêt à Voter (Bismark et al., 2009) randomizes the candidate order on the ballot, further research should investigate how this affects voters in elections where Prêt à Voter is used.

Most of the research found in the literature surveyed focused on the voter perspective, and in this book chapter, we also focus on the voter perspective. However, Chisnell et al. (2009a, 2009b), Goggin (2008), and Claasen, Magleby,

Monson and Patterson (2008) specifically focus on poll workers. Further research should investigate usability issues for this group of stakeholders.

CONCLUSION

In this book chapter we have focused on the usability of e-voting systems, and have reviewed and summarized lessons learned regarding interface design principles, user studies, and usability criteria. We have made recommendations ideally for three groups of researchers: those interested in designing e-voting system interfaces, those interested in carrying out user studies in e-voting, and those seeking for further research in this field. We have, in addition, indicated open research questions in the field of usability and e-voting that researchers can carry out to extend knowledge in the field.

This work shows that a lot of research has been done on DREs and traditional voting methods, but that in comparison, research on the usability of cryptographically-verifiable e-voting systems, and Internet-voting systems is wanting. However, both are more challenging; cryptographically-verifiable e-voting systems use terms that many voters may not be familiar with, and in Internet voting there is no poll worker available who the voter can ask for help or instructions. Research efforts need to be geared in this direction.

With the adoption of verifiable and partially-verifiable e-voting systems, for example in Norway (Stenerud & Bull, 2012), usability is an important consideration to avoid disenfranchising voters. So far, we see research carried out to reduce the technical complexity of verifiability, for example, in Eperio (Essex, Clark, Hengartner & Adams, 2010). This is one step in improving the usability of e-voting systems, making it possible for all who choose to cast their vote by electronic means to do so without undue difficulty. More and more e-voting systems, however, need to focus on and improve usability.

REFERENCES

Adida, B. (2008). Helios: Web-based open-audit voting. In *Proceedings of the 17th Conference on Security Symposium*, San Jose, CA (pp. 335–348). Berkeley, CA: USENIX Association.

Adida, B., & Rivest, R. L. (2006). Scratch & vote: Self-contained paper-based cryptographic voting. In *Proceedings of the 5th ACM Workshop on Privacy in the Electronic Society, WPES '06* (pp. 29–40). New York, NY: ACM.

Alvarez, R. M., & Hall, T. E. (2008). *Electronic elections: The perils and promises of digital democracy*. Princeton, NJ: Princeton University Press.

Arzt-Mergemeier, J., Beiss, W., & Steffens, T. (2007). The digital voting pen at the Hamburg elections 2008: Electronic voting closest to conventional voting. In Alkassar, A., & Volkamer, M. (Eds.), *VOTE-ID 2007 (Vol. 4896*, pp. 88–98). Lecture Notes in Computer Science Berlin, Germany: Springer-Verlag.

Bär, M., Henrich, C., Müller-Quade, J., Röhrich, S., & Stüber, C. (2008). *Real world experiences with bingo voting and a comparison of usability*. Paper presented at the Workshop on Trustworthy Elections (WOTE 2008), Leuven, Belgium.

Bederson, B. B., Lee, B., Sherman, R. M., Hernson, P. S., & Niemi, R. G. (2003). Electronic voting system usability issues. In *Proceedings of the SIGCHI Conference on Human Factors in Computing Systems CHI 2003* (pp. 145 - 152). New York, NY: ACM.

Benaloh, J. (2006). Simple verifiable elections. In *Proceedings of the USENIX/Accurate Electronic Voting Technology Workshop 2006 EVT'06*. Berkeley, CA: USENIX Association.

Bismark, D., Heather, J., Peel, R. M. A., Ryan, P. Y. A., Schneider, S., & Xia, Z. (2009). Experiences gained from the first prêt à voter implementation. *First International Workshop on Requirements Engineering for E-voting Systems RE-VOTE '09* (pp. 19-28) Washington, DC: IEEE Computer Society.

Bohli, J., Müller-Quade, J., & Röhrich, S. (2007). Bingo voting: Secure and coercion-free voting using a trusted random number generator. In A. Alkassar & M. Volkamer (Eds.), *Proceedings of the 1st International Conference on E-voting and Identity* (pp. 111-124). Berlin, Germany: Springer-Verlag.

Brooke, J. (1996). SUS: A 'quick and dirty' usability scale. In Jordan, P. W., Thomas, B., Weerdmeester, B. A., & McClelland, A. L. (Eds.), *Usability evaluation in industry*. London, UK: Taylor and Francis.

Burmeister, O. K. (2000). Usability testing: Revisiting informed consent procedures for testing internet sites. In J. Weckert (Ed.) *Selected Papers from the Second Australian Institute Conference on Computer Ethics CRPIT '00* (pp. 3-9). Darlinghurst, Australia: Australian Computer Society, Inc.

Byrne, M. D., Greene, K. K., & Everett, S. P. (2007). Usability of voting systems: Baseline data for paper, punch cards, and lever machines. In *Proceedings of the SIGCHI conference on Human Factors in Computing Systems CHI 2007* (pp. 171–180). New York, NY: ACM.

Campbell, B. A., & Byrne, M. D. (2009a). Now do voters notice review screen anomalies? A look at voting system usability. In *Proceedings of the 2009 Conference on Electronic Voting Technology/Workshop on Trustworthy Elections EVT/WOTE'09*. Berkeley, CA: USENIX Association.

Campbell, B. A., & Byrne, M. D. (2009b). Straight-party voting: What do voters think? *IEEE Transactions on Information Forensics and Security, 4*(4), 718–728. doi:10.1109/TIFS.2009.2031947

Carback, R., Chaum, D., Clark, J., Conway, J., Essex, A., & Herrnson, P. S. ... Vora, P.L. (2010). Scantegrity II municipal election at Takoma park: The first E2E binding governmental election with ballot privacy. In *Proceedings of the 19th USENIX Conference on Security USENIX Security'10*. Berkeley, CA: USENIX Association.

Center for American Politics and Citizenship (CAPC). (2006). *A study of vote verification technology conducted for the Maryland state board of elections. Part II: usability study*. Retrieved August 3, 2012, from http://www.cs.umd.edu/~bederson/voting/verification-study-jan-2006.pdf

Chin, J. P., Diehl, V. A., & Norman, K. L. (1988). Development of an instrument measuring user satisfaction of the human-computer interface. In *Proceedings of the SIGCHI Conference on Human Factors in Computing Systems CHI '88* (pp. 213-218). New York, NY: ACM.

Chisnell, D., Bachmann, K., Laskwoski, S., & Lowry, S. (2009b). Usability for poll workers: A voting system usability test protocol. In Jacko, J. A. (Ed.), *Human Computer Interaction, Part IV, HCII 2009 (Vol. 5613*, pp. 458–467). Lecture Notes in Computer Science Berlin, Germany: Springer-Verlag. doi:10.1007/978-3-642-02583-9_50

Chisnell, D., Becker, S., Laskowski, S., & Lowry, S. (2009a). Style guide for voting system documentation: Why user-centered documentation matters to voting security. In *Proceedings of the 2009 Conference on Electronic Voting Technology/Workshop on Trustworthy Elections (EVT/WOTE 2009)*. Berkeley, CA: USENIX Association.

Claassen, R. L., Magleby, D. B., Monson, J. Q., & Patterson, K. D. (2008). At your service: Voter evaluations of poll worker performance. *American Politics Research, 36*, 612. doi:10.1177/1532673X08319006

Cohen, J. D., & Fischer, M. J. (1985). A robust and verifiable cryptographically secure election scheme. In *Proceedings of the 26th Annual Symposium on Foundations of Computer Science SFCS '85* (pp. 372-382). Washington, DC: IEEE Computer Society.

Conrad, F. G., Bederson, B. B., Lewis, B., Peytcheva, E., Traugott, M. W., & Hanmer, M. J. (2009). Electronic voting eliminates hanging chads but introduces new usability challenges. *International Journal of Human-Computer Studies*, *67*(1), 111–124. doi:10.1016/j.ijhcs.2008.09.010

Cook, T. D., & Campbell, D. T. (1979). *Quasi-experimentation: Design and analysis issues for field settings*. Boston, MA: Houghton Mifflin Company.

Cross, E. V. II, McMillian, Y., Gupta, P., Williams, P., Nobles, K., & Gilbert, J. E. (2007). Prime III: A user centered voting system. In *CHI '07 Extended Abstracts on Human Factors in Computing Systems CHI EA '07*. New York, NY: ACM. doi:10.1145/1240866.1241006

De Jong, M., van Hoof, J., & Gosselt, J. (2008). Voters' perceptions of voting technology: Paper ballots versus voting machine with and without paper audit trail. *Social Science Computer Review*, *26*(4), 399–410. doi:10.1177/0894439307312482

Demirel, D., Frankland, R., & Volkamer, M. (2011). *Readiness of various eVoting systems for complex elections*. Technical Report, Technische Universität Darmstadt.

Election Assistance Commission (EAC). (2005). *Voluntary voting systems guidelines volumes 1 and 2*. Retrieved April 16, 2011 from, http://www.eac.gov/testing_and_certification/voluntary_voting_system_guidelines.aspx#VVSG%20Version%201.1

Essex, A., Clark, J., Hengartner, U., & Adams, C. (2010). Eperio: Mitigating technical complexity in cryptographic election verification. In *Proceedings of the 2010 Conference on Electronic Voting Technology/Workshop on Trustworthy Elections EVT/WOTE 2010*. Retrieved August 3, 2012, from http://static.usenix.org/events/evtwote10/tech/full_papers/Essex.pdf

Everett, S. P. (2007). *The usability of electronic voting machines and how votes can be changed without detection*. Doctoral dissertation, Rice University, Houston, Texas.

Everett, S. P., Byrne, M. D., & Greene, K. K. (2006). Measuring the usability of paper ballots: Efficiency, effectiveness and satisfaction. In *Proceedings of the Human Factors and Ergonomic Society 50th Annual Meeting* (pp. 2547 - 2551).

Everett, S. P., Greene, K. K., Byrne, M. D., Wallach, D. S., Derr, K., Sandler, D., & Torous, T. (2008). Electronic voting machines versus traditional methods: Improved preference, similar performance. In *Proceedings of the SIGCHI Conference on Human Factors in Computing Systems CHI '08* (pp. 883 - 892). New York, NY: ACM.

Gerlach, J., & Gasser, U. (2009). *Three case studies from Switzerland: E-voting*. Technical Report. Berkman Center Research Publications, Berkman Center for Internet & Society.

German Federal Court Decision. (2009). *Bundesverfassungsgricht. Urteil des Zweiten Senats. German Federal Court Decisions 2 BvC 3/07*, (pp. 1-163).

Goggin, S. N. (2008). Usability of election technologies: Effects of political motivation and instruction use. *The Rice Cultivator*, *1*, 30–45.

Goggin, S. N., & Byrne, M. D. (2007). An examination of the auditability of voter verified paper audit trail (VVPAT) ballots. In *Proceedings of the USENIX Workshop on Accurate Electronic Voting Technology EVT '07*. Berkeley, CA: USENIX Association.

Greene, K. K. (2008). *Usability of new electronic voting systems and traditional methods: Comparisons between sequential and direct access electronic voting interfaces, paper ballots, punch cards, and lever machines.* Unpublished Master's Thesis, Rice University, Houston, Texas.

Greene, K. K., Byrne, M. D., & Everett, S. P. (2006). A comparison of usability between voting methods. In *Proceedings of the 2006 USENIX/ACCURATE Electronic Voting Technology Workshop.* Vancouver, BC: USENIX Association.

Herrnson, P. S., Bederson, B. B., Lee, B., Francia, P. L., Sherman, R. M., & Conrad, F. G. (2005). Early appraisals of electronic voting. *Social Science Computer Review, 23*(3), 274–292. doi:10.1177/0894439305275850

Herrnson, P. S., Bederson, B. B., Niemi, R. G., Conrad, F. G., Hanmer, M. J., & Traugott, M. (2007). *The not so simple act of voting: An examination of voter errors with electronic voting.* Retrieved August 3, 2012, from http://www.bsos.umd.edu/gvpt/apworkshop/herrnson2007.pdf

Herrnson, P. S., Niemi, R. G., Hanmer, M. J., Bederson, B. B., Conrad, F. G., & Traugott, M. (2006). The importance of usability testing of voting systems. In *Proceedings of the USENIX/Accurate Electronic Voting Technology Workshop 2006 EVT '06.* Berkeley, CA: USENIX Association.

Herrnson, P. S., Niemi, R. G., Hanmer, M. J., Francia, P. L., Bederson, B. B., Conrad, F., & Traugott, M. (2005b). *The promise and pitfalls of electronic voting: results from a usability field test.* Paper presented at the Annual meeting of the Midwest Political Science Association. Chicago, IL.

Herrnson, P. S., Niemi, R. G., Hanmer, M. J., Francia, P. L., Bederson, B. B., Conrad, F. G., & Traugott, M. W. (2008). Voters' evaluations of electronic voting systems: Results from a usability field study. *American Politics Research, 36*(4), 580–611. doi:10.1177/1532673X08316667

International Organization for Standardization. ISO 9241 – 11. (1998). *Ergonomic requirements for office work with visual display terminals (VDT) – Part 11: Guidelines on usability.* Geneva, Switzerland: ISO.

Jones, D. W. (2001). *A brief illustrated history of voting.* University of Iowa, Department of Computer Science. Retrieved August 1, 2012, from http://homepage.cs.uiowa.edu/~jones/voting/pictures/

Jones, D. W. (2003). The evaluation of voting technology. In Gritzalis, D. A. (Ed.), *Secure electronic voting* (pp. 3–16). Kluwer Academic Publishers. doi:10.1007/978-1-4615-0239-5_1

Karayumak, F., Kauer, M., Olembo, M. M., Volk, T., & Volkamer, M. (2011b). User study of the improved Helios voting system interface. In *1ˢᵗ Workshop on Socio-Technical Aspects in Security and Trust* (STAST) (pp. 37 – 44). doi: 10.1109/STAST.2011.6059254

Karayumak, F., Kauer, M., Olembo, M. M., & Volkamer, M. (2011a). Usability analysis of Helios - An open source verifiable remote electronic voting system. In *Proceedings of the 2011 USENIX Electronic Voting Technology Workshop/Workshop on Trustworthy Elections (EVT/WOTE 2011).* Berkeley, CA: USENIX Association.

Kimball, D. C., & Kropf, M. (2005). Ballot design and unrecorded votes on paper-based ballots. *Public Opinion Quarterly, 69*(4), 508–529. doi:10.1093/poq/nfi054

Kincaid, J. P., Fishburne, R. P., Rogers, R. L., & Chissom, B. S. (1975). *Derivation of new readability formulas (Automated readability index, fog count, and Flesch reading ease formula) for navy enlisted personnel.* Research Branch Report 8-75. Chief of Naval Technical Training: Naval Air Station Memphis.

Kirakowski, J., & Corbett, M. (1993). SUMI: The software usability measurement inventory. *British Journal of Educational Technology, 24*, 210–212. doi:10.1111/j.1467-8535.1993.tb00076.x

Laskowski, S. J., Autry, M., Cugini, J., Killam, W., & Yen, J. (2004). *Improving the usability and accessibility of voting systems and products.* (NIST Special Publication 500 – 256).

Laskowski, S. J., & Redish, J. (2006). Making ballot language understandable to voters. In *Proceedings of the USENIX/Accurate Electronic Voting Technology Workshop EVT'06*. Berkeley, CA: USENIX Association.

Lazar, J., Feng, J. H., & Hochheiser, H. (2010). *Research methods in human computer interaction.* Wiley Publishing.

Loeber, L. (2008). E-voting in the Netherlands: From general acceptance to general doubt in two years. In R. Krimmer & R. Grimm (Eds.), *Proceedings of the 3rd International Conference on Electronic Voting*, Bregenz, Austria (pp. 21 – 30).

MacNamara, D., Carmody, F., Scully, T., Oakley, K., Quane, E., & Gibson, J. P. (2010). Dual vote: A novel user interface for e-voting systems. *IADIS International Conference Interfaces and Human Computer Interaction 2010, IHCI10,* Freiburg, Germany, (pp. 129-138).

MacNamara, D., Gibson, J. P., & Oakley, K. (2012). *A preliminary study on a DualVote and Pret a Voter hybrid system*. In 2012 Conference for E-Democracy and Open Government (Ce-DEM12). Danube University, Krems.

MacNamara, D., Scully, T., Carmody, F., Oakley, K., Quane, E., & Gibson, J. P. (2011b). (in press). DualVote: A non-intrusive e-voting interface. [IJCISIM]. *International Journal of Computer Information Systems and Industrial Management Applications.*

MacNamara, D., Scully, T., Gibson, J. P., Carmody, F., Oakley, K., & Quane, E. (2011a). *DualVote: Addressing usability and verifiability issues in electronic voting systems*. In 2011 Conference for E-Democracy and Open Government (Ce-DEM11). Danube University, Krems.

Madise, U., & Martens, T. (2006). E-voting in Estonia 2005. The first practice of country-wide binding internet voting in the world. In A. Prosser & R. Krimmer (Eds.), *2nd International Workshop on Electronic Voting 2006, Lecture Notes in Informatics vol. 47,* (pp. 83-90).

Mercuri, R. T. (2001). *Electronic vote tabulation checks and balances.* Doctoral Dissertation. University of Pennsylvania, Philadelphia.

Miller, G. A. (1956). The magical number seven, plus or minus two: Some limits on our capacity for processing information. *Psychological Review, 63*(2), 81–97. doi:10.1037/h0043158

Moran, T., & Naor, M. (2006). Receipt-free universally-verifiable voting with everlasting privacy. In C. Dwork (Ed.), *Proceedings of the 26th Annual International Conference on Advances in Cryptology CRYPTO '06,* (pp. 373-392). Berlin, Germany: Springer-Verlag.

Nielsen, J. (1993). *Usability engineering.* San Francisco, CA: Morgan Kaufmann Publishers.

Niemi, R. G., & Herrnson, P. S. (2003). Beyond the butterfly: The complexity of U.S. ballots. *Perspectives on Politics, 1*(2), 317–326. doi:10.1017/S1537592703000239

Norden, L., Creelan, J. M., Kimball, D., & Quesenbery, W. (2006). *The machinery of democracy: Usability of voting systems.* Brennan Center for Justice, New York University School of Law.

Norden, L., Kimball, D., Quesenbery, W., & Chen, M. (2008). *Better ballots.* Brennan Center for Justice, New York University School of Law.

Oostveen, A., & Van den Besselaar, P. (2009). Users' experiences with e-voting: A comparative case study. *International Journal of Electronic Governance*, 2(4). doi:10.1504/IJEG.2009.030527

Patrick, A. (2009). *Ecological validity in studies of security and human behaviour*. Key note talk: ISSNet Workshop. Retrieved April 16, 2012, from http://www.andrewpatrick.ca/cv/Andrew-Patrick-ecological-validity.pdf

Popoveniuc, S., & Hosp, B. (2010). An introduction to punchscan. In Chaum, D., Jakobsson, M., Rivest, R. L., Ryan, P. Y. A., & Benaloh, J. (Eds.), *Towards trustworthy elections* (pp. 242–259). Berlin, Germany: Springer-Verlag. doi:10.1007/978-3-642-12980-3_15

Prosser, A., Schiessl, K., & Fleischhacker, M. (2007). E-voting: Usability and accpetance of two-stage voting procedures. In Wimmer, M. A., Scholl, H. J., & Grönlund, A. (Eds.), *EGOV 2007* (*Vol. 4656*, pp. 378–387). Lecture Notes in Computer Science Berlin, Germany: Springer-Verlag.

Roth, S. K. (1998). Disenfranchised by design: Voting systems and the election process. *Information Design Journal*, 9(1), 29–38. doi:10.1075/idj.9.1.08kin

Rubin, J., & Chisnell, D. (2008). *Handbook of usability testing: How to plan, design, and conduct effective tests* (2nd ed.). Indianapolis, IN: Wiley Publishing.

Selker, T., Hockenberry, M., Goler, J., & Sullivan, S. (2005). *Orienting graphical user interfaces reduce errors: The low error voting interface*. VTP Working Paper #23. Retrieved on August 3, 2012 from www.vote.caltech.edu.

Selker, T., Rosenzweig, E., & Pandolfo, A. (2006). A methodology for testing voting systems. *Journal of Usability Studies*, 2(1), 7–21.

Sharp, H., Rogers, Y., & Preece, J. (2007). *Interaction design: Beyond human-computer interaction*. John Wiley and Sons.

Slaughter, L., Norman, K. L., & Shneiderman, B. (1995). Assessing users' subjective satisfaction with the information system for youth services (ISYS). In *Proceedings of the Third Annual Mid-Atlantic Human Factors Conference* (pp. 164-170).

Staggers, N., & Norgio, A. F. (1993). Mental models: Concepts for human computer interaction research. *International Journal of Man-Machine Studies*, 38(4), 587–605. doi:10.1006/imms.1993.1028

Stenerud, I. S. G., & Bull, C. (2012). When reality comes knocking: Norwegian experiences with verifiable electronic voting. In M. J. Kripp, M. Volkamer, & R. Grimm. (Eds.) *5th International Conference on Electronic Voting EVOTE 2012* (pp. 21-33). Bregenz, Austria.

Storer, T., Little, L., & Duncan, I. (2006). An exploratory study of voter attitudes towards a pollsterless remote voting system. In D. Chaum, R. Rivest & P. Y. A. Ryan (Eds.), *IaVoSS Workshop on Trustworthy Elections (WOTE 06) Pre-Proceedings* (pp. 77–86).

Traugott, M. W., Hanmer, M. J., Park, W., Herrnson, P. S., Niemi, R. G., Bederson, B. B., & Conrad, F. G. (2005). *The impact of voting systems on residual votes, incomplete ballots, and other measures of voting behavior*. Paper presented at the Annual Meeting of the Midwest Political Science Association, Chicago, IL.

Van Hoof, J. J., Gosselt, J. F., & De Jong, M. D. T. (2007). T*he reliability and usability of the Nedap voting machine*. University of Twente, Faculty of Behavioural Sciences. Retrieved August 3, 2012, from http://wijvertrouwenstemcomputersniet.nl/images/c/ca/UT_rapportje_over_nedap.pdf

Venkatesh, V., Morris, M. G., Davis, G. B., & Davis, F. D. (2003). User acceptance of information technology: Toward a unified view. *Management Information Systems Quarterly*, 27(3), 425–478.

Weber, J., & Hengartner, U. (2009). *Usability study of the open audit voting system Helios*. Retrieved August 3, 2012, from http://www.jannaweber.com/wp-content/uploads/2009/09/858Helios.pdf

Winckler, M., Bernhaupt, R., Palanque, P., Lundin, D., Leach, K., & Ryan, P. ... Strigini, L. (2009). Assessing the usability of open verifiable e-voting systems: A trial with the system Prêt à Voter. In *Proceedings of ICE-GOV,* (pp. 281 - 296).

World Wide Web Consortium. (W3C). (2008). *Web content accessibility guidelines (WCAG)*. Retrieved August 10, 2012, from http://www.w3.org/TR/WCAG20/

ADDITIONAL READING

Adida, B., De Marneffe, O., Pereira, O., & Quisquater, J. (2009). Electing a university president using open-audit voting: Analysis of real-world use of Helios. In *Proceedings of the 2009 Conference on Electronic Voting Technology/Workshop on Trustworthy Elections EVT/WOTE'09*. Berkeley, CA: USENIX Association.

Benaloh, J. (2006). Simple verifiable elections. In *Proceedings of the USENIX/Accurate Electronic Voting Technology Workshop 2006* EVT'06. Berkeley, CA: USENIX Association.

Brewer, M. B. (2000). Research design and issues of validity. In Reis, H. T., & Judd, C. M. (Eds.), *Handbook of research methods in social and personality psychology* (pp. 3–16). Cambridge University Press.

Chaum, D. (2004). Secret-ballot receipts: True voter-verifiable elections. *Security & Privacy, 2*(1), 38–47. doi:10.1109/MSECP.2004.1264852

Chaum, D., Carback, R., Clark, J., Essex, A., Popoveniuc, S., & Rivest, R. L. ... Sherman, A. T. (2008). Scantegrity II: End-to-end verifiability for optical scan election systems using invisible ink confirmation codes. In *Proceedings of the USENIX/Accurate Electronic Voting Technology Workshop on Electronic Voting Technology EVT '08*. Berkeley, CA: USENIX Association

Clarkson, M. R., Chong, S., & Myers, A. C. (2008). Civitas: Toward a secure voting system. *IEEE Symposium on Security and Privacy,* (pp. 354-368).

Dillman, D. A. (2000). *Mail and internet surveys: The tailored design method*. New York, NY: John Wiley and Sons.

Guzdial, M., Santos, P., Badre, A., Hudson, S., & Gray, M. (1994). *Analyzing and visualizing log files: A computational science of usability*. Presented at HCI Consortium Workshop. Retrieved August 3, 2012, from http://smartech.gatech.edu/jspui/bitstream/1853/3586/1/94-08.pdf

Heather, J., Ryan, P. Y. A., & Teague, V. (2010). Pretty good democracy for more expressive voting schemes. In D. Gritzalis, B. Preneel, & M. Theoharidou. (Eds.), *Proceedings of the 15th European Conference on Research in Computer Security ESORICS '10* (pp. 405-423). Berlin, Germany: Springer-Verlag.

Juels, A., Catalano, D., & Jakobsson, M. (2005). Coercion-resistant electronic elections. In *Proceedings of the 2005 ACM Workshop on Privacy in the Electronic Society WPES '05* (pp. 61-70) New York, NY: ACM.

Kimball, D., & Kropf, M. (2007). *Do's and don'ts of ballot design*. AEI Brookings: Election Reform Project. Retrieved August 3, 2012, from http://www.electionreformproject.org/Resources/314eaad7-6dc9-4dad-b6e4-5daf8f51b79e/r1/Detail.aspx

Kish, L. (1965). *Survey sampling*. New York, NY: Wiley.

Krug, S. (2005). *Don't make me think: A common sense approach to the web* (2nd ed.). Thousand Oaks, CA: New Riders Publishing.

Lohr, S. L. (2010). *Sampling: Design and analysis* (2nd ed.). Pacific Grove, CA: Duxbury Press.

National Center for the Study of Elections of the Maryland Institute for Policy Analysis & Research. (2006). *A study of vote verification technologies, part 1: Technical study*. Retrieved August 3, 2012, from http://www.elections.state.md.us/pdf/Vote_Verification_Study_Report.pdf

Neff, C. A. (2001). A verifiable secret shuffle and its application to e-voting. In *Proceedings of the 8th ACM Conference on Computer and Communications Security CCS '01* (pp. 116-125). New York, NY: ACM.

Norman, D. A. (2002). *The design of everyday things*. New York, NY: Basic Books.

Pieters, W. (2006). *What proof do we prefer? Variants of verifiability in voting*. In Workshop on Electronic Voting and e-Government in the UK, Edinburgh, UK.

Puiggali, J., & Morales-Rocha, V. (2007). Remote voting schemes: A comparative analysis. In A. Alkassar & M. Volkamer (Eds.) *Proceedings of the 1st International Conference on E-voting and Identity VOTE-ID '07* (pp. 16-28). Berlin, Germany: Springer-Verlag.

Quesenbery, W. (2004). *Defining a summative usability test for voting systems*. A report from the UPA 2004 workshop on voting and usability.

Quesenbery, W. (2004). *Oops! They forgot the usability: Elections as a case study*. Second Annual Usability and Accessibility Conference.

Reinhard, K., & Jung, W. (2007). Compliance of POLYAS with the BSI protection profile - Basic requirements for remote electronic voting systems. In A. Alkassar & M. Volkamer (Eds.), *Proceedings of the 1st International Conference on E-voting and Identity VOTE-ID '07* (pp. 62-75). Berlin, Germany: Springer-Verlag.

Ryan, P. Y. A. (2004). *A variant of the chaum voter-verifiable scheme*. Technical Report CS-TR 864, University of Newcastle.

Sampigethaya, K., & Poovendran, R. (2006). A framework and taxonomy for comparison of electronic voting schemes. *Computers & Security*, *25*(2), 137–153. doi:10.1016/j.cose.2005.11.003

Sampigethaya, K., & Poovendran, R. (2006). A survey on mix networks and their secure applications. In *Proceedings of the IEEE, 94*(12), 2142-2181. doi = 10.1109/JPROC.2006.889687

Sherman, A. T., Gangopadhyay, A., Holden, S. H., Karabatis, G., Koru, A. G., & Law, C. M. … Zhang, D. (2006). An examination of vote verification technologies: Findings and experiences from the Maryland study. In *Proceedings of the USENIX/Accurate Electronic Voting Technology Workshop 2006 EVT'06*. Berkeley, CA: USENIX Association.

Smith, W. D. (2005). *Cryptography meets voting*. D.O.I = 10.1.1.58.9599

U.S. Election Assistance Commission (EAC). (2011). *A survey of internet voting: Testing and certification technical paper # 2*. Retrieved August 10, 2012, from http://www.eac.gov/assets/1/Documents/SIV-FINAL.pdf

Wallschlaeger, C., & Busic-Synder, C. (1992). *Basic visual concepts and principles for artists, architects and designers*. Dubuque, IA: William C. Brown Publishers.

Wolchok, S., Wustrow, E., Halderman, A. J., Prasad, H. K., Kankipati, A., & Sakhamuri, S. K. … Gonggrijp, R. (2010). Security analysis of India's electronic voting machines. In *Proceedings of the 17th ACM Conference on Computer and Communications Security* (pp. 1–14). New York, NY: ACM.

KEY TERMS AND DEFINITIONS

DRE Voting Machines: A voting machine with mechanical or electrical components for recording votes. DRE voting machines are used at the polling station, where the voter can cast an electronic vote, but in a supervised location, given the presence of election officials.

E-Voting: A means of voting where either the vote casting processes or the tallying of the results are carried out using electronic means. As an example, this can involve using electronic voting machines, or Internet voting systems for vote casting, and optical scanning techniques for vote capture and tallying.

Mental Models: This is considered a person's cognitive internal representation of how something works in the real world. Users create these conceptual models to explain how they understand an object to work, and their interaction with it. These models are not necessarily an accurate reflection of how the object actually works.

Remote Electronic Voting: This also refers to Internet voting. In this case, the voter is able to cast his vote in an unsupervised location, whether from his home or office. Given the unsupervised nature of voting, there are concerns of vote buying or selling.

Usability Criteria: Measures used to assess whether an e-voting system meets usability goals. These criteria include efficiency (speed of performance), effectiveness (rate of errors) and subjective satisfaction. Other criteria are time to learn and retention over time.

Usability Evaluation: Studying the extent to which an e-voting system is fit for use and meets the goals of the voters. This is a process by which designs are assessed and systems tested to ensure they behave as expected, and meet voter expectations. Usability evaluation can be carried out by expert analysis or by user participation.

User Studies: Unlike what the name might suggest, these are tests carried out to determine how users perform with an e-voting system. The goal of user studies is to test if the e-voting system can be used by voters to achieve their desired goals.

Verifiable E-Voting Systems: These are e-voting systems (whether polling station-based or remote) that allow the voter to check that his vote is received and recorded correctly by the voting system, and that it is included in the final tally.

ENDNOTES

[1] Typically the voting age is set at 18 years, however some countries set it at 16 years, and others at 21 years. There is no maximum voting age requirement.

[2] Note that Voter Verifiable Audio Audit Transcript Trails (VVAATT) have been proposed. We do not consider them here as they aim to provide verification to visually-impaired voters.

[3] The voter has not made more than the allowable number of selections for any race.

[4] The voter has not made less than the allowable number of selections for any race.

[5] Roll-off is the failure to cast votes for some offices on a ballot (Bederson et al., 2003).

[6] Residual vote rate is the difference between the number of ballots cast and the number of valid votes cast in a particular contest.

[7] Straight party voting (SPV) allows a voter, by a single choice, to select all the candidates of a party on a given ballot

Section 3
Case Studies on Human-Centered E-Government

Chapter 12

Designing Online Information Systems for Volunteer–Based Court Appointed Special Advocate Organizations:
The Case of Florida Guardian *ad Litem*

Charles C. Hinnant
Florida State University, USA

Jisue Lee
Florida State University, USA

Lorri Mon
Florida State University, USA

ABSTRACT

For public organizations, the ability to harness web-based Information and Communication Technologies (ICT) to make information and services directly available to the public has become an important goal. Simultaneously, the use of volunteers by public organizations has become a crucial component of service delivery within the US. Court Appointed Special Advocate (CASA) programs rely heavily upon volunteers to advocate for neglected children. While there is no doubt variation exists across specific CASA programs, their generally ubiquitous reliance on volunteers indicates a need for recruitment, training, and coordination to successfully achieve program goals. While the discussion of User-Centered Design (UCD) factors illustrates issues for consideration, the case study of Florida's Guardian ad Litem (GAL) program more concretely illustrates how a state-level CASA can begin to harness online ICT to achieve programmatic goals. This chapter discusses key information design characteristics needed for online systems to effectively deliver required information to both volunteers and staff.

DOI: 10.4018/978-1-4666-3640-8.ch012

INTRODUCTION

A key foundational element to the design and effective operation of any organization is the ability to match the capabilities and requirements for information processing (Daft and Lengel, 1986). In most complex organizations, this has come to mean a sizeable investment in the use of Information and Communication Technologies (ICT). As the adoption of web-based ICT has increased over the past fifteen years, the ability to design and implement such systems to effectively deliver information to both internal and external stakeholders has become crucial for an organization's success. For local, state, and federal government organizations, the ability to harness the use of distributed networks to make information and services directly available to the public has become an important goal. So-called electronic government, or *e-government*, has become a foundational component of broader government reforms to improve programs by increasing access to programmatic information and services for citizens and stakeholders (Hinnant and O'Looney, 2007; Kraemer and King, 2003; Watson & Mundy, 2001).

While government has sought to make use of ICT to increase efficiencies and effectiveness, many governments have also begun to rely more heavily upon large numbers of volunteers to effectively carry out key mission activities. The use of volunteers by public organizations has become a crucial component of service delivery in the United States. As government budgets are reduced in times of economic austerity, public programs are forced to rely more on voluntary labor to deliver key services. Over a decade ago, Brudney and Kellough (2000) reported approximately 25-30% of all volunteer labor was directed toward government activities. While recent economic difficulties have sharply impacted the ability of agencies to deliver many services and increased the need for volunteer labor, there is also an increased need for improving the ability

to effectively prepare and manage such a labor resource (Brudney, 1999).

The increased reliance on both ICT and voluntary labor has brought about the need to effectively design information systems that facilitate access to key information resources by a broader set of stakeholders. There is also a concomitant need for the volunteers, as primary users, to not only access, but to effectively use information that is necessary for them to carry out assigned activities. This chapter discusses key information design characteristics needed for online systems to effectively deliver required information to both volunteers and staff. Particular focus is given to a user-centered process for assessing the information needs and capabilities of Court Appointed Special Advocate (CASA) volunteers who may not have the requisite knowledge to advocate for, or represent, children within the court and/or social service systems (Schneiderman and Plaisant, 2005). Furthermore, the chapter will highlight and describe multiple design considerations emanating from the diverse needs of staff who manage a CASA program's internal operations, the volunteers who actually serve as advocates, as well as other stakeholders such as financial donors and the children themselves. The chapter will use information obtained, and lessons learned, from an ongoing research and service project to redesign the online information systems for a state-level CASA, the Florida Guardian ad Litem program (GAL).

VOLUNTARY LABOR AND CASA PROGRAMS

Voluntary Labor in the USA

Voluntary labor has become an important factor in the ability of many organizations to fulfill their primary goals. The value of annual volunteer time in the U.S. is astounding. Calculations performed by the National Center for Charitable Statistics

(NCCS) on the Bureau of Labor Statistics' (BLS's) 2010 American Time Use Survey indicates that of the 62.8 million volunteers in that year contributed roughly 15 billion hours of work worth approximately 8.8 million full-time equivalent (FTE) employees or $283.845 billion (Roeger, Blackwood, and Pettijohn, 2011). A more recent BLS Report released in February 2012 reporting the findings from a supplement to the Current Population Survey (CPS) indicates that approximately 64.3 million people volunteered with an organization between September 2010 and September 2011. Approximately 29.9 percent of women and 23.5 percent of men volunteered their time with an organization. Similarly, young people in their twenties were less likely to volunteer (19.4 percent) when compared to older adults aged 35-44 (31.8%) and 45-54 (30.6%), respectively. With regards to racial groups, whites volunteered at a higher rate (28.2%) than did blacks (20.3%), Asians (20.0%), or Hispanics (14.9%). Married individuals tended to volunteer more than did persons who had never married. Educational status tended to influence both the likelihood of volunteering with 42.4% of college graduates volunteering in comparison to 18.2% of those with only a high school degree (Bureau of Labor Statistics, 2012).

Along with the immense amount of voluntary contributions nationwide, there is also a wide variety of organizational types and institutional sectors dependent on such contributions. Overall, more people volunteered with religious organizations (33.2 percent), then educational or youth service organizations (25.7%), followed by social or community service organizations (14.3%). In addition, the type of organizations individuals volunteered with seems to vary with age and educational attainment. For example, of those who volunteered with educational and youth service organizations, 26.7% had a college degree while 22.7% had only a high school diploma. Volunteers over 65 years old were more likely to volunteer with religious organizations

(44.9%) than were younger volunteers below 24 years of age (26.5%) (Bureau of Labor Statistics, 2012). These statistics indicate that volunteers vary greatly in regards to both demographic characteristics and experience. Organizations that rely on voluntary labor must be aware of how such differences may impact the way that vital activities are carried out.

Court Appointed Special Advocate (CASA) Programs

As with many public programs, CASA programs often depend on a mix of volunteer and permanent staff to carry out program activities in the pursuit of their primary goal to support children by providing court appointed advocates. Typically, this means that when a child has been abused, neglected, or when there is a custody disagreement between his/her parents, a judge may assign an advocate to represent the best interests of the child. CASA's can also be appointed in circumstances involving mentally challenged adults.

Enacted in 1974, the Child Abuse Prevention and Treatment Act (CAPTA) serves as the foundation for the current system of child protection and has served to solidify the role of CASA and Guardian ad Litem (GAL) programs across the country (Boumil et al, 2011; Pub. L. No. 93-247, 88 Stat, 4, 1974). Despite the impetus from a single federal law, there are meaningful legal differences between such programs at the state-level. For example, some states refer to court appointed volunteers that represent abused or neglected children as CASAs and guardian ad litem (GALs) synonymously or interchangeably. Other states differentiate between CASA and GAL programs, with only volunteer advocates referred to as CASAs and attorneys appointed to represent children in custody disputes or mentally challenged adults as GALs. Regardless of nomenclature and variations in program implementation across the states, a primary goal of all such programs is to provide court appointed volunteers to advocate for dependent children or mentally challenged adults within the courts and broader social service system.

Overall, there are over 900 CASA programs in the U.S. In 2010, 75,087 volunteers worked to represent and assist over 240,000 children (National Court Appointed Special Advocate Association, 2010). By 2011, approximately 77,000 people were assisting approximately 234,000 children (National Court Appointed Special Advocate Association, 2011). While state-level CASA programs may vary due to differences in state laws and culture, they typically all provide training and certification to volunteers who serve as advocates for children within the court and social service system. As such, attracting, training, supervising, and retaining volunteer advocates are key mission goals for all CASA programs. Designing, implementing, and managing the information resources required by staff and volunteers is crucial for the successful operation of such programs.

DESIGNING USER-CENTERED ONLINE INFORMATION SYSTEMS FOR CASA PROGRAMS

The overall number of state-level CASA programs, coupled with the variation in how such programs are operated, highlights the complexity in understanding how such organizations may effectively implement ICT to address key programmatic goals. While the development of online systems, such as websites for providing program stakeholders key information, has become a key mechanism for facilitating more effective and efficient operations for organizations, more holistic approaches to understanding the relationships between the successful use of online ICT and subsequent efficiency and effectiveness gains have only recently received broader awareness. As organizations have become more dependent on their online interfaces and systems to directly interact with key stakeholders, more attention has been focused on how to better understand the user experience and, more importantly, how to design and build more intuitive and usable systems. Mao, Vrendenburg, Smith, and Cary define such user-centered design (UCD) as a "multidisciplinary design approach based on the active involvement of users to improve the understanding of user and task requirements, and the interaction of design and evaluation" (2005, p. 105). This approach also typically includes the consideration of user perspectives during every phase of the system design process and designs are also tied closely to specific user characteristics, the tasks they must perform, and the organization's environment (Schneiderman and Plaisant, 2005).

Bertot and Jaeger (2006) indicated that there are three primary dimensions to assess e-government websites: functionality, usability, and accessibility. They believe that in order to improve these three dimensions of e-government systems, governments should employ user-centered design practices, engage in continued evaluation activities to assess these characteristics, devote more resources to evaluation and assessment of online services, understand the necessity of assistive technologies for many users, and understand that user demands, needs, and requirements are not static (Bertot and Jaeger, 2006).

Design

Design is at the core of the normative foundations of most human-centered professional work. As Simon (1996) indicates, "The engineer, and more generally the designer, is concerned with how things ought to be --- how they ought to be in order to *attain goals*, and to *function*." (pp. 4-5). Design can be defined in a variety of ways. When designing ICT, a design can refer to the actual end product and the processes that lead to the product. As Preece et al (1994, p. 352) summarizes, "....'design' refers to both the process of developing a product, artifact or system and the various representations (simulations or models) of the product that are produced during the design process." Design performance is often considered to be multi-dimensional and, therefore, success may be measured in a variety of ways. In some cases pertaining to online ICT like websites, success may be defined as the number of users who visit the site, user satisfaction, information quality, system quality, likelihood of return visit, or some combination of these measures (Palmer, 2002). In order to design a successful ICT system, UCD processes must take a socio-technical approach that considers the ultimate programmatic purposes and goals for the system, as well as a mix of social and technological factors associated with the users, program, and broader institutional environment (Eason, 1988). With regards to developing an online interface for CASAs, one must consider the ultimate goals of each organization coupled in relation to the ICT users, the online tasks that they need to perform, and the broader environment.

Program Goals, User Characteristics, Online Tasks, and Environment

As with many public programs, CASA programs typically have numerous goals that they seek to accomplish. While their ultimate institution-level goal may be to assist and represent abused and neglected children within the judicial system of their particular state or legal jurisdiction, they must also focus on achieving numerous operational-level goals that enables them to achieve high levels of performance. When designing ICT systems for CASA programs, a key activity is the gathering and analysis of the functional requirements that the system must meet. This is often done by collecting and analyzing a variety of information about the specific program's goals, operations, and activities. For example, in order to appropriately represent the children in need of advocates, CASA programs must at-

tract and recruit suitable volunteer advocates (i.e. Guardian ad Litem), as well as volunteers to provide legal services, program administrative support, and to serve as intermediaries linking the program to other individuals and organizations that may provide assistance within the broader community.

In addition to recruiting a variety of volunteers, programs must also provide required program-oriented information, training and oversight to volunteers, staff, as well as to the public. Therefore, a CASA program may seek to determine the online information and processes necessary for recruiting and retaining volunteers by interviewing, surveying, or otherwise collecting information from key stakeholders. Such information would be a crucial part of the design process for the program's online ICT systems. These systems would not only serve to facilitate attracting potential volunteers, but also address the crucial goals of preparing both volunteers and staff for carrying out organizational tasks in acceptable ways. Furthermore, they address the important goal of volunteer retention, since such programs necessarily depend on retaining personnel once they have been suitably trained and socialized to meet the program's expectations.

User Characteristics

CASA programs have a diverse set of potential ICT users such as volunteer advocates, volunteer staff, paid staff, and other potential stakeholders such as attorneys, members of the judiciary, social service workers, family of dependent children, and the children themselves. Such users not only vary with regards to their profession positions, but they often have numerous differences with regards to their backgrounds and knowledge of basic online ICT use. Users' familiarity with online ICT systems or services may be highly varied.

For example, some users may be late adopters or otherwise have limited knowledge of online ICT that is often required to operate interactive websites or make use of media, such as audio or video files. A recent report from the Pew Research Center indicates that while 82% of Americans over the age of 18 use the internet or email on an occasional basis, the percentage drops to 53% when looking only at those who are 65 years old or older. However, of those in this older age group who do use the internet, 70% use it daily (Zickuhr and Madden, 2012).

Differences are also evident between educational and racial groups. For example, a 2011 Pew Research Center survey indicates that 43% of respondents with less than a high school degree use the internet compared to 71% of those with a high school diploma, 88% of those with some college education, and 94% of those with at least a college degree. Eighty percent of non-Hispanic whites used the internet compared to 71% of non-Hispanic blacks, and 68% of Hispanics (Zickuhr and Smith, 2012).

There may also be differences in how different income groups use the internet. For example, a 2010 survey of higher income households indicates that 55% of households with incomes over $75,000 per year used the internet or email several times a day. This is compared with 42% of households with incomes of $50-74,999, 44% with incomes $30-49,999, and 31% of households with incomes less than $30,000 per year (Jansen, 2010).

Understanding the potential for differences across user groups may help understand how online ICT should be designed in order to more effectively deliver pertinent information to stakeholder groups. In fact, continually assessing the characteristics of users and inclusion of the users within the design process is paramount to providing relevant information and services that are required to interact with the program or carryout necessary program activities.

Online Tasks

Understanding the online tasks that must be completed is an important component of designing ICT systems for CASA programs as well as any other organizations. However, such tasks and the actual work that can be achieved through technological designs and artifacts are necessarily linked to the program's goals, work process analysis, the intended users' characteristics, the actual online technology artifact, and how such processes and technology interacts with, and is accepted by, the program's broader socio-technical system (Preece et al, 1994; Eason, 1988). Once a CASA program's high-level goals are identified, they can be systematically decomposed into their subcomponents and, ultimately, into online activities completed by the users. Such a process necessarily serves to connect the more abstract organizational-level goals with lower-level activities that can be connected to measureable interactions with the ICT interface and the underlying need for specific types of information (Preece et al, 1994).

Task analysis is usually hierarchical in nature. For example, while a goal may be to recruit and retain CASA volunteers, this goal must be broken down into finite activities such as providing potential volunteers program information regarding dependency issues. This means that the design process must determine how users will use the ICT to accomplish specific information-related activities. It cannot be taken for granted that all users will be able to effectively find the necessary information to accomplish a particular online activity in a successful manner. With regards to a potential goal such as volunteer recruitment, this means not only understanding the specific dependency information needed by specific stakeholders, but also knowing how such stakeholders will be able to interact successfully with the online ICT system to acquire and make use of such information.

Determining such relationships is necessarily specific to particular user groups. For example, pro bono attorneys may have at least a basic understanding of the legal and judicial system irrespective of whether they have formal training in dependency issues. However, while attorneys may have a relatively high level of overall education and legal subject matter expertise, this does not mean that all attorneys have a high level of technical expertise with regards to ICT. As a result, existing information regarding specific user groups can be helpful. In a survey of attorneys conducted by the American Bar Association in 2010, 46% of respondents indicated that their law firms provided computer (or website)-based legal training. In addition, 67% indicated that they attended live webcasts to fulfill Continuing Legal Education (CLE) requirements but only 36% indicated that they used archived webcasts for training (Johnson, 2012). Such information may be helpful when designing a CASA's online ICT system; however more specific usability studies to evaluate a program's specific users' perceptions regarding a system's technological and informational design features is a crucial aspect of UCD. Furthermore, assessing users' perceptions of how specific technological features contribute to successful task completion is an important aspect of the design process.

Organizational Environment

Closely tied to the above issues surrounding the assessment of the fit and performance between a particular task or activity, the actual technological artifact, and the specific users is the impact of the broader organizational environment. Like other public organization's, CASA programs do not operate in social or technological isolation. The operations of such programs are directly impacted by the actions of other organizations and external stakeholders within its broader environment. For example, a legislature may enact laws regarding how CASA programs must rep-

resent dependent children within a given state or how such a program may be funded. Such actions will necessarily influence a program's organizational purpose and goals. Therefore, this can influence not only what a program is permitted to do and required to do by law, but also the amount of resources that are available to carry out tasks. Similarly, such programs are influenced by a wide array of actors such as the courts, social service agencies, and community organizations within their broader environment.

The program's technological environment influences how ICT systems may be designed, as well as various performance factors. CASA programs that lack sufficient in-house technical expertise to design or redesign online ICT systems may need to acquire outside assistance to accomplish such activities. The ability to acquire such assistance may be in-and-of itself limited if sufficient financial resources are unavailable or if such expertise cannot be acquired through other means. Such issues must also be taken into consideration when planning to provide technical assistance to potential users of the ICT systems. While ideally online ICT should be designed to facilitate usability by the targeted user group(s) without the need for extensive additional assistance, assistance or technical training may still be necessarily to build efficacy within some user groups who may not have sufficient ICT knowledge to use new or redesigned systems.

With regards to CASA programs, it is important to remember that their online systems, such as websites, are often meant to address a variety of purposes such as information dissemination, recruitment, training, and acquisition of new resources (e.g. financial donations, needed professional expertise). It is also important to remember that the design process is a necessarily continuous process given the dynamic relationships between a program's goals, user characteristics, information-related tasks, and the broader socio-technical environment.

Similarly, the so-called design space, or the set of possible design options, may differ for new system designs in comparison with redesigning an existing system (Preece et al, 1994). Redesigning an existing system may necessarily be more limited or even incremental in nature because of prior design decisions, as well as existing task and functionality requirements. Conversely, designing a new system may provide more opportunities for making more decisions without consideration of as many existing factors.

A focus on UCD assists in better understanding the influence of user characteristics, tasks and activities, technology features, information resources and broader environmental factors on the usage of specific online systems and, ultimately, how such usage translates into efficiency gains for the organization (Bertot and Jaeger, 2006; Preece et al, 1994; Wang, Bretschneider, and Gant, 2005). Consideration of these factors is crucial for developing ICT that satisfies the needs of key users and, ultimately, influences how ICT may be used to accomplish key program goals. To this end, examining how a specific state-level CASA program employs online ICT can be used to highlight both specific design and implementation issues that such programs should consider as well as possible methods for facilitating a more effective overall use of such technologies. A case study of the Florida Guardian ad Litem (GAL) provides the specific details needed to illustrate key issues that impact the effective design of ICT to facilitate the attainment of child advocacy goals.

The Case of Florida Guardian *ad Litem* (GAL)

The Florida GAL program is a partnership organization that operates as a network of permanent professional staff and thousands of volunteers that work to represent and advocate for abused and neglected children that are involved in court proceedings. The state's Justice Administration Commission administratively serves the Florida GAL program and the state's governor appoints the program's director (Justice Administrative Commission, 2011; Florida Department of Children and Families, 2010).

Section 39.820(1) of the Florida statutes defines GAL as:

a certified guardian ad litem program, a duly certified volunteer, a staff attorney, contract attorney, or certified pro bono attorney working on behalf of a guardian ad litem or the program; staff members of a program office; a court-appointed attorney; or a responsible adult who is appointed by the court to represent the best interests of a child in a proceeding as provided for by law, including, but not limited to, this chapter, who is a party to any judicial proceeding as a representative of the child, and who serves until discharged by the court (2011).

In an effort to meet its legislative mandate to represent abused or neglected children within the state court system, the state-level program partners with 21 local GAL programs within 20 judicial circuits in Florida. The program works closely with private foundations, corporations, faith-based organizations, and the local bar association who assist in recruiting volunteers and pro bono attorneys, as well as providing resources to assist in meeting the program's goals (Local Programs, 2012).

In 2012, the Florida legislature appropriated $31,656,928 from the state's general revenue to fund the program operations in the next fiscal year. The program also received $320,249 in funds from its trust fund. These funds will assist a program consisting of 539 positions. The program is considered cost effective within the state government primarily because of its heavy reliance on volunteers, pro bono attorneys, and partnership with nonprofits. The program was one of 16 public programs (out of 533 nominations) that won a state-wide productivity award, the Davis Productivity Eagle Award, in 2012 (Abramowitz, 2012a; Abramowitz, 2012b).

Florida GAL: Tasks and Activities

Section 39.822(1) of the Florida Statutes dictates that a Guardian ad Litem (GAL) is assigned to the child at the earliest time possible in court proceedings dealing with the abuse or neglect of a child (2011). When a volunteer is needed to represent a child, the program seeks to accept the case as long as resources are available. A volunteer advocate typically carries out four primary roles in working with a child's case. These include: (1) investigating case background, the environment, relationships between the involved parties, and the needs of the child; (2) identifying resources and services that are available to the child and making sure that the stakeholders collaborate to work in the child's best interest; (3) representing the child within the courts and other public agencies; and (4) making sure that court orders and specified plans are carried out. In addition to these broad tasks, the advocate must keep records and file all necessary reports regarding the case as well as be present at all court hearings (Florida Guardian ad Litem Program, 2006).

A volunteer advocate is typically appointed as part of team. These volunteer advocates then work closely with case coordinators to make sure that the case is handled appropriately and the child is properly represented within the state's dependency system. Volunteers and program staff are also supported by the program attorneys or pro bono attorneys who attend hearings, depositions, negotiate with involved parties, and handle appeals. Volunteers, staff and legal counselors become knowledgeable about the child's case and circumstances in order to make recommendations to the court to bring about a safe and stable environment for the child (Florida Guardian ad Litem Program, 2012). Therefore, the program is necessarily dependent on not only

volunteers who advocate on behalf of child but on other stakeholders such as volunteer attorneys and staff who have the relevant expertise necessary to successfully manage a case.

Florida GAL: Volunteer and *Pro Bono* Attorney Qualifications

As with many volunteer-based organizations, the program is fundamentally based on the use of volunteer child advocates and legal services from local attorneys in order to advocate and represent children within the state court system. Volunteers who serve as child advocates do not need specialized legal training. However, they must but be at least 19 years old and complete a review and interview that covers their educational, employment, and criminal background. They must also complete 30 hours of initial certification training, 6 hours of recertification in each subsequent years, and 10 hours per month working on actual cases. Volunteers are supervised and reviewed on an annual basis and must complete training as necessary on the job (Florida Guardian ad Litem Program, 2006). The topics that must be covered during certification are listed in Table 1.

While the program does review all applicants' backgrounds and qualifications before they are assigned to a case, there are no specific minimal educational or background requirements. A person's certification is based on the discretion of the program and the findings from the review process.

GAL also relies heavily on attorneys for pro bono legal work associated with cases. Volunteer attorneys must be a member in good standing with the state bar association, successfully complete any necessary training pertaining to the cases for which they volunteer, and abide by the program's policies (Florida Guardian ad Litem Program, 2006). Each GAL circuit office works closely with local stakeholders such as the local bar association, the judicial branch, and members of

Table 1. Guardian ad Litem volunteer certification training topics

Program History and Structure	Communication Skills and Interviewing Techniques
Dynamics of Child Abuse and Neglect	Confidentiality
Best Interests of the Child	Record Keeping Practices
Children and Permanence	Report Writing
Role and Responsibilities of GAL Program Staff and the Volunteer	GAL Program Standards of Operations
Juvenile Court Process	Ethical Obligations
Department of Children and Families	Community Agencies and Resources
Cultural Diversity	Court Observation

(Florida Guardian *ad Litem* Program, 2006, p. 10)

their GAL non-profit board in order to assist in recruiting attorneys (Florida Guardian ad Litem Program, 2012).

As of May 2010, the Florida GAL program had 7,900 certified volunteers that served as advocates for 22,800 children per year (Frequently Asked Questions, 2012). The program provides advocates for 70% of the children in the state's dependency system. However, program leaders indicate that they will need approximately 10,000 staff and volunteers by the end of 2012 in order for all such children to be represented within the courts (Florida Guardian ad Litem Program, 2012).

Florida GAL User-Centered Website Redesign

The Florida GAL program recently began an effort to make better use of online ICT in order to increase the provision of information and information-based services. Furthermore, the enhanced use of online technologies, such the program's website and social media sites (e.g. Facebook), brings the potential for improving operational efficiency and effectiveness with a relatively limited financial investment (Florida Guardian ad Litem Program, 2012). Such operational improvements include (but are not limited to): the recruitment of new volunteers, training and retention of current volunteers, dissemination of program information, and enhanced networking with other organizations such as social service

organizations, community groups, professional associations, and government institutions.

The authors partnered with the Florida GAL program beginning in the spring of 2011, in an effort to assist the program in more effectively using online resources to provide information to a wide variety of stakeholders. This assistance included an evaluation of the program's current website, as well as follow-up assistance in developing prototypes of potential future website designs. Data from a survey of website users, as well as program archival materials collected by the authors, is used here to provide examples for key points regarding UCD issues.

The authors collected information regarding the website users by conducting an online survey in order to gather information about users' characteristics and perceptions of the GAL website. Program stakeholders were recruited to participate in the study through online sources such as the program's electronic newsletter, the program's official website and Facebook page. The survey was administered from August to November 2011 and a total of 426 respondents participated in the online survey. However, not all respondents answered all of the survey questions.

Demographics

Of 426 total respondents, 259 respondents provided information regarding their role in the GAL program, 125 (48.3%) indicated that they were volunteers and 97 (37.5%) indicated that they were program staff. Seventeen (6.6%) indicated "other" and the remaining respondents indicated that they were either pro bono attorneys (n=2; .8%), non-pro bono attorneys looking for dependency information (n=5; 1.9%), a corporate partner (n=1; .4%), a judge (n=1; .4%), an individual financial contributor (n=1; .4%), staff or expert volunteers (n=3; 1.2%), social workers (n=6; 2.3%), or an educator (n=1; .4%).

No respondents were less than 18 years old and most were over fifty years old. There were 15 (5.9%) respondents between 18 and 29 years old, 77(30.4%) between 30 and 49 years old, 99 (39.1%) between 50 and 64 years old, and 62 (24.5%) that were 65 years old or older. With regards to overall educational attainment, no respondents had less than a high school diploma, 44 (17.4%) had a high school diploma or equivalent, 96 (37.4%) had a bachelor's degree, 54 (21.3%) had a Masters degree, and 59 (23.3%) had a doctoral degree. Overall, the demographic questions indicated that the majority of respondents were typically female, over fifty years old, and possessed at least a bachelor's degree. Furthermore, a question regarding the amount of hours that respondents volunteered per year indicated that 96 (55.2%) had volunteered at least 75 hours.

It is important to highlight that 63.6% of the survey respondents were over the age of 50 and highly educated since linkages have been shown to exist between age, education, and internet use. As previously mentioned, a research study conducted by Pew has shown that older adults exhibit lower rates of internet use. However, those that do use online technologies have a relatively high rate of use. Similarly, increased education levels have also been shown to be related to increased levels of internet use (Zickuhr and Madden, 2012; Zickuhr and Smith, 2012).

Types of Information

Determining what online information users find most important is a key activity in any design or redesign process. An open-ended question that asked respondents to list the three best features of the GAL website elicited and ranked features related to general volunteer forms (26 mentions), legal case summaries (18 mentions), the volunteer application process and forms (15 mentions), and general legal information located on the main legal page (11 mentions) as the best features on the site.

Close-ended survey questions also asked respondents to rate a number of different information resources as to how important they were for the GAL website. Resources were rated on a scale of 1 indicating "Not at all important" to 7 indicating "Extremely Important." The middle value of 4 indicated "Neutral". Examining the overall responses in Table 2 indicates that several information resources received a median rating of at least 6, or "Very Important," by the overall group of respondents. These information resources include several pertaining to the program's administrative processes and operations (e.g. application process, staff information, online forms, program information, FAQs, and announcements), information related to specific stakeholder groups such as teens or volunteers (e.g. information for teens, practice aids, and links to other information sources), and information about legal resources (e.g. case law, recent legal case summaries, and legislative information).

Student's *t*-tests were carried out to investigate the differences in the mean responses between GAL volunteers and program staff. The sample was composed of respondents from these two groups (125 volunteers and 97 staff). Relatively few respondents belonged to other professional groups such as pro bono attorneys, judges, social workers, and educators. Therefore, the analysis was limited to volunteers and staff. The results of independent sample *t*-tests indicate a statistically

Table 2. Importance of information resources

How important do you consider the following types of information resources for the Guardian ad Litem website?	Descriptive Statistics for Overall Sample						Independent Sample T-Tests (Volunteers and Staff)			
	Median	Mean	Std. Dev.	Min	Max	N	T	df	p-value	Mean Difference
Information regarding being a pro bono attorney	5.000	4.630	1.905	1	7	395	-2.860	218.887	**0.005**	-0.689
Information regarding Guardian ad Litem legal issues	6.000	5.640	1.308	1	7	391	-0.131	217.000	0.896	-0.021
Information regarding the Guardian ad Litem application process	6.000	5.680	1.458	1	7	393	-2.265	217.765	**0.024**	-0.398
Information regarding the state agency initiative	5.000	5.100	1.401	1	7	393	0.339	218.000	0.735	0.061
Online forms (e.g. report to the court form, visitation form)	6.000	6.020	1.272	1	7	396	3.256	220.000	**0.001**	0.538
Archived resources(e.g. archived briefs, bulletins)	5.000	4.990	1.480	1	7	393	-0.525	217.000	0.600	-0.095
Information about/from children	6.000	5.740	1.294	1	7	392	0.729	217.000	0.467	0.121
Information about/from pro bono attorneys	5.000	4.960	1.587	1	7	393	0.593	219.000	0.554	0.116
Information about/from Guardians ad Litem	6.000	5.800	1.256	1	7	387	0.282	214.000	0.778	0.045
Information about/from other related organizations	5.000	5.020	1.413	1	7	393	1.783	218.000	0.076	0.320
Information about/from program staff	6.000	5.330	1.368	1	7	393	2.164	217.000	**0.032**	0.380
Legislative information (i.e contact information)	6.000	5.380	1.461	1	7	392	0.527	218.000	0.599	0.099
FAQs (Frequently Asked Questions)	6.000	5.550	1.289	1	7	379	-0.562	218.000	0.575	-0.093
Current event news	6.000	5.390	1.281	1	7	380	0.042	220.000	0.966	0.007
Announcements (i.e. upcoming events such as conferences)	6.000	5.660	1.220	1	7	381	1.446	219.000	0.150	0.218
Resources available by specific topic	6.000	5.840	1.122	1	7	379	2.024	218.000	**0.044**	0.264
Resources developed for teens	6.000	5.490	1.353	1	7	382	2.114	220.000	**0.036**	0.364
Materials regarding independent living for teens	6.000	5.580	1.279	1	7	382	1.825	220.000	0.069	0.298
Practice aids	6.000	5.400	1.362	1	7	378	-1.417	220.000	0.158	-0.246
Checklists regarding Guardian ad Litem issues	6.000	5.620	1.231	1	7	381	-0.187	220.000	0.852	-0.030
Links to other relevant resources	6.000	5.560	1.197	1	7	379	0.678	220.000	0.499	0.106
Information regarding how the website treats user information (i.e. privacy policy)	5.000	5.070	1.679	1	7	379	1.021	219.000	0.308	0.228
Legal briefs	5.000	5.070	1.558	1	7	374	-1.334	218.000	0.184	-0.262
Practice bulletins	5.000	5.080	1.520	1	7	373	-3.111	217.000	**0.002**	-0.589

continued on following page

Table 2. Continued

How important do you consider the following types of information resources for the Guardian ad Litem website?	Descriptive Statistics for Overall Sample						Independent Sample T-Tests (Volunteers and Staff)			
	Median	Mean	Std. Dev.	Min	Max	N	T	df	p-value	Mean Difference
Relevant case law	6.000	5.470	1.455	1	7	373	-0.967	218.000	0.334	-0.180
Recent legal case summaries	6.000	5.450	1.416	1	7	373	-1.695	218.000	0.092	-0.304
Legislative information (i.e. potential laws pertaining to dependency issues)	6.000	5.690	1.318	1	7	372	-0.091	217.000	0.927	-0.016
Legal statute information	6.000	5.650	1.347	1	7	373	-0.152	219.000	0.880	-0.027

t-values in *italic* indicate that equal variances were not assumed

P-values in **bold** when less than the alpha value (.05)

Mean Difference = Mean of Volunteers - Mean of Staff

reliable difference between the mean responses of volunteers and staff for several items at an alpha level of .05. For example, in the case of *information regarding being a pro bono attorney*, *information regarding the Guardian ad Litem application process, online forms*, and *information about/from program staff*. The mean differences between volunteers and staff on these items possibly indicate that staff may be more concerned with the ability to place information about the formal application process on the website while volunteers may find specific information regarding program staff (i.e. potential informational contacts) and the practical activities associated with actually applying to the program (i.e. online form).

There are also statistically significant differences for *resources available by specific topic*, *resources developed for teens*, and *practice bulletins* at an alpha level of .05. Volunteers seem to find the information for teens and the ability to find information resources for specific topics more important than did the program staff. Staff seemed to place more value on practice bulletins than did volunteers. These findings indicate that volunteers find information helpful when it facilitates their work with dependent children while staff members value the ability to provide

program information in an efficient manner via the program's website.

As a volunteer-based organization, online training materials should be important to the stakeholders of a CASA such as Florida GAL since such information resources would facilitate training or retraining volunteers. Table 3 shows that respondents to the survey indicated that general training information (e.g. training information for Guardians ad Litem, conference information) and specific training materials (e.g. training and practice manuals, life instruction information for teens, and training videos) all received median responses of at least six, "Very Important."

The mean responses for GAL volunteers and program staff were found to be statistically significant at $\alpha = .05$ for several items associated with training. These included *instructional information for teens, attorney training calls, training Power Point presentations for attorneys*, and *practice manuals (e.g. dependency practice manual)*. Volunteers seem to value the instructional information for teens more than did the staff respondents since they may need to directly assist teens who are leaving the dependency system and who may need information about available services. Staff members seem to value the training resources for attorneys more

Table 3. Importance of training resources

How important do you consider the following types of training resources for the Guardian ad Litem website?	Descriptive Statistics for Overall Sample						Independent Sample T-Tests (Volunteers and Staff)			
	Median	Mean	Std. Dev.	Min	Max	N	t	df	p-value	Mean Difference
General training information for Guardians ad Litem	6.000	6.060	1.118	1	7	362	1.769	219.000	0.078	0.245
Training manuals (i.e. Guardians ad Litem, support volunteers, staff)	6.000	6.050	1.103	1	7	362	0.360	219.000	0.719	0.050
Instructional information for teens (i.e. regarding health, life skills, employment, housing, or education)	6.000	5.620	1.393	1	7	360	*2.524*	180.344	**0.012**	0.447
Training conference information (i.e. CLE conference)	6.000	5.560	1.255	1	7	361	1.501	219.000	0.135	0.240
Attorney training calls	6.000	4.910	1.629	1	7	360	-2.367	218.000	**0.019**	-0.497
Training videos (i.e. for attorneys, teens, or volunteers)	6.000	5.570	1.379	1	7	362	-1.144	220.000	0.254	-0.211
Training Power Point presentations for attorneys	6.000	4.850	1.671	1	7	358	-3.917	218.000	**0.000**	-0.817
Practice manuals (e.g. dependency practice manual)	6.000	5.460	1.472	1	7	360	*3.245*	217.753	**0.001**	-0.574

t-values in *italic* indicate that equal variances were not assumed
P-values in **bold** when less than the alpha value (.05)
Mean Difference = Mean of Volunteers - Mean of Staff

than volunteers. This may indicate that staff members value the efficiency gains associated with putting such resources on the website where they can easily be reached by legal professionals. As with any CASA organization, efficiently training volunteers to work directly with dependent children and providing the necessary legal information for volunteers from the legal profession are both key components that enable a CASA organization like GAL to provide dependent children with adequate assistance within the court and dependency system.

User Preferences for Online Information and Technology

In order to make decisions regarding the redesign of the program's website, it is also important to understand more about users' technology and information preferences. One fundamental aspect

of online ICT is the ability of users to connect via a high-speed connection. In the case of these respondents, 86.4% indicated that they access the GAL website via a high-speed internet connection.

In addition to the type of broadband connection, it is also important to understand what online technologies users of the GAL website employ in circumstances other than their interaction with the GAL website. This may assist in the design process by informing designers about which online technologies may be feasible to adopt in order to increase user efficacy with regards to accomplishing online tasks. To facilitate a better understanding of GAL users' proficiency with online ICT, respondents were asked to rate a set of online technologies with regards to how useful they were to them in obtaining online information. Online technologies were rated on a scale of 1 indicating "Not at all important" to 7 indicating "Extremely Important." Table 4 shows that phone

Table 4. Importance of technologies and information formats

Please indicate how important each of the following technologies are to you in obtaining useful online information.	Descriptive Statistics for Overall Sample						Independent Sample T-Tests (Volunteers and Staff)			
	Median	Mean	Std. Dev.	Min	Max	N	t	df	p-value	Mean Difference
Text Messages (i.e. short text messages to computer or phone)	5.000	4.440	1.924	1	7	261	-2.125	218.000	**0.035**	-0.547
Email listservs (i.e. automated email mailing lists)	5.000	4.370	1.744	1	7	256	1.098	213.000	0.274	0.206
Weblog (i.e. a "blog" or a website that has regular entries of commentary, descriptions, of events, or other material in textual, graphical, or video formats)	4.000	3.670	1.674	1	7	260	-0.261	217.000	0.794	-0.059
Wiki (i.e. a collaborative website that can be directly edited by anyone who has access to it)	4.000	3.210	1.727	1	7	256	-0.720	213.000	0.472	-0.165
RSS (i.e. Really Simple Syndication or a family of web feed formats used to publish frequently updated websites such as blogs)	4.000	3.110	1.588	1	7	257	-0.331	214.000	0.741	-0.071
Microblogs (e.g. Twitter, Tumblr)	3.000	2.840	1.574	1	7	257	-1.426	214.000	0.155	-0.304
Social networking websites (e.g. Facebook, LinkedIn)	4.000	3.790	1.891	1	7	259	-3.997	216.000	**0.000**	-0.976
Podcasts (i.e. audio and/or video made available for download via a website)	4.000	3.860	1.842	1	7	256	-1.233	213.000	0.219	-0.306
Video sharing site (e.g. Youtube, Google Video)	4.000	3.810	1.798	1	7	257	-1.270	215.000	0.205	-0.302
Phone technologies (i.e. through phone conversations with knowledgeable people)	6.000	5.520	1.445	1	7	260	-0.942	216.000	0.347	-0.174
Email (i.e. email messages from knowledgeable people)	6.000	6.170	0.966	1	7	260	0.290	217.000	0.772	-0.033
Search engines (e.g. Google search, Bing search, Yahoo search)	6.000	5.910	1.114	1	7	263	-0.671	219.000	0.503	-0.095
How desirable are the following information formats in regards to effectively communicating online information to you?										
Text (i.e. narrative information)	4.000	4.050	0.987	1	5	257	-0.731	217.000	0.465	-0.101
Audio (e.g. MP3s, WAV)	3.000	3.350	1.030	1	5	253	-0.942	213.000	0.347	-0.133
Video (e.g. MOV, MPEG)	4.000	3.500	1.049	1	5	255	-1.387	215.000	0.167	-0.198
Images (i.e. digital photographs)	4.000	3.760	0.895	1	5	256	-1.348	216.000	0.179	-0.162
Data (i.e. numbers and statistics)	4.000	3.730	0.893	1	5	255	1.567	215.000	0.119	0.196
Presentation slides (i.e MS Power Point)	4.000	3.680	0.928	1	5	256	-0.802	217.000	0.424	-0.101
Interactive forms (i.e. Adobe fillable forms and web-based forms)	4.000	3.880	1.041	1	5	256	1.440	216.000	0.151	0.207

continued on following page

Table 4. Continued

	Descriptive Statistics for Overall Sample						Independent Sample T-Tests (Volunteers and Staff)			
Static graphic illustrations (i.e. a flow chart)	4.000	3.470	0.935	1	5	254	1.517	216.000	0.131	0.195

t-values in *italic* indicate that equal variances were not assumed
P-values in **bold** when less than the alpha value (.05)
Mean Difference = Mean of Volunteers - Mean of Staff

conversations with knowledgeable people, emailing knowledgeable people, and the use of online search engines as all having median ratings of six, or "Very Important."

When examining the mean responses for GAL volunteers and program staff for such technologies, *t*-tests indicate differences in the mean responses provided by the two groups concerning *text messages (i.e. short text messages to computer or phone)* and *social network websites (e.g. Facebook, LinkedIn)*. Program staff respondents seem to value both technologies more than do volunteers. Program staff may place a high level of importance on such technologies because they allow them to communicate with large numbers of stakeholders. In the case of Florida GAL, they are actively using social media applications such as Facebook to communicate with their stakeholders.

Perceptions of various information formats were also measured on a five-point scale that ranged from 1—Very Undesirable to 5—Very Desirable with 3 indicating Neutral. Information formats such as *text (i.e. narrative information)*, *video (e.g. MOV, MPEG)*, *images (i.e. digital photographs)*, *data (i.e. numbers and statistics)*, *presentation slides (i.e. MS Power Point)*, *interactive forms (i.e. Adobe fillable forms and web-based forms)*, and *static graphic illustrations (i.e. a flow chart)* all received a rating of at least 4, "Desirable." Only audio formats received a lower rating of "Neutral." As Table 3 shows, no differences between the means of volunteers and program staff are evident based on the use of Student's *t*-tests for information format.

Overall User Perceptions and Satisfaction

Overall satisfaction with the GAL website is addressed in a question which asked respondents to rate the website against six design criteria on a seven-point scale, ranging from 1 – "Poor" to 7 –"Terrific." As Table 5 shows, *effective organization*, *offers user customization*, and *provides significant user interaction* each had a median response of 4 or the mid-point on the scale. The website received slightly more positive ratings for the type and variety of information that it presented with *provides useful information*, *presents a variety of information*, and *provides information such as FAQs (Frequently Asked Questions)* each receiving a median rating of 5. Therefore, it would seem that the GAL website is perceived by users to provide valuable information although it may not present it in an overwhelmingly interactive or customizable manner.

With regards to the evaluation criteria, *t*-tests indicate statistical differences in the mean responses for GAL volunteers and program staff for *effective organization*, *provides information such as FAQs (Frequently Asked Questions)*, *offers user customization*, and *provides significant user interaction*. Volunteers provide higher ratings than do program staff members for these criteria. It is also important to highlight that, there is no statistically different difference in how volunteers and program staff view the type and variety of information. This means that volunteers place a high value on the way in which information is organized, how they can tailor their own online

Table 5. Perceptions of the Guardian ad Litem website

Please rate the Guardian ad Litem website on the following criteria	Descriptive Statistics for Overall Sample						Independent Sample T-Tests (Volunteers and Staff)			
	Median	Mean	Std. Dev.	Min	Max	N	t	df	p-value	Mean Difference
Effective organization	4.000	4.470	1.444	1	7	304	2.194	209.000	**0.029**	0.420
Provides useful information	5.000	4.990	1.379	1	7	308	0.887	211.000	0.376	0.167
Presents a variety of information	5.000	4.910	1.410	1	7	305	1.899	209.000	0.059	0.369
Provides information such as FAQs (Frequently Asked Questions)	5.000	4.610	1.494	1	7	307	2.562	212.000	**0.011**	0.502
Offers user customization	4.000	3.460	1.668	1	7	301	3.585	208.000	**0.000**	0.772
Provides significant user interaction	4.000	3.630	1.607	1	7	300	2.527	207.000	**0.012**	0.528

t-values in *italic* indicate that equal variances were not assumed
P-values in **bold** when less than the alpha value (.05)
Mean Difference = Mean of Volunteers - Mean of Staff

experience, and level of interaction that the website provides. Program staff may be more interested in providing information rather than retrieving information.

Analyses were also conducted to evaluate the possible relationships between demographic variables such as education, age, and the amount of time volunteered with the same variables from the survey. While Spearman's rank correlations did indicate some statistically significant relationships, those correlational relationships were weak. For example, respondent age is negatively correlated to the perceived importance of social networking websites in assisting respondents in obtaining online information. The correlation is statistically significant at $\alpha = .05$ ($p = .000$), but is negative and relatively weak ($r_s = -.309$). This negative correlation between age and the perceived importance of social networking sites is the strongest relationship indicated by the analysis. Even if statistically significant, such bivariate correlations are weak and, therefore, are not discussed further in this chapter.

Recommendations for Redesigning Online ICT for CASA Programs

Findings from Florida GAL may not be fully generalizable to all other state-level CASA programs. CASA programs exhibit a great amount of variation as they are subject to different state laws and other institutional factors. However, GAL does share common issues with many other CASA programs. As discussed in more depth earlier in the chapter, such organizations rely on a broad set of volunteers to achieve program goals and this impacts how they should develop and utilize ICT in an efficient and effective manner. GAL like other CASA programs must consider the variation within its potential pool of voluntary labor when designing online ICT. In addition, this need for voluntary labor leads to a constant need for recruitment, training, and coordination activities in order to successfully achieve program goals. Examination of the Florida GAL program highlights how a state-level CASA program can begin to harness online ICT to achieve programmatic goals and allows for general recommendations to be made regarding the design or redesign of online ICT for CASA programs. These recommendations include the following.

Examining the Use of Online Information Resources by User Groups

CASA organizations such as Florida GAL should conduct regular systematic evaluations of online information resources in order to make sure that such resources meet users' needs in completing tasks that are required to meet organizational goals. As seen in the case of the GAL survey, users of online resources may have different perceptions regarding the importance of information resources. These different perspectives may be associated with various user characteristics or roles within the program's operations. For example, program staff seemed to place importance on the ability to disseminate information regarding the program and its processes online while volunteers seemed to place importance on information that assisted in the advocacy of children. Not surprisingly, perceptions of specific online information seemed to match the users' specific information needs in addressing their own activities. Each user's perception of the ICT system may be valid and each must be understood in order to design and implement a successful system. By understanding the information resources needed by specific users, CASA programs will ultimately increase efficiency and effectiveness of their operations.

Facilitate Online Recruitment, Training, and Retention

The case of Florida GAL illustrates the importance of using online ICT recruit new volunteers to serve not only as child advocates but also as legal experts, community organizers, and program benefactors must be a core goal. Using online ICT as a way to facilitate the program's application process for volunteers and pro bono

attorneys was important for both volunteers and staff. Volunteers placed more importance on design features such as the availability of forms which facilitated their own ability to undertake their activities. Program staff seemed to place more importance on putting application and contact information online. Therefore, both types of stakeholders seemed to value using the online system in a manner that would benefit them as they undertook their role-based activities. In a similar vein, stakeholder groups placed importance on the types of training materials that were available. Volunteers found information resources that provided training focused on instructional materials for teens while staff seemed to place more importance on training materials that would probably provide them heightened levels of efficiency when dealing with legal professionals. Overall, stakeholders felt that online training materials were important. Like Florida GAL, CASA programs must determine which training materials should be made available online and also the format with which such materials should be provided. UCD methods provide a better understanding of the types of training resources that are required by different user groups and also provide insight into how those materials can most effectively be disseminated and used online.

Determine Users' Preferences for Using ICT to Obtain Online Information

An important feature of UCD approaches is developing an understanding of the technology efficacy and needs of the actual users rather than relying solely on the beliefs or perceptions of internal program stakeholders or technology designers. As the Florida GAL survey indicated, users do have preferences in regards to which online technologies they use to obtain information. For example, the GAL survey indicated that relatively standard technologies (e.g. text messages, email listservs, phone technologies, email, and search engines) were highly rated by the program's stakeholders. Most of the preferred technologies existed well before the present-day reliance on social media. This most likely means that the program's stakeholders are not necessarily heavy adopters of cutting-edge online technologies at least in regard to the information they obtain for their GAL-related work. However, program staff did prefer to use social media at higher levels than did volunteers. Such findings may be indicative of the program's growing move toward using social media or it may mean that program staff believe the use of social media applications is an easier means of disseminating information. Such differences may be an artifact of other factors such as volunteers—who may be older—having different needs. Such differences are not necessarily problematic, as long as the program is aware of them.

While these recommendations are by no means exhaustive, they do highlight the need for CASA programs like Florida GAL to better understand the relationships between user characteristics, their preferences, and the activities in which they engage. The use of UCD methodologies to understand such relationships is paramount if such programs truly seek to address the socio-technical nature of the ICT design/redesign process within a highly dynamic institutional environment.

FUTURE RESEARCH DIRECTIONS

This chapter provides a general overview of the issues surrounding the design of online ICT systems for CASA programs. While it provides background to issues surrounding the use of volunteers, user-centered design (UCD), and the broader issues of child advocacy organizations, it has several limitations that future research will seek to address. The primary limitation is the use

of data from only one CASA program, Florida Guardian ad Litem (GAL). The use of this case study is helpful in providing sufficient contextual information to easily highlight some issues of designing ICT for CASA programs. However, the study findings must be considered exploratory in nature since they cannot be fully generalized to the greater population of CASA programs. Future research should attempt to collect information from a larger number of CASA programs in order to gain a better, and more generalizable, understanding of how they can effectively employ ICT to achieve program goals.

In addition to the limitations associated with studying one program, the chapter relies on data collected from only one cross-sectional survey. Typically, UCD-oriented studies include several research techniques that focus on different elements of the design process. For example, usability studies employing experiments may be used to examine specific aspects of interface design, while surveys and interviews may be used to collect data about the user community. Such multi-method approaches generally provide the ability to develop a more holistic understanding of the socio-technical dimensions of the design process. Future steps in this study will expand the data collection by undertaking focus groups with stakeholders of the Florida GAL program.

Finally, this chapter uses survey data to gain a relatively simple understanding of users' perceptions of the programs' website and their own online behavior. Descriptive statistics were primarily employed, coupled with the use of some bivariate techniques to determine the impact of demographic variables. Future studies will use more sophisticated modeling techniques to better account for what may be a more complicated set of multivariate relationships between the sociotechnical dimensions of ICT design and use.

CONCLUSION

As public programs are forced - within the current environment of uncertain budgets - to do more work with fewer financial resources, they often look towards volunteer labor to accomplish core goals. CASA programs, such as Florida GAL, have a significant amount of experience in employing large numbers of volunteers to carry out their primary goal of representing abused and neglected children within the court systems. Such programs also have significant challenges with regards to recruiting, training, and retaining the very volunteers on which they rely. Many programs are looking to online ICT, such as websites or social media sites, to assist in efficiently carrying out activities such as disseminating program mission information, training materials, and assisting with communication between the program staff and other stakeholders. User-centered design (UCD) models provide an inclusive and participatory way of designing and redesigning online ICT systems for CASA programs. Such inclusion allows for a broader understanding of a program's goals and activities, user characteristics, and the organizational environment. More importantly, it provides a means of considering how users interact with ICT design to ultimately accomplish the core goals of such programs.

REFERENCES

Abramowitz, A. (2012a). *2012 Davis Productivity Eagle Award Winner!!!* Retrieved May 5, 2012 from http://gal2.org/2012/04/abramowitz-on-the-davis-productivity-eagle-award/

Abramowitz, A. (2012b). *Statewide Guardian ad Litem program announces overview of 2012 legislative results.* Retrieved May 2, 2012 from http://gal2.org/2012/03/

Bertot, J. C., & Jaeger, P. T. (2006). User-centered e-government: Challenges and benefits for government Web sites. *Government Information Quarterly, 23*(2), 163–168. doi:10.1016/j.giq.2006.02.001

Boumil, M., Freitas, C., & Freitas, D. (2011). Legal and ethical issues confronting Guardian ad Litem practice. *Journal of Law & Family Studies, 13*, 43–80.

Brudney, J. L. (1999). The effective use of volunteers: Best practices for the public sector. *Law and Contemporary Problems, 62*(4), 219–255. doi:10.2307/1192274

Brudney, J. L., & Kellough, J. E. (2000). Volunteers in state government: Involvement, management, and benefits. *Nonprofit and Voluntary Sector Quarterly, 29*(1), 111–130. doi:10.1177/0899764000291007

Bureau of Labor Statistics. (2012). *Volunteering in the United States—2011.* Retrieved from http://www.bls.gov/news.release/volun.nr0.htm

Child Abuse Prevention and Treatment Act of 1974, Pub. L. No. 93-247, 88 Stat., 4 (1974).

Daft, R. L., & Lengel, R. H. (1986). Organizational information requirements, media richness, and structural design. *Management Science, 32*(5), 554–571. doi:10.1287/mnsc.32.5.554

Eason, K. D. (1988). *Information technology and organizational change.* London, UK: Taylor & Francis.

Fla. Stat. §39.820(1) (2011)

Fla. Stat. §39.822(1) (2011)

Florida Department of Children & Families. (2010). *Abramowitz appointed to executive director of the statewide Guardian Ad Litem office.* Retrieved April 3, 2012, from http://www.dcf.state.fl.us/newsroom/pressreleases/20101230_GuardianAdLitem.shtml

Florida Guardian ad Litem Program. (2006). *Standards of operation*. Retrieved July 25, 2012, from http://www.guardianadlitem.org/training_docs/February11/StandardsofOperation.pdf.

Florida Guardian ad Litem Program. (2012). *2011 annual report*. Retrieved April 2, 2012, from http://www.guardianadlitem.org/training_docs/February11/StandardsofOperation.pdf

Frequently Asked Questions. (2012). Retrieved April 2, 2012, from http://www.guardianadlitem.org/vol_faq.asp

Hinnant, C. C., & O'Looney, J. (2007). IT innovation in local government: Theory, issues, and strategies. In David Garson, G. (Ed.), *Public information technology: Policy and management issues* (2nd ed.). Hershey, PA: Idea Group Publishing.

Jansen, J. (2010). *Use of the internet in higher-income households*. Pew Research Center's Internet & American Life Project, Nov. 24, 2010. Retrieved from http://pewinternet.org/Reports/2010/Better-off-households.aspx

Johnson, T. (2012). *Disruptive legal technologies, part 2*. Retrieved from http://www.americanbar.org/groups/departments_offices/legal_technology_resources/resources/articles/youraba0710.html

Justice Administrative Commission. (2011). *Long-range program plan: FY2012-13 through 2016-17*. Retrieved April 4, 2012, from http://floridafiscalportal.state.fl.us/PDFDoc.aspx?ID=6157

Kraemer, K. L., & King, J. L. (2003, September). *Information technology and administrative reform: Will the time after e-government be different?* Paper prepared for the Heinrich Reinermann Schrift fest, Post Graduate School of Administration, Speyer, Germany. Retrieved July 30, 2006, from http://www.crito.uci.edu/publications/pdf/egovernment.pdf

Local Programs. (2012). Retrieved April 2, 2012, from http://www.guardianadlitem.org/partners_main.asp

Mao, J., Vredenburg, K., Smith, P. W., & Carey, T. (2005). The state of user-centered design practice. *Communications of the ACM, 48*(3), 105–109. doi:10.1145/1047671.1047677

National Court Appointed Special Advocate Association. (2010). *2010 annual report*. Retrieved May 2, 2012, from http://nc.casaforchildren.org/files/public/site/communications/AnnualReport2010.pdf

National Court Appointed Special Advocate Association. (2011). *2011 annual report*. Retrieved on August 1, 2012, from http://nc.casaforchildren.org/apps/annualreport/by-the-numbers.html

Palmer, J. W. (2002). Web site usability, design, and performance metrics. *Information Systems Research, 13*(2), 151–167. doi:10.1287/isre.13.2.151.88

Preece, J., Rogers, Y., Sharp, H., Benyon, D., Holland, S., & Carey, T. (1994). *Human computer interaction*. Edinburgh Gate, UK: Addison-Wesley Longman Limited.

Roeger, K. L., Blackwood, A., & Pettijohn, S. L. (2011). *The nonprofit sector in brief: Public charities, giving, and volunteering*. Washington, DC: The Urban Institute.

Schneiderman, B., & Plaisant, C. (2005). *Designing the user interface: Strategies for effective human computer-interaction* (4th ed.). Reading, MA: Addison-Wesley.

Simon, H. A. (1996). *The sciences of the artificial* (3rd ed.). Cambridge, MA: MIT Press.

Wang, L., Bretschneider, S., & Gant, J. (2005). Evaluating Web-based e-government services with a citizen-centric approach. *Proceedings of the 38ᵗʰ Hawaii International Conference on System Sciences (HICSS'05)*, Vol. 5, (p. 129b).

Watson, R. T., & Mundy, B. (2001). A strategic perspective of electronic democracy. *Communications of the ACM, 44*(1), 27–31. doi:10.1145/357489.357499

Zickuhr, K., & Madden, M. (2012). Older adults and internet use: *For the first time, half of adults ages 65 and older are online.* Pew Research Center's Internet & American Life Project, June 6, 2012. Retrieved from http://pewinternet.org/Reports/2012/Older-adults-and-internet-use.aspx

Zickuhr, K., & Smith, A. (2012). *Digital differences.* Pew Research Center's Internet & American Life Project, April 13, 2012. Retrieved from http://pewinternet.org/Reports/2012/Digital-differences.aspx

KEY TERMS AND DEFINITIONS

Court Appointed Special Advocate: A volunteer who is certified to represent and advocate for a child within the courts or the social service system.

Guardian *ad Litem*: Someone who is appointed to represent the interests of another individual who is unable to do so because of age, disability, or impairment.

Information and Communication Technologies (ICT): Audio/video networks, computers, computerized networks, or the broader technology systems that facilitate electronic communication and information processing.

***Pro bono* Attorney:** An attorney who provides legal services without compensation.

User Centered Design (UCD): A design process that focuses on gathering and analyzing information from technology users in an attempt to better understand the tasks and activities for which they will employ the eventual technology artifact.

Volunteer: An individual who undertakes work for a community or public organization without payment.

Chapter 13
Social Media and Citizen Engagement:
Two Cases from the Philippines

Charlie E. Cabotaje
Center for Leadership, Citizenship and Democracy (CLCD), University of the Philippines, Philippines

Erwin A. Alampay
Center for Leadership, Citizenship and Democracy (CLCD), University of the Philippines, Philippines

ABSTRACT

Increased access and the convenience of participation to and through the internet encourage connectivity among citizens. These new and enhanced connections are no longer dependent on real-life, face-to-face interactions, and are less restricted by the boundaries of time and space (Frissen, 2005). In this chapter, two cases from the Philippines are documented and assessed in order to look at online citizen engagement. The first case looks at how people participate in promoting tourism in the Philippines through social media. The second case involves their use of social media for disaster response. Previous studies on ICTs and participation in the Philippines have looked at the role of intermediaries (see Alampay, 2002). Since then, the role of social media, in particular that of Facebook and Twitter, has grown dramatically and at times completely circumvents traditional notions of intermediation. The role of Facebook, in particular, will be highlighted in this chapter, and the authors will analyze its effectiveness, vis-à-vis traditional government channels for communication and delivery of similar services. By looking at these two cases and assessing the abovementioned aspects, it is hoped that the use of social media can be seen as an integral part of e-governance especially in engaging citizens to participate in local and national governance.

DOI: 10.4018/978-1-4666-3640-8.ch013

INTRODUCTION

Information and communication technologies (ICTs) provide a new arena for citizen engagement. More specifically, the Internet has provided a new media for people to get information and be involved in issues affecting society. Frissen (2005) argued for this specific role of the Internet in promoting new and diverse forms of citizen engagement. As such, the use of the Internet provides a venue for people with the same advocacies and interests to get connected and facilitate their collective actions toward the same objectives. There is increasing use of information and communication technologies (ICTs) for people to interact and transact with government. In the Philippines, this need to encourage greater use of ICTs for dealing with citizens was recognized nearly a decade ago with the passage of the e-Commerce Act in 2000. This Act called for government institutions to make services available online. Through the use of ICTs, services will be faster, more reliable, and can be more accessible, aside from other benefits.

This chapter looks into how citizens used a form of ICT-enabled interactive social media, the Facebook, to amplify and support government programs and services. In particular, two cases from the Philippines are documented to look at the citizen engagement through Facebook. The first case deals with the tourism program of the Philippine government that promotes the country to be "more fun". The second case describes how citizens were able to mobilize for disaster response. Through these cases, the facilitative role of Facebook is highlighted and compared with traditional government channels.

This chapter used literature review and content analysis of user generated information and material in the Internet during specific periods when the government campaigns or interventions occurred. Review of existing literature on e-government and citizen participation provides the background and foundation for the discussions

on this chapter. Content analysis looked at online discussions pertaining to the two cases in order to gather themes about citizen's comments and opinions. Specific themes revolved around the positive and negative feedbacks of internet users in the tourism program, and "calls" for assistance and subsequent actions generated by different users in the case of disaster response.

LITERATURE SURVEY

This section tackles literature survey on the roles of ICTs, the concept of governance, citizen participation, and intermediaries and their roles. The specific role of ICTs in providing information and its importance in promoting citizen participation and improving the process of governance is highlighted. The different roles of the intermediaries in harnessing ICTs in service provision are likewise given significance. Under the broad concept of e-governance, this literature survey also dove-tailed the concepts of citizen participation, intermediaries, and ICTs in looking at a new paradigm of improving government service provision.

ICTs, Governance, and Citizen Participation

Information and communication technologies (ICTs) touch many aspects of society including how society is governed. In particular, ICT's role in the aspect of governance is gaining ground. ICTs have been used to affect changes in and transform how government works. As identified by Malkia, Anttiroiko, and Savolainen (2004), one of the major factors that affect transformations in governance is the utilization of new ICTs, along with the changing role of knowledge, and forms of social organization and co-operation, along with globalization. These factors interplay with each other, with ICT as a catalyst in providing relevant knowledge and information among the citizenry and other stakeholders, creating new dynamics in virtual social organizations, and expanding networks across different locations.

ICTs, specifically the Internet, serve as an important source of information sharing and distribution in the information society and one of the driving forces that makes citizens informed and involved in issues of interest to them. The Internet functions as the backbone in promoting new and diverse forms of citizen engagement. Its use provides a venue for people, who share the same advocacies and interests, to get connected and facilitates collective actions toward specific issues and concerns. Access and the participation to and through the Internet encourage connections that are not dependent on real-life, face-to-face interactions, and are less limited by time and space (Frissen, 2005).

Citizens, government and other stakeholders are now more capable of accessing and providing information through ICTs on matters that affect the society including the provision of social services through more e-Government services. E-government can be broadly defined as "not merely confined to use of the web and/or Internet-based applications in government, instead it encompasses all use of digital information technology (primarily computers and networks) in the public sector" (Heeks, 2006, p.2). With proper information, citizens are made aware of developments on social services that affect their day-to-day life. As such, people can provide inputs and feedbacks on matters that are of their interest at a more opportune time. Atoev and Duncombe (2011) highlighted the importance of empowered participative citizens in the success of e-government. In Asia, e-Government has progressed dramatically, such that it is already at a stage, at least in developed ICT markets, where the availability of e-government services is a given (Sreenivisan & Singh, 2009).

In the realm of democratic e-governance, Malkia, Anttiroiko, and Savolainen (2004) identified ICTs as facilitators of interaction, communication and decision-making processes and that ICTs

have the potential to strengthen the democratic aspects of governance. This is important since citizen engagement depends on the robustness of the democratic environment that a government presents. In a robust democratic government, citizens experience genuine participation in governance as compared against token participation. Bautista (2006), for instance, showed that genuine participation generates the most satisfaction and commitment among the citizens. It motivates them to see through to completion the programs/projects/activities. With enough access to information, citizens can genuinely participate in the different phases of government programs/projects/activities (e.g., information gathering, processing/analysis of data, implementation, and monitoring and evaluation). Supporting this view is Anttiroiko's (2004) definition of democratic governance which emphasized the interactions between citizens, political representatives, and administrative machinery, and giving opportunities to citizens in influencing and participating in policy-making, development, and service processes. Through the use of ICTs, citizens are able to access information which in return can empower them and increase the possibilities of their participation in service provision.

Citizen's participation forms part of democratic governance. Bautista (2006) argued that citizen participation in government enriches transparency because people get to see directly who are involved in key decision-making processes and who could even be held accountable for the decisions they make. With ICTs, citizens participate in issues that are of their interest and that they have direct inputs which they consider responsive to the problems. With the use of ICTs, citizens can directly monitor whether their contributions have been put to use. Through citizen participation and ICT, crafted decisions through different online discussions are responsive or relevant to citizens needs since discussions and decisions are grounded on realities and information that are ongoing in the community. The legitimacy of information can be achieved since citizens who are directly affected by a problem or an issue can provide up-to-date information through the medium. Concerted citizen participation can bring about more impact in the community as compared to individual participation (Bautista, 2006). Through the use of ICTs, and social media, citizens can better leverage the convergence of their interests and their inputs can then generate more impact.

On the other hand, citizen's participation on the administrative side of the government entails several elements. In particular, two important elements should be identified when participation is applied in the context of ICTs. One is the use of information which can be turned into knowledge and the other is the changing forms of organization that cater to the needs of ICT users.

On the use of information, Malkia, Anttiroiko and Savolainen (2004) viewed that the administrative side of the government requires regular and continuing input from the public in order for government to operate effectively and responsibly. Citizens can then provide inputs and interact more with their government if they are better informed and more knowledgeable on issues that they want to engage into. Information prompts people to act, organize, and devise solutions (Mackinnon, 2012). Access to information will be considered increasingly for citizens to participate not just for convenience but also as a basic human right to be informed (Malkia & Savolainen, 2004). Through proper information, citizens can be empowered to make decisions, inspired to take actions, and educated and included in the realms of the society. Public involvement can be enhanced further with the impact of the information revolution through ICTs which include access to more information, faster access, easier interaction, irrelevance of distance and location and low cost (Thomas, 2004). However, easy access to information can also lead to information over load which can negatively affect the decision making process of individuals and organizations (e.g., delays and confusions). Also, the reliability of information

will always be a threat and an issue with citizen engagement through ICTs. This also suggests the importance of citizens trusting their information intermediaries.

On the changing forms of organization, it has been seen that changes in communication technologies are tightly linked to changes in organizations (Fulk and Desanctis, 1999). Some forms of organizations shift from centralized, authoritarian, unipolar, machine-like organizations to decentralized, democratic, multi-polar, organistic modes of interaction (Malkia and Savolainen, 2004). They are not guided by formal hierarchy, but instead by rules, convention and agreements that are coded in an information infrastructure in which they participate, craft and form (Homburg, 2008). Such organizational forms are based on respect and trust (Malkia and Savolainen, 2004).

Citizen's participation rest on the interest and motivation of people to engage. As such, citizens engage on subjects that matter to them, which are both local and relevant (Jones, Hackney, & Irani, 2007). As such, examples of how ICTs encourages citizen participation by generating interest and self-motivations are explored in the succeeding two cases. They illustrate how partnership and inclusion among and with citizens are made to help government deliver services, and solve problems.

Intermediaries and Their Roles

More specific aspects of governance in which ICTs can make a difference, are the following: (1) Information and knowledge processes; (2) Communication and learning; (3) Organization and management; (4) Social capital; and (5) Democratic control (Anttiroiko 2004). As the medium for information sharing and access, ICT can efficiently distribute information to citizens and other stakeholders. In the process, this information builds a larger stock of information and knowledge which are readily available to the citizens. These stocks of information from previous situations can be used as the foundation for providing services in similar situations (e.g., disaster response, program awareness). The limitation to this however is the validity and reliability of information. Since anybody can have access and can provide information through the use of ICTs, there is a possibility that some information might be invalid and unreliable hence the need for highly trusted intermediaries. This also pre-supposes that people are able to access data and have the ability to assess if they are useful and applicable to their situation, which would also require additional resources (e.g. social, economic, etc), for them to truly act on the opportunity (Alampay, Heeks, & Soliva, 2003).

The convenience that ICTs provide also comes with a corresponding responsibility from the individual/user. However, regulating responsible use is not easy to do in the real world and leads to the need for intermediaries. Intermediaries are the "go-betweens" who help bridge the information divide and can be real (e.g., government, non-government organizations, community groups and religious societies) or virtual (e.g., websites or Internet links) that can push and retrieve information originating from government to citizens and vice-versa (Alampay, 2002). Since they serve as bridges, they also simultaneously serve as validators and filters of information. It is of the interest of these intermediaries that the information that go in and out of their "organization" are verified and valid and that they provide control on the quality and responsiveness of the information that flows through them. Otherwise, these intermediaries lose their credibility and trust of their users.

Intermediaries, as "go-betweens" and validators and filters of information, are valuable contributors in providing better access to information. Since citizen participation through ICTs entails resources (e.g., time, effort, financial, etc.) from the users, it is important that people trust their intermediaries and the information that they provide before they commit their resources. People put their trust on intermediaries that can provide them with reliable and timely information. Having a better source of information inspires and builds a sense of inclusion and trust for citizens to get involved and participate (Malkia and Savolainen, 2004), whether it be in markets or with government.

In performing their roles, intermediaries and citizen users creates a new form of discussion (on-line). Discussion needs to be structured and it should follow certain protocols (Malkia & Savolainen, 2004). As providers of information and the arena where information sharing takes place, intermediaries need to manage these new forms of discussion and develop rules and regulations that pertain to information sharing among its users. These conditions are needed for meaningful and non-fragmented on-line discussions. Fragmented and unstructured on-line discussions tend to discourage people to get involved and support the initiatives. It is important that citizens are not reluctant to utilize these information, engage in discussions and take initiatives so that potentials of ICTs are fully utilized towards citizen engagement. Otherwise, facilities and management support systems cannot offer much help if people are reluctant to use them and to rely on them when coordinating their activities (Anttiroiko, 2004). They could also lead to fragmented and biased participation that are limited to a certain sector of society.

Intermediaries must be able to project and prove to the ICT users that they provide alternative and improved means of information access, and in return a better and more accessible way of service provision. This is not to say however that intermediaries need to completely abandon or compete with the existing traditional structure of service provision but to be able to complement or supplement them. As Anttiroiko (2004) points out, institutional mediation tools of democratic governance are needed to facilitate civic involvement with a fundamental aim of supplementing the representative system that is considered by many to be too hierarchical, inflexible, and distant from the point of view of ordinary citizens. Intermediaries can therefore, also facilitate citizen's participation.

CASE STUDIES

Situating citizens' participation through ICT, two cases are presented in this paper. Specifically, the use of the social networking application of Facebook is the focus of these two case studies.

Since its establishment in 2004, Facebook users worldwide continue to grow with 901 million monthly active users at the end of March 2012 (Facebook, 2012). In the Philippines, there are 27,088,320 Facebook users[1]. This makes the Philippines the 8th largest group on Facebook (http://www.socialbakers.com/facebook-statistics/philippines).[2] The growth in number of Facebook users is a testament to Facebook's popularity as the preferred social media of people around the globe. People use it for different purposes but it is mainly used by people to stay connected with family and friends, to know what's going on in the world and to share and express what matters to them (Facebook, 2012). Facebook, as a Web 2.0 application, encourages more user interaction on the web. Kamel Boulus and Wheelert (2007) described Web 2.0 technologies as enabler of collaboration, flexibility (where users act as readers and writers), participation, and interactivity

between users. It is in this light that people may also find the use of Facebook as an opportunity to advance their interests and personal advocacies.

The first case tackles the Philippine government's tourism campaign dubbed "It's More Fun in the Philippines." The second case focuses on the use of Facebook by citizens' for disaster response to identify needs and match this with contributions from citizens.

It's More Fun in the Philippines

The Philippine government's Department of Tourism launched its new campaign, "It's More Fun in the Philippines", on January 2012. The campaign deliberately aimed to encourage people to share their views on why it was more fun in the Philippines and take advantage of the fact that Filipinos are one of the most active demographics on Facebook. Hence, through various memes, people were able to project their own version of the slogan. The DoT campaign started with just three photo studies and there are now over 12,000 versions of the material (Garriga, 2012). Figure 1 presents one of the three photographs that jumpstarted the campaign. The photograph shows the Banaue Rice Terraces, which is considered as one of the eight wonders of the world. While showing the terraces, the campaign simply says "Going upstairs. (is) more fun in the Philippines."

Figure 1. One of the three photographs that was used in the campaign (Source: www.itsmorefuninthephilippines.com)

Over a span of a few days, the campaign went viral on Facebook and citizens began posting their own memes of the campaign. Figure 2 shows one of the thousand memes created by individuals that are circling the net and Facebook accounts. The picture reflects the Filipino culture of "Bayanihan"- Filipino version of volunteerism wherein people are helping a neighbour in transferring a house. This activity was creatively described through the picture that changing address is being more fun in the Philippines.

As such, the campaign that was actively launched through Facebook harnesses the creativity of Filipinos, while taking advantage of their active presence in the social network. With the great number of Facebook users worldwide and specifically in the Philippines, the campaign was able to penetrate the consciousness of not just the local tourists but also the foreigners. The tourism slogan is prominent in Facebook and in Twitter. Cruz (2012) described that there has been an incredible onslaught of memes in the digital community since the campaign was launched. It is only recently that a video advertisement was shown in an international news channel. But still, the video advertisement features various memes that were donated by several local photographers (Arnaldo, 2012).

The accessibility of internet and the popularity of using Facebook contributed to the fast proliferation of the slogan, with a very minimal cost to the government. Citizens were motivated to get involved in the campaign since inputs (e.g., pictures and personal information) directly came from them. Interesting contributions from others, were then re-sent, and kept the campaign alive. Information is also available and that they have an arena to contribute their creations. Several websites (e.g., abuggedlife.com and morefunmaker. com), which serve as intermediaries, provide the arena on how citizens can make their own memes and contribute in promoting the slogan. The slogan also gave some Facebook users the idea and avenue on how to share their personal experiences in the Philippines.

Figure 2. One of the Memes created by individual net users depicting the spirit of "Bayanihan"- Filipino's version of volunteerism (Source: www.spot.ph)

The proliferation of the slogan circumvents the notion that promoting government programs is the sole responsibility of the government. The campaign shows that given the access, proper information and right motivation, citizens will be willing to participate in promoting and supporting a government program.

As with any other government programs, the slogan has received positive and negative reviews and effects. Foremost among the positive review is that the slogan was coined to be used not just for usual media (e.g., TV, print and radio) but for promotion through the social network (e.g., Facebook and Twitter). Below are some of the comments posted in the discussion board of designpinoy.com (2012):

It's certainly a a big improvement from the previously failed, Pilipinas Kay Ganda (Philippines is Beautiful), campaign... The best part of this campaign was having social media in mind.

I have much hopes for this campaign, since it calls to challenge everyone to prove why #itsmorefun-inthephilippines

...At first, I thought it was hyped too much. But when you join the fun, you believe the tagline.

...the simple wordings can easily connect and be understood by any nationality who wishes to know more about the Philippines.... We had fun with it by making our own slogans from previous travels in the Philippines

As posted in the Designpinoy (2012) the slogan is easier to understand, modern looking and designed for social media participation. Other Facebook users believed that by giving the slogan a chance and joining in the fun, they are already contributing something for the country. These reviews project Bautista's (2006) argument that citizen participation in government enriches transparency and ownership. In this case, citizens were able to give their inputs freely and monitor the government initiative since they were also users of the system, which is the Facebook.

On the other hand, the use of the Facebook also has some limitations. For one, non-internet savvy people and those who do not have Facebook account may feel left out and may not be able to contribute and support the program. The information flow for those who have no Facebook account are limited to reviews from other media and other people. In this aspect it is important that this slogan/program finds its way through other forms of media (such as TV) and that other people who have no Facebook account will fell included and be able to contribute. The other limitation for this kind of ICT-based government program is that it gives opportunity for other people to mock and post negative memes that will defeat the purpose of promoting tourism in the Philippines. As can be seen in Figure 3, some users were able to make fun of the campaign. Instead of promoting tourism in the Philippines, which is the main objective of the slogan, the user projects the bad traffic condition in the country with a person crossing a main highway with no regards to pedestrian and traffic rules.

All in all, however, they were still in the spirit of fun, and fits the theme of the campaign. While one cannot totally control what is in the minds of users, by opening contributions to everyone, means government can access all ideas, including the best ones for free.

Disaster Response

The second case deals with citizens' participation in responding to a disaster caused by the tropical storm (TS) Sendong (international name Washi). TS Sendong ravaged Mindanao, the southern part of the Philippines, and left more than a thousand dead people. It affected 29 municipalities, 278 barangays and 53,240 families with hundreds of people reported missing (Philippine National Red Cross, 2012).

Figure 3. One of the citizens generated memes that spoofs the campaign (Source: http://www.creativeguerrillamarketing.com/viral-marketing/its-more-fun-in-the-philippines-crowd-sourced-marketing/)

The severity of the damage cause by TS Sendong was aggravated with the location of the affected communities. Being at the southern part of the country, transport of relief goods was an issue. Some areas were isolated for a time after the storm, and there was little if no information reported in the mainstream media (i.e., television, radio and newspaper) in some of these areas. On these isolated areas, some people relied on first hand accounts transmitted via SMS by people who were directly affected. People who received these information then relayed them through Facebook and other ICTs.

The roles of Facebook in this particular case, is that it became a medium for people to relay information (with corresponding pictures and captions), integrate the initiatives of concerned citizens (e.g., donations and volunteer work), and identify the neglected communities that required immediate help. Citizens were able to aggregate reports by making dedicated Facebook pages for

the relief efforts. It was open to everyone, and this allowed people with Facebook accounts to report first hand information and allowed those who were interested to help to identify what needs to be contributed for whom, and how this can be sent. Below are some of the actual posts from the Sendong Relief Operations Facebook account (2012):

missing child in CDO*, pls. see link above.

saon mag send ug mga clothes kung wala ka sa cdo? and kung asa e send – (How can I send clothes if I'm not in CDO and where can I send them?)**

Anyone in Cavite going to Manila tonight? We have a lot of boxes, (relief goods, medicines and what not) from surgeons waiting to be picked up. If you know anyone please comment here asap. Thank you

" am looking for a place where I can volunteer myself since my money is not that many

hmm.. taga makati po ako.. san po b malapit na pwedeng idonate ang mga lumang damit pra sa mga sendong victims?.." – (I am from Makati. Where is the nearest site that I can donate used clothes for Sendong victims?)*

As such, Facebook became a venue for matching available volunteer resources with the immediate needs for responding to the on-going disaster in real time. It also became a channel for people to call on the attention of government agencies to address their pressing concerns. Nongovernment, private organizations and even individuals, became intermediaries for relaying information and aggregating donations. Through the use of Facebook and through the intermediaries, some

immediate concerns of the communities were identified. Figure 4 reflects the immediate needs of evacuees such as cooking utensils, instant food items, medicines, etc., which were relayed through the Facebook account of the student publication of a local university (http://www.facebook.com/thecrusaderpublication).

This kind of service properly targeted the areas with corresponding services that they immediately need. This is in contrast with the traditional models of disaster response where relief goods usually consist of food and clothes. Through information sharing on Facebook, citizens, government agencies, and non-government and private organizations realized that some food and clothes were already provided, and the more immediate needs of some communities were medicines and drinking water.

Figure 4. A call for specific needs of the evacuees as posted through the Facebook account of the student publication of a local university (Source: http://www.facebook.com/thecrusaderpublication)

The response generated through Facebook created new forms of organizational structure for service provision. Cash donations needed the arrangements among the intermediaries and financial institutions (e.g., bank transfer). In-kind donations needed an organizational structure that involved the donor, collection points and logistical transport services. For those who wanted to volunteer, systems had to be developed to properly organize their efforts and optimize their contributions.

The use of Facebook greatly contributed to the access and relay of information coming from many diverse sources. It became an effective media to transport timely information. With its multi-media capability, it became a powerful tool to elicit and organize support from ordinary citizens. Pictures and videos of affected people have prompted more people to act and do their own share in responding to the catastrophe caused by TS Sendong, which would not have as easily been seen considering that the disaster occurred far from Manila. On the other hand, the reliability and validity of information remains a threat for effective disaster response. Unscrupulous individuals could also take advantage of the situation and post information for their own gain. As Mackinnon (2012) points out, power in digital world should be constrained, balanced and held accountable in order to prevent manipulation and abuse. Also with the relay of information from person to person, the validity of the information might be compromised.

CONCLUSION

The two cases has shown how social media can be harnessed by governments to engage citizens in the delivery of their programs and services. By leveraging the popularity among its citizens to use social media like Facebook, the government is able to harness the ideas, knowledge, and skills of a pool of people much larger than the limited bureaucracy under its direct control. The irony, however, is that some government agencies, remain ambivalent about social media, that it also prohibit its use in the workplace, and highlights the contradictions on the value of ICTs in organizations (Alampay, Olpoc, & Hechanova, 2012).

The two cases demonstrated that ICTs, in particular the use of Facebook, facilitated the interaction, communication, and decision-making processes which are key elements in fostering democratic e-governance. The promotion of transparency in the two cases is evident that people are able to monitor the progress of a program (It's more fun campaign) as well as implications or effects of their inputs (disaster response).

The intermediaries have proven their key roles in these two cases. They created and managed a new form of discussion and organization which they used to identify pressing needs and allocate available resources (in the case of disaster response) and promote a government program using the initiatives and creativity of the people (in the case of the tourism campaign). Indeed, intermediaries were able to prove that they can provide alternative and improved provision of information and a more accessible way of service provision. The ability of the intermediaries to complement and supplement the initiatives of the government is a critical factor in improving service provision.

As such, policies and strategies for encouraging more citizenship participation through new media, need to be developed by government. This would require a paradigm shift in how services are delivered. It may require more coordination and less planning as it becomes more networked and less centrally controlled. Policies should encourage these emerging innovations as it can only encourage and motivate more people to go online, and make them engaged participants in the governance of their own affairs.

REFERENCES

Alampay, E. (2002). People's participation, consensus-building and transparency through ICTs: Issues and challenges for governance in the Philippines. *Kasarinlan, 17*(2), 273–292.

Alampay, E., Heeks, R., & Soliva, P. (2003). *Bridging the information divide: A Philippine guidebook on ICTs for development.* Quezon City, Philippines: NCPAG, IDPM.

Alampay, E. A., Olpoc, J. C., & Hechanova, R. M. (2012). Competing values regarding internet use in "free" Philippine social institutions. In Deibert, R., Palfrey, J., Rohozinski, R., & Zittrain, J. (Eds.), *Access contested: Security, identity and resistance in Asian cyberspace* (pp. 115–132). Cambridge, MA: MIT Press.

Anttiroiko, A. (2004). Introduction to democratic e-governance. In Malkia, M., Anttiroiko, A., & Savolainen, R. (Eds.), *E-Transformation in governance: New directions in government and politics* (pp. 22–50). Hershey, PA: Idea Group Publishing.

Arnaldo, M. S. F. (2012). *Aquino admininistration to spend P63M for CNN ads.* Retrieved May 2, 2012, from http://www.interaksyon.com/article/30764/aquino-admininistration-to-spend-p63m-for-cnn-ads

Atoev, A., & Duncombe, R. (2011). E-citizen capability development. *Proceedings of the 5th International Conference on Theory and Practice of Electronic Governance,* Estonia.

Bautista, V. A. (2006). Introduction and rationale of the study. In Bautista, V. A., & Alfonso, O. M. (Eds.), *Citizen participation in rural poverty alleviation* (pp. 1–10). Quezon City, Philippines: Center for Leadership, Citizenship and Democracy.

Cruz, X. (2012). *It's more fun in the Philippines crowd sourced marketing.* Retrieved May 25, 2012, from http://www.creativeguerrillamarketing.com/viral-marketing/its-more-fun-in-the-philippines-crowdsourced-marketing/

DesignPinoy. (2012). *It's more fun in the Philippines-Filipino tourism.* Retrieved April 24, 2012, from http://designpinoy.com/its-more-fun-in-the-philippines-filipino-tourism/

Facebook. (2012). *Newsroom.* Retrieved May 11, 2012, from http://newsroom.fb.com/content/default.aspx?NewsAreaId=22

Frissen, V. A. J. (2005). The E-mancipation of the citizen and the future of e-government: Reflections on ICT and citizens' partnership. In Khosrow-Pour, M. (Ed.), *Practicing e-government: A global perspective* (pp. 163–178). Hershey, PA: Idea Group Publishing. doi:10.4018/978-1-59140-637-2.ch008

Fulk, J., & Desanctis, G. (1999). Articulation of communication technology and organization form. In De Sanctis, G., & Fulk, J. (Eds.), *Shaping organization form, communication, connection and community.* Thousand Oaks, CA: Sage.

Garriga, N. (2012). *DOT's "It's More Fun in the Philippines" campaign goes global.* Retrieved May 14, 2012, from http://ph.news.yahoo.com/dots-more-fun-philippines-campaign-goes-global-105213492.html

Heeks, R. (2006). *Implementing and managing e-government: An international text.* London, UK: Sage Publications Ltd.

Homburg, V. (2008). *Understanding e-government: Information systems in public administration.* London, UK: Routledge.

It's More Fun in the Philippines. (2012). Retrieved May 18, 2012, from http://itsmorefuninthephilippines.com/

Jones, S., Hackney, R., & Irani, Z. (2007). Towards e-government transformation: Conceptualising citizen engagement (A research note). *Transforming Government: People. Process and Policy, 1*(2), 145–152.

Kamel Boulos, M. N., & Wheeler, S. (2007). The emerging Web 2.0 social software: An enabling suite of sociable technologies in health and health care education. *Health Information and Libraries Journal, 24*, 2–23. doi:10.1111/j.1471-1842.2007.00701.x

Mackinon, R. (2012). The netizen. *Development, 55*(2), 201–204. doi:10.1057/dev.2012.5

Malkia, M., Anttiroiko, A., & Savolainen, R. (2004). Background of the project. In Malkia, M., Anttiroiko, A., & Savolainen, R. (Eds.), *E-transformation in governance: New directions in government and politics* (pp. vii–xiv). Hershey, PA: Idea Group Publishing.

Malkia, M., & Savolainen, R. (2004). Etransformation in government, politics and society: Conceptual framework and introduction. In Malkia, M., Anttiroiko, A., & Savolainen, R. (Eds.), *Etransformation in governance: New directions in government and politics* (pp. 1–21). Hershey, PA: Idea Group Publishing.

Philippine National Red Cross. (2012). *Donate now*. Retrieved May 11, 2012, from http://www.redcross.org.ph/donatenow

Sendong Relief Operations Facebook Account. (2012). Retrieved May 10, 2012, from www.Facebook.com/pages/Sendong-Relief-Operations

Socialbakers. (2012). *Facebook statistics*. Retrieved May 11, 2012, from http://www.socialbakers.com/facebook-statistics/philippines

Spot.ph. (2012). *"Its More Fun in the Philippines" meme: Top 30 fun photos on the web*. Retrieved May 25, 2012, from http://www.spot.ph/featured/50181/its-more-fun-in-the-philippines-meme-top-30-fun-photos-on-the-web

Sreenivisan, R., & Singh, A. (2009). An overview of regulatory approaches to ICTs in Asia and thoughts on best practices for the future. In Akhtar, S., & Arinto, P. (Eds.), *Digital review of Asia Pacific 2009-2010* (pp. 15–24). Haryana, India: Orbicom, IDRC and Sage Publications.

The Crusader Publication Facebook Account. (2012). Retrieved June 11, 2012, from http://www.facebook.com/thecrusaderpublication

Thomas, J. C. (2004). Public involvement in public administration in the information age: Speculations on the effects of technology. In Malkia, M., Anttiroiko, A., & Savolainen, R. (Eds.), *Etransformation in governance: New directions in government and politics* (pp. 67–84). Hershey, PA: Idea Group Publishing.

KEY TERMS AND DEFINITIONS

Citizen Participation: Process where people are able to engage in programs, such as government service provision, that will improve their own and/or others well being.

Democratic E-Governance: Emphasized the role of ICTs in promoting the interactions between citizens, political representatives, and administrative machinery, and giving opportunities to citizens in influencing and participating in policy-making, development, and service processes.

E-Government: Encompasses all use of digital information technology (primarily computers and networks) in the public sector with the aim of improving service provision.

Intermediaries: "Go-betweens" who help bridge the information divide and can be real (e.g., government, non-government organizations, community groups and religious societies) or virtual (e.g., websites or Internet links) that can push and retrieve information originating from government to citizens and vice-versa.

Meme: An idea, behavior, style, or usage that spreads from person to person within a culture

Social Media: Form of electronic communication where online communities exist for purpose of sharing information.

Web 2.0: Online applications that serve as media for facilitating online interactive experiences among users.

ENDNOTES

[1]   As of the latest posting from the website www.socialbakers.com

[2]   Women constitute 52% of these users and majority of Facebook users in the Philippines are within the age range 18-24 (39%) and 25-34 (24%) (Socialbakers 2012). In terms of penetration, the total number of Facebook users is 27.12% compared to Philippines population and 91.21% in relation to number of internet users (Socialbakers 2012).

*   CDO stands for Cagayan de Oro City, one of the worst hit areas in the Southern Philippines by TS Sendong

**   Translations provided by the authors

Chapter 14
Low Cost and Human-Centered Innovations in Healthcare Services:
A Case of Excellence in Italy

Emanuele Padovani
University of Bologna, Italy

Rebecca L. Orelli
University of Bologna, Italy

Vanni Agnoletti
Morgagni-Pierantoni Hospital Forlì, Italy

Matteo Buccioli
Morgagni-Pierantoni Hospital Forlì, Italy

ABSTRACT

This chapter focuses on a change effort for introduction of an e-governance innovation in the operating room management of a medium-sized Italian hospital, which led to higher levels of efficiency and effectiveness at once. The innovative project has made all the stages of the surgical process transparent, highlighting where there is an opportunity to improve overall performance via the introduction of organizational and process innovations. New techniques implemented and the specific factors that led to the hospital's success in achieving improved outcomes at lower costs are discussed. The chapter concludes by highlighting that low cost and human-centricity are amongst the key characteristics of success of this innovation.

DOI: 10.4018/978-1-4666-3640-8.ch014

1. INTRODUCTION

Implementing innovation is difficult in almost any organization, but it is especially so in those where the change effort must overcome the resistance of professionals. Professionals often have deeply entrenched values that are not necessarily consistent with — and often are in direct opposition to — the goals of the organization's senior management team.

Nowhere is this dilemma more evident than in the healthcare sector, where there is considerable body of evidence to suggest that physicians (and, to a lesser extent, nurses) have an agenda that often is in total contrast to that of non-clinical managers (Young & Saltman, 1985; Kitchener, 1999; Young, 2008).

We examined a change effort for introduction of e-governance innovation in the operating rooms of a medium-sized hospital. This hospital — Forlì Hospital, in Forlì Italy — had recently received an award from the European Institute for Public Administration[1] as a public healthcare organization in Europe that represents the best case of *Smart Public Service Delivery in a Cold Economic Climate*.

The project developed by the hospital made all the stages of the surgical process transparent, highlighting where there were opportunities to improve overall performance via the introduction of organizational and process changes. The project did not involve a significant technological innovation; indeed, it followed an e-governance approach that had been demonstrated to be effective elsewhere (Young, 2004). More importantly, the goal was to give managers (both clinical and non-clinical) the tools they needed to better plan and manage all available resources, as well as to reduce clinical risks, by measuring – and thus managing – performances. In fact, as it is discussed below this innovation refers more to its human-centered philosophy.

The chapter is organized as follows. The second section refers to the fundamentals of measuring and managing performance in healthcare organizations. The third section presents the factors that influence the implementation of successful innovations in healthcare. Section number four introduces the case that is analyzed in the following two sections, in terms of the basic ideas – i.e. the context of change – in section five, and the factors of change and how they were handled, in section six. In the seventh section some relevant conclusions are presented.

2. MEASURING AND MANAGING PERFORMANCE IN A HEALTHCARE ENVIRONMENT

The development of tools to better measure performance, improve efficiency, and increase accountability for results is on the agenda of many public sector organizations (Pollitt 2008; Van Dooren et al., 2010; Pollitt & Bouckaert, 2011). The goal of improving efficiency and effectiveness is not only a matter of managerial rationality but is also a political issue in many OECD countries. Indeed, more than two-thirds of OECD countries include non-financial performance data in the documentation available to managers and policy makers (internal use), and also provide reports on performance to the public (external use) (OECD, 2005).

From an external-use perspective, transparency has become a widespread symbol of "good governance" in many different contexts (O'Neill, 2008). It is a key element for public sector organizations that wish to become more accountable to their stakeholders, giving them the possibility to be involved in decision making (Pasquier & Villeneuve, 2012). Moreover, the collection of information through performance measurement can assist these organizations to move toward an improved allocation of resources through management control systems (Padovani & Young, 2012).

Internally, much of the recent focus has been on key performance indicators (KPIs). Designed properly, KPIs not only measure performance, but create incentives that help to align individual goals with the objectives of the organization, provide valuable feedback on the progress towards these objectives, and form the basis for internal and external accountability (Cavalluzzo & Ittner, 2004; Kravchuk & Schack, 1996).

Clinicians (both physicians and nurses) often are an impediment to the development of KPIs in healthcare organizations (Young, 2008). In part, this is because clinicians' decision-making processes typically do not include efficiency evaluations. Rather, their procedures for monitoring and evaluating clinical practice are patient centered. As a result, they may perceive the introduction of KPIs aimed at *economic thinking* as distant from their values (Kurunmaki, 1999) and perhaps even as a direct attack on the their profession norms (Young & Saltman, 1985; Maddock & Morgan, 1998). As a result, their patient centered thinking is likely to conflict with the implementation of KPIs.

3. CHALLENGING INNOVATION

An issue of great importance in a change effort in healthcare organizations is what Pettigrew and Lapsley (1994) call the "context" for change. This context encompasses the key factors influencing change, especially in organizations facing similar environmental and policy pressures. Whipp et al. (1987) agree, arguing that it is important to examine not only the *process* of change but the *context* in which it occurs. This is one of the underlying principles of Pettigrew et al.'s (1992) study of the British National Health Service.

In Pettigrew and Lapsley's (1994) model, context refers to the "why" and "when" of change, including influences of both the outer context (such as economic, social and/or political events) and the inner context of each specific organization. This latter context involves an examination of how

the needed changes are formulated, by whom, and how the change effort itself is managed. Such an approach distinguishes between "receptive" and "non-receptive" contexts for change.

A receptive context is one in which all parties are favorably disposed toward the change. When the reverse – a non-receptive context – exists, some parties seek to block the change. One result is an attempt to connect features of context and action to rates of adoption and change, but also to posit a relationship between capabilities for change and differences in the competitive performance of firms (Smith & Grimm, 1987; Pettigrew & Whipp, 1991).

The Pettigrew and Lapsley (1994) model consists of a set of eight factors that must be considered in a change effort in a healthcare organization:

1. Identification of a high quality and coherent policy.
2. Availability of key people to lead the change.
3. Existence of long-term environmental pressures.
4. Presence of a supportive organizational culture.
5. Development of effective managerial and clinical relations.
6. Existence of cooperative inter-organizational networks.
7. Articulation of simple and clear goals.
8. Stipulation of a change agenda and its locale.

These factors are linked and represent a pattern of association rather than a simple line of causation. As a result, they must be seen as a series of loops rather than a path between independent and dependent variables (Pettigrew et al., 1992, p.276). The underlying idea is that any innovation, amongst which electronic governance innovations, need to carefully consider a set of human-centered perspectives if they are to succeed.

4. INTRODUCING A LOW COST HUMAN-CENTERED INNOVATION IN A REAL ENVIRONMENT

The Forlì (Italy) Hospital is a facility that serves a population of about 150 thousand people, with 34 specialties. The change effort consisted of the implementation and use of an information system that gathered data on KPIs with the goal of assisting clinicians and managers to improve the Operating Room Department (ORD)'s efficiency and effectiveness.

In 2004, the hospital had moved to a new facility. An important aspect of the move was the consolidation of its eight, previously separate, operating rooms into a single location. The move meant that surgeons – and all the actors involved in operating rooms – with different specialties needed to collaborate and share all organizational and managerial arrangements.

Typically, the surgical path of each patient starts from the ward when he or she leaves it, then goes through the operating room, into recovery, then back to the ward. Along this process, many actors are involved creating the potential for mistakes and delays. The result is not only inefficiency (and higher costs than necessary) but also a possible deterioration of the quality and safety of the patient's care. Before the implementation of this project, the situation of Forlì Hospital's ORD was described by as many points of view as there were people involved in the surgical process. Moreover, there was no way to link clinicians' beliefs about the process to what actually was taking place.

In late 2005, senior managers created a multidisciplinary project team chaired and coordinated by the healthcare directorate and composed of nurses, surgeons, anesthetists, engineers, and managers. The team's mandate was to evaluate the system in place and identify opportunities to improve the level of patients' and operators' safety. A related mandate was to ensure an efficient and fair distribution of hospital resources.

The first experiment using specific IT hardware devices (Personal Digital Assistants, or PDAs) and the first generation of software was carried out in the second semester of 2006. In 2007, an analysis carried out by a new data manager with specific management engineering skills, identified two problems with the 2006 effort. First, the lack of an analytical study of the surgical process meant that information about start and end times for critical activities during the surgical path were missing. Second, the misuse of the PDAs led to an absence of data or poor quality data.

The project team discussed these problems and identified solutions to them. The solutions were communicated to all individuals involved in the process, and several tutorial sessions were organized. The renovated system was then initiated and progressively adjusted through 2008.

In January 2009, the system became institutionalized. Every month the project team analyzed the data and shared the results with the heads of the surgical departments and the other individuals involved.

In September 2010, and again in mid 2011, the data manager presented improved versions of the system which allowed the managers, surgeons, anesthetists, engineers and nurses involved in the patient flow path to view more sophisticated, relevant and current KPIs, according to the most up-to-date scientific literature (Dexter, 2005; Macario, 2006). This led to additional operational improvements. A timeline is shown in Table 1.[2]

5. THE CHANGE CONTEXT

The change process revolved around four basic ideas: (1) an inclusive approach; (2) ergonomics; (3) the absence of a specific point of view, and (4) the reuse of infrastructures and devices already available:

- **Inclusive Approach:** All of the various ORD actors were involved in the devel-

Table 1. Innovation timeline

Time	Main events
2004	The hospital moves to a new facility; consolidation of previously separate operating rooms into a single location
2005	Multidisciplinary project team with the mandate to improve the level of patients' and operators' safety and efficient and fair distribution of hospital resources is created
2006, 2nd semester	First experiment of ORD's processes analysis using specific IT hardware devices (PDAs) is carried out
2007	First data analysis shows problems related to the first experiment: critical information are missing and absence of data or poor quality data because of PDAs misuse
2008	After discussion within the project team and tutorial sessions to individuals involved into the ORD's processes, a new and continuously improved processes analysis system is implemented
2009	The ORD's processes analysis system becomes institutionalized
2010, September	Upgrade of ORD's processes analysis system
2011, June	Further upgrade of ORD's processes analysis system, with more sophisticated, relevant, and current KPIs

opment of the process and, at each step, information was shared with all of them. The inclusive process was enhanced with feedback about each component of the system. Team collaboration and the analysis of each process reengineering experiment represented some of the most important features of the project.

- **Ergonomics:** While the inclusive approach helped to create an organizational-inclusive approach, IT was introduced to deal with the large quantity of data needed by the project. The underlying logic used for IT implementation was not to interfere with the daily activities in the operating rooms. Rather, the project team set up "system ergonomics development" as its primary strategy of implementation of the new information system for the operating room.

Ergonomics, in this context, meant that methodologies, equipment and devices were designed in a way that fit the human body, its movements, and its cognitive abilities. In essence, the goal was to simplify the procedures needed to track the surgical path.

- **Absence of a specific point of view:** The project began with a rough analysis of all the flows of movements of the surgical patient throughout the ORD. The aim was to capture all the steps of surgical process and the time between each step and the next. The project team found that the flow was free of conflicts of interest. The person who carried out this task was the anesthesia nurse.

- **Reuse:** The PDA device was already available within the wards for drug administration, and was easily adapted to track the different elements (individuals involved, micro-activities of the process, their start and end times, and the locations where the activities were carried out). The project no incremental costs other than the development of the software.

6. DEALING WITH CHANGE

The eight factors of change in the Pettigrew and Lapsley (1994) model can now be discussed in conjunction with activities that took place in the ORD over the 2006 to 2011 period. As this section indicates, there were some initial problems in addressing these factors that subsequently were overcome.

6.1. Identification of a High Quality and Coherent Policy

The policy applied by the project is summarized by the four basic ideas presented in the case study: (1) an inclusive approach; (2) ergonomics; (3) absence of a specific point of view, and (4) reuse. While the concepts articulated in this policy are well established today, they were not during the first years of the project.

At the beginning of the project the ORD clinical manager and nurses started to perform a tracking system for the patient flow path to abide by the new national law, but they did not know where the project could lead.

When the data manager became part of the project in 2007, he discovered the incoherence in the policy. Nurses and all the personnel involved were not using the new system because it was time consuming and difficult. The project team then decided to undergo a thorough analysis of the patient flow path involving all its actors after which a reengineering of the entire process was introduced successfully.

6.2. Availability of Key People to Lead the Change

There were two key people leading the change. At the initiation of the project, the ORD clinical manager acted as a facilitator for the introduction of the new information system.

The PDA system was introduced without any thorough tailored analysis of the context where it was introduced. Nor did the people in charge concern themselves with the implementation of the project. Therefore, the PDA system was introduced but limited to the fulfillment of legal requirements concerning risk management.

After the first two years of testing, the ORD clinical manager felt that the tracking system needed a specialist to address the problems. There-fore, a new data manager with skills in the area of IT and clinical management engineering was

hired. This was the person who did the thorough analysis of the patient flow path and improved the system to the current level.

6.3. Existence of Long-Term Environmental Pressures

In terms of environmental pressures, the project was begun in order to comply with a 2006 law on risk management in operating rooms. The law's aim was to reduce the risk of incorrect patient treatment. However, beginning in 2007, the project evolved to a broader goal of ORD process improvements both in terms of risk management and effectiveness.

Starting from the first application of the improved system in 2009, nurses realized that the system could really improve the quality of the surgical procedures beyond the safety requirements of the 2006 law, and lead to more about efficiency and costs reduction as well. This refinement of the project toward greater attention on costs is especially evident in the award statement by the EPSA Commission at the European Institute for Public Administration: "national health service costs are exploding and runaway health spending is become more critical by the day in an ageing Europe and are of high public concern . . . [The project shows] evidence of increased efficiency of the process and a positive [qualitative] impact."

In short, the external pressures resulted in a change from simply abiding by the terms of the risk management law to a focus on efficiency and cost savings. This was the outcome of a slow process of education and communication associated with participation and involvement of all actors involved, starting from 2007.

6.4. Presence of a Supportive Organizational Culture

During the first two years of the project, it was not clear who had operational responsibility for the patient flow path and the tracking system. In this organizational "vacuum," clinicians took the lead given that they were responsible the surgical procedures.

At the outset, therefore, the organizational culture was not supportive. The consensus among the surgeons was that they knew what was best for each patient and no quantitative data could help to improve the surgical path. Similarly, nurses were seen as staff members who simply did what surgeons asked them to do. In short, the organizational culture acted in opposition to the introduction of the tracking system.

The ORD clinical manager, based on a proposal by the data manager, used two methods to deal with this resistance. First, he gave the responsibility of tracking the patient flow paths to nurses instead of surgeons. Surgeons did not resist to this change because of the perceived guarantee of objectivity by the tracking system. Second, the ORD clinical manager increased the frequency of the project team meetings in order to get more regular feedback from nurses and surgeons in order to directly address any friction between the two groups. These meetings where organized by the data manager with the aim to present and discuss the dashboards periodically produced by the informative system.

This overturning of the traditional hierarchy, so as nurses now control surgeons, led to evidence available to make decisions, overcoming the traditional environment were decisions were taken based on "nurse legends" instead of on using reliable information.

6.5. Development of Effective Managerial and Clinical Relations

An effective relationship between these two groups was in place at the beginning of the project. Clinical managers were responsible for effectiveness, and non-clinical managers were responsible for efficiency. No relevant variations in these roles

took place during the project life, and therefore no methods of dealing with resistance to change were needed.

6.6. Existence of Cooperative Inter-Organizational Networks

Cooperation with external actors emerged in May 2011, after the project received attention by the European Institute for Public Administration, and a special committee visited the hospital. As a result, media coverage increased considerably and the project became well known among the most important regional and national entities (Ministry of Health, Regional Healthcare Agency, Regional, Provincial and Municipal Councils).

The award at the European level and the extended media coverage which accompanied it, produced the right "organizational climate" to allow a network of external organizations to support the project. At the beginning, the decision of presenting the candidature for the award competition came from the project team leader, passing almost unnoticed by the top management. The project had a good potential to be selected because of its evidence of improved ORD performance. When the project was selected as finalist, the top management started to be engaged in the process and the actors involved in the project benefitted in term of highest visibility and acknowledgement throughout the organization.

In summer 2011, the ORD clinical manager accepted an offer by a private software house to develop, at no charge, some new business intelligence software that would obtain more frequent and more reliable KPIs from the data gathered using the tracking system. In November 2011, this partnership was extended to the department of management at the University of Bologna with the aim to provide improved information for cost management.[3]

6.7. Articulation of Simple and Clear Goals

The project's goals had always been simple and clear, although they changed over time. At the beginning, they were to collect the time, activities and to manage the risk of surgical procedures. As the project evolved, the goals were expanded, first to include a focus on efficiency, and then to address overall cost management. Moreover, clearly understandable KPIs provided elements to better understand the work in relation to colleagues' work.

In addition, the continuous communication of results through the above-mentioned disclosure of information on dashboards attached to the ORD hall walls, and discussed during project team meetings, allowed a better understanding of the project goals. This facilitated the acceptance of the innovation by all the individuals involved.

6.8. Stipulation of the Change Agenda and its Locale

The project was initiated following the hospital's relocation to a new facility and the consolidation of the operating rooms. This acted as window of opportunity during the project's first years of implementation of the project, pushing the actors to cooperate to develop new ways to manage the patient flow path within the new facility.

7. RESULTS

By the end of 2011, the system was able to track the entire patient flow from beginning to end. It also was able to identify the individuals who were involved in each step, and the locations where the different activities took place. The relevant KPIs were reported in a simple but comprehensive scorecard, which was posted regularly on the walls of the ORD. An Example is contained in Figure 1.

While maintaining the same level of complexity of surgical procedures performed, the ORD

Figure 1. Sample KPI report

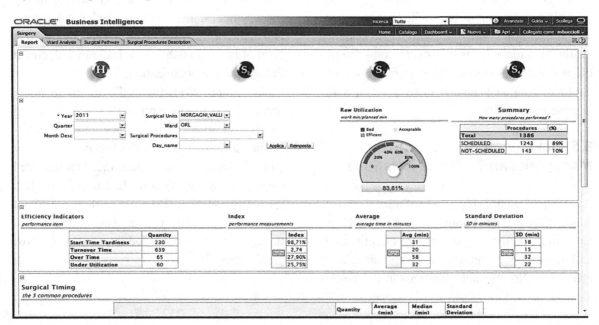

was able to improve both its efficiency and effectiveness. Relevant information is contained in Table 2 and Figure 2.

The project also proved to be cost effective since, apart from its start up expenditure of €26,500 for the PDAs and data manager's activities, its average cost per year was €14,700 (data manager's activities), or 0.2% of the ORD's fixed costs.

8. CONCLUSION

Two of the main priorities of the political system are to reduce public debt and to streamline the healthcare services costs. Consequently, health managers need to implement methods within their managerial structures such that all available resources are better planned and managed in terms of effectiveness and efficiency, without compromising – if not even improve – the quality of patient care.

The case described in this chapter has demonstrated to have successfully implemented an information system that gathers data on KPIs with the goal of assisting clinicians and managers to improve the ORD's efficiency and effectiveness. The analysis corroborates Pettigrew & Lapsley's model (1994) about the factors that describe the context influencing change in a public healthcare organization and expands this model to two supplementary perspectives.

On one hand, the low-cost of this innovation allowed the management to invest a minimum amount of money and personnel resources so as to increase the probability to have a good return on investment ratio. This also allowed the team project to avoid requests of financial resources to top management and thus have high level of freedom in managing the project. On the other hand, most of the factors which led to innovation success were human centered: from the introduction of specific skills within the organization, to the improvement of the organizational culture so as to create mechanisms that incentive performance improvements; from the retention of a good climate between managers and clinicians and the opening towards inter-organizational networks, to the involvement of all actors in the definition of goals and discussion of results.

In summary, this case proves that while electronic governance has become a minimum standard in many innovations and can now be accessible at a very low costs, humans remain at the center of success and require a strategic – and, to some extent, time consuming – approach to succeed.

Table 2. Evidence from the ORD before and after the innovation was introduced

KPI	2009	2010	Difference
Operating room occupancy	71%	79%	+11%
Number of unscheduled procedures (1)	25%	15%	-36%
Overtime working hours expenditure	€524,000	€497,000	-21%
Patient's safety	No incorrect surgery or near misses from 2009		
Turnover-time (minutes) (2)	30 ± 20	32 ± 21	
Over-time (minutes) (3)	80 ± 59	77 ± 54	

Notes:

1. In operating rooms the clinical risk for the patient is always very high and the first rule to decrease is to increase the level of standardization of the process; when the surgical team works in an unscheduled way there is a lower level of standardization and thus the clinical risk increases

2. Delay from the scheduled time to end the surgery of the last procedures of the day and the real time; mean ± standard deviation

3. Time for cleaning and setting up of the operating room between two patients - benchmark in literature is 25 min

Figure 2. Evidence from the ORD before and after the innovation was introduced (a snapshot from the business intelligence software reports)

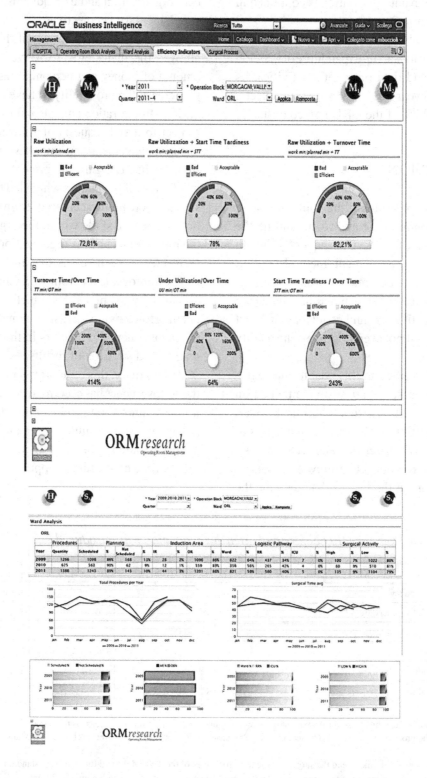

REFERENCES

Cavalluzzo, K. S., & Ittner, C. D. (2004). Implementing performance measurement innovations: Evidence from government. *Accounting, Organizations and Society, 29*(2-3), 243–267. doi:10.1016/S0361-3682(03)00013-8

Dexter, F., Ledolter, J., & Wachtel, R. E. (2005). Tactical decision making for selective expansion of operating room resources incorporating financial criteria and uncertainty in subspecialties' future workloads. *Anesthesiology and Analgesics, 100*, 1425–1432. doi:10.1213/01.ANE.0000149898.45044.3D

Kitchener, M. (1999). All fur coat and no knickers: Contemporary organizational change in United Kingdom hospitals. In Brock, D., Powell, M., & Hinings, C. R. (Eds.), *Restructuring the professional organization: Accounting, health care and law* (pp. 183–199). London, UK: Rutledge. doi:10.4324/9780203018446.ch9

Kravchuk, R. S., & Schack, R. W. (1996). Designing effective performance measurement systems under the government performance and results act of 1993. *Public Administration Review, 56*(4), 348–358. doi:10.2307/976376

Kurunmaki, L. (1999). Making an accounting entity: The case of the hospital in Finnish health care reforms. *European Accounting Review, 8*(2), 219–237. doi:10.1080/096381899336005

Macario, A. (2006). Are your hospital operating rooms "efficient"? *Anaesthesiology, 105*, 237–240. doi:10.1097/00000542-200608000-00004

Maddock, S., & Morgan, G. (1998). Barriers to transformation: Beyond bureaucracy and the market conditions for collaboration in health and social care. *International Journal of Public Sector Management, 11*(4), 234–251. doi:10.1108/09513559810225807

O'Neill, O. (2008). Transparency and the ethics of communication. In Hood, C., & Heald, D. (Eds.), *Transparency: The key for better governance?* (pp. 75–91). Oxford, UK: Oxford University Press.

OECD. (2005). *Modernising government: The way forward.* Paris, France: OECD.

Padovani, E., & Young, D. W. (2012). *Managing local governments. Designing Management control systems that deliver value.* London, UK: Routledge.

Pasquier, M., & Villeneuve, J.-P. (2012). *Marketing management and communications in the public sector.* London, UK: Routledge.

Pettigrew, A., Ferlie, E., & McKee, L. (1992). *Shaping strategic change.* London, UK: Sage.

Pettigrew, A., & Lapsley, I. (1994). Meeting the challenge: Accounting for change. *Financial Accountability and Management, 10*(2), 79–92. doi:10.1111/j.1468-0408.1994.tb00146.x

Pettigrew, A., & Whipp, R. (1991). *Managing change for competitive success.* Oxford, UK: Basil Blackwell.

Pollitt, C. (2008). *Time, policy, management. Governing with the past.* Oxford, UK: Oxford University Press.

Pollitt, C., & Bouckaert, G. (2011). *Public management reform: A comparative analysis - New public management, governance, and the neo-Weberian state.* Oxford, UK: Oxford University Press.

Smith, K. G., & Grimm, C. M. (1987). Environmental variation, strategic change and firm performance: A study of railroad deregulation. *Strategic Management Journal, 8*(4), 363–376. doi:10.1002/smj.4250080406

Van Dooren, W., Bouckaert, G., & Halligan, J. (2010). *Performance management in the public sector.* London, UK: Routledge.

Whipp, R., Rosenfeld, R., & Pettigrew, A. (1987). Understanding strategic change process: Some preliminary British findings. In Pettigrew, A. M. (Ed.), *The management of strategic change* (pp. 14–55). Oxford, UK: Basil Blackwell.

Young, D. W. (2004). Improving operating room financial performance in a center-of-excellence. *Healthcare Financial Management, 58,* 70–74.

Young, D. W. (2008). *Management accounting in health care organizations.* San Francisco, CA: Jossey-Bass.

Young, D. W., & Saltman, R. B. (1985). *The hospital power equilibrium: Physician behavior and cost control.* Baltimore, MD: The Johns Hopkins Press.

ADDITIONAL READING

Anthony, R. N., & Young, D. W. (2003). *Management control in nonprofit organizations.* New York, NY: McGraw-Hill Irwin.

Behn, R. D. (2003). Why measure performance? Different purposes require different measures. *Public Administration Review, 63*(5), 586–606. doi:10.1111/1540-6210.00322

Bouckaert, G. (1993). Measurement and meaningful management. *Public Productivity & Management Review, 17*(1), 31–43. doi:10.2307/3381047

Bouckaert, G., & Peters, B. G. (2002). Performance measurement and management: The Achilles' heel in administrative modernization. *Public Performance & Management Review, 25*(4), 359–362.

Callahan, D. (1999). *False hopes: Overcoming the obstacles to a sustainable, affordable medicine.* New Brunswick, NJ: Rutgers University Press.

Czarniawska, B. (2009). Emerging institutions: Pyramids or anthills? *Organization Studies, 30,* 423–441. doi:10.1177/0170840609102282

Czarniawska, B., & Joerges, B. (1996). Travels of ideas. In B. Czarniawska & G. Sevon (Eds.), *Translating organizational change* (pp. 13-48). Berlin, Germany: de Gruyter.

De Bruijn, H. (2002). *Managing performance in the public sector.* London, UK: Routledge.

De Lancer Julnes, P., & Holzer, M. (2001). Promoting the utilization of performance measures in public organizations: An empirical study of factors affecting adoption and implementation. *Public Administration Review, 61*(6), 693–708. doi:10.1111/0033-3352.00140

Dexter, F., & Traub, R. (2002). How to schedule elective surgical cases into specific operating rooms to maximize the efficiency of use of operating room time. *Anesthesiology and Analgesics, 94,* 933–942. doi:10.1097/00000539-200204000-00030

Dexter, F., Wachtel, R. E., & Epstein, R. H. (2011). Event-based knowledge elicitation of operating room management decision-making using scenarios adapted from information systems data. *BMC Medical Informatics and Decision Making, 11*(1), 2–13. doi:10.1186/1472-6947-11-2

Halachmi, A. (2005a). Performance measurement is only one way of managing performance. *International Journal of Productivity and Performance Management, 54*(7), 502–516. doi:10.1108/17410400510622197

He, B., Dexter, F., Macario, A., & Zenios, S. (2012). The timing of staffing decisions in hospital operating room: incorporating workload heterogeneity into the newsvendor problem. *Manufacturing & Service Operations Management, 14,* 99–114. doi:10.1287/msom.1110.0350

Heichlinger, A. (Ed.). (2011). *EPSA trends in practice. Driving public sector excellence to shape Europe for 2020.* Maastricht, The Netherlands: European Institute of Public Administration.

Komashie, A., Mousavi, A., & Gore, J. (2007). *A review of historical developments of quality assessment in industry and in healthcare.* Paper presented at the 10th International Conference on Quality Management and Organisational Development, Helsingborg, Sweden.

McLaughlin, C. P., & Kaluzny, A. D. (2006). *Continuous quality improvement in health care: theory, implementations, and applications.* Sudbury, MA: Jones & Bartlett Learning.

Orelli, R. L., Padovani, E., & Del Sordo, C. (2012). From e-government to e-governance in Europe. In Islam, M. M., & Ehsan, M. (Eds.), *From government to e-governance: Public administration in the digital age* (pp. 195–206). Hershey, PA: IGI Global. doi:10.4018/978-1-4666-1909-8.ch011

Osborne, S., & Brown, K. (2005). *Managing change and innovation in public service organizations.* London, UK: Routledge. doi:10.4324/9780203391129

Pollitt, C. (2003). *The essential public manager.* Maidenhead Berkshire, UK: Open University Press.

Rogers, E. (2003). *Diffusion of innovations* (5th ed.). New York, NY: Free Press.

Talbot, C. (2010). *Theories of performance. organizational and service improvement in the public domain.* Oxford, UK: Oxford University Press.

Taylor, J. (2009). Strengthening the link between performance measurement and decision making. *Public Administration*, *87*(4), 853–871. doi:10.1111/j.1467-9299.2009.01788.x

Van Dooren, W. (2005). What makes organisations measure? Hypotheses on the causes and conditions for performance measurement. *Financial Accountability & Management*, *21*(3), 363–383. doi:10.1111/j.0267-4424.2005.00225.x

Young, D. W. (2012). *Management control in nonprofit organizations.* Cambridge, MA: The Crimson Press.

Young, D. W., Kenagy, W., McCarthy, S. M., Barrett, D., Kenagy, J. W., & Pinakiewicz, D. C. (2001). Toward a value-based healthcare system. *The American Journal of Medicine*, *110*(2), 158–163. doi:10.1016/S0002-9343(00)00703-8

Young, D. W., McCarthy, S. M., Barrett, D., Kenagy, J. W., & Pinakiewicz, D. C. (2001). Beyond health care cost containment: Creating collaborative arrangements among the stakeholders. *The International Journal of Health Planning and Management*, *16*, 207–222. doi:10.1002/hpm.631

KEY TERMS AND DEFINITIONS

Change Management: Change management provides tools and methods to recognize and understand organizational change, to manage conservatism and resistance, and to minimize the negative human impacts of a transition.

Dashboard: A dashboard is a simplified representation of a multidimensional performance. It provides a clear "at-a-glance" understanding of which are the strengths and the weaknesses to the intended audience.

E-Governance: Expanding upon the idea of e-government as first form of extensive usage of IT in the public sector, e-governance concerns the use of information and communication technologies to support "good governance" in the public sector, so as to foster plural and pluralist complexities. More discussion in Orelli et. al. (2012).

Innovation: When an idea is translated into a good, service, or process that creates value and can be replicable at an economical cost, it is called innovation. It decreases the gap between customers/users' needs and organizational performance.

Performance Management: It is the use of information derived from performance measurement to take action.

Performance Measurement: It refers to the activity of quantifying the different aspects of performance as the result of a production process.

Transparency: It is related to the publication of performance information to foster accountability. Transparency is not limited to the right-of-access to information, but requires that performance information is easily accessible and understandable by the intended target.

ENDNOTES

[1]   The European Institute of Public Administration (EIPA) EIPA is EIPA is the leading centre of European learning and development for the public sector, with a board of governors composed of representatives of EU Member States and the Institute's associated members. These representatives form a permanent link between EIPA and national public administrations, and also guarantee a strong institutional network at the highest level of public administration throughout Europe. More information at www.eipa.eu.

[2]   For updated information about the project, visit www.operatingroommanagement.org.

[3]   The University of Bologna and the authors of this chapter belonging to the University of Bologna were not involved in the presentation of the candidature by the Morgagni-Pierantoni Hospital to the EPSA award and received no financial support for the project on cost management information system improvement.

Chapter 15
The Community Manager and Social Media Networks:
The Case of Local Governments in Spain

Manuel Pedro Rodríguez Bolívar
University of Granada, Spain

Carmen Caba Pérez
University of Almería, Spain

Antonio Manuel López Hernández
University of Granada, Spain

ABSTRACT

Local governments are increasingly embracing Web 2.0 technologies to encourage the use of means of bidirectional communication to change how they interact with stakeholders, thus providing the greater accountability demanded. Nonetheless, to make Web 2.0 tools efficient, there must be qualified people to operate and supervise the Web 2.0 and social network technologies implemented by local governments. These people, called "Community Managers," play a key role in the implementation of social networks in local government, successfully or otherwise. In this chapter, the authors analyse whether the training and education of community managers in Spanish local governments is associated with the successful use of social networks by these local governments in their interaction with the public. Their empirical study of local government in Spain shows that the position of community manager is mostly held by men who are aged 25-45 years and have a university degree in journalism, performing in addition, tasks such as updating the municipal website or running the press office.

DOI: 10.4018/978-1-4666-3640-8.ch015

1. INTRODUCTION

The implementation of Web 2.0 technologies has favoured reforms and modernisation within public administrations, changing the nature of political and public dialogue (Osimo, 2008) and encouraging participation by citizens, allowing them greater involvement in public affairs and enabling public managers to create more affordable, participatory and transparent models of public sector management (McMillan et al., 2008). The application of popular Web 2.0 technologies such as social networking (Facebook, MySpace), wikis, blogs, microblogs (Twitter), mash-up and multimedia sharing (YouTube, Flickr) can facilitate interactive information sharing, interoperability and collaboration (United Nations, 2010), thus promoting open and user-driven governance (Bertot et al., 2010a, b, c; Millard, 2009).

In the local government context, which has long been a prime focus of public sector reforms (Christiaens, 1999; Mussari, 1999; Ter Bogt and Van Helden, 2000; Pallot, 2001; Smith, 2004), social networks are becoming increasingly relevant as public sector entities are pressured by demands for improvements in information transparency and public sector services delivery, as part of the accountability required of municipalities (Gibson, 2010). Accordingly, local governments are increasingly embracing Web 2.0 technologies to encourage the use of means of bidirectional communication to change how they interact with stakeholders and to become more efficient in their response to stakeholders' demands, thus providing the greater accountability demanded (Redell and Woolcock, 2004; Leighninger, 2011).

In Spain, administrative structures were reformed in the 1990s (Gallego and Barzelay, 2010) and a process of managerial devolution was undertaken (Bastida and Bernardino, 2006). In addition, legislation promoted the modernisation of local governments by the implementation of new technologies. Thus, the Information Society Services and E-Commerce Act (No. 34/2002)

guaranteed access to government information, while the Local Government Modernisation Act (No. 57/2003) promoted the use of new technologies in order to: a) enhance participation and communication with citizens; b) enable administrative procedures to be carried out online; c) publish basic legislation on the Internet; and d) enhance interaction with municipal authorities. Finally, the definitive impetus to the inclusion of information and communication technologies (ICTs) at different levels of government – national, regional and local – in terms both of their relations with citizens and in their internal management, was given by the Electronic Access to Public Services Act (No. 11/2007), which guaranteed this right to all citizens. In view of these developments, this analysis is focused on the introduction of Web 2.0 technologies into local governments in Spain.

As a result both of legislation and of demands from the population, local governments in Spain are becoming more transparent and allowing greater public participation in municipal management, thus providing greater accountability. In this respect, social media could be an appropriate tool with which local governments could enhance their interaction with the public. Nonetheless, to make Web 2.0 tools efficient, there must be qualified people to operate and supervise the Web 2.0 and social networks technologies implemented by local governments. These people are known as community managers.

Community managers are the invisible face of local governments but they are responsible for interacting with local citizens and for creating the conditions that will underpin this relationship. Therefore, they play a key role in the success or otherwise of the implementation of social networks in local government. Accordingly, they require specialised skills, education and training to do their job well. Studies have shown that the existence of community managers and their level of professional competence are key factors in enhancing communication by municipal administrations (Cheng, 2009).

However, to the best of our knowledge, to date no research has been undertaken regarding either the competence of these community managers or the nature of their tasks. The background of the people required to perform these duties is diverse: some have studied Masters Programmes, while others have only graduate qualifications, or even less, and so there exists considerable heterogeneity in the training and education of community managers. To extend our knowledge and understanding of this question, in this study we analyse whether the training and education of community managers in Spanish local governments is associated with the successful use of social networks by these local governments in their interaction with the public.

The rest of this chapter is organised as follows. The second section is focused on community managers and their role in managing Web 2.0 tools within public administrations. We then present an empirical study made of their education, training and performance within Spanish local governments. This study is based on a prior analysis of public policies and the legal framework for Web 2.0 technologies in local and regional public administrations in Spain. In the following section, the results obtained are analysed and discussed, after which some conclusions are drawn.

2. PUBLIC ADMINISTRATIONS, WEB 2.0, AND COMMUNITY MANAGERS

Increasingly connected citizens and stakeholders are demanding that governments be more transparent and deliver services more rapidly and efficiently. Ready access to information of public value, increased transparency in government operations and a greater willingness to listen to citizens and secure their involvement: these are pivotal requirements for efficient, open and responsive government (OECD, 2010).

In recent decades, governments around the world have been faced with rapidly growing challenges on how to make public service and administration transparent, effective and efficient. The implementation of ICTs has helped meet this demand and has favoured reform and modernisation within public administrations (Chan and Chow, 2007), enabling greater accessibility to public information and services (Martins, 1995), together with greater interaction and individual participation in public management (Dunleavy et al., 2006; Taylor et al., 2007) and greater information transparency (Rodríguez et al., 2010).

Nonetheless, e-government initiatives over the past decade have been based mainly on first-generation web-based resources (including web sites, pages and services), which were based on HTML, a relatively primitive, static page mark-up technology that simply outlines what a page should look like onscreen. Governments must now strengthen their capacity to assess the needs of users (both private and commercial) and involve user groups through the use of second generation web technologies (Web 2.0) in order to listen, to engage users in the design of services and in the production of policies and to forge collective initiatives and interaction (OECD, 2010). In this regard, Web 2.0 technologies have the potential to change how government delivers services and to revolutionise its relationship with the public. It differs from first-generation web-based resources in at least three significant ways: it is participatory, it is pervasive and it is integrated (Mintz, 2008). Web 2.0 has the potential to transform public administration services, enabling the development of better policies and eliminating data silos (Klein, 2008).

Various popular Web 2.0 technologies, such as social networking (Facebook, MySpace), wikis, blogs, microblogs (Twitter), mash-up and multimedia sharing (YouTube, Flickr), facilitate interactive information sharing, interoperability and collaboration (United Nations, 2010) and can promote open, user-driven governance (Bertot et al., 2010a, b, c; Millard, 2009). Web 2.0 makes users the primary focus in the development of solutions (Kovac & Decman, 2009) because visible mistakes are rapidly and loudly protested (Klein, 2008). In fact, in the Web 2.0 era, it is no longer appropriate to conceive of users as 'end-users', as they have moved into the heart of the value chain (Tuomi, 2002). The specific benefit of users taking a proactive role is that government becomes simpler and more user-oriented, transparent, accountable, participative, inclusive, joined-up and networked (Osimo, 2008).

Therefore, public administrations can take advantage of the emergence of this new participative culture to attract attention toward municipal management, engage citizens in local decision-making and improve government-to-citizen relationships (Bonsón et al., 2012). In this regard, local governments can make use of Web 2.0 technologies in at least four important aspects. Firstly, in order to disclose a greater volume of information to a wider range of citizens, even if this distribution only takes place in a single direction. Secondly, to open up debates and engage in corporate dialogue with citizens. Thirdly, to obtain citizens' cooperation in the design of public sector services. Finally, and as in many local governments around the world, to make use of IT in back-office processes (Sandoval-Almazan & Gil-García, 2012).

The first two of these aspects concern the transparency and visibility of local government actions, while the second, in addition, favours more participative management (Kovac & Decman, 2009). For example, one Spanish municipality, taking into account the current economic and financial crisis in Spain, inquired the opinion of the local population to decide how to invest €15,000 (on a job creation scheme or on a bullfighting event). The third aspect could make public sector services more user-centred and personalised, and thus more responsive and effective (Huijboom et al., 2009, Caba et al., 2012). Finally, the use of IT in back-office processes could make local governments more efficient. This aspect is also directly related to the information published by local governments, because the provision of up-to-date, accurate information reflects internal efficiency and an appropriate organisational design (Gould et al., 2010).

Notwithstanding the above positive aspects, the implementation of Web 2.0 technologies also presents some risks, such as low rates of participation, or participation restricted to an elite, the low quality of contributions, the additional data noise created, possible loss of control due to excessive transparency, destructive behaviour by users, the manipulation of content by interested parties and privacy infringements (Osimo, 2008). In this latter respect, a permission-based system can enable citizens to exercise informational self-determination by applying their online privacy preferences, both as consumers and as creators (Cavoukian, 2009). Furthermore, legal frameworks are coming under increasing pressure as the development of legislation fails to match the pace of content creation (Huijboom et al., 2009).

In the last few years, in order to overcome the dangers associated with the implementation of Web 2.0 technologies, the figure of community manager has been defined, as a specialist in the management and control of Web 2.0 and social networks. The implementation of these technologies in local governments relies upon community managers to make Web 2.0 tools efficient, and thus favour high rates of participation in public sector management and in the activities organized by local governments, such as social events, public campaigns, political events and e-voting.

In the private sector, according to the Spanish Association of Online Community Managers (AERCO, 2009), community managers maintain and supervise online relations between the community and the firm and constitute the link between the customers' needs and the firm's capacities. As has been observed, "A community manager is the voice of the company externally and the voice of the customers internally. The value lies in the community manager serving as a hub & having the ability to personally connect with the customers (humanize the company), & providing feedback to many departments internally (development, PR, marketing, customer service, tech support, etc" (Bensen, 2009).

In the local government context, the community manager would perform similar functions, linking citizens' needs and municipal capacities, helping local government prioritise its resources and combating the above-mentioned risks of Web 2.0 technologies (Osimo, 2008).

Therefore, it is of some interest to determine whether local governments are in fact hiring community managers to manage and control their Web 2.0 tools. If so, we would expect to find improved communication and participation and thus greater public involvement in municipal management, which in turn would confer greater legitimacy on government decisions and better satisfy citizens' needs. To date, very little research has been undertaken on this subject with respect to the public sector, and so this chapter provides new information regarding the impact of community managers in the context of local government.

3. EMPIRICAL RESEARCH ON LOCAL GOVERNMENTS IN SPAIN

3.1. Public Policies and Legal Framework for Web 2.0 Technologies in Public Administration in Spain

According to the results of the UN eGovernment Survey 2010 (United Nations, 2010), governments involve their citizens for feedback and consultation via their websites, and most such sites contain polls, surveys, comment buttons or other means of reaction. However, this is believed to be just the tip of the iceberg, and Web 2.0 could enable citizens to have a more direct impact on public administrations.

The survey figures show that Europe accounts for 51% of the countries with the highest uptake of e-participation, followed by Asia with 29%, the

USA with 14% and Oceania with 6%. South Korea is the leading country in the e-participation index, followed by Australia, Spain and New Zealand.

In the framework of the European Union, many initiatives have been taken to regulate and coordinate the actions of Member States in order to facilitate digital convergence and to meet the challenges of the Information Society. The implementation of the first European eGovernment Action Plan (2006-2010) has led to the governments of all EU Member States exchanging information on good practice, and has resulted in a number of large-scale pilot projects which are developing concrete solutions for rolling out cross-border eGovernment services (ICT PSP from PIC, 2011). The second eGovernment Action Plan (2010-2015) is intended to meet the targets set out at the 5th Ministerial eGovernment Conference, in the Malmö Declaration. According to this ambitious vision, by 2015 European public administrations will be "recognized for being open, flexible and collaborative in their relations with citizens and businesses. They will use eGovernment to increase their efficiency and effectiveness and to constantly improve public services in a way that caters for users' different needs and maximizes public value, thus supporting the transition of Europe to a leading knowledge-based economy".

Let us now focus on the situation in Spain. As described in the 2010 eGovernment survey (United Nations, 2010), Spanish public administrations are beginning, albeit slowly, to make use of interactive tools to promote dialogue and receive feedback and input from citizens, as well as to provide information and services online. However, while Spain is ahead of many other countries in this respect, e-participation has been less fully developed, and is mainly limited to providing information, responding to queries and, to a much lesser extent, decision taking.

Beyond doubt, the definitive impetus to the inclusion of information and communication technologies (ICTs) in the various levels of government – national, regional and local – in terms of relations with citizens and in internal management, was given by the Electronic Access to Public Services Act (LAECSP) of 2007, which guaranteed this right to all citizens.

LAECSP is a fundamental measure in the promotion of eGovernment. The most interesting aspect of this legislation is that, apart from the general principles, the legal framework, and the rules and criteria set out, certain specific rights of individual citizens are established, and these automatically become an obligation for the administration. Therefore, the various levels of public administration will have to develop a wide range of Web-delivered services.

Nevertheless, the full implementation of LAECSP – scheduled for 2010 – has been uneven, because not all authorities have put the same emphasis on developing ICTs and because in the current economic climate, it is hard to provide the financial resources for the technological infrastructure needed.

Through the LAECSP, Spain has achieved some of the goals of the first European eGovernment Action Plan (2006-2010), and now a new strategy is needed for the next few years. This is expressed in the "National Strategy 2015" document, aimed at boosting the process of government transformation to meet the goals of the Digital Agenda and to address the three main challenges facing Europe in the short term: economic crisis, environmental degradation and an ageing population.

Table 1 presents the legislation in this respect that has been approved in the Spanish regions (autonomous communities), in most cases following the entry into force of the LAECSP. Nevertheless, not all these regional governments have met the legal requirements and there is a risk of a digital gap between citizens being created depending more on their geographical location than on the communications infrastructure installed.

3.2. Methodology

In this study, we analyse different aspects of the municipal officials responsible for issues concerning social media networks, employed by Spanish municipalities with more than 50,000 inhabitants, based on a survey carried out among these officials. The survey addresses the questions in Table 2.

The study is descriptive in nature. As indicated by Hernandez et al. (2003), this type of study aims to specify the main properties, characteristics and profiles of individuals, groups, communities or any other phenomenon to be analysed. In summary, it enables the data collected to be measured, such that the characteristics of the phenomenon under study can be systematically described, analysed and interpreted with respect to the reality of the scenario in question.

3.2.1. Population

Public administration in Spain is basically constituted of a central administration, the 17 regions (autonomous communities) and over 8,000 municipalities. Each level of this structure has its own responsibilities, but the ultimate goal in every case is to provide the population with high quality services. The dimensions of these local entities are very heterogeneous. Their distribution in terms of population is as seen in Table 3.

Obviously, the budget volume and administration is very different in a local government like that of Madrid, with over 3,000,000 inhabitants, from one of less than 5,000 inhabitants, even if the social media they use could well be identical, as these web technologies are readily available to any organisation at low cost.

This study includes all Spanish municipalities with over 50,000 inhabitants, comprising 149 municipalities, accounting for over 50% of the Spanish population. We were aware, from an initial internet search, that some of these municipalities did not have a community manager, but nevertheless a copy of the questionnaire was sent to all 149

Table 1. Legislation by Spanish regional governments on ICTS and eGovernment

Region	Legislation	Description
Andalusia	Act No. 9/2007, of 22 October 2007	Incorporates the principles governing the relations of the different agencies of the Andalusian Government with the public and with other authorities, through open communication networks. Gives legal coverage to judicial calls for the establishment of eGovernment as an alternative means of communication between public administration and citizens.
Aragon	Order of 29 July 2009	Approves the e-Government Plan of the Autonomous Community of Aragon, providing a roadmap for the modernisation of the regional administration in the coming years and complying with LAECSP.
Asturias	Decree No. 115/2008, of 20 November 2008.	Amends previous regulations on the electronic provision of documents.
	Resolution dated 9 January 2009.	Publishes procedures adapted for the automatic electronic transfer of data for national ID Cards and Residence Certificates.
Balearic Isles	Act No. 4/2011, of 31 March 2011, on effective administration and governance in the Balearic Isles.	Sets out the general principles for the use of ICTs.
Canary Isles	Act No. 5/2010, of 21 June 2010.	Regulates citizens' right to e-participation, and promotion of the latter.
	Decree No. 19/2011, of 10 February 2011	Regulates the use of ICTs by the regional authorities.
Cantabria	Decree No. 110/2006, of 9 November 2006.	Regulates the e-registration of data by the regional administration, and the provision of electronic notifications and certificates.
Castile-La Mancha	Decree No. 12/2010, of 16 March 2010	Regulates the use of ICTs by the regional authorities.
Castile and Leon	Resolution No. 29/2009, of 12 March 2009.	Approves the 2009-2011 eGovernment Introduction Plan
Catalonia	Decree No. 56/2009, of 7 April 2009	In compliance with LAECSP, promotes the implementation of eGovernment in the regional administration.
	Act No. 29/2010, of 3 August 2010	Regulates the use of ICTs by the regional authorities of Catalonia and in relations between the public sector and citizens in Catalonia. Defines the instruments to be used in developing and promoting these relations by electronic means.
Extremadura	Resolution dated 26 February 2008	Approves the 2008-2011 Plan to Advance Ongoing Improvement and Technological Modernization.
Galicia	Decree No. 198/2010, of 2 December 2010	Regulates the use of ICTs by the regional authorities of Galicia and its agencies.
La Rioja	Decree No. 57/2006, of 27 October 2006, regulating the administration's relations with citizens in La Rioja.	Sets out the principles regulating the regional administration's presence online.

continued on following page

in the view that there might exist municipal staff who, though not described explicitly as 'community managers', performed such tasks in addition to those corresponding to their job description; these would normally be officials associated with the municipality's press office.

The data were obtained by sending a copy of the e-survey to the Communications Office of all the local authorities studied, via email. The contact details were obtained from the Spanish central government's website. When no such office existed, the questionnaire was sent to the

Table 1. Continued

Region	Legislation	Description
Madrid	Decree No. 62/2009, of 25 June 2009.	Regulates the use of ICTs in public procurement by the Madrid regional authorities.
Navarre	Act No. 11/2007, of 4 April 2007.	Promotes the implementation of effective eGovernment to better serve citizens through the progressive incorporation of technical, electronic, computer and Internet communications for the performance of administrative procedures.
Basque Country	Decree No. 232/2007, of 18 December 2007.	Regulates the use of ICTs in administrative procedures.
	Decree No. 72/2008, of 29 April 2008.	Creates and regulates electronic data transfer within the general administration of the Basque Country and that of its agencies.
Valencia	Act No. 3/2010, of 5 May 2010.	Develops the right of citizens to interact electronically with the public administrations of Valencia and regulates the legal status of eGovernment and electronic administrative procedures.

Table 2. Survey of municipal officials

ASPECTS	DESCRIPTION
A. Personal characteristics	The first module considers the respondent's personal data, i.e., gender and age.
B. Employment situation	This module describes the respondent's current work situation: tenured civil servant, non-tenured civil servant, temporary staff, etc. Other questions ask about the worker's seniority and the level of income obtained.
C. Background	The third module compiles information on the worker's educational background and any training received in the field of social networks, and if so, the duration of this training.
D. Job conditions	This module inquires about the respondent's status with respect to his/her current job as head of social networks. This includes issues related to the exclusive dedication (or otherwise) to the management of social networks and to the sole responsibility (or otherwise) within the local authority for managing social networks.
E. Social networks and workload	The final module contains information about the social networks managed by the respondent and the workload this entails in terms of the number of persons dedicated to this activity.

Table 3. Population distribution

Population Strata	Nº of Municipalities	%	Population	%	% accumulated
> 1,000,000 inhabitants	2	0.02	4,678,380	10.83	10.83
500,001 to 1,000,000 inhabitants	4	0.05	2,676,465	6.19	17.02
100,001 to 500,000 inhabitants	60	0.74	10,130,473	23.45	40.47
50,001 to 100,000 inhabitants	83	1.02	4,853,517	11.24	**51.71**
20,001 to 50,000 inhabitants	207	2.55	6,131,649	14.19	65.9
5,001 to 20,000 inhabitants	877	10.81	8,525,702	19.74	85.64
<= 5,000 inhabitants	6,893	84.78	6,058,828	14.03	99.67
Total	8,107	99.98	43,055,014	99.67	
Special Autonomy Status	2	0.02	142,670	0.33	
Total	**8,109**	**100.00**	**43,197,684**	**100.00**	**100.00**

Source: Spanish National Institute of Statistics (2012)

mayor's secretary with the request that it be re-directed to the person responsible for managing the municipality's social networks.

Of the 149 municipalities that comprised the survey sample, 36 replies were received from persons responsible for social networks management. Therefore, to date, the response rate is 34.2%, of which 10% stated that the municipality had not yet introduced communication channels such as social networks, and thus no human resources were dedicated to this area.

The cut-off date for the survey data is May 2012; the survey was sent out in March 2012 and replies were received continuously during these three months.

3.2.2. Characteristics of Community Managers: The Case of Local Administrations in Spain

A. Personal Characteristics

Beginning with the question of gender, the data show that large Spanish towns and cities are more inclined to employ men than women as community managers; this job is held in 61% of cases by men, compared to 39% by women. This contrasts with the findings of "Community Manager Report 2012" on the business sector, where 65% of posts were held by women compared to 35% by men.

With respect to age, Heras (2010) observed that this need not be a deciding factor, although some specialists in the field advise that community managers should be aged around thirty years. Nevertheless, according to other specialists there is no ideal age, but it is necessary to take into account the age group of the public to which the social network is directed.

In our case, almost 90% of the municipalities analysed have a community manager aged 25-45. Specifically in 50% of these, the official is aged 25-35 years and in the remainder, 36-46 years (see Figure 1). In Spanish municipalities, thus, very few community managers are aged under 25 or over 46 years. This is consistent with the findings of the SocialFresh report (2012), in which over 50% of the professionals surveyed were aged between 25 and 40 years.

Correlation of the respondents' age and gender shows that, in general, the younger the community manager, the more likely this person is to be female. The coefficient of correlation in this case is moderately strong, at 0.49.

B. Employment Situation

To better understand the employment situation of community managers in Spanish municipalities, it is useful to clarify some of the legal concepts applicable to those employed in the Spanish public sector. There are three main figures:

Figure 1. Gender and age of community managers in Spanish local administrations

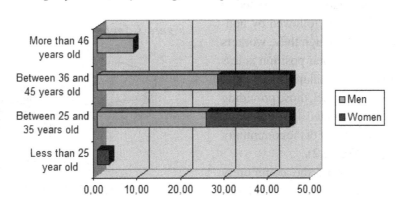

1. A tenured civil servant is someone who performs salaried work for the Spanish public administration of Spain, in a professional relationship determined by statute (laws and regulations) and governed by administrative law.
2. Non-tenured staff work for the public administration and do not belong to the group of tenured staff or that of temporary staff. They have employed status, and their activities, functions and rules and regulations are governed by the General Workers Statute and other applicable labour law provisions.
3. Temporary senior staff are those who are appointed to hold a post in the public administration. Their employment relationship is of an administrative nature; the posts thus held are not reserved for tenured staff; they are employed on a temporary basis, to work in specific organisations or authorities, in positions of special responsibility or with advisory status. In local government, they may also perform certain functions corresponding to senior officials. They may be freely dismissed by the authority that appointed them, and in any event this takes place automatically on the termination or conclusion of the mandate of the employing authority.

Our results show that the community managers in local government in Spain are mainly tenured staff. Specifically, 41% of these posts are held by staff who are permanently contracted to the Administration. These are followed by non-tenured staff, with 31%. Although these workers are, in practice, permanent, their position is less guaranteed than that of the tenured personnel. Finally, the temporary personnel, those providing political support, with very limited operational and advisory functions, occupy 28% of the community manager positions (see Figure 2).

Therefore, almost 72% of the community manager positions are held by staff whose salaries

are quite rigidly determined, fundamentally by the applicable regulations, and who enjoy a high level of job security within the administration. Thus, Spanish local governments tend to make use of existing human resources to perform community manager functions. This situation contrasts with that discussed by Castellanos (2012), who observed that in the private sector 60% of community managers work as company employees and 40% are freelance or independent.

With regard to their seniority in the position, over 50% of community managers have had this job for less than three years (see Figure 3). This is because in many local administrations in Spain, social networks have only been introduced very recently, and thus the need for a specific staff position in this respect is equally recent. Those who have been performing such duties for over three years have mainly been responsible for maintaining the information published on the municipal website, with social networking responsibilities being added subsequently.

Figure 2. Employment situation

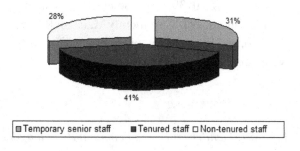

Figure 3. Seniority in the position

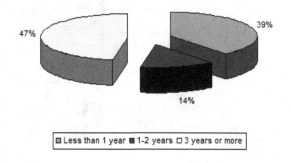

In regard to salary, in the case of the community manager of a private-sector company, this would certainly be influenced by factors such as the person's ability, training and experience, as well as the job functions performed, the type of business and the number of subordinates. Heras (2010) found salaries to range from 25,000 to 30,000 euros for more junior positions and 30,000 to 60,000 euros for more senior ones, for community managers generating multi-million euro sales figures. Cunningham (2012) reported that 28% of those working for companies in the field of social networks earn between 18,000 and 25,000 euros, while 23% are paid over 30,000 euros. However, in public administration the situation is very different, as the employees' goal is not to maximize profits but rather to improve public services and to enable citizens to participate in the organisation's activities.

As an example of community manager salaries in public administration elsewhere, in April 2011 the Telegraph newspaper stated that the British civil service had published a job offer on its website seeking a "Digital Executive Director." According to the announcement, the work of the person appointed to the post would be to adapt the government's communications and public messages to new systems of communication, and the salary offered was 161,497 euros. In Spain, the Basque government announced in April 2011

that it was looking for a community manager, to be hired on an administrative contract. The value of the tender was 80,000 euros. Moreover, in June 2012 it offered a grant of 24 months for specialist training in open government, to a value of 17,000 euros.

In this context, the salary paid to 39% of the staff responsible for maintaining the social networks of Spanish municipalities ranges from 20,000 to 30,000 euros, with only 22% of these community managers receiving more than 40,000 euros. If we consider these wages in the context of the employment situation, we find that within the higher salary range, the community managers are mainly temporary staff. However, the opposite extreme also occurs, in which it is the temporary staff who receive the lowest salaries (see Figure 4).

C. Training

When organisations refer to the role of a Community Manager, it is always described as a multidisciplinary profile, for which the ability to multitask is a prerequisite. Since it is still a relatively novel concept, and as yet there is no specific training for the post, in the selection process, both technical and social skills must be assessed. Among the technical skills, according to *AERCO y Territorio Creativo* (2009), a Community Manager should possess:

Figure 4. Salaries according to employment situation

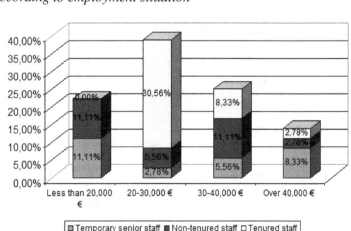

- Knowledge of the business sector, to strengthen the company's credibility and reputation.
- Knowledge of marketing, advertising and corporate communications, to understand business objectives and to align his/her activities accordingly.
- Good writing skills.
- A passion for new technologies, the Internet and Web 2.0.
- Creativity, to generate a share of attention among the overabundance of information.

Studies on the training background of community managers in the private sector have reported that 37% have studied communication and 24% marketing (Castellanos, 2012). In the first phase of the present study, the respondents were asked about their level of educational qualifications, and only 5.5% had no more than secondary school education; all the rest had studied at university.

Focusing on the university education of these staff in large Spanish municipalities, we observed that the most common qualification is that of a degree in journalism, held by 43% of community managers, followed by a degree in law, held by 21.5%, and one in advertising, marketing and public relations, by 14.3% (see Figure 5).

Furthermore, it seems logical that those responsible for social networks, many of whom previously distributed information about the organisation through a press office or similar, should change track and complement their training to undertake this new task. To do so, they may attend workshops or specialised courses, to receive new ideas, and to observe how other social media professionals work and resolve problems. Our survey revealed that 17% of these community managers had received no specific training in this field, while 19% had attended courses with a total duration of over 100 hours. The largest group (39%) was comprised of those who had received training, but for a duration of less than 50 hours (see Figure 6).

D. Workload

Turning to the community managers' present responsibilities, we wished to know whether they dealt exclusively with the management of social networks or whether they also had other duties within the administration. The questionnaire replies showed that only 11.4% dedicated their time exclusively to this task. In most cases, they also had other responsibilities, such as updating the website and running the press office. Let us recall that in the local governments analysed,

Figure 5. University degrees

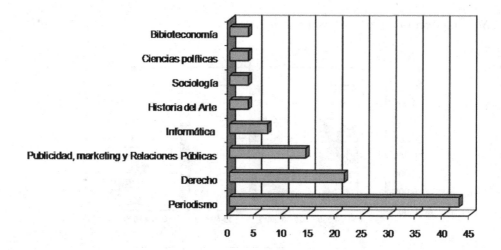

Figure 6. Complementary training via courses

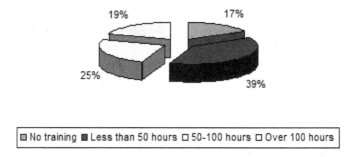

most of the community managers were originally trained and hired as journalists, and that this function was subsequently complemented with that of taking charge of the municipal social network (see Figure 7).

With respect to whether they managed only the network owned by the municipality in question or whether they also ran those of other authorities or companies, 89% stated that their job was only to manage the social networks of their own local authority.

E. Social Networks and Workload

Finally, we inquired about the respondents' workload, regarding the number of social networks managed. As shown in Figure 8, the most common situation is for a community manager to control three social networks, followed by those who manage five.

In some cases, up to ten networks are managed. The most commonly managed networks are Facebook, Twitter and YouTube. Other networks, although present, are much less commonly found (see Figure 9).

Figure 7. Workload

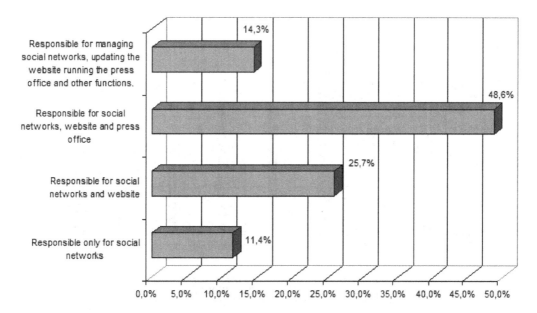

In 42% of the cases, the community manager is the only person responsible for running the municipality's social networks. On the other hand, in 25% of municipalities, the workload is shared within a department of four persons or more, and the community manager is responsible for supervising each of the networks used (see Figure 10).

CONCLUSION AND DISCUSSION

ICTs and social networks are useful tools to improve citizens' access to the public administration. Among other reasons, they provide practical information, increase the transparency and visibility of local government actions, promote participation in the design of public services, streamline the provision of specific services and provide management with useful information for planning and decision making.

However, it should be noted that the introduction of these instruments also involves risks, such as scant public interest, reflected in low levels of participation or that of only a limited group

of people; the breach of personal data protection legislation; the low quality of feedback; or inappropriate behaviour by users. Furthermore, in applying these new media, care must be taken not to create a digital breach by reason of age, gender or economic status, which could inhibit participation by certain groups.

Thus, a potential risk of the municipal use of social networks is that of low participation, and so it is important that there should be a municipal official to manage and promote the virtual community and to convey to the appropriate authorities the information provided by citizens – this official is the community manager.

Our empirical study of local government in Spain shows that the position of community manager is mostly held by men (61%). And in 90% of the municipalities analysed, the person responsible for managing social networks is aged 25-45 years. However, when age is correlated with gender, we find that, in general, the younger the community manager, the more likely this person is to be female. In our opinion, this fact could be related to the educational background of those

Figure 8. Number of social networks controlled

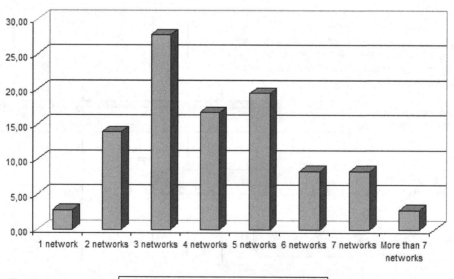

Figure 9. Social networks most often present

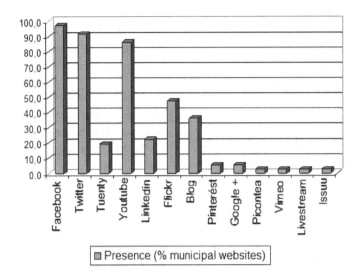

carrying out this task; most come from the area of communication, a sector that is being feminised and which, in recent years, has seen more women than men complete university studies.

Our analysis shows that almost 72% of the community manager positions are held by staff whose salaries are severely constrained by legal regulations and who enjoy a high degree of job security, which shows that town and city halls make use of their own human resources to implement these new activities. However, over 50% of community managers have been performing their duties for less than three years, which shows that the implementation of social networks is very recent.

The results of our study show that 43% of community managers have a university degree in journalism, 21.5% have a degree in law and 14.3%, a degree in advertising, marketing and public relations. In most cases, they perform, in addition, tasks such as updating the municipal website or running the press office. Facebook, Twitter and YouTube are the social networks most often used by community managers. We believe it is important that the social networks used by the municipality should be selected taking into the population to which the interaction is addressed, and the message that is to be conveyed.

Figure 10. Network supervisors

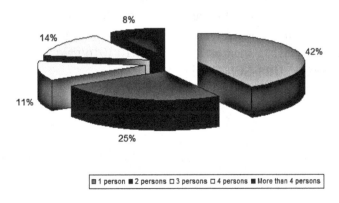

This study reveals that the use of social networks can provide the opportunity for local governments to improve their communication with the public, promoting transparency and participation and facilitating access to services. They constitute a channel for rapid communication, one that reaches a large number of people, at low cost, in comparison with print media, and allow near real-time feedback. However, it should be borne in mind that once a communication channel enabling interaction has been opened up, any reversal of such an initiative would be very counterproductive to the image of the institution.

In our opinion, in the field of corporate communication, the use of Web 2.0 and social networks provides an additional instrument for developing a comprehensive communication strategy. Accordingly, their use should be incorporated into the planning of municipal communications, setting out the objectives to be attained by the use of these channels, the audiences to be targeted and the type of information to be transmitted. A specialist in the creation, management and advancement of a community based on social networks, i.e., the community manager, should form part of the municipality's communications structure.

ACKNOWLEDGMENT

The research reported in this chapter has been supported by the Andalusia Regional Government through the excellence projects "Improving the management, transparency and participation in Local Governments through the Web 1.0 and Web 2." (P10-SEJ-06628) and "Efficiency and transparency in the municipal management under an economic crisis framework" (P11-SEJ-7700).

REFERENCES

AERCO y Territorio Creativo. (n.d.). *La función del community manager*. Retrieved June 1, 2012, from https://www.box.com/shared/pgur4btexi

Bastida, F. J., & Benito, B. (2006). Financial reports and decentralization in municipal governments. *International Review of Administrative Sciences*, 72(2), 223–238. doi:10.1177/0020852306064611

Bensen, C. (n.d.). *Community manager job description*. Retrieved June 1, 2012, from http://conniebensen.com/2008/07/17/community-manager-job-description/

Bertot, J. C., Jaeger, P. T., & Grimes, J. M. (2010a). Crowd-sourcing transparency: ICTs, social media, and government transparency initiatives. In *Proceedings of the 11th Annual International Conference on Digital Government Research*, Puebla, Mexico, May 17–20.

Bertot, J. C., Jaeger, P. T., & Grimes, J. M. (2010b). Using ICTs to create a culture of transparency: E-government and social media as openness and anti-corruption tools for societies. *Government Information Quarterly*, 27(3), 264–271. doi:10.1016/j.giq.2010.03.001

Bertot, J. C., Jaeger, P. T., Munson, S., & Glaisyer, T. (2010c). Social media technology and government transparency. *Computer*, 43(11), 53–59. doi:10.1109/MC.2010.325

Bonsón, E., Torres, L., Royo, S., & Flores, F. (2012). Local e-government 2.0: Social media and corporate transparency in municipalities. *Government Information Quarterly*, 29(2), 123–132. doi:10.1016/j.giq.2011.10.001

Caba Pérez, C., Rodríguez Bolívar, M. P., & López Hernández, A. M. (2012). The use of Web 2.0 to transform public services delivery: The case of Spain. In Reddick, C. (Ed.), *Web 2.0 technologies and democratic governance* (pp. 41–61). New York, NY: Springer. doi:10.1007/978-1-4614-1448-3_4

Castellanos, D. (n.d.). *El perfil del community manager: Funciones, herramientas, salario y actitudes*. Retrieved June 5, 2012, from http://www.inca-trade.com/blog/marketing-en-redes-sociales-2-0/el-perfil-del-community-manager-funciones-herramientas-salario/

Cavoukian, A. (2009). *Privacy and government 2.0: The implications of an open world*. Ontario, Canada: Information & Privacy Commissioner.

Chan, H. S., & Chow, K. W. (2007). Public management policy and practice in Western China: Metapolicy, tacit knowledge, and implications for management innovation transfer. *American Review of Public Administration*, *37*(4), 479–497. doi:10.1177/0275074006297552

Cheng, Y. L. (2009). On the professional competence-building of management personnel in urban community neighborhood committees. *Proceedings of the 2009 International Conference on Public Economics and Management (ICPEM), Vol. 4: Econometrics*, (pp. 433-437).

Christiaens, J. R. (1999). Financial accounting reform in Flemish municipalities: An empirical investigation. *Financial Accountability & Management*, *15*(1), 21–40. doi:10.1111/1468-0408.00072

Dunleavy, P., Margetts, H., Bastow, S., & Tinkler, J. (2006). New public management is dead - Long live digital-era governance. *Journal of Public Administration: Research and Theory*, *16*, 467–494. doi:10.1093/jopart/mui057

eGovernment Action Plan. (2006). *eGovernment action plan 2006, COM 2006/173 of 25.04.2006*. Retrieved June 10, 2012, from http://ec.europa.eu/information_society/activities/egovernment/library/index_en.htm

Gallego, R., & Barzelay, M. (2010). Public management policymaking in Spain: The politics of legislative reform of administrative structure, 1991-1997. *Governance: An International Journal of Policy, Administration and Institutions*, *23*(2), 277–296. doi:10.1111/j.1468-0491.2010.01479.x

Gibson, A. (2010). *Local by social: How local authorities can use social media to achieve more for less*. London, UK: NESTA.

Gould, E., Gomez, R., & Camacho, K. (2010). Information needs in developing countries: How are they being served by public access venues? Paper presented at the AMCIS. Retrieved from http://aisel.aisnet.org/81625amcis2010/9.

Heras, M. (2010). Community manager, ese gran desconocido. *Revista de Comunicación*, *13*, 16.

Hernández Sampieri, R., Fernández Collado, C., & Baptista Lucio, P. (2003). *Metodología de la investigación* (3rd ed.). México: McGraw-Hill.

Huijboom, N., Van den Broek, T., Frissen, V., Kool, L., Kotterink, B., Nielsen, M., & Millard, J. (2009). *Public services 2.0: The impact of social computing on public services*. Luxembourg: Institute for Prospective Technological Studies, Joint Research Centre, European Commission, Office for Official Publications of the European Communities.

ICT PSP from PIC. (n.d.). *ICT policy support programme*. Retrieved June 1, 2012, from http://ec.europa.eu/information_society/activities/egovernment/implementation /ict_psp/index_en.htm

Klein, P. (2008). Web 2.0: Reinventing democracy. *CIO Insight*, 30-43.

Kovac, P., & Decman, M. (2009). Implementation and change of processual administrative legislation through an innovative Web 2.0 solution. *Transylvanian Review of Administrative Sciences, 28E*, 65–86.

Leighninger, M. (2011). *Using online tools to engage – And be engaged by – The public.* Washington, DC: IBM Center for The Business of Government, IBM Center for The Business of Government.

Martins, M. R. (1995). Size of municipalities, efficiency, and citizens participation: A cross-European perspective. *Environment and Planning. C, Government & Policy, 13*(4), 441–458. doi:10.1068/c130441

McMillan, P., Medd, A., & Hughes, P. (n.d.). *Change the world or the world will change you: The future of collaborative government and Web 2.0.* Retrieved June 5, 2012, from www.deloitte.com

Millard, J. (2009). Government 1.5: Is the bottle half full or half empty? *European Journal of ePractice, 9*(1), 35-50.

Mintz, D. (2008). Government 2.0 –Fact or fiction? *Public Management, 36*(4), 21–24.

Mussari, R. (1999). Some considerations on the significance of the assets and liabilities statement in Italian local government reform. In Capperchione, E., & Mussari, R. (Eds.), *Comparative issues in local government accounting* (pp. 175–190). Norwell, MA: Kluwer.

OECD. (n.d.). *Denmark: Efficient e-government for smarter service delivery.* OECD Publishing. Retrieved June 1, 2012, from http://dx.doi.org/10.1787/9789264087118-en

Osimo, D. (2008). *Web 2.0 in government: Why and how?* Luxembourg: Joint Research Centre. Institute for Prospective Technological Studies, Office for Official Publications of the European Communities.

Pallot, J. (2001). Transparency in local government: Antipodean initiatives. *European Accounting Review, 10*(3), 645–660.

Redell, T., & Woolcock, G. (2004). From consultation to participatory governance? A critical review of citizen engagement strategies in Queensland. *The Australian Journal of Public Administration, 63*(3), 75–87. doi:10.1111/j.1467-8500.2004.00392.x

Rodríguez, M. P., Alcaide, L., & López, A. M. (2010). Trends of e-government research: Contextualization and research opportunities. *International Journal of Digital Accounting Research, 10*, 87–111. doi:10.4192/1577-8517-v10_4

Sandoval-Almazan, R., & Gil-García, J. (2012). Are government internet portals evolving towards more interaction, participation, and collaboration? Revisiting the rhetoric of e-government among municipalities. *Government Information Quarterly, 29*(2), S72–S81. doi:10.1016/j.giq.2011.09.004

Smith, K. A. (2004). Voluntary reporting performance measures to the public: A test of accounting reports from U.S. cities. *International Public Management Journal, 7*(1), 19–48.

SocialFresch. (n.d.). *Community manager report 2012.* Retrieved June 10, 2012, from http://socialfresh.com/community-manager-report-2012/

Taylor, J., Lips, M., & Organ, J. (2007). Information-intensive government and the layering and sorting of citizenship. *Public Money and Management, 27*(2), 161–164. doi:10.1111/j.1467-9302.2007.00573.x

Ter Bogt, H. J., & Van Helden, G. J. (2000). Management control and performance measurement in Dutch local government. *Management Accounting Research, 11*(2), 263–279. doi:10.1006/mare.2000.0132

Tuomi, I. (2002). *Networks of innovation: Change and meaning in the age of the internet.* Oxford, UK: Oxford University Press.

United Nations. (2010). *E-government survey 2010: Leveraging e-government at a time of financial and economic crisis.* New York, NY: United Nations.

KEY TERMS AND DEFINITIONS

Blogs: A blog (a portmanteau of the term web log) is a discussion or information site published on the World Wide Web consisting of discrete entries ("posts") typically displayed in reverse chronological order so the most recent post appears first. Until 2009 blogs were usually the work of a single individual, occasionally of a small group, and often were themed on a single subject.

Community Managers: Persons that hold the collective vision, create and manage relationships and manage collaborative processes. The community manager builds and monitors multiple communities generated in blogs, forums, social networks, etc. S/he becomes the authorised voice of the company.

E-Government: It is a term used to refer to the use of information and communication technology to provide and improve government services, transactions and interactions with citizens, businesses, and other arms of government.

E-Participation: e-participation represents the expansion, transformation and greater involvement of citizens in public life and consultation processes.

Information transparency: It is defined as the availability of online governmental information and access to this information.

Stakeholders: A corporate stakeholder is a party that can affect or be affected by the actions of the business as a whole. The stakeholder concept was first used in a 1963 and it was defined as those groups without whose support the organization would cease to exist.

Web 2.0: Web 2.0 is a concept that takes the network as a platform for information sharing, interoperability, user-centered design, and collaboration on the World Wide Web.

Chapter 16
Kenya E–Participation Ecologies and the Theory of Games

Vincenzo Cavallo
Cultural Video Foundation, Kenya

ABSTRACT

An e-Participation ecology is composed of five elements—actors, contents, traditional culture of participation, existing media skills and practices, and discourses in conflicts (establishment vs. antagonists)—and three macro-dimensions—cultural/traditional, political, and socio-technological—with which the five elements are interacting (Cavallo, 2010). Game theory can be used to understand how a certain actor or a group of actors can develop a successful strategy in/for each one of the three dimensions. Therefore, the concept of Nash equilibrium (Nash Jr., 1950), developed in physics and successfully applied in economy and other fields of study, can be borrowed also by e-Participation analysts/project managers to develop "Win-Win" scenarios in order to increase e-Participation projects' chances of success and consequently reduce e-Participation's "risk of failures," especially in developing countries where they usually occur more frequently (Heeks, 2002). The Kenyan e-Participation platform, Ushahidi, generated a techno-discourse about the rise of African Cyberdemocracy and the power of crowd-sourcing that is probably more relevant than the real impact that these e-Participation platforms had or will have on the lives of normal citizens and media activists.

INTRODUCTION

In this chapter we will explore the possible applications of game theory to the field of e-Participation in Kenya. The term e-Participation is used to define a specific field of study that attempts to investigate how ICTs can be applied to improve citizen/community participation and consequently their capacity to influence government decision-making processes. In this respect, e-Participation projects are considered strategies to influence power in a specific direction.

It is important to specify that in the context of this chapter, e-Participation is not considered a neutral space in which different actors communicate amongst themselves.

DOI: 10.4018/978-1-4666-3640-8.ch016

Game theory should be applied only to specific concepts of e-Participation: those that imply the existence of opposite and complementary interests of actors, fighting and collaborating between each other to reach their objectives. In this context e-Participation can be defined as: the electronic space/dimension in which different actors are competing or cooperating to support a specific or different "discourse/s" (to: gain, expand or to maintain their power) (Cavallo, 2010).

Furthermore, in this chapter we will investigate e-Participation in relation to international development and the history of "North-South" relations in Sub-Saharan Africa from colonialism up to now.

BACKGROUND

Citizen participatory deliberation systems imply devolution of power for the government. Why would bureaucrats - and in most cases also politicians - be open and willing to delegate part of their power to the people? While in some cases politicians are ready to do so, in order to increase their consensus among voters, often even if they open the doors to participation and participatory deliberation, it does not mean that they will keep them open forever. Furthermore, bureaucratic power is based on knowledge of institutional mechanisms. If citizens gain access to this type of information and understand these mechanisms, bureaucrats will swiftly begin to lose their power. As a result, bureaucrats need to keep the different "information spheres" separate from each other in order to maintain their sphere of influence (Meyrowitz, 1985).

Communication can generally be understood as an act of power. Power, understood in this case as the capacity to influence other people's decisions to support the values, the interests and the will of the people who hold power (Castells, 2009).

In the field of e-Participation, researchers have not always been able or willing to focus on power dynamics. This may occur because in some cases

it is difficult to develop specific frameworks to do so and in others because they decide to avoid the topic deliberately, in order to prevent conflicts or please their funders/donors - who in most cases, are the same institutions that are financing the project that they should be assessing or researching.

In this context power dynamics are not ignored but represent instead the core of the enquiry to demonstrate how it is possible to develop frameworks, strategies and actions based on the analysis of opposite and complementary power relationships.

These power dynamics have a direct or indirect impact on e-Participation studies and therefore should not be ignored by researchers. Instead, they should become one of the most important subjects/elements of any scientific analysis of e-Participation, in both the so-called developing and developed countries.

MAIN FOCUS OF THE CHAPTER

Power Dynamics and Analytical Frameworks

The emancipation of a social actor cannot be separated from its capacity to gain power over other actors unless we want to accept a naïve image of a reconciled human community, a normative utopia, that historical observation debunks. (Castells 2009, p. 5)

The main focus of this chapter is on e-Participation's power dynamics and how to analyse them to support the development of an e-Participation project strategy in either a developing or a developed country. These dynamics can be classified as follows:

- **Discourse-Influence:** The system of thoughts composed by ideas, attitudes, courses of action, beliefs and practices, that systematically construct the subjects and the worlds (Foucault, 1972). For the scope of this chapter, it is therefore necessary to consider how power dynamics influence the way in which "developing countries" are studied by the "developed countries" from which most of the funding for research projects originates. For this reason, it is very important to consider, the influence of dominant theories of development studies, such as the "modernization theory" influence – in our case the ones created by "ICT for development" experts who did not questioned the "technological determinism" behind the dominant theoretical frameworks. These theories are directly linked to "discourses" embedded in western societies and discussed during international events such as the World Summit on Information Society (WSIS). Many concepts, values and ideas on which these discourses are based should be deconstructed to understand the "discourses influences" on e-Participation projects.

Most of the above-mentioned internationally developed frameworks and discourses have been proved to be functional for the ruling elites who produced and used them to expand their power by influencing local/national and international policies before and after colonialism. Both participatory methodologies and ICTs for development projects have been used, and are still used today, within these very frameworks. This implies that a critical approach to studying both participation as a way of managing resources and the use of ICTs within an informational system (Castells, 1998) is needed before starting any type of research or project in the field of e-Participation - especially but not exclusively in the context of the so called "developing countries."

A possible solution could be to use the approach proposed by Olivier de Sarden: a "*socio-anthropologie du changement social*" (an anthropology of social change), in order to overcome the dialectics

between the development anthropologist and the anthropologist of development (de Sarder, 1995) to create an anthropological approach capable of analyzing and deconstructing external/oppressive "discourse influences" in order to develop new ones aimed at supporting the uprising of marginalized communities.

This chapter tries to demonstrate the need and feasibility of shifting the established approach in the study of e-Participation and in general of ICT-4Development. The new approach should be used to deconstruct "non-participatory" discourses, in order to construct new "antagonistic discourses" made by and for local communities that wish to use an e-Participation project to influence government policies.

- **Techno-Semantic-Influence:** Technocrats behind semantic technologies, especially in the legislative field (the so called "legal semantic web"), decide how different concepts and bodies of information are linked to each other, thus influencing the sense of an event and the relationships between different episodes of political relevance. As a result, technocrats are often able to influence the way legislative acts are discussed, approved or rejected. Episodes occurring outside legislative bodies but related to what happens inside them can be easily ignored or manipulated by omitting a link to specific information or by selecting certain types of information while ignoring others. These "techno-semantic-influences" may represent a threat to democracy.

For example: a hypothetic minister that is proposing a certain law is in fact guilty of a related crime. However, if the law in question does go through, the crime of which he or she is guilty will no longer be consider a crime. By controlling the way a semantic "mark up" is applied, technocrats influenced by governments may decide that the criminal record of a minister should not be linked to the laws he or she is proposing and discussing in parliament. In this way, it will become more difficult for journalists, civil society or other members of parliament to discover possible conflicts of interests or other factors that may negatively influence policy making. The importance of using open standards such as XML is directly linked to the right of social movements, journalists and members of parliament or other legislative bodies, to mark up in a participatory manner, important information in order to "democratize" the legislative semantic web. Examples of semantic "participatory mark ups" such as the http://www.theyworkforyou.com/ website in the UK.

- **Actors-influences:** These influence the way projects start, develop and end. An e-Participation project can be defined as a coordinate action, organized by a group of media-actors related to social movements, or/and political parties, public or private institutions, local communities, who decide to use digital media in order to gain or to maintain political power. These groups of actors objective is to influence decision makers and in general the public opinion. The on-line dimension/spaces developed by these different groups for and during their actions/campaigns are defined as e-Participation platforms/spaces. An e-Participation project may be started as a result of a conflict between two or more actors. In other words, it may exist because a specific conflict occurred and one of the actors decided to use e-Participation as a tool to fight against the opponents. An e-Participation project may end because the conflict from which it was generated has been solved. In other cases e-Participation projects may be kept alive by citizens and civil society as an act of resistance. An e-Participation project may also start because an international/national development agency decides to target ICTs and eGovernance as

a field of technological assistance through which it is possible to influence a country's policy-making processes. A social actor or a group of activists may use e-Participation to gain the attention of public opinion and influence civil society at international, national and local levels.

Actors's motivations are considered the strongest influences of power, capable of determining the success or failure of a given e-Participation project.

In this context, the so called "creative class" (Florida 2003; McKenzie 2004; Aronowitz 2006; Formenti 2008) and its capacity of developing and influencing e-Participation narratives, technologies, projects and strategies in both developed and developing countries is very important.

A "creative class" interested in supporting a specific political agenda should be able to influence all the above mentioned dynamics, therefore the "creative class" is a key player that can become both an essential partner of the establishment class or of the "multitude" (Hardt and Negri, 2004). The "creative class" may therefore play a key role in the "informational society" (Castells, 2000), in order to contribute to the development of either an *"informational establishment"* or an *"informational multitude."* Therefore e-Participation projects should all be considered as actions made by specific private or public actors who aim to influence other public/private actors in order to reach specific objectives. These actors may start e-Participation projects to defend themselves from others, or may start e-Participation projects to gain power over others. Within these ongoing conflicts, new actors emerge and alliances are formed. In some cases these alliances may generate *"informational multitudes"* able to change the political systems and overthrow the establishments. In other cases new and old enemies continue to fight against each other in order to gain power within the already existing political system/framework.

By understanding the history of conflicts and alliances, it should be possible to draw future scenarios and develop new strategies. Nevertheless to accomplish this, a new framework based on the historical interactions between media actors and political actors is needed in order to help us understand the process by which the failure and success of e-Participation projects can be influenced.

The famous game theory became an interdisciplinary research field after Theory of Games and Economic Behavior was published in 1944 (Neumann and Morgenstern, 1944). Different games have been invented and used by social scientists - mostly political scientists and economists - in order to understand international conflicts and trade. Consequently, other games can be used to understand e-Participation and the conflicts that are behind the failure or the success of different projects. This approach requires a deep historical analysis of the interactions between actors to be effective.

The Win-Win e-Participation model I present in this chapter focuses on how different media and already existing participatory practices/process, which are part of a specific communicative ecology (Foth & Hearn, 2007), can be utilized in an e-Participation project to reduce the risks of failure associated to technological and cultural barriers.

The "Win-Win" terms stands for relationships or transactions in which both or all parties gain something. The model is based on the assumption that a "Win-Win" relationship is happening when all the actors involved in a communication process are benefiting from a specific media mix strategy, used to pursue a common objective.

The methodological proposal is based on a new concept: the "e-Participation ecology" (Cavallo, 2010) in which three transverse dimensions co-exist: cultural/traditional, political and socio-technological.

The idea of creating an e-Participation ecology has been developed following the already existing communicative ecology concept (Tacchi, Slater & Hearn, 2003) and the cooperative or non-cooperative games.

An *e-Participation ecology* is an analytical framework (see Figure 1) necessary to map a specific network composed of actors, contents, cultures and traditions, media skills and "dis-

Figure 1. E-participation ecologies

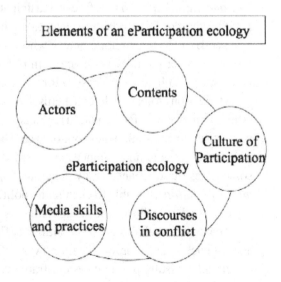

courses" interested/involved in and by communication practices/modalities aimed to influence a specific group of policy makers.

These actors are not necessarily sharing the same physical spaces, however their objective is to influence policy-making processes in specific geographic/administrative areas. These actors are able to actively influence such policies being involved at local, national or international levels. During the course of the field research and thanks to the last interviews and social network analysis I discovered that young people living in Nairobi are actively involved in local debates concerning their constituencies of origin. In the same way, the influence of Kenyans abroad through the use of social networks such as Facebook drastically increased during the last election and is becoming a new source of opinion, able to influence the decision of many Kenyan voters. The most evident example is Ory Okolloh, a human rights activist behind two of the most relevant e-Participation projects developed in Kenya Ushahidi and Mzalendo. Okolloh studied and still lives abroad, despite being undoubtedly one of the most influential human rights/political activist in Kenya.

Therefore, e-Participation ecologies are borrowing the concept of *field* developed by the *Network Ethnography* (NE). Following this approach the meaning of '*field sites*' is adapted, and instead of choosing territorial field sites, the researcher has to choose a perceived community and select the important nodes in the social network as field sites. Indeed, the field site may not be a socially significant physical place at all (Howards, 2002).

For this reason, e-Participation ecologies are not only composed of specific geographic areas, in which policies are discussed and implemented, but also networks of actors, who do not necessarily live in these specific geographic areas, but that are able to use their networks to influence policy-making processes occurring in one or more geographic areas.

An e-Participation ecology is composed by five elements: actors, contents, traditional culture

of participation, existing media skills and practices, discourses in conflicts (establishment vs. antagonists) and three macro-dimensions: cultural/traditional, political and socio-technological, in which the five elements are interacting between each other (Cavallo, 2010). In these three dynamic dimensions, different actors interact with each other. The result is a complex framework, whose analysis should support project managers in assessing/predicting the actions of other actors that may wish to influence their project results and outcomes.

In conclusion, the Win Win e-Participation model based on the e-Participation ecology framework should be considered as one comprehensive model, capable of assessing both the accessibility and power dynamics that belong to a specific e-Participation ecology.

Kenya Cross-Media Models and Power Dynamics

In developing countries, Community Multimedia Centers (CMCs) can be considered fertile grounds where many low cost cross-media innovations/solutions have been developed and experimented. CMCs are usually located in marginalized urban areas or remote rural areas to improve the life conditions of a specific community through the use of ICTs (BBC Trust, 2006). In such situations the main problem is accessibility. Radio is considered the most used and accessible media in Africa (Ibid.) and should therefore be the main communication channel between the community that has no access to the Internet and the CMCs that should be able to access Internet and provide the know-how to use it.

One of the characteristics that allowed radio to become so popular is the capacity and adaptability of this media. For example, using phones to call radio presenters or to take part in a public debate transformed radio into an interactive media. Additionally information for radios can be produced and distributed locally, but also distributed worldwide through a decentralized interactive network of stations distributed in different areas.

Radio structure and content are perfectly compatible with the emerging culture of Internet and social networks (Girard, 2003). Examples such as radio browsing programs, in which the radio presenters gather information in response to listeners' needs and queries from reliable sites on the Internet, have demonstrated the potential of this cross-media approach in the field of education, health and participatory development (Collin, F. and Restrepo, E., 2001). Radio browsing has also contributed to disseminating values and ideals embedded in the western "discourse" on ICT4Development (Soriano, 2007).

Over the last few years, developing countries have been also the protagonists of an incredible mobile phone market expansion. Mobile Penetration rates have been incredibly high in Africa, Asia and Latin America changing forever the media landscapes of developing countries and the use of other media such as the Internet and the Radio (UNCTAD, 2010).

The impact of this mobile penetration in Kenya has been tremendous. In just five years, this country has become an innovation hub thanks to mobile applications and cross-media solutions that have been developed and experimented by local media activists/developers in and for the local market. These innovations are having a huge impact at both national and international levels.

Field research on how different actors communicate between each other on issues related to Community Development Founds and local/national policies in 16 Kenyan constituencies (Cavallo, 2009) confirmed most of the above mentioned trends about the use of radios and mobile phones.

The overall aim of the field research was to advise policy makers, project managers and media activists working in the field of eDemocracy/e-Participation in Kenya on how to avoid failures and provide, at project completion, ROI (return of investment) data for all the relevant stakeholders.

The concrete output of this study is "a model" developed to support and enhance the implementation of accessible technologies used to support existing/traditional information systems supporting existing/traditional participatory practices: the "Win Win e-Participation model" (Ibid).

The "Win Win e-Participation model" (Ibid) is based on three analytical micro-dimensions (see Figure 2) developed to assess one of the tree dimensions that are forming part of an e-Participation ecology, the "socio-technological dimension":

- **Organizational:** Consists of a pre-implementation and post-evaluation analysis of the project in relation to the traditional participatory policy practices and cycle - e.g. Identify an already existing/traditional practice and evaluate the impact of the project on it.
- **Socio technological:** Consists of a pre-implementation and post-evaluation analysis of the project in relation to how citizens and their representatives already use the existing media (radio, video, press) and how the e-Participation platform can empower and extend their networking ability - e.g., how the e-Participation platform could empower the use of local radio, mobile phones etc. in order to support communication between citizens and their representatives.
- **Socio-political:** Consists of an overall pre-implementation and post-evaluation analysis in relation to the policy cycle and the achievement of public participation goals, e.g. political vision, policy information, decision making, implementation, impact on the final decision of the Parliament (International Association of Public Participation).

The field research, consisted of 16 interviews with personal assistants (PAs) of Kenyan members of parliament (MPs), 8 constituency officers (COs) and 2 community radio journalists (CRJs).

Figure 2. Win-win e-participation model

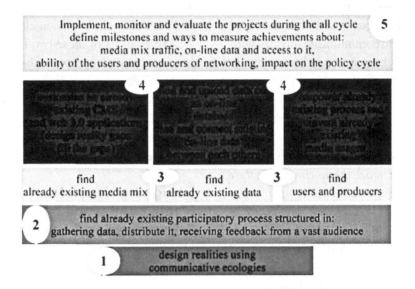

The main research focus areas were:

- The use of different media to communicate between (Pas, Mps, Cos & CRJs) citizens, civil society and journalists.
- **The use of the media mix:** New and old media, strategies and tools to inform and engage citizens in public debates.
- Perceptions and prioritization of different technologies usage and finally socio technological ideas and scenarios for the future.

The main results of this research are summarized as follows.

Traditional forums such as mabaraza_can constitute the main source of information to understand citizens problems and opinions about political issues.

The baraza (pl. mabaraza) is a feature of Zanzibar's "public sphere." In organizational terms, a baraza may represent different degrees of formality and informality, institutionalization, and abstractness. A baraza might be a simple (informal) "meeting" of people, but it could also be a "council," or, in historical times, the "audience" of the Sultan of Zanzibar. Finally, it could refer to a vast range

of clubs, unions or associations.In spatial terms, a Baraza is a public or semi-public space where people meet to chat, communicate, quarrel, sit, similar to a "Piazza" in Italy "Agorà" in Greece or the "Majlis" in Arabia (Loimeier, 2005).

The "baraza" appears to be at the same time: a place, an event, and a forum, therefore it is "an essential node in the social network" (Cavallo, 2009 p.9).

On 16 PAs interviewed 13 defines baraza has the most used way to communicate with the citizens, furthermore in the perception of most of the PAs and COs baraza is also the most effective way of communicate with the constituents.

From a transcription of the interviews:

Our policies? We use a lot of ways to communicate. The most effective is a baraza. We also have printouts of vision.... (Wajir East Constituency).

Usage/Media mix: mobile phones and community radios are used to organize mabaraza and communicate their contents to a vast audience, the Internet is used to send information from Nairobi to the constituencies, once in the constituencies they are distributed in different formats: radio and print. Radio and mobile phones are mostly used to mobilize people while the Internet is seen mostly as a fundraising and a business-to-business tool rather than a media to communicate directly with citizens by most of the constituencies officers.

From a transcription of the interviews:

We have a representative per village, so this representative has a mobile phone, so when the MP wants to communicate with them he calls the representatives, they organize the meeting, then they speak ......we also use posters then we put it in the markets and churches" (Rarieda Constituency).

We also have this public address system mounted on vehicles, so we prefer announcing our meetings ... We also have Musii FM which we also use to communicate our meetings" (Kibwezi Constituency).

For projects that we are trying to start, we take photos then we put them in those newsletters, is about projects that have not been completed by the former MP, so after we evaluate the situation then we send the information. The objectives of the newsletter is to inform the constituents about projects we intend to initiate and we urge them to prioritize every village should prioritize which projects are fundamental, the newsletter is sent from Nairobi to the constituency office by email, then they photo copy it and they distribute an hard copy version of it in local churches, schools and different public spaces (Kitutu Chache Constituency).

Already existing data: radio have an enormous amount of information about citizens opinions and polls in their on-line database, constituency staff members are video recording mabaraza and store the videos off-line.

Once assessed the socio-technological dimension to understand what type of media and participatory practices are mostly used successfully in the 16 interested constituencies, the e-Participation ecology can be enriched by developing the Cultural and the Political dimension also.

The analysis of these two dimensions should help analysts to complete their e-Participation ecologies assessments by tracing the genesis of the latest tech innovations to understand who developed them, why and how they were developed and therefore who will probably develop new ones in the near future.

Designing Kenya New-Media Landscape by Using E-Participation Ecologies

The genesis of Ushahidi - in Swahili witness - can be explained by using the tree fundamental e-Participation ecologies categories (Discourses – Actors – Conflicts). This open-source software was invented during the 2007 post election violence. The Kenyan government (Discourse 1 - Actor 1) decided to obscure live programs on TV and to limit media coverage of daily episodes of violence (Conflict). Therefore, a human right activist, Ory Okolloh, a blogger, Erik Hersman and a developer, David Kobia (Discourse 2 - Actor 2), managed over the course of three days, to develop a software that could be used to map in a participatory manner episodes of violence around the country. They created an e-Participation platform/project.

They achieved this by using existing software: Google Maps and FrontlineSMS[1] a software allowing to manage SMSs with the help of a simple laptop. The main objective of Actor 2 was to create a platform through which Kenyan citizens could inform each other about what was happening around the country. In this way, citizens who needed to move from one place to another for emergency reasons, could be informed about road blocks and other dangers in order to avoid them.

As we already mentioned, radios and mobile phones are the most accessible media in Africa, therefore the on-line platform had to be developed according to the specific needs of Kenyan users. FrontlineSMS was synced to be used with Ushahidi. Once installed, the program enabled users to send and receive text messages with other groups of people through mobile phones while concurrently volunteers and bloggers were mapping information on-line using Google Maps. Consequently, other media such as CRs began using the Ushahidi platform as one of their main sources to inform citizens. In just a few days Ushahidi, a newborn e-Participation platform, became a credible source of information for both citizens and mainstream media. The platform had 45,000 users in Kenya during this time of turbulence. Radio deejays read some of the reports on air.

A study done by the Harvard Humanitarian Initiative (Meier, Patrick and Kate Brodock, 2008) analysed data reported on the site and compared it to reports from mainstream media outlets. The study concluded that Ushahidi was better at reporting incidents as they started (rather than just the deaths resulting from incidents) and reports covered a broader geographical area than those coming from mainstream media.

FrontlineSMS or Google Maps could be used for different purposes, and of course existed before Ushahidi did. Nevertheless their combined use with Ushahidi in that specific situation, introduced a new dimension which made Ushahidi famous around the all world. Soon after its initial use in Kenya, the Ushahidi software was used to create a similar site to track anti-immigrant violence in South Africa, to map violence in eastern Congo, to track pharmacy stockouts in Malawi, Uganda, and Zambia, to monitor elections in Mexico and India, to collect eyewitness reports during the 2008-2009 Gaza War by Al Jazeera, to develop a crisis information system in support of aid workers during the earthquake in Haiti and Chile, to map blocked roads and other information in US by the Washington Post during the wake of winter storms, to set up a "map of help" for voluntary workers needed after a wildfires in Russia. This software allowed pro-democracy demonstrators across the Middle-East to organize and communicate what was happening around them in early 2011. It has been used also in Italy, in Japan, in Australia, in and in the Balkans.

An innovation coming from the "South" of the world is now used to solve problems in the "North." However this does not necessarily mean that "North-South" power dynamics have been radically changed by these experiences. In reality we are witnessing to an appropriation of both "discourses" (Human Rights and Modernization) and "technologies" (Crowdsourcing/Social Media) by the emerging African "creative class" and the new political elites.

The idea of Mzalendo, an e-Participation project that began at the end of 2005, came to Ory Okolloh (Actors 1) a few years before Ushahidi. The slogan of Mzalendo is "keep an eye on the Kenyan Parliament" (Discourse 1), and came about after the website for Kenya's Parliament was shut down (Conflict) following protests by some MPs (Actors 2) who were embarrassed about their CVs being published online (Discourse 2). The initial goal of Mzalendo, then, was to provide the basic information that otherwise would have been available on the official parliamentary website. Kenya's parliament website is now back online - and much improved since its former 2005 incarnation – but the activists behind this e-Participation project continue to feel that they still have an important role to play in using online tools to hold Kenyan MPs more accountable. Therefore the Mzalendo project is still going on following the model of the British "TheyWorkForYou" project[2].

The case of Mzalendo, which means "patriot" in Swahili, also confirm the validity of the analytical framework based on the "Discourses – Actors – Conflicts" model based on the e-Participation ecologies framework.

The Kenyan e-Participation ecology is strongly influenced by all the actors mentioned above. They support antagonists discourses that are generating conflicts, themselves generating e-Participation projects. Kenyan politicians believe that the media should not be totally free to report about sensitive issues, especially during internal crisis, such as tribal clashes, but also that MPs CVs should not be public. Kenyan human rights activists believe that crowd-sourcing can be a strategic resource to face all types of emergencies and that the Kenyan Parliament like the British one should provide citizens open access to all types of data. Different values and ideologies are embedded in both discourses.

The same analytical framework can be used to analyse the genesis of other technological innovations that are not directly linked to the political sphere but may have an impact on it in the next future.

The following analysis on the M-Pesa case may help us to understand more about how a successful tech business innovation has been developed and how it may be used in the next future to organize and/or support political actions.

M-Pesa, ("M" for mobile, "Pesa" Swahili slang-word for money) is the product name of a mobile-phone based money transfer service for Safaricom, which is a Vodafone affiliate. Therefore it is the result of a partnership between different organizations which represent a significant example of how mobile low-cost technologies can be used in creative ways to improve the life conditions of the populations of developing countries. This innovation started as a "development project" (Discourse1-Influence) financed by the Vodafone Foundation and the UK-based Department for International Development (DFID) (Actor 1-Influence) trying to solve a problem that most Kenyans (Actor 2-Influence) have to face: credit.

The initial concept of M-Pesa was to create a service that allowed micro-finance borrowers to conveniently receive and repay loans using the network of Safaricom airtime resellers. This would enable micro-finance institutions (MFIs) to offer more competitive loan rates to their users, as there is a reduced cost of dealing in cash. The users of the service would gain by being able to track their finances more easily (Discourse 1 made by Actor

1). However when the service was eventually set up for user-trials, it was discovered that customers adopted the service for a variety of alternative uses (Discourse 2 made by Actor 2). After this complications arose with Faulu, the partnering micro-finance institution (MFI) (Conflict).

M-Pesa was re-focused and launched with a different value proposition: sending remittances home across the country and making payments. This conflict of interests between the vision of Actor 1 and Actor 2 generated a new "discourse" from which both Actors (1 & 2) could benefit from, while Faulu decided to leave when the project lost the micro-finance component.

M-Pesa is now a branchless banking service, meaning that it is designed to enable users to complete basic banking transactions without the need to visit a bank branch. The continuing success of M-Pesa in Kenya is due to the creation of a highly popular, affordable payment service with only limited involvement of a bank. The system was developed and ran by Sagentia (U.K. based company) from initial development to the 6 million customer mark. The service has now been transitioned to be operationally run by IBM Global Services on behalf of Vodafone (U.K. based company). The initial 3 markets (Kenya, Tanzania & Afghanistan), are hosted between Rackspace and Vodafone.

In conclusion what began as a "development project," partially financed by international aid public funds and implemented by a "western private company," thanks to the users "appropriation" of the mobile media, was transformed into one of the most profitable business ever invented for the African market, considering that by 2012 mobile financial systems in developing countries will create a market of about 5 million of US dollars (CGAP and GSMA, 2009).

The appropriation is an ongoing transformation of use continuously brought about by interactions with other users and by interactivity with equipment and software. Appropriation is a concept that helps us get out of a naïve prediction, built exclusively on technical possibilities. To think in terms of appropriation necessarily entails introducing social representations/perceptions of the potential users in their contexts/networks (Flichy, 1995).

Therefore the appropriation process that caused the rise and success of the M-Pesa case in Kenya may continue and provoke other changes. For example in the political context, systems such as M-Pesa may change the ways political fundraising campaigns could be organized by new candidates running for presidential and local elections.

This article was published on the web-portal of the most important Kenyan news publisher, the Standard Group: "The burial of his wife Virginia Nyakio, a fortnight ago, provided the perfect occasion for Njenga to demonstrate his fundraising skills — another critical aspect that makes him ideal for presidential seekers. Although he sent out an appeal for Sh5 million to offset mortuary fees, he got in excess of Sh9 million. He says thousands of his supporters contributed small amounts via M-Pesa." (Standard Group, 2012).

The impact of these changes has not been assessed yet, however based on the previous analysis we should be able to develop possible scenarios.

In this chapter e-Participation ecologies have been used to analyse different cases and to demonstrate how different projects were generated and developed from/around specific conflicts. These conflicts are all generated by different actors's discourses (Foucault, 1972). Even in the case of M-Pesa, that should not be considered as an e-Participation project, analytical frameworks such as the e-Participation ecology can be useful to understand why and how a tech-innovation emerged from a specific conflict and how tech-innovations can be "*appropriate*" by users to be implemented in totally different contexts. Therefore we may conclude that by understanding or predicting possible conflicts and users "*appropriations*" we should be able to understand tech-innovations and even predict their impact in some cases.

There is also another interesting aspect related to the psycho-social impact of these techno-innovations that are able to generate new or reinforce already existing techno-discourses.

For example Geo-local applications and in general crowd-sourcing platforms such as Ushahidi and Mzalendo are also able to respond to a collective psychological need by giving to citizens a chance to overcome their sense of impotence by actively engaging in an event of social and political relevance and at the same time being able to disclose and share the realities ignored by mainstream media. It is a first step to overcome a psycho-social sense of impotence against the establishment's power.

Users can share information and connect places by mapping them, creating in this way a cognitive map of the problems that is able to delegitimize the mainstream media version of reality. In this sense Ushahidi maps can be defined as Heterotopia (Foucault, 1967) a concept elaborated to describe places and spaces that function in non-hegemonic conditions. A parallel space that contains undesirable bodies to make a real utopian space possible.

Mzalendo and Ushahidi's capacity of generate a techno-discourse about African Cyberdemocracy and the power of Crowd-sourcing is probably more relevant than the real impact that these two e-Participation platforms had or will have on the life of citizens and media activists.

This techno-discourse can be seen as the main African contribution to the development of a new idea of antagonist actor that is based on concepts such as crowd-sourcing and in general the capacity of independent users of reprogramming networks and generate data in a collective manner.

Two concepts, the "multitude" formulated by Negri and Hardt (2004) and the "informational society" further developed by Manuel Castells in "Communication and Power" (2009) fits into this Ushahidi techno-discourse.

The "informational-multitude" may therefore represent the emerging techno-discourse that will be able to influence the next generation of media activists living in both the developed and the developing countries. Groups of users, producers, bloggers, human rights activists, common citizens, that may start to consider themselves as a whole of singularities, always productive and always *in motion*.

Participatory social mapping for these actors may start to represent a sort of constitutional process "momentum" through which the "informational-multitude" appears/manifest itself like a sort of techno-spirit, to disclose and publicly display problems in order to solve them or just to discredit the mainstream media. A new Cyber ritual that can be performed by different actors during a crisis such as the already mentioned post election violence in Kenya.

Other examples such as the participatory social mapping events in occasion of the 2010 Kenya constitutional referendum, during which hundreds of Ushahidi volunteers physically met to map data, may demonstrate the emerging of these new socio-techno-rituals to overcome our sense of impotence and fear that is generated everyday by mainstream media to support the establishments discourses and political agenda.

Ushahidi has been used to map spaces in which negative episodes/events occurred, in order to destroy utopias and impose heterotopias using an on-line off-line mobilization. At the same time, Ushahidi has been used to display utopian spaces by mapping areas in which positive episodes/events have occurred, in order to balance a negative image created by mainstream media.

In conclusion, mapping is directly connected to the need of social movements to counterbalance heterotopias or utopians imposed by the establishment. Once again we find that not only is technology determined by use, but that the reverse is also true and the two are actually co-determined.

SOLUTIONS, RECOMMENDATIONS AND CONCLUSION

The recommendations formulated after obtaining the field research results were the following ones.

An information system to engage citizens in Kenya should focus mainly on:

- Inform citizens about incoming mabaraza, using mobile phones and radios
- Record mabaraza contents using video and audio
- Distribute contents and generate debates using radio browsing
- Get feedback from citizens using mobile phones and store it online to record and keep track of citizens opinions using relational database systems
- Connect different radio databases between each others to have a clear picture of citizen opinions and their trends at local and national level.
- MPs should allow their staff to upload all the contents about mabaraza on line so that radio stations, TV and single citizens could have access to them. Policy makers should use the data gathered to understand priorities and needs of the citizens.

The above-mentioned recommendations should be integrated with the results and recommendation coming from the analysis of the other two Kenyan e-Participation ecology macrodimensions (the Political and the Cultural).

The genesis analysis of two e-Participation platforms Mzalendo and Ushahidi should have helped us to understand: who, why and how an e-Participation project may start and develop in Kenya.

In both cases the projects are direct consequences of a mainstream/institutional media censure wanted by the political establishment, the Parliament in the case of Mzalendo and the Government in the case of Ushahidi.

In both cases the answer organized by the media activists in form of e-Participation project has not been suppressed, this level of freedom allowed the "informational-multitude" to generate a techno-logical innovation and the rise of an indigenous techno-discourse. A new techno-narrative based on a common history and rituals.

All the obtained information can be used to develop payoff matrices and win-win scenarios using the game theory approach (see Figure 3).

A win-win game is a game that is designed in such a way that all participants can profit from it in one-way or another. In conflict resolution a win-win strategy is a conflict resolution process that aims to accommodate all disputants.

In e-Participation, a *Win-Win e-Participation scenario* is a situation in which a specific group, in conflict with other groups, is able to reach its priorities/objectives reducing at the same time both the ostracism of other groups of opponents,

Figure 3. Cultural, political, and social participation models

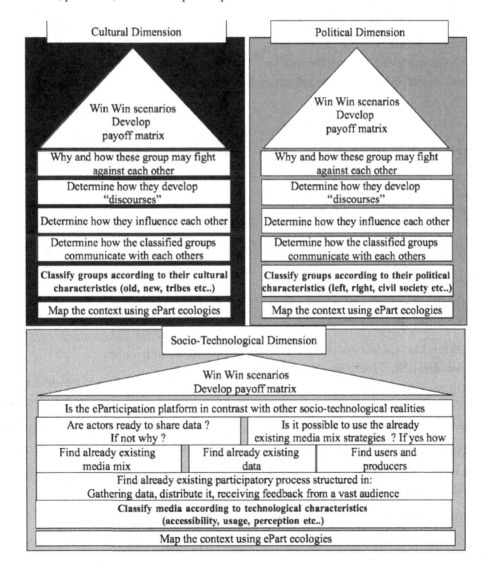

and the technological "reality-gaps." A *Win-Win e-Participation scenario* can be created by planning *ad hoc* strategies based on payoff matrix/tables developed using cooperative non-cooperative game theories.

In conclusion, the *e-Participation ecology*, based on ideas such as the *communicative ecology* and the *networking ethnography*, together with the latest studies on *non-cooperative* and *cooperative games*, developed in the last 50 years of research, can offer to e-Participation analysts/project managers a new set of useful analytical tools to understand e-Participation and to develop successful projects by reducing risks of failure. The development of such analytical tools that are focusing on power dynamics rather than on pure technical assessments should improve the quality and the success rate of e-Participation projects/initiatives.

REFERENCES

Aronowitz, S., & Cutler, J. (1997). *Post-work: The wages of cybernation*. New York, NY: Routledge.

Castells, M. (1998). *End of millennium, the information age: Economy, society and culture (Vol. III)*. Oxford, UK: Blackwell.

Castells, M. (2009). *Communication and power*. Oxford, UK: Oxford University Press.

Cavallo, V. (2009). *The win win eparticipatione-participation model*. LNCS Journals, Academic Books & Online Media | SpringerElectronic Participation First International Conference, ePart 2009 Linz, Austria, August 31–September 4, 2009.

Cavallo, V. (2009). *The win win eparticipatione-participation model*. Paper presented at the IST Africa Kampala Uganda, 6 May 2009

Cavallo, V. (2010). *eParticipatione-Participation and the theory of games*. Unpublished doctoral dissertation, IULM University, Milan.

Collin, F., & Restrepo, E. (2001). *Community radio handbook*. Paris, France: UNESCO.

de Sardan, J. O. (2006). *Anthropology and development: Understanding contemporary social change*. London, UK: Zed Books.

Donal, J. (2009). *Mobile Money Summit 2009. Accelerating the Development of Mobile Money Ecosystems*. IFC and the Harvard Kennedy School.

Edong, K. (2007). *Radio Browsing*. Technobiography Blog. Retrieved from http://www.techno-biography.com/edongs-dreams/radio-browsing/

Flichy, P. (1995). *Dynamics of modern communication: The shaping and impact of new communication technologies*. London, UK: Sage Publications.

Florida, R. (2004). *The rise of the creative class*. New York, NY: Basic Books.

Florida, R. (2005). *The flight of the creative class: The new global competition for talent* (1st ed.). Toronto: HarperBusiness.

Focault, M. (1984). Utopie: Eterotopie. In *Des Espace Autres*. Architecture /Mouvement

Formenti, C. (2008). *Cybersoviet, Utopie post-democratiche and new media*. Milan, Italy: Raffaello Cortina Editore.

Foucault, M. (1970). *The order of things. An archaeology of the human sciences*. London, UK: Tavistock Publications.

Foucault, M. (1972). *The archaeology of knowledge*. London, UK: Tavistock Publications.

Foucault, M. (1977). *Discipline and punish: The birth of the prison*. London, UK: Penguin Books.

Girard, B. (Ed.). (2003). *The one to watch: Radio new ICT and interactivity*. Geneva, Switzerland: FES.

Hardt, M., & Negri, A. (2001). *Empire*. Harvard University Press.

Hardt, M., & Negri, A. (2004). *Moltitude: War and democracy in the age of the empire.* New York, NY: The Penguin Press.

Haugerud, A. (1995). *The culture of politics in modern Kenya.* Cambridge, UK: Cambridge University Press. doi:10.1017/CBO9781139166690

Hearn, G. N., & Foth, M. (2007). Communicative ecologies: Editorial Preface. *Electronic Journal of Communication, 17*(1-2).

Heeks, R. (2006). *Implementing and managing egovernment.* London, UK: Sage Publications.

Howards, P. (2002). *Network ethnography and the hypermedia organization: new media, new organizations, new methods.* London, UK: Sage Publication.

Loimeier, R. (2005). The Baraza: A grassroots institution. *International SIM Review for the Study of Islam in the Modern World, 16,* 26–27.

McKenzie, W. (2004). *A hacker manifesto.* Cambridge, UK: Harvard University Press.

Meier, P., & Brodock, K. (2008). *Crisis mapping Kenya's election violence: Comparing mainstream news, citizen journalism and Ushahidi.* Harvard Humanitarian Initiative, HHI, Harvard University: Boston). Retrieved from http://irevolution.net/2008/10/23/mapping-kenyas-election-violence

Meyrowitz, J. (1985). *No sense of place: The impact of electronic media on social behaviour.* Oxford, UK: Oxford University Press.

Nash, J. F. Jr. (1950). Equilibrium points in n-person games. *Proceedings of the National Academy of Sciences of the United States of America, 36,* 48–49. doi:10.1073/pnas.36.1.48

Nash, J. F. Jr. (1951). Non-cooperative games. *The Annals of Mathematics, 54,* 286–295. doi:10.2307/1969529

Neumann, J., & Morgenstern, O. (1944). *Theory of games and economic behavior.* Princeton University Press.

Standard Digital. (2012, July 22). Suddenly, everybody loves Maina Njenga. *Standard Digital.* Retrieved from http://www.standardmedia.co.ke/?articleID=2000003272&pageNo=2

Trust, B. B. C. (2006). *Africa media development initiatives.* AMDI.

UNCTAD. (2010). *Informational economy report.* Geneva, Switzerland: Author.

ADDITIONAL READING

Barabasi, A.-L. (2003). *Linked.* Plume Books.

Castells, M. (1997). *The power of identity, the information age: Economy, society and culture (Vol. II).* Cambridge, MA: Blackwell.

Castells, M. (1998). *End of millennium, the information age: Economy, society and culture (Vol. III).* Cambridge, MA: Blackwell.

Castells, M., & Himanen, P. (2002). *The information society and the welfare state: The Finnish model.* Oxford, UK: Oxford UP. doi:10.1093/acprof:oso/9780199256990.001.0001

Castells, M. F.-A., Linchuan Qiu, J., & Sey, A. (2006). *Mobile communication and society: A global perspective.* Cambridge, MA: MIT Press.

Cernea, M. (1992). The building blocks of participation: Testing a social methodology. In Bhatnagar, B., & Williams, A. C. (Eds.), *Participatory planning and the World Bank: Potential directions for change* (pp. 96–108). Washington, DC: World Bank.

Chaveau, J. P. (1994). *Participation paysanne et populisme bureaucratique: Essai d'histoire et de sociologie de la culture du developpement.*

Colin, W. (1973). *Anarchy in action*. New York, NY: Allen & Unwin, Freedom Press.

Crouch, C. (2004). *Post-democracy*. Cambridge, UK: Polity Press.

Escobar, A. (1984). Discourse and power in development: Michel Foucault and the relevance of his work to the third world. *Alternatives*.

Goffman, E. (1974). *Frame analysis: An essay on the organization of experience*. Northeastern University Press.

Hardt, M., & Negri, A. (2009). *Commonwealth*. Belknap Press of Harvard University Press.

Haugerud, A. (1995). *The culture of politics in modern Kenya*. Cambridge, UK: Cambridge University Press. doi:10.1017/CBO9781139166690

Lessig, L. (2008). *Making art and commerce thrive in the hybrid economy*. Bloomsbury Academic. doi:10.5040/9781849662505

Misuraca, G. (2007). *E-governance in Africa, from theory to action*. IDRC.

Morawczynski, O. (2009). Exploring the usage and impact of 'transformational' mobile financial services: The case of M-PESA in Kenya. *Journal of Eastern African Studies*, *3*(3), 509–525. doi:10.1080/17531050903273768

Mulgan, G. (1991). *Communication & control: Networks & the new economies of communication*. New York, NY: Guilford Press.

Tacchi, J., Slater, D., & Hearn, G. (2003). *Ethnographic action research: A user's handbook*. New Delhi, India: UNESCO.

KEY TERMS AND DEFINITIONS

E-Participation Ecologies: An e-Participation ecology is an ethonographic model based on the concept of "Communicative ecologies."

Following this analytical framework there are five elements and tree dimensions that should be analyzed and then mapped to obtain enough data for the development of an e-Participation strategy. The elements to analyze and then map are: actors, contents, traditional culture of participation, existing media skills and practices, discourses in conflicts (establishment vs. antagonists). Finally we should analyze the three macro-dimensions in which the elements are interacting between each other: cultural/traditional, political and socio-technological. Through the historical analysis of the above mentioned interactions, we should be able to understand more about possible future scenarios.

Informational Multitude: Two concepts, the "multitude" formulated by Michael Hardt and Antonio Negri (2004) and the "informational society" further developed by Manuel Castells in "Communication and Power" (2009) fits into the idea/discourse of an emerging antagonist actor. Groups of users, producers, bloggers, human rights activists, common citizens, that may start to consider themselves as a whole of singularities, always productive and always *in motion*. The "informational-multitude" seams to represent the new emerging techno-discourse influencing this generation of media activists living in both the developed and the developing countries.

The Win-Win E-Participation Model: The" Win-Win e-Participation model" is based on the assumption that: the concept of "e-Participation ecology" is the most appropriate one for understanding the reality in which an hypothetical e-Participation project should be implemented, in order to "design realities" and reduce the risks of failures caused by the "reality gaps"; the "Win-Win" terms stands for relationship or transaction in which both or all parties gain from. The model is based on the assumption that a "Win-Win" relationship is happening when all the actors involved in a communication process are benefiting from a specific media mix strategy, used to pursue a common objective. The model is based on three analytical dimensions: Organizational: consists of a pre-implementation and post-evaluation analysis of the project in relation to the traditional participatory policy practices and cycle - e.g. Identify an already existing practice (like the baraza) and evaluate the impact of the project on it. Socio technological: consist of a pre-implementation and post-evaluation analysis of the project in relation to how citizens and their representatives already use the existing media (radio, video, press) and how the e-Participation platform can empower and extend their networking ability - e.g., how the e-Participation platform could empower the use of local radio, mobile phones etc.. in order to support communication between citizens and their representatives. Socio-political: consists of an overall pre-implementation and post-evaluation analysis in relation to the policy cycle and the achievement of public participation goals, e.g. political vision, policy information, decision making, implementation, impact on the final decision of the Parliament (International Association of Public Participation).

Win Win E-Participation Scenarios: A *Win-Win e-Participation scenario* is a situation in which a specific group, in conflict with other groups, is able to reach its priorities/objectives reducing at the same time both the ostracism of other groups of opponents, and the technological "reality-gaps." A *Win-Win e-Participation scenario* can be created by planning *ad hoc* strategies based on payoff matrix/tables developed using cooperative non-cooperative game theories. To develop a Win Win e-Participation scenario it is important to analyze the five elements (actors, contents, traditional culture of participation, existing media skills and practices, discourses in conflicts - establishment vs. antagonists) interacting in their tree macro-dimensions (cultural/traditional, political and socio-technological).

ENDNOTES

[1] http://www.frontlinesms.com/
[2] http://www.theyworkforyou.com/

Chapter 17
The Effect of Trust on the Continuance Intention of E-Filing Usage:
A Review of Literatures

T. Santhanamery
Universiti Teknologi Mara Malaysia, Malaysia

T. Ramayah
Universiti Sains Malaysia, Malaysia

ABSTRACT

Research on e-government is taking a new phase nowadays, with researchers focusing more to evaluate the continued usage intention by the citizens rather than the initial intention. Continuance intention is defined as a person's intention to continue using, or long term usage intention of a technology. Unlike initial acceptance decision, continuance intention depends on various factors that affect the individual's decision to continue using a particular system, with trust being one for the most important factors. Therefore, this case study aims to examine the role of trust, particularly trust in the system, on continuance usage intention of an e-filing system by taxpayers in Malaysia. The primary discussion in this case study concerns the e-filing system in Malaysia, followed by the strategies for successful adoption of e-government services and the benefits of e-government adoption, concluding with future research directions.

INTRODUCTION

Advances in Information and Communication Technologies (ICT) have challenged governments all over the world to innovates their traditional structures and consider e-enabled approaches for the implementation of effective public service delivery and for improved performance within public administration (Adeshara et al, 2004). Interest in e-government has expanded over the past 10 years as government has viewed e-government as a lever for changing outmoded bureaucracies, making improvements in the efficiencies and effectiveness

DOI: 10.4018/978-1-4666-3640-8.ch017

in public services, enhancing service to citizens and businesses and promoting participation and democracy (Rowley, 2011). According to Bhatnagar (2009), governments are spending billions of dollars to build online service delivery portals and United Nations E-Government Survey (UNPAN) (2010) reports that high income countries enjoy the top rankings in the e-government development index as they have the financial resources to develop and rollout advanced e-government initiatives and create a favourable environment for citizen engagement and empowerment.

Electronic government or e-government refers to the government that makes use of ICT to work more effectively, to share information and deliver better services to the public more efficiently and to increase the speed of delivery of services combined with reduction in costs' (Chadwick & May, 2003). According to Fang (2000), the term e-government refers to the use of information technology by government agencies, such as web-based networks, the internet and mobile computing, that have the ability to transform relations with citizens, businesses and other arms of government.

Malaysia's strategic shift into the information and knowledge era were guided by the Vision 2020 whereby, Malaysia has embarked on an ambitious plan by launching the Multimedia Super Corridor (MSC) in August 1996 which aimed to accelerate the country's entry into Information Age and is executed in three phases from 1996–2020. The vision of e-Government was to transform administrative process and service delivery through the use of ICT and multimedia (Lean et al., 2009). Seven specific flagship applications were identified as the pioneering MSC projects which includes e-government flagship. Under the e-government flagship, seven pilot projects of E-Government Flagship Application were identified such as Electronic Procurement (EP), Project Monitoring System (PMS), Electronic Services Delivery (E-Services), Human Resource Management Information System (HRMIS), Generic Office Environment (GOE), E-Syariah and

Electronic Labor Exchange (ELX) ((Muhammad Rais & Nazariah, 2003). Besides these seven main projects, several government agencies have taken initiatives to introduce online services for the public which includes e-filing of income tax payment (Ambali, 2009).

As such, the objective of this paper is to evaluate the progress of the e-filing system and examine the effect of trust on the continuance intention of e-filing system among tax payers in Malaysia.

BACKGROUND

E-Filing System in Malaysia

Traditionally, Malaysian taxpayers filed their tax manually by completing their BE and B (Resident Individual), M (Non-Resident Individuals), PE (Others), C, R and CP204 (Companies) forms, do a self-calculation on their tax, attach together all the payment receipts and submit it over in person or by mail to the Inland Revenue Board of Malaysia (IRBM) branches. IRBM later sends the confirmation on the tax payment amount to be settled by the taxpayers. However, a new paradigm has taken place with the introduction of the e-Filing system or online tax filing in 2006 and ever since has undergone a progressive improvement with a more robust engine promised to the users'. E-filing system as a whole integrates tax preparation, tax filing and tax payment, which serves as a major advantage over the traditional manual procedure (Ambali, 2009). Since its introduction in 2006, e-filing has evolved each year in order to provide better service to the taxpayers. Figure 1 shows the progress of e-filing system since its introduction.

Currently there are two major methods of tax filing in Malaysia: Manual and e-filing. Taxpayers are free to choose their preferred way of filing the tax. The submission via e-filing has shown a tremendous increase since its launching in 2006 particularly for individual taxpayers. The number of submission grew from 186,271 (2006) to

Figure 1. Evolution of e-filing (Adapted from: Hasmah, 2009)

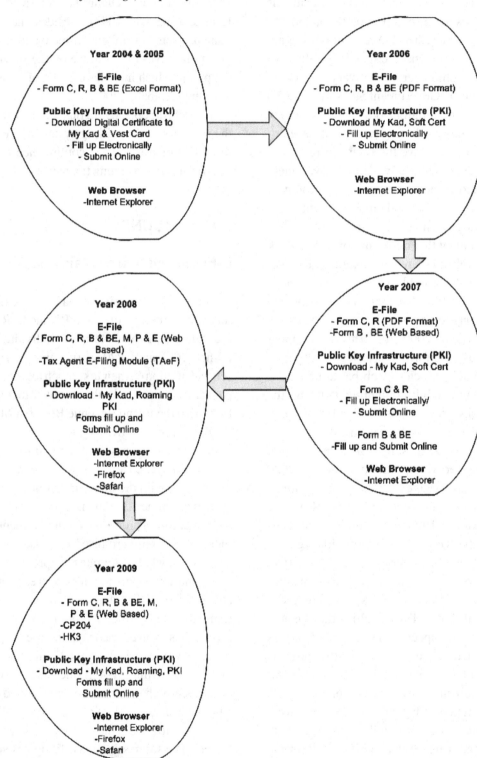

873,095 (2007) (Annual Report IRBM, 2007) and from 1,171,105 (2008) to 1,466,507 (2009) (Annual Report IRBM, 2009). This shows that 30% of the total registered individual taxpayers (4,785,452) have filed their income taxes via e-filing in 2009 (Annual Report IRBM, 2009). The number of submission increases further to 1,666,242 (2010) and 1,597,113 (as of 30th April, 2011) (Abdul Manap, 2011). Various advantages gained from the e-filing system such as time saving, cost effective, accurate, easy and able to increase productivity and security are highlighted (see Figure 2) (Hasmah, 2009).

LITERATURE REVIEW

Technology Adoption

Previous studies have proven various reasons affecting the technology adoption. Survey done by Anuar and Othman (2010) on the taxpayers and e-Bayaran system in Malaysia had found

that subjective norm is the strongest predictor of intention to use online e-Bayaran system in Malaysia followed by perceived usefulness and self-efficacy. However, the perceived ease of use and the perceived credibility and amount of information were found to have no significant influence on taxpayers' intention to use. Contradictorily, Hussein et al. (2010) found that perceived ease of use had the strongest impact on taxpayers' intention to use e-filing system in Malaysia. This was followed by the perceived usefulness and trust of the government. This finding is supported by Zakaria et al. (2009) whose study had also revealed that taxpayers in Malaysia perceived the e-filing system as easy to use (ease of use).

Further, Illias, Suki and Yasoa (2008) found that one's attitude plays an important role in influencing taxpayer's intention to use e-filing system in Malaysia. Their study also posits that attitude can be changed based on first experience of handling e-filing system. Similarly, a study on the intention to use e-government services in Malaysia also found that attitude has a significant

Figure 2. Submission via e-filing (Adapted from: Annual Report IRBM 2009, Abdul Manap, 2011)

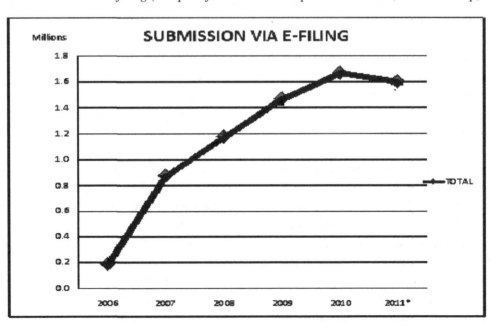

* as at 30th April 2011

positive impact on intention to use (Suki & Ramayah, 2010). Furthermore, based on a study done by Illias, Razak and Yasao (2009) on taxpayers' attitude, they found that education background of taxpayers plays an important role in encouraging the attitude of taxpayers to use e-filing system in Malaysia.

In addition, a study done by Ramayah, Ramoo and Ibrahim (2008) disclosed that one of the main reason that prevents the taxpayers in Malaysia to adopt the e-filing system was the security and privacy issue, whereby the taxpayers were sceptical over the security and privacy of the data transmitted through the web.

Similarly, various studies had also found that a major concern in intention to adopt e-government services in Malaysia is extended to trust issues. Berdykhanova, Dehghantanha and Hariraj (2010) found that in an e-tax prospective, citizen's decision to use online system is influenced by their willingness to trust the environment and the agency involved. Correspondingly, Lean et al. (2009) also found that trust has a significant positive relationship with citizens' intention to use e-government services in Malaysia.

Based on the previous studies explained above, it is concluded that most of the studies undertaken previously in Malaysia particularly on the e-government services were all focused on the intention to use or to adopt and had less focus on the continuance usage intention

Continuance Intention

Continuance intention is defined as ones intention to continue using or long term usage intention of a technology (Bhattacherjee, 2001; Bhattarcherjee & Premkumar, 2004). The research on continuance usage intention actually had been examined variously in the past decade with various terms such as "implementation" (Zmud, 1982), "post adoption" (Karahanna, Straub & Chervany, 1999) and infusion (Meister & Compeau, 2002). Research on information system (IS) continuance have been

explored both at the organizational and individual level of analysis (Limayem, Hirt & Cheung, 2007). For example, Zmud (1982) found that in order for an organizational innovation to be successful, the member of the organizational must accept and utilize it. The individual level of analysis assumes that IS continuance behaviour is the continued usage of IS by adopters, which is follows an initial acceptance decision (Kim, Chan & Chan, 2007). However, unlike initial acceptance decisions, IS continuance depends on various factors that affect the individuals' decision to continue using a particular system (Limayem, Hirt & Chin, 2001) with one of the most important factor that is the trust. Table 1 summarizes the local studies done in Malaysia in the context of continuance intention.

Similarly, there are also various researches which have been carried out in the Western context on the continuance usage or repurchase intention. Table 2 summarizes the western studies done in the context of continuance intention.

Continuance Intention and E-Government

Previous research on e-government had focused mainly on the intention to use the e-government services by the citizens. Nevertheless, the research on e-government is taking a new phase with new researches focusing more on the evaluation of the continued usage intention by the citizens rather than the initial intention.

Wangpipatwong, Chutimaskul and Papasratorn (2007) in their study had examined the role of Technology Acceptance Model (TAM)'s belief (perceived usefulness and perceived ease of use) and computer self-efficacy to predict the citizen's continuance intention to use e-government websites. They found that all the three variables (perceived ease of use, perceived usefulness and computer self-efficacy) positively influence citizen's continuance intention to use e-government websites but out of all three variables, perceived usefulness has the greatest influence on continu-

Table 1. Local studies in Malaysia with continuance intention

Authors/Year	Independent Variable	Dependent Variable
Ahmad, Omar and Ramayah (2010)	Lifestyle	Continuance Intention
Mohamed, Hussein and Zamzuri (2010)	Perceived Usefulness, Satisfaction, Preferences, Consumer Addiction, Personality, Lifestyle	Re-purchase Intention in Electronic Commerce Environment
Ramayah, Ahmad and Lo (2010)	System Quality, Information Quality, Service Quality	Continuance Intention of e-Learning System
Aziz and Wahiddin (2010)	Service Excellence, Service Quality and Customer Satisfaction	Intention to Re-purchase in Hotel Service Environment
Bojei and Hoo (2010)	Brand Awareness, Brand Association, Relative Advantage, Ease of Use	Re-purchase Intention of Smartphones
Yahya, Arshad and Wahab (2009)	Perceived Organizational Benefit, Perceived Usefulness	Organizational IS Continuance
Ambali (2009)	Perceived Security, Facilitating Conditions, Perceived Usefulness, Perceived Ease of Use	Users' Retention in E-Filing System
Zamzuri, Mohamed and Hussein (2008)	Customer Satisfaction	Re-purchase Intention in Electronic Commerce Environment

ance intention. This finding was supported by another study of Wangpipatwong, Chutimaskul and Papasratorn (2008) when they investigated the fundamental factors influencing the citizen's continuance intention to use e-government web-sites and discovered that the perceived usefulness, the perceived ease of use and the computer self-efficacy significantly influence the citizen's continuance intention to use e-government websites.

Table 2. Western studies done in the context of continuance intention

Authors/Year	Independent Variable	Dependent Variable
Boakye, Prybutok and Ryan (2012)	Perceived ease of use, Perceived Usefulness, Perceived Service Quality, Perceived Satisfaction in Web Access Quality	Intention of Continued Usage
Rose, Clark, Samouel and Hair (2012)	Perceived Control, Cognitive Experiential State (CES), Affective Experiential State (AES), Satisfaction, Trust	Online Re-purchase Intention
Lawkobit and Speece (2012)	Perceived Usefulness, Confirmation, Satisfaction, Structural Fairness, Social Fairness	IS Continuance Intention in Cloud Computing
Olsen, Tudoran, Brunso and Verbeke (2012)	Satisfaction, Loyalty Intention, Habit Strength	Loyalty Behaviour (Repeat Patronage)
Lopez, Miguel and Pradas (2012)	Value, Quality, Satisfaction, Confirmation, Perceived Usefulness, Purchasing Habits	Re-purchase Intention
Hernandez, Jimenez and Martin (2010)	Acceptance of Internet, Internet Use Frequency, Satisfaction with the Internet, Perceived Self-Efficacy, Perceived Ease of Use, Perceived Usefulness, Attitude	Future Purchase Intention over the Internet
Hume and Mort (2010)	Core Service Quality, Peripheral Service Quality, Appraisal Emotion, Perceived Value, Customer Satisfaction	Re-purchase Intention in Performing Arts
Theodorakis, Koustelios, Robinson & Barlas (2009)	Team Identification, Service Quality	Re-purchase Intention Among Spectators of Professional Sports

In another study, Chai Herath, Park and Rao (2008) explored the influence of perceived performance of the government websites, the satisfaction and the perceived confidentiality on the users' intention to repeat using the e-government websites. They revealed that users' intention to continue using e-government websites was positively influenced by users' satisfaction, perceived performance of the websites and the requirement for confidential information. Subsequently, research by Ambali (2009) on the continuance usage intention of e-filing system in Malaysia had also found a strong relationship between the perceived usefulness, ease of use as well as the security and facilitating condition towards users' retention of the system.

Trust in the System

Consumer's trust in the service provider is considered as an important antecedent for the continuation of exchange relationships, where such belief depends on the perceived risk with the service provider by the consumers' (Palvia, 2009). In electronic commerce, trust is important when there is a risk of negative outcomes especially when financial transaction or personal information is involved (Kini & Choobineh, 1998). However, the role of trust in e-government services is even more important and crucial because citizens who are using e-government websites are unable to find the alternative websites which serve the same purpose. In the absence of sufficient trust in the e-government websites, users may be motivated to revert to the traditional way of interacting with the government (Teo, Srivastava & Jiang, 2008). Further, citizens' trust in e-government has been identified as a unique dimension, due to the unfriendly nature of the online environment, the widespread use of technology and natural uncertainty using an open infrastructure. These together with extensive media coverage on privacy, security and fraud on the internet have actually increases the hesitation of citizens in adopting e-government (Al-adawi, Yousafzai & Pallister,

2005). Table 3 summarizes the research done on trust in e-government studies in general.

Trust in the system is defined as the perception of the operation of the system which will display availability, fault tolerance and its security and correctness which are guaranteed together with stability in system response time (Papadopoulou, Nikolaidou & Martakos, 2010). Hung, Chang & Yu (2006) has studied the effect of trust in the system of the online tax and payment. In this study, the researchers found that trust is one of the major determinants towards the attitude and acceptance of online tax filing and payment system. Similarly, Schaupp and Carter (2010) and Schaupp, Carter and McBride (2010) also studied the effect of trust in the system of e-filing on the intention to use directly and indirectly through perceived risk. The study had found that e-filer will have a negative effect on perceived risk and positive effect on intention to use the tax system. The dimensions of the trust in the system as listed by Papadopoulou et al. (2010) are adopted in this case study as it is more relevant in the context of the e-filing tax payment system. *Correctness* is defined as the assurance that the system works properly and produces correct output. *Availability* is defined as the assurance that the system is up and running, is fully functional whenever needed and is protected from denial of service. *Security* is the assurance that the system is protected against intrusion threats. *Failure* is the assurance that the system is protected against loss of user data in case of failure and *Accountability* is the assurance that actions of an entity are traced (auditing) to allow for non-repudiation, intrusion detection and prevention and legal action. *Response Time* means the system responds to requests within a short and acceptable time period. Lastly, *System Support* is defined as the mechanical and tailored support to access the needed information without problems. It includes help desks, online support services, customized support and other facilities (Cho, Cheng & Lai, 2009b).

Table 3. Research done on trust in e-government studies in general

Author/Year	Topic of Analysis	Constructs	Results
Akkaya, Wolf and Krcmar (2012)	Factors Influencing Citizen Adoption of E-Government Services: A cross-cultural comparison (Research in Progress)	Trust in Internet, Trust in Government, Perceived Risk, Relative Advantage, Complexity, Compatibility and Subjective Norm.	Trust in Internet and Trust in Government are the important barriers in using e-government services.
Schooley, Harold, Horan and Burkhard (2011)	Citizen Perspectives on Trust in a Public Online Advanced Traveler Information System	Trust (usefulness, dependability, reliability, compatibility and timeliness)	All the five elements of trust are found to be the important elements for citizens in using e-government services.
Lee, Kim and Ahn (2011)	The willingness of e-Government service adoption by business users: The role of offline service quality and trust in technology	Offline Service Quality (Timeliness, Empathy, Responsiveness, Tangible, Assurance, Satisfaction, Promptness, Service Quality) and Trust in the technology	Trust in technology does not have any significant impact on the willingness of e-government service adoption. Also, trust in technology does not significantly moderate the effect of service quality on the willingness to adopt e-Government
Berdykhanova, Dehghantanha and Hariraj (2010)	Trust Challenges and Issues of E-Government: E-Tax Prospective	Trust in environment, Trust in agency	Trust emerges as the key component in e-government initiatives particularly in e-taxation. Further trust only occurs when proper security is guaranteed. As such, this study proposes Trusted Platform Module (TPM) to enhance the security robustness of e-tax services
Colesca (2009)	Understanding Trust in e-Government	Age, Gender, Education, Income, Internet experience, Propensity to Trust, Trust in technology, Perceived organizational trust, Privacy, Risk Perception, Perceived Quality, Perceived Usefulness	Privacy concern has the greatest influence on trust in e-government. Followed by age, propensity to trust, Internet experience, Perceived organizational trust, Trust in technology, Perceived Quality, Perceived Usefulness which also shows a significant influence on trust in e-government. However, Perceived risk, Gender, Education and Income do not show any significant relationship
Grimsley and Meehan (2007)	E-Government information systems: Evaluation-led design for public value and client trust	Public value (service provision and service outcomes), Trust	Trust is related to the extent to which people feel that an e-Government service enhances their sense of being well-informed, gives them greater personal control and provides them with a sense of influence or contingency
Horst, Kuttschreuter, Gutteling (2007)	Perceived usefulness, personal experiences, risk perception and trust as determinants of adoption of e-government services in The Netherlands	Perceived Usefulness, Personal Experience, Trust, Risk, Perceived Behaviour Control, Subjective Norm	Respondents trust in the e-government is found to be high. Perceived usefulness of electronic services is the main determinant of the intention to use e-government services while trust in e-government is the main determinant of the perceived usefulness of e-government services.
Avgerou, Ganzaroli, Poulymenakou and Reinhard (2007)	ICT and citizens' trust in government: lessons from electronic voting in Brazil	System Trustworthiness, Trust towards ICT and Perceived Trustworthiness of Tribunal Superior Eleitoral (TSE) and Tribunal Electoral Regional (TRE)	E-Government requires active trust formation in ICT and government institutions, rather than overcoming their perceived trustworthiness deficiencies.

As we can see, when we want to examine the effect of trust on the intention to use or continually use of a particular system, it will be improper to measure the effect trust based on the trust on the service provider, trust on the internet, trust on the government, trust on the vendor, trust in the third parties and more. This is because all this trust can be considered as an external trust;

which does not reflects the trust on the particular system internally. So, what is most important is the system itself; the functions of the system that will make people take it on trust.

In fact, previous researches had argued that designing high quality websites not only improves the functional efficacies but also helps to build citizens trust towards public e-service (Tan, Benbasat, Cenfetelli, 2008). Similarly, Cyr, Kindra and Dash (2008) had found that website design is important and it has an impact on the users trust, satisfaction and e-loyalty. Ou and Sia (2010) also revealed that website design attributes have an important effect in shaping users trust and distrust.

As such, compared to external features, the design of the particular system will have a more powerful effect in building the taxpayers trust in the system. With this knowledge the Inland Revenue Board can plan their strategies to enhance or improve that particular design or dimension, which will encourage more and more taxpayers to use the e-filing system and continue using it.

MAIN FOCUS OF THE CHAPTER

Malaysia's ranking for e-government development index has improved from 43rd position (2005) to 34th position (2008) and recently was ranked at number 32 (2010) (UNPAN, 2010). In South East Asia Category, Malaysia's ranking has also improved from 3rd position (2005) to 2nd position in 2008 and 2010 but with an improved index (UNPAN, 2010). In terms of e-Participation Index, there has been a tremendous improvement for Malaysia where its ranking has improved from 41st position (2008) to 12th position (2010) (UNPAN, 2010). In reality, all these rankings actually show the initiatives and technology developments made by the Malaysian Government to ensure that the e-government services reach the citizens. These initiatives truly resemble the "supply side" of the e-government development delivery (Gauld, Goldfinch & Horsburgh, 2010). What about

the "demand side"? To what extent would the citizens use or continually use these particular services had remained as an important question to be answered because a country can be at high position in e-government rankings which focus more on technology developments but a system is still considered failed if the intended recipients do not use or continually use them (Gauld et al., 2010). Thus, the underlying problem now is not the problem of design but utilization. The key to the successful e-government depends very much on the utilization of the implemented systems (Economist, 2008).

Previous studies have verified that the intention of continuing usage is vital because it guarantees the long term revenue of an organization (Parthasarathy & Bhattacherjee, 1998), lowers the operational cost by retaining customers (Ndubisi, 2004). It is also important for the persistent exist of many customers based on electronic commerce firms (Bhattacherjee, 2001). Parthasarathy and Bhattacherjee (1998) also stressed that acquiring new customers is more expensive than to retain existing ones given the cost involved in identifying new customers, setting up account and initiating customers. Although initial acceptance is important in recognizing the success of an information system (Bhattacherjee, 2001) but continued usage is even more significant in ensuring the long-term viability of technology innovations (Premkumar & Bhattacherjee, 2008). Furthermore, Devaraj and Kohli (2003) argued that the long term usage of a technology will enhance the financial and quality performance of an organization. Similarly, Zmud (1982) argues that for an innovation to be successful, members of the organization must accept and utilize the technology. Thus, it is important for businesses to accentuate on continuance usage as the key for long term growth.

In this vein, investigating the continuance intention of e-government particularly in e-filing system is deemed to be important because as more citizens use e-filing services, the more operation and management costs are reduced (Wangpipatwong et al., 2008). In Malaysia, the IRBM has saved millions of ringgits annually by reducing the cost of printing, imaging, postal and storage through their e-filing system. In 2009, a total cost of RM9, 162,845.92 have been saved via the e-filing submission of tax. The number has gradually increased since its introduction in 2006 (RM1, 302,590.40), 2007 (RM4, 876,564.64) and 2008 (RM8, 187,144.96) (Hasmah, 2009) as depicted in Figure 3.

Apart from that, Bhatnagar (2009) reveals that governments are spending millions of dollars to build online service delivery portals in terms of hardware, software, training and maintenance and communication infrastructure. For example, in building the online tax system in Thailand, the government invested USD55.8 million, whereas in building the e-Lanka in Sri Lanka, the project cost USD53 million. In Malaysia, huge amount of investment has been made in developing the e-filing online service portal, especially to upgrade the agency's computer hardware and software (Bernama, 2005). Moreover, the possibility of discontinuance may occur if the system does not meet the users need even after its successful prior adoption (Limayem, Hirt & Cheung, 2003), which may incur undesirable cost or a waste of effort in developing the technology (Hong, Thong & Tam, 2006). Thus it is believed that the heavy investments in the development of e-government

online services will be a waste of effort if people do not use the services continuously in the long term. This is essential in the case of e-filing system in Malaysia because filing of income tax online is a voluntary usage, means that the taxpayers still have an option to submit their tax manually. According to IRBM, as at April 30, 2010, it has received 677,885 income tax return forms manually from individual tax payers (Shari, 2010). In fact, recent report has confirmed that the number of forms submitted manually have increased by 11.92% (Rahim, 2011) compared to 30% taxpayers who have used the e-filing system (Annual Report IRBM, 2009).

Therefore, to ensure that the heavy investments in developing the e-filing online portals will not be wasteful, identifying the factors that will motivate the continuance usage intention is crucial.

Considerable evidence has also shown that lack of trust is the main reason that obstructs citizens in adopting and using e-government services especially which involves sharing of personal information on the internet (Hussein et al., 2010; Weerakkody & Choudrie, 2005; Navarra & Cornford, 2003). One of the major challenges faced by the IRBM in the implementation of this e-filing system, according to its Chief Executive Officer/ Director General, is to develop trust and confidence among the taxpayers and the public (including tax agents) for e-filing system. Chidambaram

Figure 3. Cost saving from e-filing submission (Adapted from: Hasmah, 2009)

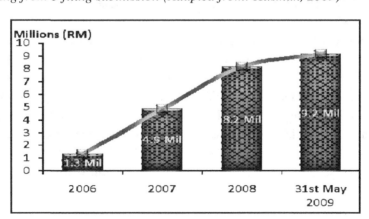

(2008) pointed out that one of the loop holes of e-filing system in Malaysia which leads to user dissatisfaction were trust and system functionality issues where someone's private and confidential information can be accessed by others in the system accidentally. In fact, Organization for Economic Cooperation and Development (OECD) (2008) also discovered that one of the main challenges for increasing user take-up in the e-government services is trust which, it is hoped, could meet citizens' requirements and also provide privacy and security by protecting their personal data. It also highlighted that the more an e-government service is user-centric, the higher will be the take-up of its services. Trust is not only time-consuming to engender, but also fragile and easily destroyed. Therefore, the process of continuous trust development deserves special attention. Continuous trust development can be in terms of improve site quality, sharpen business competence, maintain company integrity, post privacy policy, foster a virtual community, encourage communication and increase accessibility, use external auditing to monitor operations and security controls (Siau & Shen, 2003).

As such, exploring the role of trust in the context of continued usage intention of e-filing system in Malaysia is deemed to be an important task.

STRATEGIES FOR SUCCESSFUL ADOPTION OF E-GOVERNMENT SERVICES

Governments all around are investing in the development of strategies to enhance their e-government initiatives. Strategies are systematic and long term approaches to problems (Gil-Gacia & Pardo, 2005). Thus, to ensure the successful adoption of e-government services, influential strategies are essential.

Support Systems

Technical Support Systems is the foundation stone of a successful adoption programme and should be integral to e-services rollout. Any weakness in technical support systems may present a barrier to all e-government implementation stages (Alshehri, Drew & Alfarraj, 2012). Technical support which dealt with inevitable network and other problems was also found to be an important factor in acceptance of web-based distance learning course (Williams, 2002). Similarly, human resource support system such as help desks, qualified IT staff, IT governance and technical e-government experiences are the important factors that contributes towards an organization's e-government success (Alghamdi, Goodwin & Rampersad, 2011).

Transaction Transparency

A sense of trust and security develops when the process of online transaction is transparent and when the citizens are informed of every detail in a procedure. For example, in case of any incomplete transaction due to technical error such as system failure or otherwise, the service providers need to inform users about the unfinished task immediately. Sometimes if there is a situation for refund, users should be informed through email or via other communication mode, to avoid any fraudulent activity. It is observed that citizens are willing to use e-transaction services if the method is transparent, smooth and cost effective, otherwise if they are unsure of security and privacy mechanisms or do not receive proper communication about a transaction, they prefer to use alternative traditional system (Bhattacharya, Gulla & Gupta, 2012). This is supported by Welch, Hinnant and Moon (2004) whose study concluded that transaction transparency is an important factor that directly affects e-government satisfaction and indirectly affects trust.

Relative Advantage and Technical Adequacy

Relative advantage, in the context of e-government includes speedy log ins, fast download, quick upload of documents and swift reliable transactions, quality of public services, flexibility offered to the individuals, the ease of use and convenience. Similarly, technical adequacy refers to the reliability, availability, flexibility and speed of a service system of end users. Both these factors are found to be an important issue for the successful of e-government adoption and continuance usage (Bhattacharya et al., 2012). Equally, several studies have proved that relative advantage is an important factor in influencing intention to use e-government services (Carter & Campbell, 2011; Sang, Lee & Lee, 2010; Carter & Weerakkody, 2008; Carter & Belanger, 2004). Likewise, technical adequacy was also found to be positively affect users' perceived usefulness and trust towards continuance usage (Liao, Palvia & Lin, 2006) and an important measure of system quality (Aladwani & Palvia, 2002)

Privacy and Security

Privacy and Security are major concern of citizens when providing personal data online, as such citizens need to be explained why the data are needed and how the record will be protected from third party invasion. As online government is virtual and different from traditional government in the mode of operation, policy makers need to be extra careful in developing citizens' trust on the transaction website. The transaction site should take complete responsibility on the transaction performed online by integrating functions of interdependent departments instead of redirecting citizen to corresponding authorized web interfaces (Bhattacharya et al., 2012). Similarly, previous studies had also found privacy and security as an important determinant for online bank usage (Yousafzai, Foxall and Pallister, 2010), major

source of dissatisfaction (Poon, 2008), major barrier for internet shopping (Godwin, 2001) and obstacle for the adoption of online banking (Sathye, 1999).

Trust

Trust has been identified as a crucial enabling factor in almost all type of social interactions (He, Fang & Wei, 2009). The role of trust in e-government services is even more important and crucial because citizens who use e-government websites are unable to find alternative websites that serve the same purpose. In the absence of sufficient trust in the e-government websites, users may be motivated to revert to the traditional way of interacting with the government (Teo et al., 2008). As such, governments and relevant parties should put more effort in enhancing the trust factor. Strategies to enhance trust included improving quality of communication with users and minimizing any opportunistic behaviour such as failing to provide expected services or violating business promises (Li, Browne & Chau, 2006), privacy and security of the citizens data online are protected (Bhattacharya et al., 2012) and transaction transparency are uphold (Welch et al., 2004).

BENEFITS OF E-GOVERNMENT ADOPTION

Cost Reduction and Efficiency Gains

Providing services online substantially decreases the processing costs of many activities compared to manual way of handling operations (Ndou, 2004). For example, in the case of e-filing in Malaysia, Inland Revenue Board of Malaysia has saved millions of ringgit by reducing cost of printing, imaging, postal and storage. The e-filing system also eliminates the mistakes in calculation of the tax as the calculation is done automatically by the system and not manually as per the traditional method. Efficiency also attained by reducing processing activities of return forms which will speed up the process of tax refund or repayment to the taxpayers (Hasmah, 2009).

Quality of Service Delivery to Taxpayers

The usual procedures that need to be followed in the traditional model of public service delivery are long, time consuming and lack of transparency (Ndou, 2004). For example, the traditional method of tax payment in Malaysia is considered as high cost and time consuming. However, the e-filing way of submitting the tax reduces bureaucracy, offers around the clock accessibility, provides fast and convenience transaction and enhances the quality of services in terms of time, content and accessibility (Hasmah, 2009).

Transparency and Accountability

E-Government helps to increase transparency of decision making. In many cases e-government offers opportunities for citizens to directly participate in the decision making by allowing them to provide ideas and suggestion in forums and online communities (Ndou, 2004). For example, Malaysia is one of the countries that actively involve the public in the decision making by allowing the citizens to express their view and ideas in online surveys. In fact, Malaysia's ranking in e-participation index has improved from ranking number 41 (2008) to ranking number 12 (2010) (UNPAN, 2010).

The diversity of publications available in regard to the activities of the public administration, as well as economic and legislative aspects increases the transparency (Ndou, 2004). For instance, IRBM has introduced Key Performance Indicator (KPI) in 2009 to measure and evaluate the achievements of its objectives. The KPIs will focus on the main processes that are directly related to its strategic objectives. The implementation of KPIs is an ideal

step towards improving its future efficiency and productivity and they provide a basis for improving accountability and transparency to the public (Annual Report IRBM, 2009).

Enhance the Use of ICT

Continuous interaction and communication between government and its citizens contributes to the creation of awareness about the potential contribution of ICT to local communities activities (Ndou, 2004). In this way, e-Government plays a vital role in initiating the process of introducing the ICT to large number of its citizens particularly those in the rural area. For example, with the implementation of e-government services, the Malaysian government has taken more initiatives to increase the internet access particularly to citizens in the rural area. In line with this, more Medan Info Desa and Pusat Internet Desa has been built and upgraded (Mid Term Review, 9MP). Until the first quarter of 2010, a total of 217 Medan Info Desa and 42 Pusat Internet Desa in East and West Malaysia had been established (Tajudin, 2010).

Digital Divide Reduction

The evolution of ICT has brought significant changes to individuals and communities around the world. However considerable gap exist between those who has high access and those who has less or no access to the ICT which is called the digital divide (Noor, 2010). With the implementation of e-government services, the Malaysian government has taken various initiatives to reduce the digital divide with the latest development of Jendela Informasi Anda (JENii) to promote the use of communication information technology among the elderly and to bridge the age digital divide (Abdullah, Salman, Razak, Noor and Malek, 2011). Similarly, in bridging the digital divide among the urban and rural communities, various programmes has been created such as Bestari.com

(State of Terengganu info center and ICT training and learning center for rural citizens), Mobile Internet Unit (offering) training on ICT skills to rural area students, Bridging Digital Divide Initiative (to bridge digital divide and give awareness about ICT usefulness), Portal NurIta (training on computer literacy among women especially single mother) and Warga Emas Network (providing education and computer access to senior citizens to enhance their IT knowledge) (Mahbob, Sulaiman, Mahmud, Mustaffa & Abdullah, 2012).

FUTURE RESEARCH DIRECTIONS

The growing awareness that IT investment has to be measured in terms of "value" and "impact" of the investment has shifted the growing interest of research from electronic government to electronic governance (Davies, Janowski, Ojo & Shukla, 2007). While electronic government refers to government's use of ICT to work more effectively, to share information and deliver better services to the public (Chadwick & May 2003), electronic governance is the actual use of these instruments (Bea & Matotay, 2011) and how the use of ICT can change relationships between citizens and governments (Pathak, Singh, Smith & Naz, 2008). Therefore, the emerging trend is the route to e-governance. E-Governance is not only the act of automation but it implies major socioeconomic innovations and political administrative changes based on a new ICT application and developments (Saxena, 2005). Thus, future research should focus more on the important of e-governance in the area of e-filing returns of income, e-payment of taxes and computerised processing of returns and refunds (Agrawal & Agarwal, 2006).

CONCLUSION

It is our tentative conclusion that the trust in the system is an important factor that influences the continuance usage intention of e-filing system in Malaysia. In the absence of adequate trust in the e-filing system, taxpayers may choose to revert to the manual way of submitting their income tax, which will actually lead to the failure of the system and incur heavy financial loss to the government. Also, it can act as a propagation of e-government services by providing guidelines on how other e-government service providers should address trust in the system to encourage usage and continuance usage intention of the particular system.

REFERENCES

Abdul Manap, D. (2011). *Statistik E-Filing*. Kuala Lumpur, Malaysia: Lembaga Hasil Dalam Negeri (LHDN).

Abdullah, M. Y., Salman, A., Razak, N. A., Noor, N. F. M., & Malek, I. A. (2011). Issues affecting the use of information and communication technology among the elderly: A case study on Jenii. *IEEE 10th Malaysia International Conference on Communications* (MICC) (pp. 29-32). 2nd - 5th October Sabah, Malaysia.

Adeshara, P., Juric, R., Kuljis, J., & Paul, R. (2004). A survey of acceptance of e-government services in the UK. *Journal of Computing and International Technology*, *12*(2), 143–150. doi:10.2498/cit.2004.02.10

Agrawal, P., & Agarwal, M. (2006). E-filing of returns. *The Chartered Accountant*, 1567-1573. Retrieved April 22, 2012, from http://www.icai.org/resource_file/101991567.pdf

Al-Adawi, Z., Yousafzai, S., & Pallister, J. (2005). Conceptual model of citizen adoption of e-government. *Proceedings of the Second International Conference on Innovations in Information Technology (IIT)*, (pp. 1-10).

Aladwani, A. M., & Palvia, P. C. (2002). Developing and validating an instrument for measuring user-perceived web quality. *Information & Management, 39*(6), 467–476. doi:10.1016/S0378-7206(01)00113-6

Alghamdi, I. A., Goodwin, R., & Rampersad, G. (2011). E-government readiness assessment for government organizations in developing countries. *Computer and Information Science, 4*(3), 3–17. doi:10.5539/cis.v4n3p3

Alshehri, M., Drew, S., & Alfarraj, O. (2012). A comprehensive analysis of e-government services adoption in Saudi Arabia: Obstacles and challenges. *International Journal of Advanced Computer Science and Applications, 3*(2), 1–6.

Ambali, A. R. (2009). E-government policy: Ground issues in e-filing system. *European Journal of Soil Science, 11*(2), 249–266.

Annual Report Inland Revenue Board of Malaysia. (2007). Retrieved from http://www.hasil.gov.my/pdf/pdfam/AR2007_2.pdf

Annual Report Inland Revenue Board of Malaysia. (2009). Retrieved from http://www.hasil.gov.my/pdf/pdfam/AR2009_2.pdf

Anuar, S., & Othman, R. (2010). Determinants of online tax payment system in Malaysia. *International Journal of Public Information System, 1*, 17–32.

Bea, G., & Matotay, E. (2011). E-government and e-governance. In J. Itika, K. de Ridder, & A. Tollenaar (Eds.), *Theories and stories in African public administration* (pp. 157-168). African Public Administration and Management series, 1, African Studies Centre/University of Groningen/Mzumbe University.

Berdykhanova, D., Dehghantanha, A., & Hariraj, K. (2010). Trust challenges and issues of e-government: E-tax prospective. *International Symposium of Informational Technology, 2*, 1015-1019.

BERNAMA, The Malaysian National News Agency. (12/5/1999). *E-government a new way to serve people says Dr. Mahathir*. Retrieved from http://www.accessmylibrary.com/article-IGI-54639942/e-government-new-way.html

Bhatnagar, S. (2009). *Unlocking e-government potential: Concepts, cases and practical insights* (1st ed.). SAGE Publications India Pvt Ltd.

Bhattacharya, D., Gulla, U., & Gupta, M. P. (2012). E-service quality model for Indian government portals: Citizens' perspective. *Journal of Enterprise Information Management, 25*(3), 246–271. doi:10.1108/17410391211224408

Bhattacherjee, A. (2001). Understanding information systems continuance: An expectation-confirmation model. *Management Information Systems Quarterly, 25*(3), 351–370. doi:10.2307/3250921

Bhattacherjee, A., & Premkumar, G. (2004). Understanding changes in belief and attitude toward information technology usage: A theoretical model and longitudinal test. *Management Information Systems Quarterly, 28*(2), 229–254.

Boakye, K. G., Prybutok, V. R., & Ryan, S. D. (2012). The intention of continued web-enabled phone service usage: A quality perspective. *Operations Management Research, 5*, 14–24. doi:10.1007/s12063-012-0062-1

Carter, L., & Belanger, F. (2004). The influence of perceived characteristics of innovating on e-government adoption. *Electronic. Journal of E-Government, 2*(1), 11–20.

Carter, L., & Campbell, R. (2011). The impact of trust and relative advantage on internet voting diffusion. *Journal of Theoretical and Applied Electronic Commerce Research, 6*(3), 28–42. doi:10.4067/S0718-18762011000300004

Carter, L., & Weerakkody, V. (2008). E-government adoption: A cultural comparison. *Information Systems Frontiers*, *10*(4), 473–482. doi:10.1007/s10796-008-9103-6

Chadwick, A., & May, C. (2003). Interaction between states and citizens in the age of the Internet: "e-government" in the United States, Britain, and the European Union. *Governance: An International Journal of Policy, Administration and Institutions*, *16*(2), 271–300. doi:10.1111/1468-0491.00216

Chai, S., Herath, T. C., Park, I., & Rao, H. R. (2008). Repeated use of e-government websites: A satisfaction and confidentiality perspective. In Norris, D. (Ed.), *E-government research: Policy and management* (pp. 158–182).

Chidambaram, N. (2008). *Evaluating electronic service quality and user satisfaction: Malaysian income tax e-filing system*. Unpublished MBA Dissertation, University Science Malaysia.

Cho, V., Cheng, T. C. E., & Lai, W. M. J. (2009). The role of perceived user-interface design in continued usage intention of self-paced e-learning tools. *Computers & Education*, *53*(2), 216–227. doi:10.1016/j.compedu.2009.01.014

Cyr, D., Kindra, G., & Dash, S. (2008). Website design, trust, satisfaction, and e-loyalty: The Indian experience. *Online Information Review*, *32*(6), 773–790. doi:10.1108/14684520810923935

Davies, J., Janowski, T., Ojo, A., & Shukla, A. (2007). Technological foundations of electronic governance. *International Conference on E-Government*, December 10-13, Macao, (pp. 5-11).

Devaraj, S., & Kohli, R. (2003). Performance impacts of information technology: Is actual usage the missing link? *Management Science*, *49*(3), 273–289. doi:10.1287/mnsc.49.3.273.12736

Economist. (2008). *The good, the bad and the inevitable: The pros and cons of e-government*. Retrieved from http://www.economist.com/node/10638105?story_id=10638105

Fang, Z. (2002). E-government in digital era: Concept, practice and development. *International Journal of the Computer, the Internet and Management, 10*(2), 1-22.

Gauld, R., Goldfinch, S., & Horsburgh, S. (2010). Do they want it? Do they use it? The demand-side of e-government in Australia and New Zealand. *Government Information Quarterly*, *27*(2), 177–186. doi:10.1016/j.giq.2009.12.002

Gil-Garcia, J. R. G., & Pardo, T. A. (2005). E-government success factors: Mapping practical tools to theoretical foundation. *Government Information Quarterly, 22*(2), 187–216. doi:10.1016/j.giq.2005.02.001

Godwin, J. U. (2001). Privacy and security concerns as a major barrier for e-commerce: A survey study. *Information Management & Computer Security*, *9*(4), 165–174. doi:10.1108/EUM0000000005808

Hasmah, A. (2009). *E-filing pays*. Retrieved from http://www.intanbk.intan.my/i-portal/nict/nict/DAY1/SESSION3PARALLEL1(SESSION3A)/Dato_hasmah_e-filing_Pays.pdf

He, W., Fang, Y., & Wei, K. K. (2009). The role of trust in promoting organizational knowledge seeking using knowledge management systems: An empirical investigation. *Journal of the American Society for Information Science and Technology*, *60*(3), 526–537. doi:10.1002/asi.21006

Hernandez, B., Jimenez, J., & Martin, M. J. (2010). Customer behavior in electronic commerce: The moderating effect of e-purchasing experience. *Journal of Business Research*, *63*(9-10), 964–971. doi:10.1016/j.jbusres.2009.01.019

Hong, S. J., Thong, J. Y. L., & Tam, K. Y. (2006). Understanding continued information technology usage behavior: A comparison of three models in the context of mobile internet. *Decision Support Systems*, *42*(3), 1819–1834. doi:10.1016/j.dss.2006.03.009

Hume, M., & Mort, G. S. (2010). The consequence of appraisal emotion, service quality, perceived value and customer satisfaction on repurchase intent in the performing arts. *Journal of Services Marketing*, 24(2), 170–182. doi:10.1108/08876041011031136

Hung, S. Y., Chang, C. M., & Yu, T. J. (2006). Determinants of user acceptance of the e-government services: The case of online tax filing and payment system. *Government Information Quarterly*, 23(1), 97–122. doi:10.1016/j.giq.2005.11.005

Hussein, R., Mohamed, N., Ahlan, A. R., Mahmud, M., & Aditiawarman, U. (2010). An integrated model on online tax adoption in Malaysia. *European, Mediterranean & Middle Eastern Conference on Information Systems (EMCIS)*, (pp. 1-16).

Ilias, A., Razak, M. Z. A., & Yasao, M. R. (2009). Taxpayers' attitude in using e-filing system: Is there any significant difference among demographic factors? *Journal of Internet Banking and Commerce*, 14(1), 1–13.

Illias, A., Suki, N. M., Yasao', M. R., & Rahman, R. A. (2008). A study of taxpayers' intention in using e-filing system: A case in Labuan F.T s. *Computer and Information Science*, 1(2), 110–119.

Karahanna, E., Straub, D. W., & Chervany, N. L. (1999). Information technology adoption across time: A cross-sectional comparison of pre-adoption and post-adoption beliefs. *Management Information Systems Quarterly*, 23(2), 183–213. doi:10.2307/249751

Kim, H. W., Chan, H. C., & Chan, Y. P. (2007). A balanced thinking-feelings model of information systems continuance. *International Journal of Human-Computer Studies*, 65(6), 511–525. doi:10.1016/j.ijhcs.2006.11.009

Kini, A., & Choobineh, J. (1998). Trust in electronic commerce: Definition and theoretical considerations. *31st Annual Hawaii International Conference on System Sciences (HICSS)*, Vol. 4, (p. 51).

Lawkobit, M., & Speece, M. (2012). Integrating focal determinants of service fairness into post-acceptance model of IS continuance in cloud computing. *Proceeding of the 11th International Conference on Computer and Information Science IEEE/ACIS, Thailand*, May 30 – June 1, (pp. 49-55).

Lean, O. K., Zailani, S., Ramayah, T., & Fernando, Y. (2009). Factors influencing intention to use e-government services among citizens in Malaysia. *International Journal of Information Management*, 29(6), 458–475. doi:10.1016/j.ijinfomgt.2009.03.012

Li, D., Browne, G. J., & Chau, P. Y. K. (2006). An empirical investigation of web site use using a commitment-based model. *Decision Sciences*, 37(3), 427–444. doi:10.1111/j.1540-5414.2006.00133.x

Liao, C., Palvia, P., & Lin, H. N. (2006). The role of habit and web site quality in e-commerce. *International Journal of Information Management*, 26(6), 469–483. doi:10.1016/j.ijinfomgt.2006.09.001

Limayem, M., Hirt, S. G., & Cheung, C. M. K. (2003). Habit in the context of IS continuance: Theory extension and scale development. In C. U. Ciborra, R. Mercurio, M. de Marco, M. Martinez, & A. Carignani (Eds.), *Proceedings of the 11th European Conference on Information Systems*, Naples, Italy, June 16-21, 2003.

Limayem, M., Hirt, S. G., & Cheung, C. M. K. (2007). How habit limits the predictive power of intention: The case of information system continuance. *Management Information Systems Quarterly*, 31(4), 705–737.

Limayem, M., Hirt, S. G., & Chin, W. W. (2001). Intention does not always matter: The contingent role of habit on IT usage behavior. *The 9th European Conference on Information Systems*, (pp. 274-286).

Lopez, A. U., Miguel, F. P., & Prasad, S. I. (2012). Value, quality, purchasing habits and repurchase intention in B2C: Differences between frequent and occasional purchaser. *Dirección y Organización, 47,* 70–80.

Mahbob, M. H., Sulaiman, W. I. W., Mahmud, W. A. W., Mustaffa, N., & Abdullah, M. Y. (2012). The elements of behavioral control in facilitating the acceptance of technological innovation on malaysia on-line government services. *Asian Social Science, 8*(5), 125–131. doi:10.5539/ass.v8n5p125

Meister, D. B., & Compeau, D. R. (2002). *Infusion of innovation adoption: An individual perspective* (pp. 23–33). Winnipeg, Manitoba: ASAC.

Mid Term Review of the Ninth Malaysian Plan. (9MP). (2006-2010). *Economic Planning Unit* (EPU). Retrieved from the http://www.btimes.com.my/Current_News/BTIMES/Econ2007_pdf/Mid-term%20Review%20of%20the%20Ninth%20Malaysia%20Plan%202006-2010

Muhammad Rais, A. K., & Nazariah, M. K. (2003). *E-government in Malaysia. Pelanduk Publications (M).* Sdn Bhd.

Navarra, D., & Cornford, T. (2003). A policy making view of e-government innovations in public governance. *Proceedings of the Ninth Americas Conference on Information Systems (AMCIS),* Tampa, Florida, (p. 103).

Ndou, V. D. (2004). E-government for developing countries: Opportunities and challenges. *The Electronic Journal on Information Systems in Developing Countries, 18*(1), 1–24.

Ndubisi, N. O. (2004). Understanding the salience of cultural dimensions on relationship marketing, it's underpinnings and aftermaths. *Cross Cultural Management, 11*(3), 70–89. doi:10.1108/13527600410797855

Noor, M. M. (2010). Assessing the impacts of rural internet center programs on quality of life in rural areas of Malaysia. *Proceedings of IPID Postgraduate Strand at ICTD* (pp. 43-48). Karlstad University Studies.

Olsen, S. O., Tudoran, A. A., Brunson, K., & Verbeke, W. (2012). Extending the prevalent consumer loyalty modelling: The role of habit strength. *European Journal of Marketing, 47*(1), 1–30.

Organization for Economic Cooperation and Development (OECD). (2008). *Public governance.* Retrieved from http://www.oecd.org/dataoecd/39/19/40556222.pdf

Ou, C. X., & Sia, C. L. (2010). Consumer trust and distrust: An issue of website design. *International Journal of Human-Computer Studies, 68*(12), 913–934. doi:10.1016/j.ijhcs.2010.08.003

Palvia, P. (2009). The role of trust in e-commerce relational exchange: A unified model. *Information & Management, 46*(4), 213–220. doi:10.1016/j.im.2009.02.003

Papadopoulou, P., Nikolaidou, M., & Martakos, D. (2010). What is trust in e-government? A proposed typology. *Proceedings of the 43rd Hawaii International Conference on System Sciences (HICSS),* (pp. 1-10).

Parthasarathy, M., & Bhattacherjee, A. (1998). Understanding post-adoption behavior in the context of online services. *Information Systems Research, 9*(4), 362–379. doi:10.1287/isre.9.4.362

Pathak, R. D., Singh, G., Smith, R. F. I., & Naz, R. (2008). *Contribution of information and communication technology in improving government coordination, services and accountability in Fiji.* Retrieved April 22, 2012, from http://www.napsi-pag.org/pdf/RAGHUVAR_PATHAK.pdf

Poon, W. C. (2008). Users' adoption of e-banking services: The Malaysian perspective. *Journal of Business and Industrial Marketing, 23*(1), 59–69. doi:10.1108/08858620810841498

Premkumar, G., & Bhattacherjee, A. (2008). Explaining information technology usage: A-test of competing models. *The International Journal of Management Science, 36*(1), 64–75.

Rahim, R. (2011, May 14). 227,851 taxpayers successfully refunded. *The STAR. Nation (New York, N.Y.)*, N3.

Ramayah, T., Ramoo, V., & Ibrahim, A. (2008). Profiling online and manual tax filers: Results from an exploratory study in Penang, Malaysia. *Labuan e-Journal of Muamalat and Society, 2*, 1-8.

Rose, S., Clark, M., Samouel, P., & Hair, N. (2012). Online customer experience in e-retailing: An empirical model of antecedents and outcomes. *Journal of Retailing, 88*(2), 308–322. doi:10.1016/j.jretai.2012.03.001

Rowley, J. (2011). E-government stakeholders – Who are they and what do they want. *International Journal of Information Management, 31*(1), 53–62. doi:10.1016/j.ijinfomgt.2010.05.005

Sang, S., Lee, J. D., & Lee, J. (2010). E-government adoption in ASEAN: The case of Cambodia. *Internet Research, 19*(5), 517–534. doi:10.1108/10662240910998869

Sathye, M. (1999). Adoption of internet banking by Australian consumers: An empirical investigation. *International Journal of Bank Marketing, 17*(7), 324–334. doi:10.1108/02652329910305689

Saxena, K. B. C. (2005). Towards excellence in e-governance. *Centre for Excellence in Information Management (CEXIM) Working Paper Series*, No. 2005-1, (pp. 1-15).

Schaupp, L. C., & Carter, L. (2010). The impact of trust, risk and optimism bias on e-file adoption. *Information Systems Frontiers, 12*(3), 299–309. doi:10.1007/s10796-008-9138-8

Schaupp, L. C., Carter, L., & McBride, M. E. (2010). E-file adoption: A study of U.S. taxpayers' intentions. *Computers in Human Behavior, 26*(4), 636–644. doi:10.1016/j.chb.2009.12.017

Shari, I. (2010, May 18). Govt forks out a bomb for taxpayers who shy away from e-filing. *The STAR*. Retrieved from http://the-star.com.my/news/story.asp?file=/2010/5/18/nation/6281573&sec=nation

Siau, K., & Shen, K. (2003). Building customer trust in mobile commerce. *Communications of the ACM, 46*(4), 91–94. doi:10.1145/641205.641211

Suki, N. M., & Ramayah, T. (2010). User acceptance of the e-government services in Malaysia: Structural equation modelling approach. *Interdisciplinary Journal of Information, Knowledge, and Management, 5*, 395–413.

Tajuddin, U. (2010). *From pilot to national initiative*. Retrieved from http://www.itu.int/ITUD/asp/CMS/ASP-CoE/2010/IRD/S4-Mr_Uzer_Tajuddin.pdf

Tan, C. W., Benbasat, I., & Cenfetelli, R. T. (2008). Building citizen trust towards e-government services: Do high quality websites matter? *Proceedings of the International Conference on System Sciences, Vancouver*, 7-10 Jan, (p. 217).

Teo, T. S. H., Srivastava, S. C., & Jiang, L. (2008). Trust and electronic government success: An empirical study. *Journal of Management Information Systems, 25*(3), 99–131. doi:10.2753/MIS0742-1222250303

Theodorakis, N. D., Koustelios, A., Robinson, L., & Barlas, A. (2009). Moderating role of team identification on the relationship between service quality and repurchase intentions among spectators of professional sports. *Managing Service Quality, 19*(4), 456–473. doi:10.1108/09604520910971557

United Nations E-Government Survey (UNPAN). (2010). *Leveraging e-government at a time of financial and economic crisis.* Retrieved from http://unpan1.un.org/intradoc/groups/public/documents/un/unpan038851.pdf

Wangpipatwong, S., Chutimaskul, W., & Papasratorn, B. (2007). The role of technology acceptance model's beliefs and computer self-efficacy in predicting e-government website continuance intention. *WSEAS Transaction on Information Science and Applications, 4*(6), 1212–1218.

Wangpipatwong, S., Chutimaskul, W., & Papasratorn, B. (2008). Understanding citizen's continuance intention to use e-government website: A composite view of technology acceptance model and computer self-efficacy. *The Electronic. Journal of E-Government, 6*(1), 55–64.

Weerakkody, V., & Choudrie, J. (2005). Exploring e-government in the UK: Challenges, issues and complexities. *Journal of Information Science and Technology, 2*(2), 25–45.

Welch, E. W., Hinnant, C. C., & Moon, M. J. (2005). Linking citizen satisfaction with e-government and trust in government. *Journal of Public Administration: Research and Theory, 15*(3), 371–391. doi:10.1093/jopart/mui021

Williams, P. (2002). The learning web: The development, implementation and evaluation of internet-based undergraduate materials for the teaching of key skills. *Active Learning in Higher Education, 3*(1), 40–53. doi:10.1177/1469787402003001004

Yousafzai, S. Y., Foxall, G. R., & Pallister, J. G. (2010). Explaining internet banking behavior: Theory of reasoned action, theory of planned behavior, or technology acceptance model? *Journal of Applied Social Psychology, 40*(5), 1172–1202. doi:10.1111/j.1559-1816.2010.00615.x

Zakaria, Z., Hussin, Z., Zakaria, Z., Noordin, N. B., Sawal, M. Z. H. B. M., Saad, S. F. B. M., & Kamil, S. B. O. (2009). E-filing system practiced by inland revenue board (IRB): Perception towards malaysian taxpayers. *Cross Cultural Communication, 5*(4), 10.

Zmud, R. W. (1982). Diffusion of modern software practices: Influence of centralization and formalization. *Management Science, 28*(12), 1421–1431. doi:10.1287/mnsc.28.12.1421

ADDITIONAL READING

Akkaya, C., Wolf, D. P., & Krcmar, H. (2010). The role of trust in e-government adoption: A literature review. *Proceeding of Americas Conference on Information Systems* (AMCIS), paper 297.

Belanche, D., Casalo, L. V., & Flavian, C. (2010). The importance of confirming citizens' expectations in e-government. *IFIP Advances in Information and Communication Technology, 341*, 103–111. doi:10.1007/978-3-642-16283-1_14

Dawes, S. S. (2008). An exploratory framework for future e-government research investments. *Proceedings of the 41ˢᵗ Hawaii International Conference on System Sciences (HICSS)*, (p. 201).

Hannas, G., Andersen, O., & Buvik, A. (2010). Electronic commerce and governance forms: A transaction cost approach. *International Journal of Procurement Management, 3*(4), 409–427. doi:10.1504/IJPM.2010.035470

Hsu, S. L., Wang, H., & Doong, H. (2010). Determinants of continuance intention towards self-service innovation: A case of electronic government services. *Exploring Service Sciences, 53*, 58–64. doi:10.1007/978-3-642-14319-9_5

Jiang, X. (2011). Enhancing users' continuance intention to e-government portals: An empirical study. *Proceeding of the International Conference on Management and Service Sciences (*MASS), (pp. 1-4).

Nan, Z., Xin, G., & Qingguo, M. (2011). *Exploring different roles between service expectation and technology expectation in citizen's e-government continuance adoption: An extended expectation-confirmation mode*l. Research in Progress. Retrieved on April 30, 2012, from http://projects.business.uq.edu.au/pacis2011/papers/PACIS2011-222.pdf

Rahman, M. M., & Rajon, A. S. A. (2011). An effective framework for implementing electronic governance in developing countries: Bangladesh perspective. *14th International Conference on Computer and Information Technology (ICCIT),* (pp. 360-365).

Schooley, B., Harold, D. A., Horan, T. A., & Burkhard, R. (2011). Citizen perspectives on trust in a public online advanced traveler information system. *Proceedings of the 44th Hawaii International Conference on System Sciences*, (pp. 1-9).

Uma, R. W. O. (2000). Electronic governance: Reinventing good governance. Retrieved from http://vallenacional.gob.mx/work/sites/ELOCAL/resources/LocalContent/1192/9/Okot-Uma.pdf

KEY TERMS AND DEFINITIONS

Continuance Intention: Continuance Intention refers to ones intention to continue using or long term usage intention of a technology (Bhattacherjee, 2001).

E-Filing System: E-Filing system which is launched in 2006 in Malaysia "allows taxpayers to submit their income tax details online and is considered as an alternative to the usual manual paper submission" (Ambali, 2009). "With the e-filing system, taxpayers and tax practitioners can file income tax returns electronically via the enabling technologies, rather than through mail or physically visiting the tax office" (Ling et al., 2005).

Electronic Government: Electronic Government or e-government refers to government's use of ICT to work more effectively, to share information and deliver better services to the public more efficiently and to increase the speed of delivery of services combined with reduction in costs' (Chadwick & May, 2003).

Information Communication Technologies (ICT): ICT are a diverse set of technological tools and resources used to communicate, and to create, disseminate, store, and manage information (Blurton, 1999).

Manual filing of Income Tax: Manual Tax Filing is the traditional submission method "either by hand or typewriter. Taxpayers usually perform complex calculations using mental arithmetic or calculator, and then the return is delivered to the tax agency through the postal service or in person. After receiving a return, the agency uses a data entry service to input the data. (Fu, Farn & Chao, 2006).

Taxpayers: Individuals who are liable to pay income tax on all of his or her income for a tax year (Melville, 2005).

Trust in the System: Trust in the system is defined as the perception of the operation of the system which will display availability, fault tolerance and its security and correctness is guaranteed together with stability in system response time (Papadopoulou, Nikolaidou & Martakos, 2010).

Chapter 18

E–Government and Public Service Delivery:
A Survey of Egypt Citizens

Hisham M. Abdelsalam
Cairo University, Egypt

Christopher G. Reddick
University of Texas at San Antonio, USA

Hatem A. ElKadi
Cairo University, Egypt

Sara Gamal
Cairo University, Egypt

ABSTRACT

This chapter aims to better understand what citizens think regarding the currently available e-government public services in Egypt. This is done through an analysis of a public opinion survey of Egyptian citizens, examining citizens' use and associated issues with usage of e-government portals. This chapter is different from existing research in that most of the studies that examine e-government and citizens focus on developed countries. This study focuses on a developing country, Egypt, as an emerging democracy, which has very unique and important challenges in the delivery of public services to its citizens. The results revealed that only gender, daily use of the internet, and the desire to convert all of the services to electronic ones were important factors that affected the use of the Egyptian e-government portal. On the other hand, age, education, trust in information confidentiality on the internet, and believing in e-government did not play any role in using e-government.

DOI: 10.4018/978-1-4666-3640-8.ch018

1. INTRODUCTION

Governments worldwide are integrating Information and Communications Technologies (ICT) in their public administrative reform programs to digitize their delivery of public services (Electronic Government for Developing Countries, 2008). Electronic government, or e-government, is used as a way of integrating ICT into public service delivery. One of the most important motivations behind such transformation stems from the potential of e-government to enhance the delivery of public services to promote greater transparency, accountability, and responsiveness to citizens (Bwalya, 2009).

Egypt is an interesting case since, being a developing country, in that it would benefit greatly from e-government to improve public service delivery (Heeks, 2002; Chen, Chen, Ching, & Huang, 2007; Hamner & Qazi, 2009). Applying e-government in a developing country is challenging to a noticeable degree, as face-to-face communication is the most preferable and dominant way to contact government entities, since only 29% of the Egyptians using the internet; according to the last estimates of the Central Agency for Public Mobilization and Statistics in Egypt. In addition, the Egyptian illiteracy rate is 29.7% of the total population in age 10 years and older, according to the Egyptian census in 2006.

This study aims to better understand what citizens think regarding the currently available e-government services using a public opinion survey of Egyptian citizens. Most of the existing research has focused on the United States (Thomas & Streib, 2003; Reddick, 2005; Streib & Navarro, 2006), and there is much less research that focuses on citizens' use of e-government in the context of developing countries.

There are two research questions examined in this study:

1. What is the current use of e-government in Egypt by its citizens?

2. What factors affect using e-government for Egyptian citizens?

The answers to these two questions will be addressed through several sections. Following the introduction section, the rest of this paper is organized as follows. Section 2 provides background information on the context of the Egyptian e-government program. Research hypotheses derived from the e-government literature are presented in Section 3. Section 4 details the research methodology of this chapter. Detailed results and hypotheses testing are presented in Section 5. Section 6 presents the conclusion, research limitations, and future research possibilities are suggested.

2. CONTEXT

Egypt is a unitary country that comprises 27 administrative sections, called governorates (or municipalities), of various sizes, populations, and resources. Governorates are administratively further divided into cities and districts, which are, in turn, divided into smaller entities called neighborhoods (in cities) and villages (in the districts).

Egypt has established its ICT strategy in 2001 in what has been known as the Egyptian Information Society Initiative (EISI). EISI was built on seven pillars; one of which was e-government. EISI initiative was put into action and, hence, the e-government program in Egypt started in 2001. In 2004, program ownership was transferred to the Ministry of State for Administrative Development (MSAD), where the former e-government Program Director was appointed as the minister. This reflects the Egyptian understanding of e-government as a natural component of administrative development and reform. Thus, the e-government program in Egypt became one of the two mandates of MSAD, the other one being public administration institutional reform. In 2010 Egypt was ranked 86th out of 195 countries in e-government development, it ranked 23rd in the online service index, and 42nd in the e-participation index (United Nations, 2010).

Initially, the e-government program consisted of four main subprograms among which came the Egyptian Local Government Development Program (ELGDP). In turn, ELGDP has three main projects: (1) service enhancement in municipalities which includes automation of services provided to citizens; (2) development of web portals for the governorates; and (3) citizen relationship management (CRM) systems. Finally, a 10-Year National Broadband Plan (e-Masr 2020) is to be announced with four key political objectives: promote economic growth nationwide; avoid an increased digital divide within Egypt; increase job opportunities and foster social cohesion; and improve quality of life for all citizens (Minister of Communications & Information Technology, 2011).

3. HYPOTHESIS

There are eight hypotheses which are used to examine e-government access by citizens in Egypt. The most common factors noted in the literature that may affect accessing e-government programs were access to the internet and socio-demographic factors (Belanger & Carter, 2006; 2009). One other key factor that affects the use of e-government services could be how citizens perceive the importance of e-government project (Titah and Barki, 2006).

Accessing e-government services in this study means using the e-government portal as it considered the most popular project that was included in the e-government program in Egypt. This examines whether citizens have access to internet technology and are able to use it effectively. In addition, the socio-demographic status of the citizens and their use of e-government were also examined.

3.1 Accessing the Internet

The following hypotheses are derived from the literature. Hypothesis 1 indicates that citizens that have daily access to the internet would have a greater ability to use it and would be more willing to go online to access e-government (Belanger & Carter, 2006; van Deursen & van Dijk, 2009). Hypothesis 2 examines whether a citizen has broadband internet access. Citizens that have this type of high-speed internet access would be more likely to access e-government (Reddick, 2010).

Hypothesis 1: Citizens that access the internet daily are likely to use e-government.

Hypothesis 2: Citizens that have broadband internet are more likely to use e-government.

3.2 Socio-Demographic Factors

Socio-demographic status of citizens is the second factor that is used to predict e-government usage. Research show that older citizens are less likely to use the internet than the younger generation (Dimitrova & Chen, 2006; Morgeson, VanAmburg, & Mithas, 2010), therefore, Hypothesis 3 is shown below. Research also shows that citizens that have a college education would be more likely to go online to engage in e-government (McNeal, Hale, & Dotterweich, 2008). This implies that citizens' that are college educated would be more knowledgeable about the internet and be able to navigate a government website as shown in Hypothesis 4. Citizens that are male would be more likely to go online and use e-government as shown in Hypothesis 5. Research shows that there is a closing of the gap of gender and e-government access, but there is little research that examines developing countries to determine whether this is the case (Al-Rababah & Abu-Shanab, 2010).

Hypothesis 3: Citizens that are older will use e-government services less.

Hypothesis 4: Citizens that have a college education are more likely to use e-government services.

Hypothesis 5: Male citizens will use e-government services more than females.

3.3. Perceived Importance of E-Government Projects

Titah and Barki (2006) identified five main research streams in the e-government studies; one of which is concerned with the influence of the 'individual beliefs' on e-government use and acceptance. Citizens who believe in the importance of e-government programs are more likely to use e-government than citizens who do not believe this, as shown in hypothesis 6. Carter and Belanger (2005) emphasize that citizen's perceptions of trustworthiness issues such as security and privacy (trust of internet and of government) can also influence the use e-government service use. Reddick (2004) also referred to privacy and security as an important issue that may limit e-government use. Hypothesis 7 examines that citizens' who trust their data confidentially are more likely to use e-government. The final hypothesis 8 examines if the desire to convert all the services to electronic ones through the portal will have a positive impact on using e-government.

Hypothesis 6: Citizens that believe in the importance of e-government programs are more likely to use this technology.

Hypothesis 7: Citizens who trust to put their own personal information on the internet are more likely to use e-government.

Hypothesis 8: Citizens who wish to convert all services to electronic ones are more likely to use e-government.

4. RESEARCH METHODS AND DATA

A questionnaire was prepared to answer the two main research questions using open and closed ended questions. Data was collected during August 10, 2010 to September 20, 2010 by interviewing 120 respondents which were randomly selected from Cairo, Giza and Dakahlia. The Decision Support and Future Studies Center – Cairo University carried out this survey examining citizens' awareness about the Egyptian e-government program and the level of interaction with such programs. The most preferred services, services channel, method of payment, services evaluation, and barriers of use were also addressed in the survey.

The data were processed and analyzed using SPSS 19. Descriptive analysis was prepared to discuss the current use of e-government services along with the perceived importance and interaction with these services. Logistic regression was also applied in order to determine which factors affected e-government services' usage in Egypt.

This study is considered an empirical pilot study (e.g., Dimitrova and Chen 2006; Dossani et al., 2005) that mainly depends on quantitative analysis to help researchers identify the extent of e-government development and administration. The convenience sample done in this study is a very suitable design for such cases (Hackett & Bambang, 2005) as it helps to achieve the required sample size in a relatively fast and inexpensive way as we are in a pilot study so we need to obtain the results as soon as possible in order to open the way for more accurate studies in the same field. Even knowing with our sample, the representativeness is less reliable, we have freedom in exploring important questions on the direction of unknown factors impacting e-government development.

5. RESULTS AND DISCUSSION

The results of this chapter are presented in two sections. The first section uses descriptive statistics that examines the sample according to the analyzed questions in order to draw on background of the respondents using the following factors: use of the internet, e-government program awareness, perceived importance, e-government service usage, services channel selection, and security and privacy. The second section contains a logistic regression model of the statistically significant variables that explained e-government usage.

5.1 Sample Characteristics

Males and females were equally distributed in the sample as they have the same percentages approximately (see Table 1). About 37% of the respondents were aged from 20 to less than 30 and about 28% from 30 to less than 40. These age ranges are realistic as they are the most potential ones to deal with e-government services and channels compared to the other age groups. The majority of the sample were living in Cairo and Giza (70.6%) and having a university degree (57.5%). As a result, the survey results are more representative of educated and younger population of Egyptian society, which are the more likely users of e-government.

5.2 Using the Internet

Table 2 shows that the majority of the sample (86.6%) accessed the internet and about 48% of them accessed it between 3-5 days per week. In addition, there were about 86% of the internet users having e-mail access. DSL connection was the most popular way to access the internet (38.3%), while an illegal connection took the second place (34.6%).

5.3 E-Government Program Awareness

The overwhelming majority of the sample was aware of the e-government program (93%). Table 3 shows that 64.3% of the aware people were aged

Table 1. Survey sample characteristics

	Percentage (%)
Gender	
Male	49.2
Female	50.8
Age	
Less than 20	12.5
From 20 to 30	36.7
From 30 to 40	28.3
from 40 to 50	17.5
Older than 50	5.0
Education	
Illiterate	3.3
High school	15.0
2-year diploma	14.2
University degree	57.5
Post graduate	10.0
Governorate	
Cairo	31.3
Giza	39.3
Dakahlia	23.2
Other	6.3

Table 2. Internet access

	Percentage (%)
Internet Usage Rate	
Between 3-5 days per week	48.2
between once to twice per week	16.1
One day every few weeks	10.7
Less than that	7.1
Does not use the internet	13.4
How did you access the Internet	
Telephone line	17.8
DSL	38.3
Mobile networks	9.3
Illegal connection	34.6
Do you use e-mail	
Yes	86.0
No	14.0

315

Table 3. The e-government program awareness according to the demographic characteristics

		% (of Yes)
Age	Less than 20	13.4
	From 20 to 30	35.7
	From 30 to 40	28.6
	from 40 to 50	17.0
	Older than 50	5.4
Gender	Male	50.9
	Female	49.1
Education	Illiterate	0.9
	High school	12.5
	2-year diploma	14.3
	University degree	61.6
	Post graduate	10.7

from 20 to less than 40, and about 51% of them were male. Finally, 61.6% had a university degree. Only 10.7% of the citizens were post graduates.

The Egyptian municipalities' portals had been launched in 2001 by the Ministry of State for Administrative Development (MSAD) in order to provide citizens with services and information through the internet. Asking the respondents if they knew about these portals, or not, we found that all respondents did not know anything about the Egyptian municipalities' portals. There was 87.3% who did not hear about it before. Therefore, the lack of using the municipalities' portals may be due to the lack of media campaign. Recently, the municipalities' portals started to launch new pages on Facebook to reach their citizens which is considered a good step towards more interaction with the citizens.

The MSAD is now working to provide a standardized design for municipal portals in order to be more user-friendly. In addition it is providing more services electronically through the portal. At the end of June 2011 the number of uniformed portals reached to 12 portals out of 26, the MSAD in on their way to making sure that all municipal portals have a uniform design.

5.4 Perceived Importance

Figure 1 shows that the majority of the respondents (77.7%) believed that e-government program was important or very important. This importance was justified by saving time and effort (37.5%), providing services quickly (16.1%), and eliminating bribery (9%).

What factors play a role in the citizens' satisfaction? Using cross tabulation the following model was obtained as it shown in Table 4. There

Figure 1. What do you think of the e-government program?

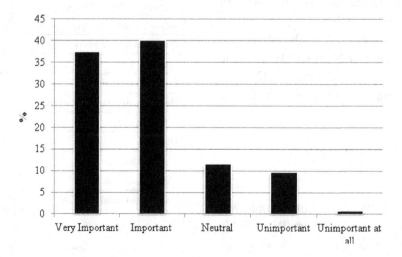

Table 4. The related variables with the importance of e-government program

	What do you think of e-government program			
	Chi-Square Tests		Gamma	
	Value	Asymp. Sig. (2-sided)	Value	Asymp. Sig. (2-sided)
Do you use Egyptian government portal	34.45*	0.000	0.45*	0.007
Do you wish to convert all governmental services to electronic ones	45.54*	0.001	0.47*	0.000
Respondents annual usage rate of the portal services	10.55**	0.032	0.37**	0.041

Notes: *significant at the 0.01 level; **significant at the 0.05 level

were three variables that significantly correlate with the citizens' view of e-government program importance.

Using the e-government portal was significantly correlated with the importance associated with e-government programs with 99% confidence level. Gamma measurement of ordinal by ordinal association assessed that there was a positive but moderate relationship between the two variables; people who use the portal were more likely to believe in the e-government importance within the 99% confidence level.

Egyptian citizens that have a strong belief in the e-government programs wish to convert all governmental services to electronic ones as a result of their convictions. Table 4 shows that there was a positive moderate significant relationship between the two variables with 99% confidence level. Finally, the increasing annual usage rate of portal services leads to more satisfaction of e-government program; there was a positive and significant relationship between the two variables with 95% confidence level.

5.5 E-Government Services Usage

The Ministry of State for Administrative Development (MSAD) and the Ministry of Communications and Information Technology (MCIT) launched the Egyptian government portal (Bawaba) in 2004. Bawaba is a gateway through which citizens can access all Egyptian government services, information, and documents. The results showed that there were about 70% of the respondents using the e-government portal (Bawaba).

The Egyptian portal provides many of services which can be requested electronically. Figure 2 shows that 50% of the portal users access the portal services 3-7 times annually, while 9% only access the services more than 7 times annually.

Figure 3 shows that 'query on the telephone bill' was the most popular services where about 87.5% of the respondents knew about this service. The electronic university admission was a good step for e-government services dissemination as it opens the door to know the portal and its services. The results revealed that there were 80.4% of the respondents who knew the university admission coordinating service, while 34.8% only knew they could renew their car license electronically. On the other side, the following services are less popular than the previous one; submission of tax returns (13.4%), booking airline tickets (12.5%), maps (5.4%) and finally complaints and inquiries services (1.8%).

5.5.1 Services Evaluation

Figure 4 shows that the most accessed e-services for Egyptian citizens were the telephone bill and coordinating university admissions. There were almost 70% of the respondents querying their telephone bill using the Egyptian government

Figure 2. The annual usage rate of portal services

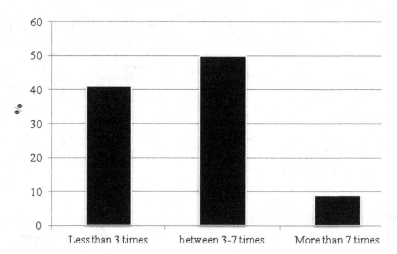

Figure 3. The most popular (known) services

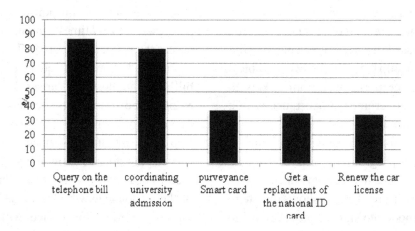

web portal, while about 49% were coordinating their university admissions through the portal.

While asking the respondents to evaluate these electronic services, the survey results reveal that 75.6% of respondents believed they were satisfied of the service level (very good and good), as Figure 5 shows, while only 4% were unsatisfied with the service level (bad and very bad).

Figure 6 shows the relation between the annual usage rate of the portal users and citizens' evaluation of the accessed services. It was very clear that the increased annual usage rate of the portal services led to decreased satisfaction level, this may be due to more ambitious users who expect more compared to first time users. For instance, when you use something for the first time you will be very impressed with it, however, when you return you might expect more leader to lower satisfaction.

Spearman correlation coefficient supported the previous result that was drawn from Figure 6; we were 95% confident that there were a signifi-

Figure 4. The most used services

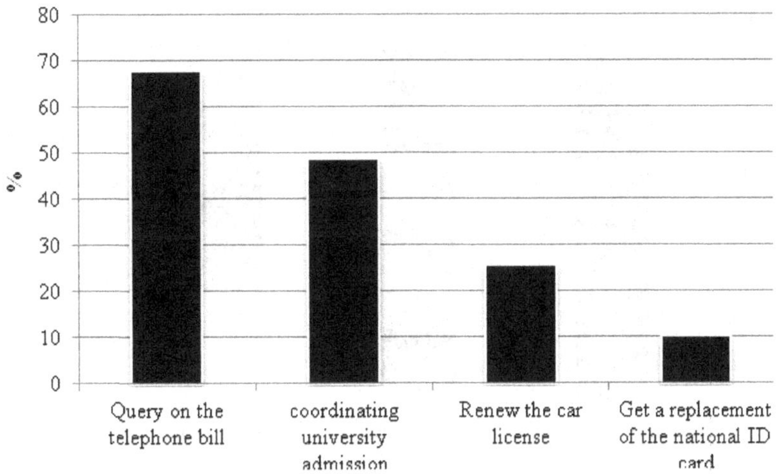

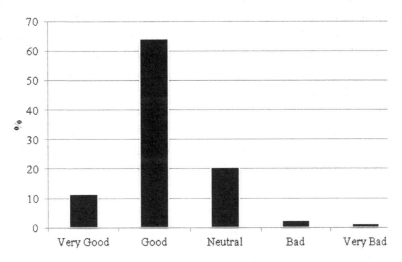

Figure 5. Services evaluation

cant weak negative linear relationship between the annual usage rate of the portal services and satisfaction level of the accessed services (Spearman Coefficient = -0.23).

5.5.2 Barriers

Barriers to e-government adoption have been the focus of a considerable number of research articles. Several factors were confirmed to play a significant role in new e-government adoptions, such as information security (Conklin, 2007), trust of internet and of government (Carter & Belanger, 2005), usability (Bwalya, 2009), and website navigability and aesthetics (Coleman, 2006).

The results of this research showed that about 69% of the respondents were unsatisfied with the current government performance while only 4% were satisfied. This is primarily due to corruption among government officials. "Freedom, democracy, and social justice" were the three governing principles called for by the Egyptian revolution in

Figure 6. The relation between the evaluation of portal services and its annual usage rate

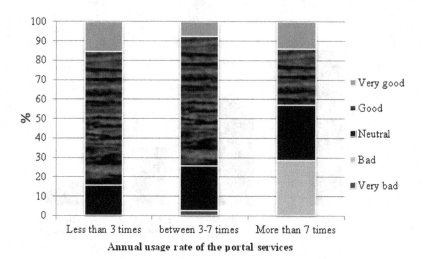

25th January, 2011. Now, the Egyptian government is seeking quick and major changes in the political regime that leads to enhancing citizens' empowerment. The results show that about 30% of the respondents did not use the Egyptian government portal as they did not need to use it (41.2%), and they cannot deal quickly with the portal (20.6%).

Figure 7 shows that there were 50% of the respondents that trust the confidentially of their data on the internet (very confident or confident)

while there were about 15% of them did not trust (didn't trust and didn't trust at all).

On the other hand, there were about 40% of the portal users that had some difficulties accessing the services online. As Figure 8 shows slow loading of the portal was the most common problem (84%), while 19.4% complained of lack of clarity of service request steps, also the same percentage complained of waiting a long time to get the service.

Figure 7. The extent of your trust in the confidentiality of your data on the internet

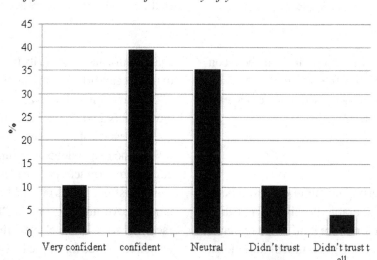

Figure 8. The most common problems during accessing the services online

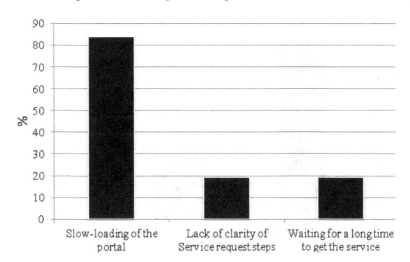

The most popular method of paying for the electronic services was paying upon delivery (89.7%) as Figure 9 shows. On the other hand, only 10.3% used credit cards and no one was used bank transfer or prepaid cards as a method of payment.

What reasons are there for the lack of using of credit cards for paying for services? Figure 10 clarified that 64.3% of the respondents preferred the upon delivery method of payment in order to make sure that they get the services, while 22.9%

of them said this was the only available way for them to pay for the service.

5.5.3 Required Services

In a survey conducted by the Center for Technology in Government (University at Albany/SUNY) in 2000, indicated that citizens' most required services to be provided online were renewing a driver's license, followed by voter registration.

In this research, asking the respondents about the services that they wish to access electronically

Figure 9. The preferred way of paying for the electronic services

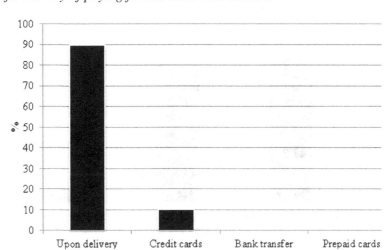

Figure 10. The reason of preferring upon delivery as a way of paying for the electronic services

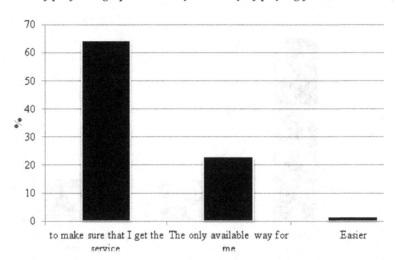

through the government portal we found that 42% saw that the existed services were enough and they did not need to add any more services on the portals. About 32% of the respondents proposed that financial transactions be available through the portal such as paying all bills and schools fees.

5.6 Services' Channel Selection

This section examines channel choice and the digital divide through a public opinion survey, which is a relatively unexplored area of research

(Pieterson & Ebbers, 2008; Reddick, 2010). Figure 11 shows that the traditional offline service provider locations were the most preferred services channels as 48.2% of the respondents depend on this service. Some citizens still prefer to receive public services in a traditional way, while in many countries governments have started to offer public services online (European, Mediterranean & Middle Eastern Conference on Information Systems 2010). The internet and mobile phones was the second services channel as 46.4% of the

Figure 11. Services' channels

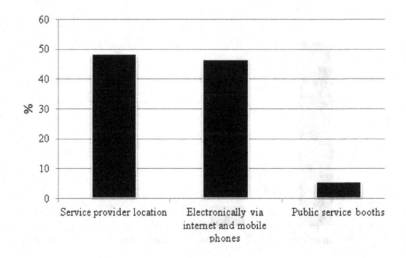

Figure 12. Why did you choose these channels?

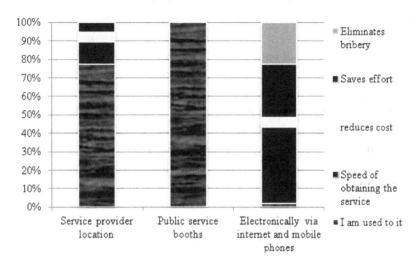

respondents prefer to obtain services via this channel.

But, what was the reason behind using such channels? This question was answered in the 2010 European, Mediterranean & Middle Eastern Conference on Information Systems: The (face-to-face) communication is an important aspect to be included in the service delivery. Many people may also not be aware of the fact that the services are provided online as well, and therefore, they need to be informed. (p. 5)

The previous results were supported by Figure 12 with 83.3% of the respondents justified their face-to-face transaction through the service provider location by saying "I am used to it". There were a variety of reasons for using the electronic services channels such as the internet and mobile phones such as obtaining services faster compared to the traditional channels (77%), saving efforts (53.8%), and eliminating bribery (42.3%).

Therefore, the government needs to be aware that citizens tend to use more than one channel; e-government is one of many channels that citizens

Figure 13. The preferred way to contact one of the governmental agencies

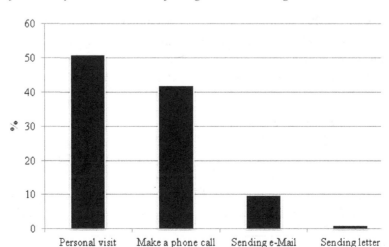

can use when they have a question or face a problem which requires face to face government contact.

Figure 13 shows that there were about 51% of the respondents who visited the government agency in person when they have a question or a problem. Making a phone call is the second preferred way to contact a government entity (42%), while only10% sent e-mails.

5.7. Logistic Regression Model

Logistic regression model was constructed to explore which variables affected using e-government. Table 4 was obtained by implementing binary logistic regression forward LR (Likelihood Ratio). The Egyptian e-government portal usage was a binary dependent variable, with a "1" representing if the citizen used the portal and "0" otherwise. The predictor variables were age, gender, education, the desire to convert all the services to electronic ones, data confidentially on the internet, the internet usage rate, the way of accessing the internet, and believing in the e-government program. Odds Ratios (OR) can be calculated in logistic regression to show the likelihood of a predictor variable impacting the dependent one. The resulted model presented in Table 5 classified 85.4% of the data correctly. In addition, Nagelkerke R-Square was 0.53 for this model.

Table 5 shows that we are 99% confident that females were 8 times more likely to use e-government portal than males which contradicted with hypothesis 5, which supposed that females

would be less likely to go online and access e-government. Descriptive results showed that 80% of females were using e-government, while 59.6% of males do this which support the logistic regression results and contradicted the previous research presented in the hypothesis.

Citizens who used the internet daily were 28.7 times more likely to use e-government with 99% confidence level. Therefore hypothesis 1 was verified in that citizens that access the internet daily will have a positive impact on the use of e-government. The desire to convert all the services to electronic ones reflects the citizens' orientation towards accessing e-government. Table 5 shows that citizens who want to convert all services to electronic ones were 2 times more likely to use e-government with 95% confidence level, which supported hypothesis 8. Finally, Age, education, trust of information confidentiality on the internet, and believing in e-government did not play a role in using e-government according to the statistical tests done.

6. CONCLUSION, LIMITATIONS, AND FUTURE RESEARCH

This study aimed to understand citizens' thoughts regarding e-government services, exploring their demands and expectations towards e-government in Egypt using a public opinion poll. The results concluded that there were some factors affected using e-government noted in the literature that did not play a role in our model such as age, education, trust information confidentiality on

Table 5. Logistic regression of using e-government portal

	Beta Coefficient	Wald Statistic	Prob. Sign.	Odds Ratio
Female	2.088	7.048*	.008	8.068
Using internet Daily	3.357	9.027*	.003	28.696
The desire to convert all the services to electronic ones	.752	4.782**	.029	2.121

Notes:*significant at the 0.01 level; **significant at the 0.05 level

the internet, and believing in e-government. On the other hand, gender gap was the opposite from what was expected. This result can be justified by assessing that the sample which was drawn from metropolitan areas that gives females the right to learn, work, and develop their self, compared to more rural regions in Egypt.

There is a digital divide in Egypt; face-to-face communication is the most preferable way to contact government entity. The second preferable channels are the internet and mobile phones as they save time and effort and eliminate bribery. After the Egyptian revolution on 25th of January, e-government could be seen as a way of restoring public trust to the government as it can enhance good governance, strengthen the democratic process, and can also facilitate access to information, freedom of expression, greater equity, efficiency, productivity, growth and social inclusion. Successful e-government initiatives can have demonstrable and tangible impact on improving citizen participation and quality of life (ADF IV, 2004). But these reforms are difficult to achieve without addressing the digital divide that exists in Egypt, and other developing countries, which makes access to e-government services difficult for a vast segment of the population.

The MSAD has provided a uniform design for the municipalities' portals to be more easy to use; in addition providing more services electronically. On the other hand, the municipalities' portals started to launch new pages on Facebook to reach citizens and propagate is an initial first step towards more interaction with the citizens.

Using a public opinion survey with a few key variables was a limitation of this study. Another limitation is the sample being composed of citizens that are younger, educated, and in an urban setting, which is not broadly representative of Egyptian society. As a result of this limitation future work could do focus groups examining channels choice and cover other factors not covered in this quantitative study. Focus groups could drill down into some of the nuisances missed in our quantitative analysis.

REFERENCES

ADF IV. (2004). *Fourth African Development Forum, Governance for a Progressing Africa*, 11-15th October 2004, Addis Ababa, Ethiopia. Retrieved from http://www.uneca.org/adf/adfiv/adf_4_report_final_sml.pdf

Al-Rababah, B., & Abu-Shanab, E. (2010). E-government and gender digital divide: The case of Jordan. *International Journal of Electronic Business Management, 8*(1), 1–8.

Belanger, F., & Carter, L. (2006). The effects of the digital divide on e-government: An empirical evaluation. *Proceedings of the 39th Annual Hawaii International Conference on System Sciences*. Retrieved April 13, 2011, from http://www.computer.org/portal/web/csdl/doi/10.1109/HICSS.2006.464

Belanger, F., & Carter, L. (2009). The impact of the digital divide on e-government use. *Communications of the ACM, 52*(4), 132–135. doi:10.1145/1498765.1498801

Bwalya, K. (2009). Factors affecting adoption of e-government in ZAMBIA. *Electronic Journal of Information Systems in Developing Countries, 38*(4), 1–13.

Carter, L., & Belanger, F. (2005). The utilization of e-government services: Citizen trust, innovation and acceptance factors. *Information Systems Journal, 15*(1), 5–25. doi:10.1111/j.1365-2575.2005.00183.x

Chen, Y., Chen, H., Cling, R., & Huang, W. (2007). Electronic government implementation: A comparison between developed and developing countries. *International Journal of Electronic Government Research, 3*(2), 45–61. doi:10.4018/jegr.2007040103

Coleman, S. (2006). *African e-governance – Opportunities and challenges.* Oxford, UK: Oxford University Press.

Cook, M. A. (2000). *What citizens want from e-government: Current practice research.* Center for Technology in Government-University at Albany/SUNY.

Dimitrova, D., & Chen, Y.-C. (2006). Profiling the adopters of e-government information and services: The influences of psychological characteristics, civic mindedness, and information channels. *Social Science Computer Review, 24*(2), 172–188. doi:10.1177/0894439305281517

Dossani, R., Jhaveri, R., & Misra, D. (2005). *Enabling ICT for rural India,* (pp. 1-75). Asia-Pacific Research Center, Stanford University and National Informatics Center, Government of India. Retrieved from http://iis-db.stanford.edu/pubs/20972/ICT_full_Oct05.pdf

Electronic Government for Developing Countries. (2008). Retrieved from www.itu.int/ITU-D/cyb/app/e-gov.html

Hackett, S., & Parmanto, B. (2005). A longitudinal evaluation of accessibility: Higher education web sites. *Internet Research, 15*(3), 281–294. doi:10.1108/10662240510602690

Hamner, M., & Qazi, R. (2009). Expanding the technology acceptance model to examine personal computing technology utilization in government agencies in developing countries. *Government Information Quarterly, 26*(1), 128–136. doi:10.1016/j.giq.2007.12.003

Heeks, R. (2002). E-government in Africa: Promise and practice. *Information Polity, 7,* 97–114.

McNeal, R., Hale, K., & Dotterweich, L. (2008). Citizen-government interaction and the internet: Expectations and accomplishments in contact, quality, and trust. *Journal of Information Technology & Politics, 5*(2), 213–229. doi:10.1080/19331680802298298

Minister of Communications & Information Technology. (2011). Keynote speech by Dr. Magued Osman in the 15th round of Cairo ICT, 25 May 2011.

Morgeson, F., VanAmburg, D., & Mithas, S. (2010). Misplaced trust? Exploring the structure of the e-government-citizen trust relationship. *Journal of Public Administration: Research and Theory, 20,* 1–27.

Pieterson, W., & Ebbers, W. (2008). The use of service channels by citizens in the Netherlands: Implications for multi-channel management. *International Review of Administrative Sciences, 74*(1), 95–110. doi:10.1177/0020852307085736

Reddick, C. (2004). Empirical models of e-government growth in local governments. *E- Service Journal, 3*(2).

Reddick, C. (2005). Citizen interaction with e-government: From the streets to servers? *Government Information Quarterly, 22*(1), 38–57. doi:10.1016/j.giq.2004.10.003

Reddick, C. (2010). Comparing citizens' use of e-government to alternative service channels. *International Journal of Electronic Government Research, 6*(2), 54–67. doi:10.4018/jegr.2010040104

Streib, G., & Navarro, I. (2006). Citizen demand for interactive e-government: The case of Georgia consumer services. *American Review of Public Administration, 36*(3), 288–300. doi:10.1177/0275074005283371

Thomas, J., & Streib, G. (2003). The new face of government: Citizen-initiated contacts in the era of e-government. *Journal of Public Administration: Research and Theory, 13*(1), 83–102. doi:10.1093/jpart/mug010

Titah, R., & Barki, H. (2006). E-government adoption and acceptance: A literature review. *International Journal of Electronic Government Research, 2*(3), 23–57. doi:10.4018/jegr.2006070102

United Nations. (2010). *United Nations global e-government survey.* Retrieved from http://egovernments.wordpress.com/2010/04/15/united-nations-global-e-government-survey-2010/

Van Deursen, A., & van Dijk, J. (2009). Improving digital skills for the use of online public information and services. *Government Information Quarterly, 26*(2), 333–340. doi:10.1016/j.giq.2008.11.002

KEY TERMS AND DEFINITIONS

Citizen Participation: The interaction of citizens within the frame-work of democratic government. This includes the facets of e-government and government to citizen (G2C) communications.

Developing Countries: Countries that have low developed industrial base, low literacy rate, high subsistence agriculture base. This is also in relation to economic development.

Digital Divide: The gap between people who have easy access to web, cell phone, and other information technology and those people who do not have access. This divide can include gender, economic, and literacy factors.

E-Government: Is the use of information technologies by governments and agencies to transform government to citizen (G2C) interactions into viable modes of citizen empowerment and efficient government management.

Gender Gap: The difference in participation by the sexes, in this case, within the political and social activity of emerging e-governance.

Information and Communications Technology (ICT): Technologies that provide access to information through telecommunications and includes internet, smartphone technology, and other communications channels.

Web Portal: An internet website that channels information from different sources; in the case of e-governance, a government managed website clearinghouse that citizens can use to obtain information, forms, and connect with policy makers from different agencies, rather than going to individual agency web sites.

Compilation of References

Abdul Manap, D. (2011). *Statistik E-Filing*. Kuala Lumpur, Malaysia: Lembaga Hasil Dalam Negeri (LHDN).

Abdullah, M. Y., Salman, A., Razak, N. A., Noor, N. F. M., & Malek, I. A. (2011). Issues affecting the use of information and communication technology among the elderly: A case study on Jenii. *IEEE 10th Malaysia International Conference on Communications* (MICC) (pp. 29-32). 2nd - 5th October Sabah, Malaysia.

Abramowitz, A. (2012). *2012 Davis Productivity Eagle Award Winner!!!* Retrieved May 5, 2012 from http://gal2.org/2012/04/abramowitz-on-the-davis-productivity-eagle-award/

Abramowitz, A. (2012). *Statewide Guardian ad Litem program announces overview of 2012 legislative results*. Retrieved May 2, 2012 from http://gal2.org/2012/03/

Abu-Shanab, E. (2012). Digital government adoption in Jordan: An environmental model. *The International Arab Journal of e-Technology, 2*(3), 129-135.

Abu-Shanab, E., Knight, M., & Refai, H. (2010). E-voting systems: A tool for e-democracy. *Management Research and Practice, 2*(3), 264–274.

Accenture. (2003). *E-government leadership: Engaging the customer*. Retrieved from http://www.accenture.com/xdoc/en/newsroom/epresskit/egovernment/egov_epress.pdf

Accenture. (2006). *Leadership in customer service: Building the trust*. Retrieved from http://www.accenture.com/xdoc/en/industries/government/acn_2006_govt_report_FINAL2.pdf

Adeshara, P., Juric, R., Kuljis, J., & Paul, R. (2004). A survey of acceptance of e-government services in the UK. *Journal of Computing and International Technology, 12*(2), 143–150. doi:10.2498/cit.2004.02.10

ADF IV. (2004). *Fourth African Development Forum, Governance for a Progressing Africa*, 11-15th October 2004, Addis Ababa, Ethiopia. Retrieved from http://www.uneca.org/adf/adfiv/adf_4_report_final_sml.pdf

Adida, B. (2008). Helios: Web-based open-audit voting. In *Proceedings of the 17th Conference on Security Symposium*, San Jose, CA (pp. 335–348). Berkeley, CA: USENIX Association.

Adida, B., & Rivest, R. L. (2006). Scratch & vote: Self-contained paper-based cryptographic voting. In *Proceedings of the 5th ACM Workshop on Privacy in the Electronic Society, WPES '06* (pp. 29–40). New York, NY: ACM.

Aelst, P. V., & Walgrave, S. (2004). New media, new movements? The role of the internet in shaping the anti-globalization movement. In W. van de Donk, et al. (Ed.), *Cyberprotest, new media, citizens and social movements*, (pp. 97-122). London, UK: Routledge.

AERCO y Territorio Creativo. (n.d.). *La función del community manager*. Retrieved June 1, 2012, from https://www.box.com/shared/pgur4btexi

Affisco, J. F., Khalid, S., & Soliman, K. (2006). E-government: A strategic operations management framework for service delivery. *Business Process Management Journal, 12*(1), 13–21. doi:10.1108/14637150610643724

Agarwal, A. (Ed.). (2007). *eGovernance case studies*. Hyderabad, India: Universities Press.

Agrawal, P., & Agarwal, M. (2006). E-filing of returns. *The Chartered Accountant*, 1567-1573. Retrieved April 22, 2012, from http://www.icai.org/resource_file/101991567.pdf

Agarwal, R., & Karahanna, E. (2000). Time flies when you're having fun: Cognitive absorption and beliefs about information technology usage. *Management Information Systems Quarterly*, *24*, 665–694. doi:10.2307/3250951

Agarwal, R., & Prasad, J. (1997). The role of innovation characteristics and perceived voluntariness in the acceptance of information technologies. *Decision Sciences*, *28*(93), 557–582. doi:10.1111/j.1540-5915.1997.tb01322.x

Agarwal, R., & Prasad, J. (1998). A conceptual and operational definition of personal innovativeness in the domain of information technology. *Information Systems Research*, *9*(2), 204–215. doi:10.1287/isre.9.2.204

Agarwal, R., & Venkatesh, V. (2002). Assessing a firm's web presence: A heuristic evaluation procedure for the measurement of usability. *Information Systems Research*, *13*, 168–186. doi:10.1287/isre.13.2.168.84

Ajzen, I. (1985). From intentions to actions: A theory of planned behaviour. In Kuhl, J., & Backmann, J. (Eds.), *Action control: From cognition to behaviour* (pp. 11–39). New York, NY: Springer-Verlag. doi:10.1007/978-3-642-69746-3_2

Ajzen, I. (1991). The theory of planned behaviour. *Organizational Behavior and Human Decision Processes*, *50*(2), 179–221. doi:10.1016/0749-5978(91)90020-T

Ajzen, I., & Fishbein, M. (1980). *Understanding attitudes and predicting social behaviour*. Upper Saddle River, NJ: Prentice-Hall.

Ajzen, I., & Madden, T. J. (1986). Predication of goal-directed behaviour: Attitude, intentions, and perceived behavioural control. *Journal of Experimental Social Psychology*, *22*, 453–474. doi:10.1016/0022-1031(86)90045-4

Al-Adawi, Z., Yousafzai, S., & Pallister, J. (2005). Conceptual model of citizen adoption of e-government. *Proceedings of the Second International Conference on Innovations in Information Technology (IIT)*, (pp. 1-10).

Al-Anie, H., Alia, M., & Hnaif, A. (2011). E-voting protocol based on public-key cryptography. *International Journal of Network Security & Its Applications*, *3*(4), 87–98. doi:10.5121/ijnsa.2011.3408

Al-Maskari, A., & Sanderson, M. (2010). A review of factors influencing user satisfaction in information retrieval. *Journal of the American Society for Information Science and Technology*, *61*(5), 859–868. doi:10.1002/asi.21300

Al-Rababah, B., & Abu-Shanab, E. (2010). E-government and gender digital divide: The case of Jordan. *International Journal of Electronic Business Management*, *8*(1), 1–8.

Aladwani, A. M., & Palvia, P. C. (2002). Developing and validating an instrument for measuring user-perceived web quality. *Information & Management*, *39*(6), 467–476. doi:10.1016/S0378-7206(01)00113-6

Alalwan, J. (2010). Can IT resources lead to sustainable competitive advantage. *Proceedings of the Southern Association for Information Systems Conference*, Atlanta, GA, USA March 26th-27th, (pp. 231-236).

Alampay, E. (2002). People's participation, consensus-building and transparency through ICTs: Issues and challenges for governance in the Philippines. *Kasarinlan*, *17*(2), 273–292.

Alampay, E., Heeks, R., & Soliva, P. (2003). *Bridging the information divide: A Philippine guidebook on ICTs for development*. Quezon City, Philippines: NCPAG, IDPM.

Alampay, E. A., Olpoc, J. C., & Hechanova, R. M. (2012). Competing values regarding internet use in "free" Philippine social institutions. In Deibert, R., Palfrey, J., Rohozinski, R., & Zittrain, J. (Eds.), *Access contested: Security, identity and resistance in Asian cyberspace* (pp. 115–132). Cambridge, MA: MIT Press.

Alavi, H. (2008). Trading up: How Tunisia used ICT to facilitate trade. *IFC SmartLessons*, June 2008.

Albers, J. (2009). When marketing merges with learning, customers profit. *T + D, 63*(6), 72–75.

Alghamdi, I. A., Goodwin, R., & Rampersad, G. (2011). E-government readiness assessment for government organizations in developing countries. *Computer and Information Science*, *4*(3), 3–17. doi:10.5539/cis.v4n3p3

Allen, M. (2002). A case study of the usability testing of the University of South Florida's virtual library interface design. *Online Information Review, 26*(1), 40–53. doi:10.1108/14684520210418374

Alshawi, S., & Alalwany, H. (2009). E-government evaluation: Citizen's perspective in developing countries. *Information Technology for Development, 15*(3), 193–208. doi:10.1002/itdj.20125

Alshehri, M., Drew, S., & Alfarraj, O. (2012). A comprehensive analysis of e-government services adoption in Saudi Arabia: Obstacles and challenges. *International Journal of Advanced Computer Science and Applications, 3*(2), 1–6.

Altun, A., & Bilgin, M. (2011). Web based secure e-voting system with fingerprint authentication. *Scientific Research and Essays, 6*(12), 2494–2500.

Aluko, B. T. (2005). Building urban local governance fiscal autonomy through property taxation financing option. *International Journal of Strategic Property Management, 9*(1), 201–214.

Alvarez, R. M., & Hall, T. E. (2008). *Electronic elections: The perils and promises of digital democracy.* Princeton, NJ: Princeton University Press.

Amato, A., Calabrese, M., Di Lecce, V., & Quarto, A. (2008). Multi agent system to promote electronic data interchange in port systems. In *Proceedings of the 21st IEEE Canadian Conference on Electrical and Computer Engineering* (pp. 729-734). IEEE.

Ambali, A. R. (2009). E-government policy: Ground issues in e-filing system. *European Journal of Soil Science, 11*(2), 249–266.

Amberg, M., Markov, R., & Okujava, S. (2005). A framework for valuing the economic profitability of e-government. *Proceedings of International Conference on E-government*, (pp. 31-41).

Anane, R., Freeland, R., & Theodoropoulos, G. (2007). E-voting requirements and implementation. *Proceedings of the 9th IEEE Conference on E-Commerce Technology, CEC '07* Tokyo, Japan, (pp. 382–392).

Annual Report Inland Revenue Board of Malaysia. (2007). Retrieved from http://www.hasil.gov.my/pdf/pdfam/AR2007_2.pdf

Annual Report Inland Revenue Board of Malaysia. (2009). Retrieved from http://www.hasil.gov.my/pdf/pdfam/AR2009_2.pdf

Antonis, K., Lampsas, P., & Prentzas, J. (2008). In Leung, H., Li, F., Lau, R., & Li, Q. (Eds.), *Adult distance learning using a Web-based learning management system: Methodology and results (Vol. 4823*, pp. 508–519). Lecture Notes in Computer ScienceHeidelberg, Germany: Springer.

Anttiroiko, A. (2004). Introduction to democratic e-governance. In Malkia, M., Anttiroiko, A., & Savolainen, R. (Eds.), *E-Transformation in governance: New directions in government and politics* (pp. 22–50). Hershey, PA: Idea Group Publishing.

Anuar, S., & Othman, R. (2010). Determinants of online tax payment system in Malaysia. *International Journal of Public Information System, 1*, 17–32.

Arnaldo, M. S. F. (2012). *Aquino admininistration to spend P63M for CNN ads.* Retrieved May 2, 2012, from http://www.interaksyon.com/article/30764/aquino-admininistration-to-spend-p63m-for-cnn-ads

Arnstein, S. R. (1969). A ladder of citizen participation. *Journal of the American Institute of Planners, 35*(1), 216–224. doi:10.1080/01944366908977225

Aronowitz, S., & Cutler, J. (1997). *Post-work: The wages of cybernation.* New York, NY: Routledge.

Arzt-Mergemeier, J., Beiss, W., & Steffens, T. (2007). The digital voting pen at the Hamburg elections 2008: Electronic voting closest to conventional voting. In Alkassar, A., & Volkamer, M. (Eds.), *VOTE-ID 2007 (Vol. 4896*, pp. 88–98). Lecture Notes in Computer ScienceBerlin, Germany: Springer-Verlag.

Asgarkhan, M. (2005). Digital government and its effectiveness in public management reform. *Public Management Review, 7*(3), 465–487. doi:10.1080/14719030500181227

Asquith, A. (1998). Non-elite employees' perceptions of organizational change in English local government. *International Journal of Public Sector Management, 11*(4), 262–280. doi:10.1108/09513559810225825

Atoev, A., & Duncombe, R. (2011). E-citizen capability development. *Proceedings of the 5th International Conference on Theory and Practice of Electronic Governance,* Estonia.

Avgerou, C., & Walsham, G. (2000). Introduction: IT in developing countries. In C. Avgerou & G. Walsham (Eds.), *Information technology in context: Studies from the perspective of developing countries* (pp. 1-8). Ashgate, UK: Aldershot.

Bagozzi, R. P. (1982). A field investigation of causal relations among cognitions, affects, intentions, and behaviour. *JMR, Journal of Marketing Research*, *19*(4), 562–584. doi:10.2307/3151727

Bahl, R. W., & Linn, J. F. (1992). *Urban public finance in developing countries*. New York, NY: Oxford University Press.

Balka, E., & Wagner, I. (2006). Making things work: Dimensions of configurability as appropriation work. [ACM.]. *Proceedings of CSCW*, *2006*, 229–238. doi:10.1145/1180875.1180912

Bangalore Mahanagara Palike. (2000). *Property tax self-assessment scheme handbook: Golden Jubilee Year 2000*. Bangalore, India: BBMP.

Bangalore Mahanagara Palike. (2007). *Assessment and calculation of property tax under the capital value system (New SAS): 2007- 2008*. Unpublished Handbook.

Banks, B. (2012). Let us recommit to our mission. *The Connection*. Retrieved February 21, 2012, from http://www.casaforchildren.org/site/c.mtJSJ7MPIsE/b.7984779/k.D31D/From_the_President.htm

Bär, M., Henrich, C., Müller-Quade, J., Röhrich, S., & Stüber, C. (2008). *Real world experiences with bingo voting and a comparison of usability*. Paper presented at the Workshop on Trustworthy Elections (WOTE 2008), Leuven, Belgium.

Bardhan, P. (1997). Corruption and development: A review of issues. *Journal of Economic Literature*, *35*, 1320–1346.

Bastida, F. J., & Benito, B. (2006). Financial reports and decentralization in municipal governments. *International Review of Administrative Sciences*, *72*(2), 223–238. doi:10.1177/0020852306064611

Batini, C., Viscusi, G., & Cherubini, D. (2009). GovQual: A quality driven methodology for e-government project planning. *Government Information Quarterly*, *26*, 106–117. doi:10.1016/j.giq.2008.03.002

Bautista, V. A. (2006). Introduction and rationale of the study. In Bautista, V. A., & Alfonso, O. M. (Eds.), *Citizen participation in rural poverty alleviation* (pp. 1–10). Quezon City, Philippines: Center for Leadership, Citizenship and Democracy.

Bea, G., & Matotay, E. (2011). E-government and e-governance. In J. Itika, K. de Ridder, & A. Tollenaar (Eds.), *Theories and stories in African public administration* (pp. 157-168). African Public Administration and Management series, 1, African Studies Centre/University of Groningen/Mzumbe University.

Bederson, B. B., Lee, B., Sherman, R. M., Herrnson, P. S., & Niemi, R. G. (2003). Electronic voting system usability issues. In *Proceedings of the SIGCHI Conference on Human Factors in Computing Systems CHI 2003* (pp. 145 - 152). New York, NY: ACM.

Belanger, F., & Carter, L. (2006). The effects of the digital divide on e-government: An empirical evaluation. *Proceedings of the 39th Annual Hawaii International Conference on System Sciences*. Retrieved April 13, 2011, from http://www.computer.org/portal/web/csdl/doi/10.1109/HICSS.2006.464

Bélanger, F., & Carter, L. (2008). Trust and risk in e-government adoption. *The Journal of Strategic Information Systems*, *17*(2), 165–176. doi:10.1016/j.jsis.2007.12.002

Belanger, F., & Carter, L. (2009). The impact of the digital divide on e-government use. *Communications of the ACM*, *52*(4), 132–135. doi:10.1145/1498765.1498801

Benaloh, J. (2006). Simple verifiable elections. In *Proceedings of the USENIX/Accurate Electronic Voting Technology Workshop 2006 EVT'06*. Berkeley, CA: USENIX Association.

Bensen, C. (n.d.). *Community manager job description*. Retrieved June 1, 2012, from http://conniebensen.com/2008/07/17/community-manager-job-description/

Benston, M. (1990). Participatory designs by non-profit groups. In *Proceedings of the Participatory Design Conference*, Palo Alto CA, 1990 (pp. 107-113).

Berdykhanova, D., Dehghantanha, A., & Hariraj, K. (2010). Trust challenges and issues of e-government: E-tax prospective. *International Symposium of Informational Technology, 2*, 1015-1019.

Berio, G., Harzallah, M., & Sacco, G. M. (2006). Portals for integrated competence management. In Tatnall, A. (Ed.), *Encyclopedia of portal technology and applications*. Idea Group Inc.

Berman, B. J., & Tettey, W. J. (2001). African states, bureaucratic culture and computer fixes. *Public Administration and Development, 21*(1), 1–13. doi:10.1002/pad.166

BERNAMA, The Malaysian National News Agency. (12/5/1999). *E-government a new way to serve people says Dr. Mahathir*. Retrieved from http://www.access-mylibrary.com/article-IGI-54639942/e-government-new-way.html

Berners-Lee, T., Hendler, J., & Lassila, O. (2001, May 17). The Semantic Web. *Scientific American*, 35–43.

Bertot, J. C., & Jaeger, P. T. (2006). User-centered e-government: Challenges and benefits for government Web sites. *Government Information Quarterly, 23*(2), 163–168. doi:10.1016/j.giq.2006.02.001

Bertot, J. C., Jaeger, P. T., & Grimes, J. M. (2010). Crowdsourcing transparency: ICTs, social media, and government transparency initiatives. In *Proceedings of the 11th Annual International Conference on Digital Government Research*, Puebla, Mexico, May 17–20.

Bertot, J. C., Jaeger, P. T., & Grimes, J. M. (2010). Using ICTs to create a culture of transparency: E-government and social media as openness and anti-corruption tools for societies. *Government Information Quarterly, 27*(3), 264–271. doi:10.1016/j.giq.2010.03.001

Bertot, J., Jaeger, P., & McClure, C. (2008). Citizen-centered e-government services: Benefits, costs, and research needs. *Proceedings of the 2008 International Conference on Digital Government Research*, Vol. 289, (pp. 137-142).

Bertot, J. C., Jaeger, P. T., Munson, S., & Glaisyer, T. (2010). Engaging the public in open government: Social media technology and policy for government transparency. *U.S. National Science Foundation Workshop Draft Report*, (pp. 1-18).

Bertot, J. C., Jaeger, P. T., Munson, S., & Glaisyer, T. (2010). Social media technology and government transparency. *Computer, 43*(11), 53–59. doi:10.1109/MC.2010.325

Beynon-Davies, P. (2005). Constructing electronic government: The case of the UK inland revenue. *International Journal of Information Management, 25*(1), 3–20. doi:10.1016/j.ijinfomgt.2004.08.002

Bhagat, R. B. (2005). Rural-urban classification and municipal governance in India. *Singapore Journal of Tropical Geography, 26*(1), 61–73. doi:10.1111/j.0129-7619.2005.00204.x

Bhagwan, J. (1983). *Municipal finance in the metropolitan cities of India: A case study of Delhi Municipal Corporation*. New Delhi, India: Concept Publishing.

Bhatia, D., Bhatnagar, S., & Tominaga, J. (2009). *How do manual and e-government services compare? Experiences from India. Information and Communications for Development* (pp. 67–82). World Bank Publications.

Bhatnagar, S. (2003). E-government: Building a SMART administration for India's states. In Howes, S., Lahiri, A., & Stern, N. (Eds.), *State-level reform in India: Towards more effective government* (pp. 257–267). New Delhi, India: Macmillan India Ltd.

Bhatnagar, S. (2003). Public service delivery: Does e-government help? In S. Ahmed & S. Bery (Eds.), *The Annual Bank Conference on Development Economics 2003* (pp. 11-20). New Delhi, India: The World Bank and National Conference of Applied Economic Research.

Bhatnagar, S. (2003). Role of government: As an enabler, regulator, and provider of ICT based services. Asian Forum on ICT Policies and e-Strategies, Asia-Pacific Development Information Programme, United Nations Development Programme.

Bhatnagar, S. (2003). *The economic and social impact of e-government*. Background technical paper for E-government, the Citizen and the State: Debating Governance in the Information Age, the proposed UNDESA publication (World Public Sector Report for 2003).

Bhatnagar, S. (2003). *Transparency and corruption: Does e-government help?* Draft paper for the compilation of the Commonwealth Human Rights Initiative 2003 Report 'Open Sesame: Looking for the Right to Information in the Commonwealth.

Bhatnagar, S. (2004). *E-government: From vision to implementation*. New Delhi, India: Sage Publications.

Bhatnagar, S. (2005). *E-government: Opportunities and challenges.* World Bank Presentation. Retrieved from http://siteresources.worldbank.org/INTEDEVELOPMENT/Resources/559323-1114798035525/1055531-1114798256329/10555556-1114798371392/Bhatnagar1.ppt

Bhatnagar, S. (2009). *Unlocking e-government potential: Concepts, cases and practical insights* (1st ed.). SAGE Publications India Pvt Ltd.

Bhattacharya, D., Gulla, U., & Gupta, M. P. (2012). E-service quality model for Indian government portals: Citizens' perspective. *Journal of Enterprise Information Management, 25*(3), 246–271. doi:10.1108/17410391211224408

Bhattacherjee, A. (2001). Understanding information systems continuance: An expectation-confirmation model. *Management Information Systems Quarterly, 25*(3), 351–370. doi:10.2307/3250921

Bhattacherjee, A., & Premkumar, G. (2004). Understanding changes in belief and attitude toward information technology usage: A theoretical model and longitudinal test. *Management Information Systems Quarterly, 28*(2), 229–254.

Bismark, D., Heather, J., Peel, R. M. A., Ryan, P. Y. A., Schneider, S., & Xia, Z. (2009). Experiences gained from the first prêt à voter implementation. *First International Workshop on Requirements Engineering for E-voting Systems RE-VOTE '09* (pp. 19-28) Washington, DC: IEEE Computer Society.

Blackmore, P. (2010). *Intranets: A guide to their design, implementation and management.* New York, NY: Routledge.

Blair, D. C., & Maron, M. E. (1985). An evaluation of retrieval effectiveness for a full-text document-retrieval system. *Communications of the ACM, 28*(3), 289–299. doi:10.1145/3166.3197

Boakye, K. G., Prybutok, V. R., & Ryan, S. D. (2012). The intention of continued web-enabled phone service usage: A quality perspective. *Operations Management Research, 5,* 14–24. doi:10.1007/s12063-012-0062-1

Bohli, J., Müller-Quade, J., & Röhrich, S. (2007). Bingo voting: Secure and coercion-free voting using a trusted random number generator. In A. Alkassar & M. Volkamer (Eds.), *Proceedings of the 1st International Conference on E-voting and Identity* (pp. 111-124). Berlin, Germany: Springer-Verlag.

Bonsón, E., Torres, L., Royo, S., & Flores, F. (2012). Local e-government 2.0: Social media and corporate transparency in municipalities. *Government Information Quarterly, 29*(2), 123–132. doi:10.1016/j.giq.2011.10.001

Bostrom, R., & Heinen, J. (1977). MIS problems and failures: A socio-technical perspective part I: The causes. *Management Information Systems Quarterly, 1*(3), 17–32. doi:10.2307/248710

Boumil, M., Freitas, C., & Freitas, D. (2011). Legal and ethical issues confronting Guardian ad Litem practice. *Journal of Law & Family Studies, 13,* 43–80.

Bouras, C., Destounis, P., Garofalakis, J., Triantafillou, V., Tzimas, G., & Zarafidis, P. (1999). A co-operative environment for local government: an Internet-Intranet approach. *Telematics and Informatics, 16,* 75–89. doi:10.1016/S0736-5853(99)00020-9

Bouras, C., Katris, N., & Triantafillou, V. (2003). An electronic voting service to support decision-making in local government. *Telematics and Informatics, 20,* 255–274. doi:10.1016/S0736-5853(03)00017-0

Boyd, d. m., & Ellison, N. B. (2008). Social network sites: Definition, history, and scholarship. *Journal of Computer-Mediated Communication, 13*(1), 210-230.

Bozinis, A., & Lakovou, E. (2005). Electronic democratic governance: Problem, challenge and best practice. *Journal of Information Technology Impact, 5*(2), 73–80.

Brady, L., & Phillips, C. (2003). Aesthetics and usability: A look at color and balance. *Usability News,* Vol. 5.1. Retrieved from http://psychology.wichita.edu/surl/usabilitynews/51/aesthetics.htm

Bresciani, P., Donzelli, P., & Forte, A. (2003). Requirements engineering for knowledge management in egovernment. *Lecture Notes in Artificial Intelligence, 2645,* 48–59.

Brooke, J. (1996). SUS: A 'quick and dirty' usability scale. In Jordan, P. W., Thomas, B., Weerdmeester, B. A., & McClelland, A. L. (Eds.), *Usability evaluation in industry.* London, UK: Taylor and Francis.

Brudney, J. L. (1999). The effective use of volunteers: Best practices for the public sector. *Law and Contemporary Problems*, *62*(4), 219–255. doi:10.2307/1192274

Brudney, J. L., & Kellough, J. E. (2000). Volunteers in state government: Involvement, management, and benefits. *Nonprofit and Voluntary Sector Quarterly*, *29*(1), 111–130. doi:10.1177/0899764000291007

Brussell, J. (2012). *Corruption and reforms in India: Public service ii the digital age.* Cambridge University Press. doi:10.1017/CBO9781139094023

Bryer, T., & Zavattaro, S. M. (2011). Social media and public administration: Theoretical dimensions and introduction to the symposium. *Administrative Theory & Praxis*, *33*(3), 325–340. doi:10.2753/ATP1084-1806330301

Brynjolfsson, E., & Smith, M. (2000). Frictionless commerce? A comparison of Internet conventional retailers. *Management Science*, *46*, 563–585. doi:10.1287/mnsc.46.4.563.12061

Buchsbaum, T. (2005). E-voting: Lessons learnt from recent pilots. *International Conference on Electronic Voting and Electronic Democracy: Present and the Future*, Seoul, Korea, March 2005, (pp. 1-22).

Buckley, J. J. (1984). The multiple judge, multiple criteria ranking problem: A fuzzy set approach. *Fuzzy Sets and Systems*, *13*, 25–37. doi:10.1016/0165-0114(84)90024-1

Buckley, J. J. (1985). Fuzzy hierarchical analysis. *Fuzzy Sets and Systems*, *17*, 233–247. doi:10.1016/0165-0114(85)90090-9

Buckley, J. J. (1985). Ranking alternatives using fuzzy numbers. *Fuzzy Sets and Systems*, *15*(1), 21–31. doi:10.1016/0165-0114(85)90013-2

Budhiraja, R. (2003). *Electronic governance: A key issue in the 21st century.* Electronic Governance Division, Ministry of Information Technology, Government of India. Retrieved from http://www.mit.gov.in/eg/article2.htm

Bui, T., Sankaran, S., & Sebastian, I. (2003). A framework for measuring national e-readiness. *International Journal of Electronic Business*, *1*(1), 3–22. doi:10.1504/IJEB.2003.002162

Bureau of Labor Statistics. (2012). *Volunteering in the United States— 2011.* Retrieved from http://www.bls.gov/news.release/volun.nr0.htm

Burmeister, O. K. (2000). Usability testing: Revisiting informed consent procedures for testing internet sites. In J. Weckert (Ed.) *Selected Papers from the Second Australian Institute Conference on Computer Ethics CRPIT '00* (pp. 3-9). Darlinghurst, Australia: Australian Computer Society, Inc.

Burne, J. (2002). *Better public services through e-government.* London, UK: The National Audit Office.

Bwalya, K. (2009). Factors affecting adoption of e-government in ZAMBIA. *Electronic Journal of Information Systems in Developing Countries*, *38*(4), 1–13.

Byrne, M. D., Greene, K. K., & Everett, S. P. (2007). Usability of voting systems: Baseline data for paper, punch cards, and lever machines. In *Proceedings of the SIGCHI conference on Human Factors in Computing Systems CHI 2007* (pp. 171–180). New York, NY: ACM.

Caba Pérez, C., Rodríguez Bolívar, M. P., & López Hernández, A. M. (2012). The use of Web 2.0 to transform public services delivery: The case of Spain. In Reddick, C. (Ed.), *Web 2.0 technologies and democratic governance* (pp. 41–61). New York, NY: Springer. doi:10.1007/978-1-4614-1448-3_4

Calero, C., Ruiz, J., & Piattini, M. (2005). Classifying Web metrics using the Web quality model. *Online Information Review*, *29*(3), 227–248. doi:10.1108/14684520510607560

Cammaerts, B., & Van Audenhove, L. (2003). *ICT-usage among transnational social movements in the networked society: To organize, to mediate & to influence.* Amsterdam, The Netherlands: ASCoR, Amsterdam Free University.

Campbell, B. A., & Byrne, M. D. (2009). Now do voters notice review screen anomalies? A look at voting system usability. In *Proceedings of the 2009 Conference on Electronic Voting Technology/Workshop on Trustworthy Elections EVT/WOTE'09.* Berkeley, CA: USENIX Association.

Campbell, B. A., & Byrne, M. D. (2009). Straight-party voting: What do voters think? *IEEE Transactions on Information Forensics and Security*, *4*(4), 718–728. doi:10.1109/TIFS.2009.2031947

Cao, M., Zhang, Q., & Seydel, J. (2005). B2C e-commerce Web site quality: An empirical examination. *Industrial Management & Data Systems*, *105*(5), 645–661. doi:10.1108/02635570510600000

Cappel, J. J., & Huang, Z. (2007). A usability analysis of company websites. *Journal of Computer Information Systems*, *48*(1), 117.

Carback, R., Chaum, D., Clark, J., Conway, J., Essex, A., & Herrnson, P. S. … Vora, P.L. (2010). Scantegrity II municipal election at Takoma park: The first E2E binding governmental election with ballot privacy. In *Proceedings of the 19th USENIX Conference on Security USENIX Security'10*. Berkeley, CA: USENIX Association.

Cardoso, G., & Neto, P. P. (2004). Mass media driven mobilization and online protest ICTs and the pro-East Timor movement in Portugal. In Van de Donk, W. (Eds.), *Cyber protest, new media, citizens and social movements* (pp. 147–163). London, UK: Routledge.

Card, S., Moran, T., & Newell, A. (1983). *The psychology of human-computer interaction*. Hillsdale, NJ: Lawrence Erlbaum Associates.

Carter, L., & Belanger, F. (2004). Citizen adoption of electronic government initiatives. *Proceedings of 37ᵗʰ Annual Hawaii International Conference on System Sciences*, Big Island, Hawaii, (pp. 5–8).

Carter, L., & Belanger, F. (2004). The influence of perceived characteristics of innovating on e-government adoption. *Electronic. Journal of E-Government*, *2*(1), 11–20.

Carter, L., & Belanger, F. (2005). The utilization of e-government services: Citizen trust, innovation and acceptance factors. *Information Systems Journal*, *15*(1), 5–25. doi:10.1111/j.1365-2575.2005.00183.x

Carter, L., & Campbell, R. (2011). The impact of trust and relative advantage on internet voting diffusion. *Journal of Theoretical and Applied Electronic Commerce Research*, *6*(3), 28–42. doi:10.4067/S0718-18762011000300004

Carter, L., & Weerakkody, V. (2008). E-government adoption: A cultural comparison. *Information Systems Frontiers*, *10*(4), 473–482. doi:10.1007/s10796-008-9103-6

Casa for Children. (2012). *About us*. Retrieved February 21, 2012, from http://www.casaforchildren.org/site/c.mtJSJ7MPIsE/b.5301303/k.6FB1/About_Us__CASA_for_Children.htm

Casselberry, R. (1996). *Running a perfect intranet*. Que Publications.

Castellanos, D. (n.d.). *El perfil del community manager: Funciones, herramientas, salario y actitudes*. Retrieved June 5, 2012, from http://www.inca-trade.com/blog/marketing-en-redes-sociales-2-0/el-perfil-del-community-manager-funciones-herramientas-salario/

Castells, M. (1998). *End of millennium, the information age: Economy, society and culture (Vol. III)*. Oxford, UK: Blackwell.

Castells, M. (2009). *Communication and power*. Oxford, UK: Oxford University Press.

Cavallo, V. (2009). *The win win eparticipatione-participation model*. LNCS Journals, Academic Books & Online Media | SpringerElectronic Participation First International Conference, ePart 2009 Linz, Austria, August 31–September 4, 2009.

Cavallo, V. (2010). *eParticipatione-Participation and the theory of games*. Unpublished doctoral dissertation, IULM University, Milan.

Cavalluzzo, K. S., & Ittner, C. D. (2004). Implementing performance measurement innovations: Evidence from government. *Accounting, Organizations and Society*, *29*(2-3), 243–267. doi:10.1016/S0361-3682(03)00013-8

Cavoukian, A. (2009). *Privacy and government 2.0: The implications of an open world*. Ontario, Canada: Information & Privacy Commissioner.

Center for American Politics and Citizenship (CAPC). (2006). *A study of vote verification technology conducted for the Maryland state board of elections. Part II: usability study*. Retrieved August 3, 2012, from http://www.cs.umd.edu/~bederson/voting/verification-study-jan-2006.pdf

Center for Technology in Government. (2009). *Exploratory social media project*. The Research Foundation of State University of New York, 2009.

Centre for Policy Research. (2001). *The future of urbanisation: Spread and shape in selected states*. New Delhi, India: Centre for Policy Research.

Cetinkaya, O., & Cetinkaya, D. (2007). Verification and validation issues in electronic voting. *Electronic Journal of E-Government*, *5*(2), 117–126.

Chadwick, A., & May, C. (2003). Interaction between states and citizens in the age of the Internet: "e-government" in the United States, Britain, and the European Union. *Governance: An International Journal of Policy, Administration and Institutions, 16*(2), 271–300. doi:10.1111/1468-0491.00216

Chae, B., Paradice, D., Courtney, J. F., & Cagle, C. J. (2005). Incorporating an ethical perspective into problem formulation: Implications for decision support system design. *Decision Support Systems, 40*(2), 197–212. doi:10.1016/j.dss.2004.02.002

Chai, S., Herath, T. C., Park, I., & Rao, H. R. (2008). Repeated use of e-government websites: A satisfaction and confidentiality perspective. In Norris, D. (Ed.), *E-government research: Policy and management* (pp. 158–182).

Chan, H. S., & Chow, K. W. (2007). Public management policy and practice in Western China: Metapolicy, tacit knowledge, and implications for management innovation transfer. *American Review of Public Administration, 37*(4), 479–497. doi:10.1177/0275074006297552

Chan, M., & Chung, W. (2002). A framework to develop an enterprise information portal for contract manufacturing. *International Journal of Production Economics, 75*(1), 113–126. doi:10.1016/S0925-5273(01)00185-2

Chan, S., & Lu, M. (2004). Understanding internet banking adoption and use behaviour: A Hong Kong perspective. *Journal of Global Information Management, 12*(3), 21–43. doi:10.4018/jgim.2004070102

Chander, S., & Kush, A. (2012). E-governance web portals assessment of two states. *International Journal of Advanced Research in Computer Science and Software Engineering, 2*(2).

Chang, D. Y. (1992).Extent analysis and synthetic decision, Optimisation techniques and applications. *World Scientific, 1,* 352-355.

Chang, D. Y. (1996). Application of extent analysis method to fuzzy AHP. *European Journal of Operational Research, 95*(3), 649–655. doi:10.1016/0377-2217(95)00300-2

Chang, J. C., & King, W. R. (2005). Measuring the performance of the information system: A functional scorecard. *Journal of Management Information Systems, 22*(1), 85–115.

Chaplinksy v. New Hampshire. (March 9, 1942). 315 U.S. 568; 62 S. Ct. 766; 86 L. Ed. 1031; 1942 U.S. Lexis 851.

Chau, P. Y. K. (1996). An empirical assessment of a modified technology acceptance model. *Journal of Management Information Systems, 13*(2), 185–204.

Chaum, D. (1988). Elections with unconditionally secret ballots and disruption equivalent to breaking RSA. In Guenther, C. G. (Ed.), *Advances in Cryptology - EUROCRYPT '88, LNCS 330* (pp. 177–182). doi:10.1007/3-540-45961-8_15

Cheema, G. S. (2005). *Building democratic institutions: Governance Reform in developing countries.* Bloomfield, CT: Kumarian Press, Inc.

Chen, B., Wang, H., Proctor, R. W., & Salvendy, G. (1997). A human-centered approach for designed world-wide-Web browsers. *Behavior Research Methods, Instruments, & Computers, 29,* 172–179. doi:10.3758/BF03204806

Chen, Y., Chen, H., Cling, R., & Huang, W. (2007). Electronic government implementation: A comparison between developed and developing countries. *International Journal of Electronic Government Research, 3*(2), 45–61. doi:10.4018/jegr.2007040103

Chen, Y., Chen, H., Huang, W., & Ching, R. (2006). E-government strategies in developed and developing countries: An implementation framework and case study. *Journal of Global Information Management, 14*(1), 23–46. doi:10.4018/jgim.2006010102

Cheng, T. C. E., David, Y. C. L., & Yeung, A. C. L. (2006). Adoption of internet banking: An empirical study in Hong Kong. *Decision Support Systems, 42*(3), 1558–1572. doi:10.1016/j.dss.2006.01.002

Cheng, Y. L. (2009). On the professional competence-building of management personnel in urban community neighborhood committees. *Proceedings of the 2009 International Conference on Public Economics and Management (ICPEM), Vol. 4: Econometrics,* (pp. 433-437).

Cheta, R. (2004). Dis@bled people, ICTs and a new age of activism: A Portuguese accessibility special interest group study. In van de Donk, W. (Eds.), *Cyber protest, new media, citizens and social movements* (pp. 207–232). London, UK: Routledge.

Cheung, C. M. K., & Lee, M. K. O. (2005). The asymmetric impact of Web site attribute performance on user satisfaction: An empirical study. *Proceedings of the 38th Annual Hawaii International Conference on System Sciences* (HICSS-38), Big Island, HI, 3-6 January (CD Rom).

Cheung, C. M. K., Zhu, L., Kwong, T., Chan, G. W. W., & Limayem, M. (2003). Online consumer behaviour: A review and agenda for future research. *16ᵗʰ Bled eCommerce Conference,* Bled, Slovenia, June 9-11, (pp. 194-218).

Chevallier, M. (2003). *Internet voting: Status, perspectives and issues.* Geneva State, Internet Voting Project. Retrieved from www.itu.int/itudoc/itu-t/workshop/e.../e-gov010.pdf

Chi, E. H., Czerwinski, M., Millen, D., Randall, D., Stevens, G., Wulf, V., & Zimmermann, J. (2011). Transferability of research findings: context-dependent or model-driven. In *Proceedings of the 2011 Annual Conference: Extended Abstracts on Human Factors in Computing Systems* (CHI 'EA 11), (pp. 651-654). New York, NY: ACM.

Chiang, L. (2009). Trust and security in the e-voting system. *Electronic Government: An International Journal,* 6(4), 343–360. doi:10.1504/EG.2009.027782

Chidambaram, N. (2008). *Evaluating electronic service quality and user satisfaction: Malaysian income tax e-filing system.* Unpublished MBA Dissertation, University Science Malaysia.

Chin, J. P., Diehl, V. A., & Norman, K. L. (1988). Development of an instrument measuring user satisfaction of the human-computer interface. In *Proceedings of the SIGCHI Conference on Human Factors in Computing Systems CHI '88* (pp. 213-218). New York, NY: ACM.

Chin, W. W., & Gopal, A. (1995). Adoption intention in GSS: Importance of beliefs. *Data Base Advances, 26,* 42–64. doi:10.1145/217278.217285

Chisnell, D., Bachmann, K., Laskwoski, S., & Lowry, S. (2009). Usability for poll workers: A voting system usability test protocol. In Jacko, J. A. (Ed.), *Human Computer Interaction, Part IV, HCII 2009* (Vol. 5613, pp. 458–467). Lecture Notes in Computer Science Berlin, Germany: Springer-Verlag. doi:10.1007/978-3-642-02583-9_50

Chisnell, D., Becker, S., Laskowski, S., & Lowry, S. (2009). Style guide for voting system documentation: Why user-centered documentation matters to voting security. In *Proceedings of the 2009 Conference on Electronic Voting Technology/Workshop on Trustworthy Elections (EVT/WOTE 2009).* Berkeley, CA: USENIX Association.

Cho, V., Cheng, T. C. E., & Lai, W. M. J. (2009). The role of perceived user-interface design in continued usage intention of self-paced e-learning tools. *Computers & Education, 53*(2), 216–227. doi:10.1016/j.compedu.2009.01.014

Chowdhury, S., Landoni, M., & Gibb, F. (2006). Usability and impact of digital libraries: A review. *Online Information Review, 30*(6), 656–680. doi:10.1108/14684520610716153

Christiaens, J. R. (1999). Financial accounting reform in Flemish municipalities: An empirical investigation. *Financial Accountability & Management, 15*(1), 21–40. doi:10.1111/1468-0408.00072

Cil, I., Alpturk, O., & Yazgan, H. R. (2005). A new collaborative system framework based on a multiple perspective approach: Inteli team. *Decision Support Systems, 39*(4), 619–641. doi:10.1016/j.dss.2004.03.007

City of Aurora. (2012). *City of Aurora, IL info.* Retrieved January 20, 2012, from http://www.facebook.com/cityofaurorail#!/cityofaurorail?sk=info

City of Aurora Police Department. (2012). *City of Aurora Police Department Facebook page.* Retrieved January 20, 2012, from http://www.facebook.com/AuroraPolice?sk=notes

City of Chicago. (2011). *Connect with the City of Chicago vis social media.* Retrieved November 15, 2011, from htt://www.cityofchicago.org/content/city/en/narr/misc/social_media.html

City of Collinsville. (March 1, 2010). *City of Collinsville notes.* Retrieved January 15, 2012, from http://www.facebook.com/cityofcollinsville#!/cityofcollinsville?sk=notes

City of Rock Island. (February 10, 2010). *City of Rock Island notes.* Retrieved January 10, 2012, from http://www.facebook.com/rockislandil?sk=info#!/rockislandil?sk=notes

City of Sterling. (November 10, 2010). *City of Sterling notes*. Retrieved January 14, 2012, from http://www.facebook.com/sterling.il?sk=notes

Claassen, R. L., Magleby, D. B., Monson, J. Q., & Patterson, K. D. (2008). At your service: Voter evaluations of poll worker performance. *American Politics Research, 36*, 612. doi:10.1177/1532673X08319006

Clay, G. W. (2001). E-government in the Asia-Pacific region. *Asian Journal of Political Science, 9*(2), 1–26.

Cohen, J. D., & Fischer, M. J. (1985). A robust and verifiable cryptographically secure election scheme. In *Proceedings of the 26th Annual Symposium on Foundations of Computer Science SFCS '85* (pp. 372-382). Washington, DC: IEEE Computer Society.

Colby, S.-S. (2001). *Anti-corruption and ICT for good governance*. Deputy Secretary-General, OECD in Anti-Corruption Symposium 2001: The Role of Online Procedures in Promoting and Good Governance.

Coleman, S. (2006). *African e-governance – Opportunities and challenges*. Oxford, UK: Oxford University Press.

Collin, F., & Restrepo, E. (2001). *Community radio handbook*. Paris, France: UNESCO.

Commonwealth Centre for E-Governance. (2002). *E-government, e-governance and e-democracy: A background discussion paper*. International Tracking Survey Report, no. 1.

Conrad, F. G., Bederson, B. B., Lewis, B., Peytcheva, E., Traugott, M. W., & Hanmer, M. J. (2009). Electronic voting eliminates hanging chads but introduces new usability challenges. *International Journal of Human-Computer Studies, 67*(1), 111–124. doi:10.1016/j.ijhcs.2008.09.010

Cook, M. A. (2000). *What citizens want from e-government: Current practice research*. Center for Technology in Government-University at Albany/SUNY.

Cook, T. D., & Campbell, D. T. (1979). *Quasi-experimentation: Design and analysis issues for field settings*. Boston, MA: Houghton Mifflin Company.

Cornwell, B., Curry, T. J., & Schwirian, K. P. (2003). Revisiting Norton long's ecology of games: A network approach. *City & Community, 2*(2), 121–142. doi:10.1111/1540-6040.00044

Courtney, J. F. (2001). Decision-making and knowledge management in inquiring organisations: towards a new decision-making paradigm for DSS. *Decision Support Systems, 31*, 17–38. doi:10.1016/S0167-9236(00)00117-2

Crabtree, A., O'Brien, J., Nichols, D., Rouncefield, M., & Twidale, M. (2000). Ethnomethodologically informed ethnography and information system design. *Journal of the American Society for Information Science American Society for Information Science, 51*(7), 666–682. doi:10.1002/(SICI)1097-4571(2000)51:7<666::AID-ASI8>3.0.CO;2-5

Cross, E. V. II, McMillian, Y., Gupta, P., Williams, P., Nobles, K., & Gilbert, J. E. (2007). Prime III: A user centered voting system. In *CHI '07 Extended Abstracts on Human Factors in Computing Systems CHI EA '07*. New York, NY: ACM. doi:10.1145/1240866.1241006

Crozier, M., & Friedberg, E. (1980). *Actors and systems*. Chicago, IL: University of Chicago Press.

Cruz, X. (2012). *It's more fun in the Philippines crowd sourced marketing*. Retrieved May 25, 2012, from http://www.creativeguerrillamarketing.com/viral-marketing/its-more-fun-in-the-philippines-crowdsourced-marketing/

Currion, P. (2006). *NGO information technology and requirements assessment report*. Emergency Capacity Building Project Obtained through the Internet. Retrieved from http://www.ecbproject.org/publications_4.htm

Curtin, G., Sommer, M., & Vis-Sommer, V. (Eds.). (2003). *The world of e-government*. New York, NY: Hayworth Press.

Cyr, D., Kindra, G., & Dash, S. (2008). Website design, trust, satisfaction, and e-loyalty: The Indian experience. *Online Information Review, 32*(6), 773–790. doi:10.1108/14684520810923935

Dada, D. (2006). The failure of e-government in developing countries: A literature review. *The Electronic Journal on Information Systems in Developing Countries, 26*(7), 1–10.

Daft, R. L., & Lengel, R. H. (1986). Organizational information requirements, media richness, and structural design. *Management Science, 32*(5), 554–571. doi:10.1287/mnsc.32.5.554

Datta, A. (1984). *Municipal finances in India*. New Delhi, India: Indian Institute of Public Administration.

Datta, A. (1999). Institutional aspects of urban governance in India. In Jha, S. N., & Mathur, P. C. (Eds.), *Decentralization and local politics* (pp. 191–211). New Delhi, India: Sage Publications.

Davidson, F. (2005). *Information and organizations/ use of indicators-governance syllabus*. Rotterdam, The Netherlands: IHS.

Davies, J., Janowski, T., Ojo, A., & Shukla, A. (2007). Technological foundations of electronic governance. *International Conference on E-Government*, December 10-13, Macao, (pp. 5-11).

Davis, F. D. (1986). *A technology acceptance model for empirically testing new end-user information systems: Theory and results*. Doctoral dissertation, Sloan school of management, Massachusetts Institute of Technology.

Davis, F. D. (1989). Perceived usefulness, perceived ease of use, and user acceptance of information technology. *Management Information Systems Quarterly*, *13*(3), 319–339. doi:10.2307/249008

Davis, F. D., Bagozzi, R. P., & Warshaw, P. R. (1989). User acceptance of computer technology: A comparison of two theoretical models. *Management Science*, *35*(8), 982–1003. doi:10.1287/mnsc.35.8.982

Dawes, S. (2008). The evolution and continuing challenges of e-governance. *Public Administration Review*, *68*(6), S86–S102. doi:10.1111/j.1540-6210.2008.00981.x

De, R. (2006). Evaluation of e-government systems: Project assessment vs. development assessment. In Wimmer, M. A. (Eds.), *EGOV 2006, LNCS 4084* (pp. 317–328). Berlin, Germany: Springer-Verlag. doi:10.1007/11823100_28

De, R. (2007). *Antecedents of corruption and the role of e-government systems in developing countries*. Paper presented at the Electronic Government 6th International Conference, EGOV 2007, Ongoing Research, Regensburg, Germany, September 3-7, 2007.

De Jong, M., van Hoof, J., & Gosselt, J. (2008). Voters' perceptions of voting technology: Paper ballots versus voting machine with and without paper audit trail. *Social Science Computer Review*, *26*(4), 399–410. doi:10.1177/0894439307312482

de Sardan, J. O. (2006). *Anthropology and development: Understanding contemporary social change*. London, UK: Zed Books.

Deep, K. (2011). Various authentication techniques for security enhancement. *International Journal of Computer Science & Communication Networks*, *1*(2), 176–185.

Demirel, D., Frankland, R., & Volkamer, M. (2011). *Readiness of various eVoting systems for complex elections*. Technical Report, Technische Universität Darmstadt.

Denton, K., & Richardson, P. (2012). Using intranets to reduce information overload. *Journal of Strategic Innovation and Sustainability*, *7*(3), 84–94.

DesignPinoy. (2012). *It's more fun in the Philippines-Filipino tourism*. Retrieved April 24, 2012, from http://designpinoy.com/its-more-fun-in-the-philippines-filipino-tourism/

Devaraj, S., & Kohli, R. (2003). Performance impacts of information technology: Is actual usage the missing link? *Management Science*, *49*(3), 273–289. doi:10.1287/mnsc.49.3.273.12736

Dexter, F., Ledolter, J., & Wachtel, R. E. (2005). Tactical decision making for selective expansion of operating room resources incorporating financial criteria and uncertainty in subspecialties' future workloads. *Anesthesiology and Analgesics*, *100*, 1425–1432. doi:10.1213/01.ANE.0000149898.45044.3D

Dimitrova, D., & Chen, Y.-C. (2006). Profiling the adopters of e-government information and services: The influences of psychological characteristics, civic mindedness, and information channels. *Social Science Computer Review*, *24*(2), 172–188. doi:10.1177/0894439305281517

Donal, J. (2009). *Mobile Money Summit 2009. Accelerating the Development of Mobile Money Ecosystems*. IFC and the Harvard Kennedy School.

Dossani, R., Jhaveri, R., & Misra, D. (2005). *Enabling ICT for rural India*, (pp. 1-75). Asia-Pacific Research Center, Stanford University and National Informatics Center, Government of India. Retrieved from http://iis-db.stanford.edu/pubs/20972/ICT_full_Oct05.pdf

Dourish, P. (2003). The appropriation of interactive technologies: Some lessons from placeless documents. *International Journal of Computer Supported Cooperative Work*, *12*(4), 465–490. doi:10.1023/A:1026149119426

Dubin, R. (1978). *Theory building*. New York, NY: The Free Press.

Dunleavy, P., & Margetts, H. (2000). *The advent of digital government: Public bureaucracies and the state in the information age*. Paper to the Annual Conference of the American Political Science Association, September 2000.

Dunleavy, P., Margetts, H., Bastow, S., & Tinkler, J. (2006). New public management is dead - Long live digital-era governance. *Journal of Public Administration: Research and Theory*, *16*, 467–494. doi:10.1093/jopart/mui057

Dutton, W. H. (1992). The ecology of games shaping telecommunications policy. *Communication Theory*, *2*(4), 303–324. doi:10.1111/j.1468-2885.1992.tb00046.x

Dutton, W. H. (1996). *Information and communication technologies: Visions and realities*. London, UK: Oxford University Press.

Dutton, W. H. (1999). *Society on the line: Information politics in the digital age*. Oxford, UK

Dutton, W. H., & Guthrie, K. (1991). An ecology of games: The political construction of Santa Monica's public electronic network. *Informatization and the Public Sector*, *1*(4), 279–301.

Eason, K. D. (1988). *Information technology and organizational change*. London, UK: Taylor & Francis.

Economist. (2008). *The good, the bad and the inevitable: The pros and cons of e-government*. Retrieved from http://www.economist.com/node/10638105?story_id=10638105

Edmiston, K. D. (2003). State and local e-government: Prospects and challenges. *American Review of Public Administration*, *33*(1), 20–45. doi:10.1177/0275074002250255

Edong, K. (2007). *Radio Browsing*. Technobiography Blog. Retrieved from http://www.technobiography.com/edongs-dreams/radio-browsing/

Edwards, A. (2004). The Dutch women's movement online Internet and the organizational infrastructure of a social movement. In van de Donk, W. (Eds.), *Cyber protest, new media, citizens and social movements* (pp. 183–206). London, UK: Routledge.

eGovernment Action Plan. (2006). *eGovernment action plan 2006, COM 2006/173 of 25.04.2006*. Retrieved June 10, 2012, from http://ec.europa.eu/information_society/activities/egovernment/library/index_en.htm

Election Assistance Commission (EAC). (2005). *Voluntary voting systems guidelines volumes 1 and 2*. Retrieved April 16, 2011 from, http://www.eac.gov/testing_and_certification/voluntary_voting_system_guidelines.aspx#VVSG%20Version%201.1

Electronic Government for Developing Countries. (2008). Retrieved from www.itu.int/ITU-D/cyb/app/e-gov.html

El-Kiki, T., & Lawrence, E. (2007). *Mobile user satisfaction and usage analysis model of m-government services*. Retrieved from http://www.mgovernment.org/resrces/euromgv022006/PDF/11_El-Kiki.pdf

Eschenfelder, K. R., & Miller, C. (2005). *The openness of government websites: Toward a socio-technical government website evaluation toolkit*. MacArthur Foundation/ALA Office of Information Technology Policy Internet Credibility and the User Symposium, Seattle, WA.

Essex, A., Clark, J., Hengartner, U., & Adams, C. (2010). Eperio: Mitigating technical complexity in cryptographic election verification. In *Proceedings of the 2010 Conference on Electronic Voting Technology/Workshop on Trustworthy Elections EVT/WOTE 2010*. Retrieved August 3, 2012, from http://static.usenix.org/events/evtwote10/tech/full_papers/Essex.pdf

Esteve, J., Goldsmith, B., & Turner, J. (2012). *International experience with e-voting*. National Foundation for Electoral Systems (IFES). Retrieved from www.IFES.org

Esteves, J., & Joseph, R. (2008). Comprehensive framework for the assessment of eGovernment projects. *Government Information Quarterly*, *25*, 118–132. doi:10.1016/j.giq.2007.04.009

Everett, S. P. (2007). *The usability of electronic voting machines and how votes can be changed without detection*. Doctoral dissertation, Rice University, Houston, Texas.

Everett, S. P., Byrne, M. D., & Greene, K. K. (2006). Measuring the usability of paper ballots: Efficiency, effectiveness and satisfaction. In *Proceedings of the Human Factors and Ergonomic Society 50th Annual Meeting* (pp. 2547 - 2551).

Everett, S. P., Greene, K. K., Byrne, M. D., Wallach, D. S., Derr, K., Sandler, D., & Torous, T. (2008). Electronic voting machines versus traditional methods: Improved preference, similar performance. In *Proceedings of the SIGCHI Conference on Human Factors in Computing Systems CHI '08* (pp. 883 - 892). New York, NY: ACM.

Facebook. (2012). *Facebook basic information.* Retrieved January 4, 2012, from http://www.facebook.com/#!/facebook?sk=info

Facebook. (2012). *Newsroom.* Retrieved May 11, 2012, from http://newsroom.fb.com/content/default.aspx?NewsAreaId=22

Fang, Z. (2002). E-government in digital era: Concept, practice and development. *International Journal of the Computer, the Internet and Management, 10*(2), 1-22.

Farbey, B., Land, F., & Targett, D. (1993). *How to assess your IT investment: A study of methods and practice.* Oxford, UK: Butterworth-Heinemann Ltd.

Farbey, B., Land, F., & Targett, D. (1995). A taxonomy of information systems applications: The benefits evaluation ladder. *European Journal of Information Systems, 4*, 41–50. doi:10.1057/ejis.1995.5

Farooq, U. (2005). Conceptual and Technical Scaffolds For End User Development: Using scenarios and wikis in community computing. In *Proceedings of the IEEE Symposium on Visual Languages and Human-Centric Computing* Los Alamitos, California, (pp. 329-330).

Farooq, U., Merkel, C. B., Nash, H., Rosson, M. B., Carroll, J. M., & Xiao, L. (2005). Participatory design as apprenticeship: Sustainable watershed management as a community computing application. *Proceedings of the 38th Annual Hawaii International Conference on System Sciences*, Hawaii, January 3-6, 2005.

Farooq, U., Merkel, C. B., Xiao, L., Nash, H., Rosson, M. B., & Carroll, J. M. (2006). Participatory design as a learning process: Enhancing community-based watershed management through technology. In Depoe, S. P. (Ed.), *The environmental communication yearbook (Vol. 3*, pp. 243–267). doi:10.1207/s15567362ecy0301_12

Ferney, S. L., & Marshall, A. L. (2006). Website physical activity interventions: preferences of potential users. *Health Education Research, 21*(4), 560–566. doi:10.1093/her/cyl013

Fine, G. A. (2000). Games and truths: Learning to construct social problems in high school debate. *The Sociological Quarterly, 41*(1), 103–123. doi:10.1111/j.1533-8525.2000.tb02368.x

Firestone, W. A. (1989). Educational policy as an ecology of games. *Educational Researcher, 18*(7), 18–24.

Fishbein, M., & Ajzen, I. (1975). *Belief, attitude, intention and behaviour: An introduction to theory and research.* Reading, MA: Addison-Wesley.

Fitsilis, P., Anthopoulos, L., & Gerogiannis, V. C. (2010). An evaluation framework for e-government projects. In Reddick, C. (Ed.), *Citizens and e-government: Evaluating policy and management* (pp. 69–90). Hershey, PA: Information Science. doi:10.4018/978-1-61520-931-6.ch005

Fla. Stat. §39.820(1) (2011)

Fla. Stat. §39.822(1) (2011)

Fla. Stat. §39.8298(1) (2011)

Flatters, F., & MacLeod, W. B. (1995). Administrative corruption and taxation. *International Tax and Public Finance, 2*, 397–417. doi:10.1007/BF00872774

Flavian, C., & Guinaliu, M. (2006). Consumer trust, perceived security and privacy policy: Three basic elements of loyalty to a Web site. *Industrial Management & Data Systems, 106*(5), 601–620. doi:10.1108/02635570610666403

Flichy, P. (1995). *Dynamics of modern communication: The shaping and impact of new communication technologies.* London, UK: Sage Publications.

Florida Department of Children & Families. (2010). *Abramowitz appointed to executive director of the statewide Guardian Ad Litem office.* Retrieved April 3, 2012, from http://www.dcf.state.fl.us/newsroom/pressreleases/20101230_GuardianAdLitem.shtml

Florida Guardian ad Litem Program. (2006). *Standards of operation.* Retrieved July 25, 2012, from http://www.guardianadlitem.org/training_docs/February11/StandardsofOperation.pdf.

Florida Guardian ad Litem Program. (2012). *2011 annual report.* Retrieved April 2, 2012, from http://www.guardianadlitem.org/training_docs/February11/StandardsofOperation.pdf

Florida, R. (2005). *The flight of the creative class: The new global competition for talent* (1st ed.). Toronto: HarperBusiness.

Focault, M. (1984). Utopie: Eterotopie. In *Des Espace Autres*. Architecture /Mouvement

Formenti, C. (2008). *Cybersoviet, Utopie postdemocratiche and new media*. Milan, Italy: Raffaello Cortina Editore.

Foucault, M. (1970). *The order of things. An archaeology of the human sciences.* London, UK: Tavistock Publications.

Foucault, M. (1972). *The archaeology of knowledge.* London, UK: Tavistock Publications.

Foucault, M. (1977). *Discipline and punish: The birth of the prison.* London, UK: Penguin Books.

Fountain, J. (2003). Prospects for improving the regulatory process using e-rule making. *Communications of the ACM, 46*(1), 43–44. doi:10.1145/602421.602445

Fountain, J. E. (2001). *Building the virtual state: Information technology and institutional change.* Washington, DC: Brookings Institution.

Fountain, J. E. (2002). A theory of federal bureaucracy. In Kamarck, E., & Nye, J. S. Jr., (Eds.), *Governance. com: Democracy in the information age* (pp. 117–140). Washington, DC: Brookings Institution.

Frequently Asked Questions. (2012). Retrieved April 2, 2012, from http://www.guardianadlitem.org/vol_faq.asp

Frissen, V. A. J. (2005). The E-mancipation of the citizen and the future of e-government: Reflections on ICT and citizens' partnership. In Khosrow-Pour, M. (Ed.), *Practicing e-government: A global perspective* (pp. 163–178). Hershey, PA: Idea Group Publishing. doi:10.4018/978-1-59140-637-2.ch008

Frøkjær, E., & Hornbæk, K. (2008). Metaphors of human thinking for usability inspection and design. *ACM Transactions on Computer Human Interaction, 14*(4). doi:10.1145/1314683.1314688

Fulk, J., & Desanctis, G. (1999). Articulation of communication technology and organization form. In De Sanctis, G., & Fulk, J. (Eds.), *Shaping organization form, communication, connection and community.* Thousand Oaks, CA: Sage.

Funilkul, S., Quirchmayry, G., Chutimaskul, W., & Traunmuller, R. (2006). *An evaluation frame work for e-government services based on principles laid out in COBIT, the ISO 9000 standard, and TAM.* 17th Australasian Conference on Information Systems.

Futuregov.asia. (2011). *Philippines e-elections miracle.* Retrieved from http://www.futuregov.asia/articles/2010/jul/30/philippines-e-election-miracle/

Galer, M., Harker, S., & Ziegler, J. (1992). *Methods and tools in user-centered design for information technology (Human factors in information technology).* Elsevier Science Ltd, North-Holland.

Gallego, R., & Barzelay, M. (2010). Public management policymaking in Spain: The politics of legislative reform of administrative structure, 1991-1997. *Governance: An International Journal of Policy, Administration and Institutions, 23*(2), 277–296. doi:10.1111/j.1468-0491.2010.01479.x

Gallegos-Garcia, G., Gomez-Cardenas, R., & Duchen-Sanchez, G. (2010). Identity based threshold cryptography and blind signatures for electronic voting. *WSEAS Transactions on Computers, 9*(1), 62–71.

Garfinkel, H. (1967). *Studies in ethnomethodology.* Englewood Cliffs, NJ: Prentice-Hall.

Garfinkel, H. (1974). On the origins of the term ethnomethodology. In Turner, R. (Ed.), *Ethnomethodology* (pp. 15–18). Harmondsworth, UK: Penguin.

Garriga, N. (2012). *DOT's "It's More Fun in the Philippines" campaign goes global.* Retrieved May 14, 2012, from http://ph.news.yahoo.com/dots-more-fun-philippines-campaign-goes-global-105213492.html

Garson, G. D. (2006). *Public information technology and e-government managing the virtual state.* Sudbury, MA: Jones & Bartlett Publishers.

Gascó, M. (2003). New technologies and institutional change in public administration. *Social Science Computer Review, 21*(1), 6–14. doi:10.1177/0894439302238967

Gauld, R., Goldfinch, S., & Horsburgh, S. (2010). Do they want it? Do they use it? The demand-side of e-government in Australia and New Zealand. *Government Information Quarterly, 27*(2), 177–186. doi:10.1016/j.giq.2009.12.002

Gefen, D., & Straub, D. W. (2003). The relative importance of perceived ease of use in IS adoption: A study of e-commerce adoption. *Journal of the Association for Information Systems*, *1*(8), 1–28.

Gehrke, D., & Turban, E. (1999). Determinants of successful Website design: relative importance and recommendations for effectiveness. *Proceedings of the 32nd Hawaii International Conference of Information* Systems, Maui, HI. Retrieved from http://ieeexplore.ieee.org/iel5/6293/16785/00772943.pdf

General Accounting Office. (2001). *Electronic government: Challenges must be addressed with effective leadership and management.* (GAO-01-959T). Retrieved from http://www.gao.gov/new.items/d01959t.pdf

Gerlach, J., & Gasser, U. (2009). *Three case studies from Switzerland: E-voting.* Technical Report. Berkman Center Research Publications, Berkman Center for Internet & Society.

Gerlach, F. (2009). Seven principles for secure e-voting. *Communications of the ACM*, *52*(2), 1–8.

German Federal Court Decision. (2009). *Bundesverfassungsgricht. Urteil des Zweiten Senats. German Federal Court Decisions 2 BvC 3/07*, (pp. 1-163).

Gibson, A. (2010). *Local by social: How local authorities can use social media to achieve more for less.* London, UK: NESTA.

Gil-Garcia, J. R. G., & Pardo, T. A. (2005). E-government success factors: Mapping practical tools to theoretical foundation. *Government Information Quarterly*, *22*(2), 187–216. doi:10.1016/j.giq.2005.02.001

Gils, D. (2002). Examples of evaluation practices used by OECD members countries to assess e-government. *Draft Point*, *9*, 1–64.

Ginige, A., & Murugesan, S. (2001). The essence of Web engineering – Managing the diversity and complexity of Web application development. *IEEE MultiMedia*, *8*, 22–25. doi:10.1109/MMUL.2001.917968

Girard, B. (Ed.). (2003). *The one to watch: Radio new ICT and interactivity.* Geneva, Switzerland: FES.

Godwin, J. U. (2001). Privacy and security concerns as a major barrier for e-commerce: A survey study. *Information Management & Computer Security*, *9*(4), 165–174. doi:10.1108/EUM0000000005808

Goggin, S. N. (2008). Usability of election technologies: Effects of political motivation and instruction use. *The Rice Cultivator*, *1*, 30–45.

Goggin, S. N., & Byrne, M. D. (2007). An examination of the auditability of voter verified paper audit trail (VVPAT) ballots. In *Proceedings of the USENIX Workshop on Accurate Electronic Voting Technology EVT '07.* Berkeley, CA: USENIX Association.

Goldstein, R. (2008). *Community informatics, electronic government and inclusion: Strategies for the consolidation of a citizens' democracy in Latin America.* Prato CIRN 2008 Community Informatics Conference: ICTs for Social Inclusion: What is the Reality? Refereed Paper.

Google. com. (2012). Ten things we know to be true. Retrieved from http://www.google.com/about/company/philosophy

Gothwal, J., Yadav, S., & Singh, R. (2011). Enhancing fingerprint authentication system using fragile image watermarking technique. *International Journal of Computer Science and Communication*, *2*(2), 459–463.

Gould, E., Gomez, R., & Camacho, K. (2010). Information needs in developing countries: How are they being served by public access venues? Paper presented at the AMCIS. Retrieved from http://aisel.aisnet.org/81625amcis2010/9.

Greene, K. K. (2008). *Usability of new electronic voting systems and traditional methods: Comparisons between sequential and direct access electronic voting interfaces, paper ballots, punch cards, and lever machines.* Unpublished Master's Thesis, Rice University, Houston, Texas.

Greene, K. K., Byrne, M. D., & Everett, S. P. (2006). A comparison of usability between voting methods. In *Proceedings of the 2006 USENIX/ACCURATE Electronic Voting Technology Workshop.* Vancouver, BC: USENIX Association.

Griffin, D., & Halpin, E. (2002). Local government: A digital intermediary for the information age? *Information Polity*, *7*(4), 217–231.

Griffths, J. R., Johnson, F., & Hartley, R. J. (2007). User satisfaction as a measure of system performance. *Journal of Librarianship and Information Science, 39*(3), 142–152. doi:10.1177/0961000607080417

Groenbaek, K., & Trigg, R. (Eds.). (1994). Hypermedia. *Communications of the ACM, 37*(2).

Guhan, S., & Paul, S. (Eds.). (1997). *Corruption in India: Agenda for action*. New Delhi, India: Vision Books.

Gupta, M. P., & Jana, D. (2003). E-government evaluation: A framework and case study. *Government Information Quarterly, 20*, 365–387. doi:10.1016/j.giq.2003.08.002

Gupta, P. (2007). Challenges and issues in e-government project assessment. *Proceedings of the 1st International Conference on Theory and Practice of Electronic Governance ACM*, USA.

Gupta, P., & Bagga, R. K. (Eds.). (2008). *Compendium of egovernance initiatives in India*. Hyderabad, India: Universities Press.

Gurstein, M. (2011). Evolving relationships: Universities, researchers and communities. *Journal of Community Informatics, 7*(3).

Hackett, S., & Parmanto, B. (2005). A longitudinal evaluation of accessibility: Higher education web sites. *Internet Research, 15*(3), 281–294. doi:10.1108/10662240510602690

Hall, D. J., & Davis, R. A. (2007). Engaging multiple perspectives: A value based decision-making model. *Decision Support Systems, 43*(4), 1588–1664. doi:10.1016/j.dss.2006.03.004

Hall, D., Guo, Y., Davis, R. A., & Cegielski, C. (2005). Extending unbounded system thinking with agent oriented modelling: Conceptualizing a multiple perspective decision-making support systems. *Decision Support Systems, 41*(1), 279–295. doi:10.1016/j.dss.2004.06.009

Hamilton, S., & Chervany, N. (1981). Evaluating information system effectiveness - Part I: Comparing evaluation approach. *Management Information Systems Quarterly, 5*(3), 55–69. doi:10.2307/249291

Hammond, A. L. (2001). Digitally empowered development. *Foreign Affairs, 80*(2), 96–106. doi:10.2307/20050067

Hamner, M., & Qazi, R. (2009). Expanding the technology acceptance model to examine personal computing technology utilization in government agencies in developing countries. *Government Information Quarterly, 26*(1), 128–136. doi:10.1016/j.giq.2007.12.003

Hand, L. C., & Ching, B. D. (2011). You have one friend request: An exploration of power and citizen engagement in local governments' use of social media. *Administrative Theory and Praxis, 33*(3), 362–382. doi:10.2753/ATP1084-1806330303

Hangen, M., & Kubicek, H. (2000). *One-stop government in Europe: Results of 11 national surveys*. University of Bremen.

Haque, S. M. (2002). E-governance in India: Its impacts on relations amongst citizens, politicians and public servants. *International Review of Administrative Sciences, 68*, 231–250. doi:10.1177/0020852302682005

Hardt, M., & Negri, A. (2001). *Empire*. Harvard University Press.

Hardt, M., & Negri, A. (2004). *Moltitude: War and democracy in the age of the empire*. New York, NY: The Penguin Press.

Hartson, H. R., Andre, T. S., & Williges, R. C. (2001). Criteria for evaluating usability evaluation methods. *International Journal of Human-Computer Interaction, 13*(4), 373–410. doi:10.1207/S15327590IJHC1304_03

Hasmah, A. (2009). *E-filing pays*. Retrieved from http://www.intanbk.intan.my/i-portal/nict/nict/DAY1/SESSION3PARALLEL1(SESSION3A)/Dato_hasmah_e-filing_Pays.pdf

Hatzilygeroudis, I., & Prentzas, J. (2006). Knowledge representation in intelligent educational systems. In Ma, Z. (Ed.), *Web-based intelligent e-learning systems: Technologies and applications* (pp. 175–192). Hershey, PA: Information Science Publishing.

Haugerud, A. (1995). *The culture of politics in modern Kenya*. Cambridge, UK: Cambridge University Press. doi:10.1017/CBO9781139166690

Haziemeh, F., Khazaaleh, M., & Al-Talafhah, K. (2011). New applied e-voting system. *Journal of Theoretical and Applied Information Technology, 25*(2), 88–97.

He, W., Fang, Y., & Wei, K. K. (2009). The role of trust in promoting organizational knowledge seeking using knowledge management systems: An empirical investigation. *Journal of the American Society for Information Science and Technology, 60*(3), 526–537. doi:10.1002/asi.21006

Hearn, G. N., & Foth, M. (2007). Communicative ecologies: Editorial Preface. *Electronic Journal of Communication, 17*(1-2).

Hearst, M. (2002). Finding the flow in web site search. *Communications of the ACM, 45*(9), 42–49. doi:10.1145/567498.567525

Hearst, M. (2006). Clustering versus faceted categories for information exploration. *Communications of the ACM, 49*(4). doi:10.1145/1121949.1121983

Heeks, R. (1998). *Information technology and public sector corruption* (Working Paper 4). Institute for Development Policy Management, University of Manchester.

Heeks, R. (1998). *Information age reform of the public sector: The potential and problems of IT for India.* (Information Systems for Public Sector Management Working Paper Series Paper No. 6). IDPM, University of Manchester.

Heeks, R. (2000). The approach of senior public officials to information technology related reform: Lessons from India. *Public Administration and Development, 20*(3), 197–205. doi:10.1002/1099-162X(200008)20:3<197::AID-PAD109>3.0.CO;2-6

Heeks, R. (2001). *Building e-governance for development: A framework for national and donor action.*

Heeks, R. (2001). *Understanding e-governance for development.* Manchester, UK: Institute for Development Policy Management.

Heeks, R. (2002). E-government in Africa: Promise and practice. *Information Polity, 7,* 97–114.

Heeks, R. (2002). i-Development not e-development: Special issue on ICTs and development. *Journal of International Development, 14*(1), 1–11. doi:10.1002/jid.861

Heeks, R. (2002). Information systems and developing countries: Failure, success and local improvisations. *The Information Society, 18,* 101–112. doi:10.1080/01972240290075039

Heeks, R. (2003). *Most egovernment-for-development projects fail: How can the risks be reduced?* (iGovernment Working Paper Series – Paper No. 14), University of Manchester.

Heeks, R. (2005). eGovernment as a carrier of context. *Journal of Public Policy, 25*(1), 51–74. doi:10.1017/S0143814X05000206

Heeks, R. (2006). *Implementing and managing e-government: An international text.* London, UK: Sage Publications Ltd.

Helander, M. G. (2000). Theories and models of electronic commerce. *Proceedings of the IEA 2000/HFES 2000 Congress,* Vol. 2, (pp. 770-3).

Henderson, A., & Kyng, M. (1991). There's no place like home: Continuing design in use. In Greenbaum, J., & Kyng, M. (Eds.), *Design at work: Cooperative design of computer systems* (pp. 219–240). Hillsdale, NJ: Lawrence Erlbaum Association.

Heras, M. (2010). Community manager, ese gran desconocido. *Revista de Comunicación, 13,* 16.

Hermana, B., & Silfianti, W. (2011). Evaluating e-government implementation by local government: Digital divide in internet-based public services in Indonesia. *International Journal of Business and Social Science, 2*(3).

Hernandez, B., Jimenez, J., & Martin, M. J. (2010). Customer behavior in electronic commerce: The moderating effect of e-purchasing experience. *Journal of Business Research, 63*(9-10), 964–971. doi:10.1016/j.jbusres.2009.01.019

Hernández Sampieri, R., Fernández Collado, C., & Baptista Lucio, P. (2003). *Metodología de la investigación* (3rd ed.). México: McGraw-Hill.

Herrnson, P. S., Bederson, B. B., Lee, B., Francia, P. L., Sherman, R. M., & Conrad, F. G. (2005). Early appraisals of electronic voting. *Social Science Computer Review, 23*(3), 274–292. doi:10.1177/0894439305275850

Herrnson, P. S., Bederson, B. B., Niemi, R. G., Conrad, F. G., Hanmer, M. J., & Traugott, M. (2007). *The not so simple act of voting: An examination of voter errors with electronic voting.* Retrieved August 3, 2012, from http://www.bsos.umd.edu/gvpt/apworkshop/herrnson2007.pdf

Herrnson, P. S., Niemi, R. G., Hanmer, M. J., Bederson, B. B., Conrad, F. G., & Traugott, M. (2006). The importance of usability testing of voting systems. In *Proceedings of the USENIX/Accurate Electronic Voting Technology Workshop 2006 EVT '06.* Berkeley, CA: USENIX Association.

Herrnson, P. S., Niemi, R. G., Hanmer, M. J., Francia, P. L., Bederson, B. B., Conrad, F., & Traugott, M. (2005). *The promise and pitfalls of electronic voting: results from a usability field test.* Paper presented at the Annual meeting of the Midwest Political Science Association. Chicago, IL.

Herrnson, P. S., Niemi, R. G., Hanmer, M. J., Francia, P. L., Bederson, B. B., Conrad, F. G., & Traugott, M. W. (2008). Voters' evaluations of electronic voting systems: Results from a usability field study. *American Politics Research, 36*(4), 580–611. doi:10.1177/1532673X08316667

Hindriks, J., Keen, M., & Muthoo, A. (1999). Corruption, extortion and evasion. *Journal of Public Economics, 74,* 395–430. doi:10.1016/S0047-2727(99)00030-4

Hinnant, C. C., & O'Looney, J. (2007). IT innovation in local government: Theory, issues, and strategies. In David Garson, G. (Ed.), *Public information technology: Policy and management issues* (2nd ed.). Hershey, PA: Idea Group Publishing.

Hix, D., & Hartson, H. R. (1993). *Developing user interfaces: Ensuring usability through product and process.* New York, NY: Wiley.

Hochstrasser, B. (1992). Justifying IT investment. *Proceedings of the Advanced Information Systems Conference; The New Technologies in Today's Business Environment*, UK, (pp. 17–28).

Homburg, V. (2008). *Understanding e-government: Information systems in public administration.* London, UK: Routledge.

Hong, S. J., Thong, J. Y. L., & Tam, K. Y. (2006). Understanding continued information technology usage behavior: A comparison of three models in the context of mobile internet. *Decision Support Systems, 42*(3), 1819–1834. doi:10.1016/j.dss.2006.03.009

Hosseini, R., & Mazinani, M. (2006). A fuzzy approach for measuring IT effectiveness of business processes. *Proceedings of the 6th WSEAS International Conference on Applied Informatics*, (pp. 1-6). Elounda, Greece.

Howards, P. (2002). *Network ethnography and the hypermedia organization: new media, new organizations, new methods.* London, UK: Sage Publication.

Hrdinová, J., & Helbig, N. (2011). *Designing social media policy for government* (pp. 1–9). Issues in Technology Innovation.

Hrdinova, J., Helbig, N., & Stollar Peters, C. (2010). *Designing social media policy for government: Eight essential elements. Center for Technology in Government.* The Research Foundation of State University of New York.

Hsu, M. H., & Chiu, C. M. (2004). Internet self-efficacy and electronic service acceptance. *Decision Support Systems, 38*(3), 369–381. doi:10.1016/j.dss.2003.08.001

Huang, T., Chen, K., Huang, P., & Lei, C. (2008). A generalizable methodology for quantifying user satisfaction. *IEICE Transaction Communication. E (Norwalk, Conn.), 91-B*(5), 1260–1268.

Huang, T., & Lee, C. (2010). Evaluating the impact of e-government on citizens: Cost-benefit analysis. In Reddick, C. (Ed.), *Citizens and e-government: Evaluating policy and management* (pp. 37–52). Hershey, PA: Information Science Reference. doi:10.4018/978-1-61520-931-6.ch003

Hughes, J. A., King, V., Rodden, T., & Andersen, H. (1994). Moving out from the control room: Ethnography in system design. [ACM.]. *Proceedings of CSCW, 94,* 429–439.

Huijboom, N., Van den Broek, T., Frissen, V., Kool, L., Kotterink, B., Nielsen, M., & Millard, J. (2009). *Public services 2.0: The impact of social computing on public services.* Luxembourg: Institute for Prospective Technological Studies, Joint Research Centre, European Commission, Office for Official Publications of the European Communities.

Hume, M., & Mort, G. S. (2010). The consequence of appraisal emotion, service quality, perceived value and customer satisfaction on repurchase intent in the performing arts. *Journal of Services Marketing, 24*(2), 170–182. doi:10.1108/08876041011031136

Hung, S. Y., Chang, C. M., & Yu, T. J. (2006). Determinants of user acceptance of the e-government services: The case of online tax filing and payment system. *Government Information Quarterly, 23*(1), 97–122. doi:10.1016/j.giq.2005.11.005

Huque, A. S. (1994). Public administration in India: Evolution, change and reform. *Asian Journal of Public Administration, 16*(2), 249–259.

Hussein, R., Mohamed, N., Ahlan, A. R., Mahmud, M., & Aditiawarman, U. (2010). An integrated model on online tax adoption in Malaysia. *European, Mediterranean & Middle Eastern Conference on Information Systems (EMCIS)*, (pp. 1-16).

Hyvönen, E., Saarela, S., & Viljanen, K. (2004). Application of ontology techniques to view-based semantic search and browsing. *Proceedings of the First European Semantic Web Symposium (ESWS 2004), LNCS 3053*, (pp. 92-106).

ICMA. (2011). *E-government 2011 survey summary.* Retrieved March 28, 2012, from http://icma.org/en/icma/knowledge_network/documents/kn/Document/302947/EGovernment_2011_Survey_Summary

ICT PSP from PIC. (n.d.). *ICT policy support programme.* Retrieved June 1, 2012, from http://ec.europa.eu/information_society/activities/egovernment/implementation / ict_psp/index_en.htm

Igbaria, M., Guimaraes, T., & Davis, G. B. (1995). Testing the determinants of microcomputer usage via a structural equation model. *Journal of Management Information Systems, 11*(4), 87–114.

Ilias, A., Razak, M. Z. A., & Yasao, M. R. (2009). Taxpayers' attitude in using e-filing system: Is there any significant difference among demographic factors? *Journal of Internet Banking and Commerce, 14*(1), 1–13.

Illias, A., Suki, N. M., Yasao', M. R., & Rahman, R. A. (2008). A study of taxpayers' intention in using e-filing system: A case in Labuan F.T s. *Computer and Information Science, 1*(2), 110–119.

International Organization for Standardization. ISO 9241 – 11. (1998). *Ergonomic requirements for office work with visual display terminals (VDT) – Part 11: Guidelines on usability.* Geneva, Switzerland: ISO.

Irani, Z., Al-Sebie, M., & Elliman, T. (2006). Transaction stage of e-government systems: Identification of its location & importance. *Proceedings of the 39th Hawaii International Conference on System Sciences*, (pp. 1-9).

Irani, Z., Love, P., Elliman, T., Jones, S., & Themistocleous, M. (2005). Evaluating e-government: Learning from the experiences of two UK local authorities. *Information Systems Journal, 15*, 61–82. doi:10.1111/j.1365-2575.2005.00186.x

Isen, A. M. (1993). Positive affect and decision-making. In Lewis, M., & Haviland, J. M. (Eds.), *Handbook of emotions* (pp. 261–277). New York, NY: Guilford.

ISO_9241-11. (1998). *Ergonomic requirements for the office works with visual display terminals (VDTs)-part-11: Guidance on usability* (No. ISO 9241-11:1998(E)). Geneva, Switzerland: International Organisation for Standardisation.

It's More Fun in the Philippines. (2012). Retrieved May 18, 2012, from http://itsmorefuninthephilippines.com/

Iwaarden, J., Wiele, T., Ball, L., & Millen, R. (2003). Applying SERVQUAL to Web sites: An exploratory study. *International Journal of Quality & Reliability Management, 20*, 919–935. doi:10.1108/02656710310493634

Iwaarden, J., Wiele, T., Ball, L., & Millen, R. (2004). Perceptions about the quality of websites: A survey amongst students at Northeastern University and Erasmus University. *Information & Management, 41*, 947–959. doi:10.1016/j.im.2003.10.002

Jaeger, P. T. (2003). The endless wire: E-government as global phenomenon. *Government Information Quarterly, 20*(4), 323–331. doi:10.1016/j.giq.2003.08.003

Jafari, S., Karimpour, J., & Bagheri, N. (2011). A new secure and practical electronic voting protocol without revealing voters identity. *International Journal on Computer Science and Engineering, 3*(6), 2191–2199.

Jalal, J. (2005). Good practices in public sector reform: A few examples from two Indian cities. In Singh, A. (Ed.), *Administrative reforms: Towards sustainable practices* (pp. 96–116). New Delhi, India: Sage Publications.

Jansen, A. (2005). *Assessing e-government progress—Why and what.* University of Oslo. Retrieved from http://www.uio.no/studier/emner/jus/afin/FINF4001/h05/undervisningsmateriale/AJJ-nokobit2005.pdf

Jansen, J. (2010). *Use of the internet in higher-income households.* Pew Research Center's Internet & American Life Project, Nov. 24, 2010. Retrieved from http://pewinternet.org/Reports/2010/Better-off-households.aspx

Janssen, D., Rotthier, S., & Snijkers, K. (2004). If you measure it they will score: An assessment of international e-government benchmarking. *Information Polity, 9*(3/4), 121–130.

Jaruchirathanakul, B., & Fink, D. (2005). Internet banking adoption strategies for a developing country: The case of Thailand. *Internet Research, 15*(3), 295–311. doi:10.1108/10662240510602708

Ji, Z. (2009). The research on the evaluation of e-government system. *International Conference on Industrial and information Systems*, (pp. 220-223).

Johansen, J. A., Olalsen, J., & Olsen, B. (1999). Strategic use of information technology for increased innovation and performance. *Information Management & Computer Security, 7*(1), 5–22. doi:10.1108/09685229910255133

John, B., & Kieras, D. E. (1996). Using GOMS for user interface design and evaluation: Which technique? *ACM Transactions on Computer-Human Interaction, 4*, 287–319. doi:10.1145/235833.236050

Johnson, T. (2012). *Disruptive legal technologies, part 2.* Retrieved from http://www.americanbar.org/groups/departments_offices/legal_technology_resources/resources/articles/youraba0710.html

Jones, D. W. (2001). *A brief illustrated history of voting.* University of Iowa, Department of Computer Science. Retrieved August 1, 2012, from http://homepage.cs.uiowa.edu/~jones/voting/pictures/

Jones, D. W. (2003). The evaluation of voting technology. In Gritzalis, D. A. (Ed.), *Secure electronic voting* (pp. 3–16). Kluwer Academic Publishers. doi:10.1007/978-1-4615-0239-5_1

Jones, S., Hackney, R., & Irani, Z. (2007). Towards e-government transformation: Conceptualising citizen engagement (A research note). *Transforming Government: People. Process and Policy, 1*(2), 145–152.

Jones, S., & Hughes, J. (2001). Understanding IS evaluation as a complex social process: A case study of a UK local authority. *European Journal of Information Systems, 10*(1), 189–203. doi:10.1057/palgrave.ejis.3000405

Jones, S., Irani, Z., Sharif, A., & Themistocleous, M. (2006). E-government evaluation: Reflections on two organizational studies. *Proceedings of the 39th Hawaii International Conference on System Sciences*, Kauai, Hawaii. January 4–7.

Jørgensen, A. H. (1990). Thinking-aloud in user interface design: A method promoting cognitive ergonomic. *Ergonomics, 33*(4), 501–507. doi:10.1080/00140139008927157

Jukic, T., Bencina, J., & Vintar, M. (2012). Multi-attribute evaluation of e-government projects: Slovenian approach. *International Journal of Information Communication Technologies and Human Development, 4*(1), 82–92. doi:10.4018/jicthd.2012010106

Justice Administrative Commission. (2011). *Long-range program plan: FY2012-13 through 2016-17.* Retrieved April 4, 2012, from http://floridafiscalportal.state.fl.us/PDFDoc.aspx?ID=6157

Kahani, M. (2005). Experiencing small-scale e-democracy in Iran. *Electronic Journal on Information System in Developing Country, 22*(5), 1–9.

Kahraman, C., Cebeci, U., & Raun, D. (2004). Multi-attribute comparison of catering servicing companies using fuzzy AHP: The case of Turkey. *International Journal of Production Economics, 87*(2), 171–184. doi:10.1016/S0925-5273(03)00099-9

Kaisara, G., & Pather, S. (2011). The e-government evaluation challenge: A South African Batho Pele-aligned service quality approach. *Government Information Quarterly, 28*(2), 211–221. doi:10.1016/j.giq.2010.07.008

Kalaichelvi, V., & Chandrasekaran, R. (2011). Design and analysis of secured electronic voting protocol. *Journal of Theoretical and Applied Information Technology, 34*(2), 151–157.

Kalaichelvi, V., & Chandrasekaran, R. (2011). Secured single transaction e-voting protocol: Design and implementation. *European Journal of Scientific Research, 51*(2), 276–284.

Kamal, M. M., Weerakkody, V., & Jones, S. (2009). The case of EAI in facilitating-government services in a Welsh authority. *International Journal of Information Management, 29*(2), 161–165. doi:10.1016/j.ijinfomgt.2008.12.002

Kamel Boulos, M. N., & Wheeler, S. (2007). The emerging Web 2.0 social software: An enabling suite of sociable technologies in health and health care education. *Health Information and Libraries Journal*, *24*, 2–23. doi:10.1111/j.1471-1842.2007.00701.x

Karahanna, E., Straub, D. W., & Chervany, N. L. (1999). Information technology adoption across time: A cross-sectional comparison of pre-adoption and post-adoption beliefs. *Management Information Systems Quarterly*, *23*(2), 183–213. doi:10.2307/249751

Karat, J. (1997). Evolving the scope of user-centered design. *Communications of the ACM*, *40*, 33–38. doi:10.1145/256175.256181

Karayumak, F., Kauer, M., Olembo, M. M., Volk, T., & Volkamer, M. (2011). User study of the improved Helios voting system interface. In *1ˢᵗ Workshop on Socio-Technical Aspects in Security and Trust* (STAST) (pp. 37 – 44). doi: 10.1109/STAST.2011.6059254

Karayumak, F., Kauer, M., Olembo, M. M., & Volkamer, M. (2011). Usability analysis of Helios - An open source verifiable remote electronic voting system. In *Proceedings of the 2011 USENIX Electronic Voting Technology Workshop/Workshop on Trustworthy Elections (EVT/WOTE 2011)*. Berkeley, CA: USENIX Association.

Kats, Y. (Ed.). (2013in press). *Upgrading, maintaining and securing learning management systems: Advances and developments*. Hershey, PA: Information Science Reference.

Kavada, A. (2005). Civil society organizations and the internet: The case of Amnesty International, Oxfam and the world development movement. In de Jong, W. (Eds.), *Global activism, global media* (pp. 208–222). London, UK: Pluto Press.

Kavada, A. (2007). *The European social forum and the internet: A case study of communication networks and collective action*. Ph.D Thesis, University of Westminster, UK.

Keeker, K. (1997). *Improving Web-site usability and appeal: guidelines compiled by MSN usability research*. Retrieved from http://msdn.microsoft.com/library/default.asp

Kelly, D., Wacholder, N., Rittman, R., Sun, Y., Kantor, P., Small, S., & Strzalkowski, T. (2007). Using interview data to identify evaluation criteria for interactive, analytical question-answering system. *Journal of the American Society for Information Science and Technology*, *58*(7), 1032. doi:10.1002/asi.20575

Khalifa, G., Irani, Z., Baldwin, L. P., & Jones, S. (2004). Evaluating information technology with you in mind. *Electronic Journal of Information Systems Evaluation*, *4*(5), 246–252.

Khan, B., Khan, M., & Alghathbar, K. (2010). Biometrics and identity management for homeland security applications in Saudi Arabia. *African Journal of Business Management*, *4*(15), 3296–3306.

Khan, H. (2010). Comparative study of authentication techniques. *International Journal of Video & Image Processing and Network Security*, *10*(4), 9–15.

Kim, H. W., Chan, H. C., & Chan, Y. P. (2007). A balanced thinking-feelings model of information systems continuance. *International Journal of Human-Computer Studies*, *65*(6), 511–525. doi:10.1016/j.ijhcs.2006.11.009

Kim, H., Pan, G., & Pan, S. (2007). Managing IT-enabled transformation in the public sector: A case study on e-government in South Korea. *Government Information Quarterly*, *24*(2), 338–352. doi:10.1016/j.giq.2006.09.007

Kimball, D. C., & Kropf, M. (2005). Ballot design and unrecorded votes on paper-based ballots. *Public Opinion Quarterly*, *69*(4), 508–529. doi:10.1093/poq/nfi054

Kincaid, J. P., Fishburne, R. P., Rogers, R. L., & Chissom, B. S. (1975). *Derivation of new readability formulas (Automated readability index, fog count, and Flesch reading ease formula) for navy enlisted personnel*. Research Branch Report 8-75. Chief of Naval Technical Training: Naval Air Station Memphis.

King, J. L., Gurbaxani, V., Kraemer, K. L., McFarlan, F. W., Raman, K. S., & Yap, C. S. (1994). Institutional factors in information technology innovations. *Information Systems Research*, *5*(21), 139–169. doi:10.1287/isre.5.2.139

King, W. R., & He, J. (2006). A meta-analysis of the technology acceptance model. *Information & Management*, *43*, 740–755. doi:10.1016/j.im.2006.05.003

Kini, A., & Choobineh, J. (1998). Trust in electronic commerce: Definition and theoretical considerations. *31ˢᵗ Annual Hawaii International Conference on System Sciences (HICSS)*, Vol. 4, (p. 51).

Kinzie, M. B., Cohn, W. F., Julian, M. F., & Knaus, W. A. (2002). A user-centered model for Web site design. *Journal of the American Medical Informatics Association*, 9, 320–330. doi:10.1197/jamia.M0822

Kirakowski, J., Claridge, N., & Whitehand, R. (1998). Human centered measures of success in Web site design. *Proceedings of the 4th Conference on Human Factors & the Web*, Basking Ridge, NJ, 5 June.

Kirakowski, J., & Corbett, M. (1993). SUMI: The software usability measurement inventory. *British Journal of Educational Technology*, 24, 210–212. doi:10.1111/j.1467-8535.1993.tb00076.x

Kitchener, M. (1999). All fur coat and no knickers: Contemporary organizational change in United Kingdom hospitals. In Brock, D., Powell, M., & Hinings, C. R. (Eds.), *Restructuring the professional organization: Accounting, health care and law* (pp. 183–199). London, UK: Rutledge. doi:10.4324/9780203018446.ch9

Klein, P. (2008). Web 2.0: Reinventing democracy. *CIO Insight*, 30-43.

Kline, R., & Pinch, T. (1996). Users as agents of technological change: The social construction of the automobile in the rural United States. *Technology and Culture, 37*(4), 763–795. doi:10.2307/3107097

Kluver, R. (2005). The architecture of control: A Chinese strategy for egovernance. *Journal of Public Policy, 25*(1), 75–97. doi:10.1017/S0143814X05000218

Kostopoulos, G. K. (2004). E-government in the Arabian Gulf: A vision toward reality. *Electronic Government: An International Journal, 1*(3), 293–299. doi:10.1504/EG.2004.005553

Koutsojannis, C., Beligiannis, G., Hatzilygeroudis, I., Papavlasopoulos, C., & Prentzas, J. (2007). Using a hybrid AI approach for exercise difficulty level adaptation. *International Journal of Continuing Engineering Education and Lifelong Learning, 17*(4-5), 256–272. doi:10.1504/IJCEELL.2007.015042

Kovac, P., & Decman, M. (2009). Implementation and change of processual administrative legislation through an innovative Web 2.0 solution. *Transylvanian Review of Administrative Sciences, 28E*, 65–86.

Kraemer, K. L., & King, J. L. (2003, September). *Information technology and administrative reform: Will the time after e-government be different?* Paper prepared for the Heinrich Reinermann Schrift fest, Post Graduate School of Administration, Speyer, Germany. Retrieved July 30, 2006, from http://www.crito.uci.edu/publications/pdf/egovernment.pdf

Kraft, P., Rise, J., Sutton, S., & Roysamb, E. (2005). Perceived difficulty in the theory of planned behaviour: Perceived behavioural control or affective attitude? *The British Journal of Social Psychology, 44*, 479–496. doi:10.1348/014466604X17533

Kral, J., & Zemlicka, M. (2008). Implementation of business processes in service-oriented systems. *International Journal of Business Process Integration and Management, 3*(3), 208–219. doi:10.1504/IJBPIM.2008.023220

Kravchuk, R. S., & Schack, R. W. (1996). Designing effective performance measurement systems under the government performance and results act of 1993. *Public Administration Review, 56*(4), 348–358. doi:10.2307/976376

Krug, S. (2006). *Don't make me think, second edition: A common sense approach to Web usability*. New York, NY: Pearson Education Inc.

Kumar, K., Kumar, N., Md, A., & Sandeep, M. (2011). PassText user authentication using smartcards. *International Journal of Computer Science and Information Technologies, 2*(4), 1802–1807.

Kumar, R., & Best, M. L. (2006). Impact and sustainability of e-government services in developing countries: Lessons learned from Tamil Nadu, India. *The Information Society, 22*, 1–12. doi:10.1080/01972240500388149

Kumar, R. L., Smith, M. A., & Bannerjee, S. (2004). User interface features influencing overall ease of use and personalization. *Information & Management, 41*, 289–302. doi:10.1016/S0378-7206(03)00075-2

Kumar, S., & Walia, E. (2011). Analysis of electronic voting system in various countries. *International Journal on Computer Science and Engineering, 3*(5), 1825–1830.

Kumar, S., & Walia, E. (2011). Analysis of various biometric techniques. *International Journal of Computer Science and Information Technologies, 2*(4), 1595–1597.

Kunstelj, M., & Vintar, M. (2004). Evaluating the progress of eGovernment development: A critical analysis. *Information Polity, 9*(3–4), 131–148.

Kurunmaki, L. (1999). Making an accounting entity: The case of the hospital in Finnish health care reforms. *European Accounting Review, 8*(2), 219–237. doi:10.1080/096381899336005

Lam, W. (2005). Barriers to e-government integration. *Journal of Enterprise Information Management, 18*(5), 511–530. doi:10.1108/17410390510623981

Langer, L., Schmidt, A., Buchmann, J., Volkamer, M., & Stolfik, A. (2010). Towards a framework on the security requirements for electronic voting protocols. *Proceedings of First International Workshop on Requirements Engineering for e-Voting Systems (RE-VOTE)*, Atlanta, USA, (pp. 61-68).

Laskowski, S. J., Autry, M., Cugini, J., Killam, W., & Yen, J. (2004). *Improving the usability and accessibility of voting systems and products.* (NIST Special Publication 500 – 256).

Laskowski, S. J., & Redish, J. (2006). Making ballot language understandable to voters. In *Proceedings of the USENIX/Accurate Electronic Voting Technology Workshop EVT'06*. Berkeley, CA: USENIX Association.

Law, E. L.-C., & Hvannberg, E. T. (2004). Analysis of strategies for estimating and improving the effectiveness of heuristic evaluation. In *Proceedings of NordiCHI*, 23-27 October, Tampere, Finland.

Lawkobit, M., & Speece, M. (2012). Integrating focal determinants of service fairness into post-acceptance model of IS continuance in cloud computing. *Proceeding of the 11th International Conference on Computer and Information Science IEEE/ACIS, Thailand*, May 30 – June 1, (pp. 49-55).

Layne, K., & Lee, J. (2001). Developing fully functional e-government: A four stage model. *Government Information Quarterly, 18*(2), 122–136. doi:10.1016/S0740-624X(01)00066-1

Lazar, J., Feng, J. H., & Hochheiser, H. (2010). *Research methods in human computer interaction*. Wiley Publishing.

Lean, O. K., Zailani, S., Ramayah, T., & Fernando, Y. (2009). Factors influencing intention to use e-government services among citizens in Malaysia. *International Journal of Information Management, 29*(6), 458–475. doi:10.1016/j.ijinfomgt.2009.03.012

Lecerof, A., & Paterno, F. (1998). Automatic support for usability evaluation. *IEEE Transactions on Software Engineering, 24*, 863–887. doi:10.1109/32.729686

Lee, C., Chang, K., & Berry, F. (2011). Testing the development and diffusion of e-government and e-democracy: A global perspective. *Public Administration Review*, (May-June): 444–454. doi:10.1111/j.1540-6210.2011.02228.x

Lee, S. M., Tan, X., & Trimi, S. (2005). Current practices of leading e-government countries. *Communications of the ACM, 48*(10), 99–104. doi:10.1145/1089107.1089112

Lefebvre, E., & Lefebvre, L. (1996). *Information and telecommunication technologies: The impact of their adoption on small and medium-sized enterprises*. International Development Research Centre. Retrieved from http://www.idrc.ca/en/ev-9303-201-1-DO_TOPIC.html

Legrain, P. (2002). *Open world: The truth about globalisation*. London, UK: Abacus.

Leighninger, M. (2011). Citizenship and governance in a wild, wired world: How should citizens and public managers use online tools to improve democracy? *National Civic Review, 100*(2), 20–29. doi:10.1002/ncr.20056

Leighninger, M. (2011). *Using online tools to engage – And be engaged by – The public*. Washington, DC: IBM Center for The Business of Government, IBM Center for The Business of Government.

Lessen, V. T., Nitzsche, J., & Leymann, F. (2009). Conversational web services: Leveraging BPEL for expressing WSDL 2.0 message exchange patterns. *Enterprise Information Systems, 3*(3), 347–367. doi:10.1080/17517570903046300

Lewis, A. (1982). *The psychology of taxation*. Oxford, UK: Martin Robertson & Company.

Lewis, C., Polson, P., Wharton, C., & Rieman, J. (1990). Testing a walkthrough methodology for theory-based design of walk-up-and-use interfaces. In *Proceedings of the SIGCHI conference on human factors in computing systems* (pp. 235–242), Seattle, WA, USA.

Li, D., Browne, G. J., & Chau, P. Y. K. (2006). An empirical investigation of web site use using a commitment-based model. *Decision Sciences*, *37*(3), 427–444. doi:10.1111/j.1540-5414.2006.00133.x

Li, W., Zheng, W., & Guan, X. (2007). Application controlled caching for web servers. *Enterprise Information Systems*, *1*(2), 161–175. doi:10.1080/17517570701243273

Li, Y., & Zhang, X. (2010). Study on development and system application of China E-port. In *Proceedings of the 3rd International Conference on Information Management, Innovation Management and Industrial Engineering* (pp. 430-433). IEEE.

Liao, C., Palvia, P., & Lin, H. N. (2006). The role of habit and web site quality in e-commerce. *International Journal of Information Management*, *26*(6), 469–483. doi:10.1016/j.ijinfomgt.2006.09.001

Liao, S., Shao, Y. P., Wang, H., & Chen, A. (1999). The adoption of virtual banking: An empirical study. *International Journal of Information Management*, *19*, 63–74. doi:10.1016/S0268-4012(98)00047-4

Lim, J. H., & Tang, S.-Y. (2008). Urban e-government initiatives and environmental decision performance in Korea. *Journal of Public Administration: Research and Theory*, *18*, 109–138. doi:10.1093/jopart/mum005

Limayem, M., Hirt, S. G., & Cheung, C. M. K. (2003). Habit in the context of IS continuance: Theory extension and scale development. In C. U. Ciborra, R. Mercurio, M. de Marco, M. Martinez, & A. Carignani (Eds.), *Proceedings of the 11th European Conference on Information Systems*, Naples, Italy, June 16-21, 2003.

Limayem, M., Hirt, S. G., & Chin, W. W. (2001). Intention does not always matter: The contingent role of habit on IT usage behavior. *The 9th European Conference on Information Systems*, (pp. 274-286).

Limayem, M., Hirt, S. G., & Cheung, C. M. K. (2007). How habit limits the predictive power of intention: The case of information system continuance. *Management Information Systems Quarterly*, *31*(4), 705–737.

Lindgaard, G. (1999). Does emotional appeal determine perceived usability of Websites? In L. Straker & C. Pollock (Eds.), *Proceedings of CybErg: The Second International Cyberspace Conference on Ergonomics,* (pp. 202-11). The International Ergonomics Association Press, Curtin University of Technology, Perth, Western Australia.

Linggang, C., & Hitoshi, I. (2005). Expressway policy-set analysis from multi-perspective view-points: Model, algorithm and application. *Journal of the Eastern Asia Society for Transportation Studies*, *6*, 4144–4159.

LINGOs. (2008). *Obtained through the Internet*. Retrieved from http://www.lingos.org

Liu, C., & Arnett, K. P. (2000). Exploring the factors associated with Website success in the context of electronic commerce. *Information & Management*, *38*, 23–33. doi:10.1016/S0378-7206(00)00049-5

Liu, J., Derzs, Z., Raus, M., & Kipp, A. (2008). Lecture Notes in Computer Science: *Vol. 5184. Egovernment project evaluation: An integrated framework* (pp. 85–97). New York, NY: Springer.

Liu, T., Liu, R., & Zhao, P. (2004). Research on e- government system assessment methods. *Journal of Wuhan Automotive Polytechnic University*, *26*(3).

Local Programs. (2012). Retrieved April 2, 2012, from http://www.guardianadlitem.org/partners_main.asp

Loeber, L. (2008). E-voting in the Netherlands: From general acceptance to general doubt in two years. In R. Krimmer & R. Grimm (Eds.), *Proceedings of the 3rd International Conference on Electronic Voting*, Bregenz, Austria (pp. 21 – 30).

Lohse, G., & Spiller, P. (1999). Internet retail store design: How the user interface influences traffic and sales. *Journal of Computer-Mediated Communication*, *5*(2).

Loimeier, R. (2005). The Baraza: A grassroots institution. *International SIM Review for the Study of Islam in the Modern World*, *16*, 26–27.

Long, N. E. (1958). The local community as an ecology of games. *American Journal of Sociology*, *64*(3), 251–261. doi:10.1086/222468

Lopez, A. U., Miguel, F. P., & Prasad, S. I. (2012). Value, quality, purchasing habits and repurchase intention in B2C: Differences between frequent and occasional purchaser. *Dirección y Organización*, *47*, 70–80.

Lorente, A. R. M., Dewhust, F., & Dale, B. G. (1999). TQM and business innovation. *European Journal of Innovation Management*, *2*(1), 12–19. doi:10.1108/14601069910248847

Luarn, P., & Lin, L. H. (2004). Towards an understanding of the behavioural intention to use mobile banking. *Computers in Human Behavior*, *21*(6), 1–19.

Lyytinen, K., Mathiassen, L., & Ropponen, J. (1998). Attention shaping and software risk—A categorical analysis of four classical risk management approaches. *Information Systems Research*, *9*(3), 233–255. doi:10.1287/isre.9.3.233

Maaten, E., & Hall, T. (2008). Improving the transparency of remote e-voting: The Estonian experience. *Proceedings of 3rd International Conference on Electronic Voting*, Austria, August 6-9, (pp. 31-43).

Macario, A. (2006). Are your hospital operating rooms "efficient"? *Anaesthesiology*, *105*, 237–240. doi:10.1097/00000542-200608000-00004

Mackinon, R. (2012). The netizen. *Development*, *55*(2), 201–204. doi:10.1057/dev.2012.5

MacNamara, D., Carmody, F., Scully, T., Oakley, K., Quane, E., & Gibson, J. P. (2010). Dual vote: A novel user interface for e-voting systems. *IADIS International Conference Interfaces and Human Computer Interaction 2010, IHCI10,* Freiburg, Germany, (pp. 129-138).

MacNamara, D., Gibson, J. P., & Oakley, K. (2012). *A preliminary study on a DualVote and Pret a Voter hybrid system*. In 2012 Conference for E-Democracy and Open Government (CeDEM12). Danube University, Krems.

MacNamara, D., Scully, T., Carmody, F., Oakley, K., Quane, E., & Gibson, J. P. (2011). (in press). DualVote: A non-intrusive e-voting interface. [IJCISIM]. *International Journal of Computer Information Systems and Industrial Management Applications*.

MacNamara, D., Scully, T., Gibson, J. P., Carmody, F., Oakley, K., & Quane, E. (2011). *DualVote: Addressing usability and verifiability issues in electronic voting systems*. In 2011 Conference for E-Democracy and Open Government (CeDEM11). Danube University, Krems.

Madden, T. J., Ellen, P. S., & Ajzen, I. (1992). A comparison of the theory of planned behaviour and the theory of reasoned action. *Personality and Social Psychology Bulletin*, *18*(1), 3–9. doi:10.1177/0146167292181001

Maddock, S., & Morgan, G. (1998). Barriers to transformation: Beyond bureaucracy and the market conditions for collaboration in health and social care. *International Journal of Public Sector Management*, *11*(4), 234–251. doi:10.1108/09513559810225807

Madigan, L. (2004). *Guide to the Illinois Open Meetings Act: 5 ILCS 120*. State of Illinois.

Madigan, L. (March 2011). *Public access counselor annual report 2010*. Illinois Attorney General. Retrieved January 12, 2012, from http://foia.ilattorneygeneral.net/pdf/Public_Access_Counselor_Annual_Report_2010.pdf

Madigan, L. (2011). *Public Access Opinion No. 11-006. State of Illinois*. Office of the Attorney General.

Madise, U., & Martens, T. (2006). E-voting in Estonia 2005. The first practice of country-wide binding internet voting in the world. In A. Prosser & R. Krimmer (Eds.), *2nd International Workshop on Electronic Voting 2006, Lecture Notes in Informatics vol. 47*, (pp. 83-90).

Madise, U., & Vinkel, P. (2011). Constitutionality of remote internet voting: The Estonian perspective. *Juridica International*, *18*(1), 4–16.

Madon, S. (1993). Introducing administrative reform through the application of computer-based information systems: A case study in India. *Public Administration and Development*, *13*, 37–48. doi:10.1002/pad.4230130104

Madon, S. (2004). Evaluating the developmental impact of e-governance initiatives: An exploratory framework. *Electronic Journal of Information Systems in Developing Countries*, *20*(5), 1–13.

Madon, S., & Bhatnagar, B. (2000). Institutional decentralised information systems for local level planning: Comparing approaches across two states in India. *Journal of Global Information Technology Management*, *3*(4), 45–59.

Madon, S., Sahay, S., & Sahay, J. (2004). Implementing property tax reforms in Bangalore: An actor-network perspective. *Information and Organization, 14*, 269–295. doi:10.1016/j.infoandorg.2004.07.002

Mahalik, D. K. (2010). Outsourcing in e-governance: A multi-criteria decision-making approach. *Journal of Administration and Governance, 5*(1), 21–35.

Mahbob, M. H., Sulaiman, W. I. W., Mahmud, W. A. W., Mustaffa, N., & Abdullah, M. Y. (2012). The elements of behavioral control in facilitating the acceptance of technological innovation on malaysia on-line government services. *Asian Social Science, 8*(5), 125–131. doi:10.5539/ass.v8n5p125

Maheswari, S. R. (1993). *Administrative reform in India.* New Delhi, India: Jawahar Publishers and Distributors.

Maier, M. (2010). *Youth Parliament Esslingen – Living e-democracy first binding election to public office over the internet worldwide.* Retrieved from http://www.jgrwahl.esslingen.de/paper.pdf

Malkia, M., Anttiroiko, A., & Savolainen, R. (2004). Background of the project. In Malkia, M., Anttiroiko, A., & Savolainen, R. (Eds.), *E-transformation in governance: New directions in government and politics* (pp. vii–xiv). Hershey, PA: Idea Group Publishing.

Malkia, M., & Savolainen, R. (2004). Etransformation in government, politics and society: Conceptual framework and introduction. In Malkia, M., Anttiroiko, A., & Savolainen, R. (Eds.), *Etransformation in governance: New directions in government and politics* (pp. 1–21). Hershey, PA: Idea Group Publishing.

Mangaraj, B. K., & Upali, A. (2008). Multi-perspective evaluation of community development programmes: A case study for a primitive tribe of Orissa. *Journal of Social and Economic Development, 10*(1), 98–126.

Manian, A., Fathi, M. R., Zarchi, M. K., & Omidian, A. (2011). Performance evaluating of IT department using a modified fuzzy TOPSIS and BSC methodology. *Journal of Management Research, 3*(2:E10), 1-20.

Manstead, A. S. R., & Parker, D. (1995). Evaluating and extending the theory of planned behaviour. In Stroebe, W., & Hewstone, M. (Eds.), *European review of social psychology* (pp. 69–96). Chichester, UK: John Wiley & Sons. doi:10.1080/14792779443000012

Mao, J., Vredenburg, K., Smith, P. W., & Carey, T. (2005). The state of user-centered design practice. *Communications of the ACM, 48*(3), 105–109. doi:10.1145/1047671.1047677

March, J. G., & Olsen, J. P. (1989). *Rediscovering institutions: The organisational basis of politics.* New York, NY: The Free Press.

Marcus, A. (2005). User interface design's return on investment: examples and statistics. In Bias, R. G., & Mayhew, D. J. (Eds.), *Cost-justifying usability.* San Francisco, CA: Morgan Kaufman. doi:10.1016/B978-012095811-5/50002-X

Margetts, H. (1998). *Information technology in government: Britain and America.* London, UK: Routledge.

Margetts, H. (2006). Transparency and digital government. In Hood, C., & Heald, D. (Eds.), *Transparency: The key to better governance?* (pp. 197–210). London, UK: The British Academy. doi:10.5871/bacad/9780197263839.003.0012

Martins, M. R. (1995). Size of municipalities, efficiency, and citizens participation: A cross-European perspective. *Environment and Planning. C, Government & Policy, 13*(4), 441–458. doi:10.1068/c130441

Mason, D. (2011). *E-government takes policy-making social.* Retrieved from http://www.publicserviceeurope.com/article/667/e-government-makes-policy-making-social

Masuku, W. (2006). *An exploratory study on the planning and design of a future e-voting system for South Africa,* (pp. 1-159). A thesis published in 2006, in the University of the Western Cape, South Africa.

Mathew, G. (2006). A new deal for municipalities. In *Proceedings of the National Seminar on Urban Governance in the Context of the Jawaharlal Nehru National Urban Renewal Mission,* India Habitat Centre, New Delhi 24th – 25th November 2006, (pp. 102–116).

Mathieson, K. (1991). Predicting user intentions: Comparing the technology acceptance model with the theory of planned behaviour. *Information Systems Research, 2*(3), 173–191. doi:10.1287/isre.2.3.173

McIver, W. (2004). Software support for multi-lingual legislative drafting. In *Proceedings of the Community Informatics Research Network Conference and Colloquium* Tuscany, Italy, 2004.

McIver, W. (2004). Tools for collaboration between tans national NGOs: Multilingual, legislative drafting. In *Proceedings of International Colloquium on Communication and Democracy: Technology and Citizen Engagement,* Fredericton, New Brunswick, Canada, 2004.

McKenzie, W. (2004). *A hacker manifesto.* Cambridge, UK: Harvard University Press.

McMillan, P., Medd, A., & Hughes, P. (n.d.). *Change the world or the world will change you: The future of collaborative government and Web 2.0.* Retrieved June 5, 2012, from www.deloitte.com

McMullen, S. (2001). Usability testing in a library Web site redesign project. *RSR. Reference Services Review, 29,* 7–22. doi:10.1108/00907320110366732

McNeal, R., Hale, K., & Dotterweich, L. (2008). Citizen-government interaction and the internet: Expectations and accomplishments in contact, quality, and trust. *Journal of Information Technology & Politics, 5*(2), 213–229. doi:10.1080/19331680802298298

McPhail, B., Costantino, T., Bruckmann, D., Barclay, R., & Clement, A. (1998). CAVEAT exemplar: Participatory design in a non-profit volunteer organisation. *Computer Supported Cooperative Work, 7*(3), 223–241.

Mechling, J. (2002). Information age governance. In Kamarck, E., & Nye, J. S. Jr., (Eds.), *Governance.com: Democracy in the information age* (pp. 171–189). New York, NY: Brookings Institution.

Meier, P., & Brodock, K. (2008). *Crisis mapping Kenya's election violence: Comparing mainstream news, citizen journalism and Ushahidi.* Harvard Humanitarian Initiative, HHI, Harvard University: Boston). Retrieved from http://irevolution.net/2008/10/23/mapping-kenyas-election-violence

Meijer, A. (2002). Geographical information systems and public accountability. *Information Policy, 7,* 39–47.

Meister, D. B., & Compeau, D. R. (2002). *Infusion of innovation adoption: An individual perspective* (pp. 23–33). Winnipeg, Manitoba: ASAC.

Mercuri, R. T. (2001). *Electronic vote tabulation checks and balances.* Doctoral Dissertation. University of Pennsylvania, Philadelphia.

Merkel, C. B., Xiao, L., Farooq, U., Ganoe, C. H., Lee, R., Carroll, J. M., & Rosson, M. B. (2004). Participatory design in community computing contexts: Tales from the field. In *Proceedings of the Participatory Design Conference*, Toronto, Canada, (pp. 1-10).

Meyrowitz, J. (1985). *No sense of place: The impact of electronic media on social behaviour.* Oxford, UK: Oxford University Press.

Mid Term Review of the Ninth Malaysian Plan. (9MP). (2006-2010). *Economic Planning Unit* (EPU). Retrieved from the http://www.btimes.com.my/Current_News/BTIMES/Econ2007_pdf/Mid-term%20Review%20of%20the%20Ninth%20Malaysia%20Plan%202006-2010

Millard, J. (2009). Government 1.5: Is the bottle half full or half empty? *European Journal of ePractice, 9*(1), 35-50.

Miller, G. A. (1956). The magical number seven, plus or minus two: Some limits on our capacity for processing information. *Psychological Review, 63*(2), 81–97. doi:10.1037/h0043158

Minister of Communications & Information Technology. (2011). Keynote speech by Dr. Magued Osman in the 15th round of Cairo ICT, 25 May 2011.

Ministry of Public Administration and Security. (2009). *G4C: Government for Citizens.* Korean Government. Retrieved from http://korea.go.kr/html/files/intro/001.pdf

Minogue, M. (2002). Power to the people? Good governance and the reshaping of the state. In Kothari, U., & Minogue, M. (Eds.), *Development theory and practice* (pp. 117–135). Basingstoke, UK: Palgrave.

Mintz, D. (2008). Government 2.0 – Fact or fiction? *Public Management, 36*(4), 21–24.

Misra, S. (2005). eGovernance: Responsive and transparent service delivery mechanism. In A. Singh (Ed.), *Administrative reforms: Towards sustainable practices* (pp. 283–302). New Delhi, India: Sage Publications.

Mitra, R. (2000). Emerging state-level ICT development strategies. In Bhatnagar, S., & Schware, R. (Eds.), *Information and communication technology in development: Cases from India* (pp. 195–205). New Delhi, India: Sage Publications.

Mitroff, I., & Linestone, H. A. (1993). *The unbounded mind*. New-York, NY: Oxford University Press.

Moore, G. C., & Benbasat, I. (1991). Development of an instrument to measure the perceptions of adopting an information technology. *Information Systems Research*, 2(3), 173–191. doi:10.1287/isre.2.3.192

Moore, G. C., & Benbasat, I. (1996). Integrating diffusion of innovations and theory of reasoned action models to predict utilisation of information technology by end-users. In Kautz, K., & Prier-Hege, J. (Eds.), *Diffusion and adoption of information technology* (pp. 132–146). London, UK: Chapman & Hall.

Moraga, A., Calero, C., & Piattini, M. (2006). Comparing different quality models for portals. *Online Information Review*, 30(5), 555–568. doi:10.1108/14684520610706424

Moran, T., & Naor, M. (2006). Receipt-free universally-verifiable voting with everlasting privacy. In C. Dwork (Ed.), *Proceedings of the 26th Annual International Conference on Advances in Cryptology CRYPTO '06*, (pp. 373-392). Berlin, Germany: Springer-Verlag.

Morgeson, F. V. (2012). E-government performance measurement: A citizen-centric approach in theory and practice. In Chen, Y., & Chu, P. (Eds.), *Electronic governance and cross-boundary collaboration: Innovations and advancing tools* (pp. 150–165). Hershey, PA: Information Science Reference.

Morgeson, F., VanAmburg, D., & Mithas, S. (2010). Misplaced trust? Exploring the structure of the e-government-citizen trust relationship. *Journal of Public Administration: Research and Theory, 20*, 1–27.

Muhammad Rais, A. K., & Nazariah, M. K. (2003). *E-government in Malaysia. Pelanduk Publications (M)*. Sdn Bhd.

Mussari, R. (1999). Some considerations on the significance of the assets and liabilities statement in Italian local government reform. In Capperchione, E., & Mussari, R. (Eds.), *Comparative issues in local government accounting* (pp. 175–190). Norwell, MA: Kluwer.

Mutula, S., & Brakel, P. (2006). An evaluation of e-readiness assessment tools with respect to information access: Towards an integrated information rich tool. *International Journal of Information Management, 26*(3), 212–223. doi:10.1016/j.ijinfomgt.2006.02.004

Nash, J. F. Jr. (1950). Equilibrium points in n-person games. *Proceedings of the National Academy of Sciences of the United States of America, 36*, 48–49. doi:10.1073/pnas.36.1.48

Nash, J. F. Jr. (1951). Non-cooperative games. *The Annals of Mathematics, 54*, 286–295. doi:10.2307/1969529

Nathan, R. J., & Yeow, P. H. P. (2009). An empirical study of factors affecting the perceived usability of websites for student internet users. *Universal Access in the Information Society, 8*(3).

Nathan, R. J., & Yeow, P. H. P. (2011). Crucial web usability factors of 36 industries for students – A large empirical study. *Journal of Electronic Commerce Research, 12*(2), 150–180.

Nathan, R. J., Yeow, P. H. P., & Murugesan, S. (2008). Key usability factors of service-oriented websites for students: An empirical study. *Online Information Review, 32*(3). doi:10.1108/14684520810889646

National Court Appointed Special Advocate Association. (2010). *2010 annual report*. Retrieved May 2, 2012, from http://nc.casaforchildren.org/files/public/site/communications/AnnualReport2010.pdf

National Court Appointed Special Advocate Association. (2011). *2011 annual report*. Retrieved on August 1, 2012, from http://nc.casaforchildren.org/apps/annualreport/by-the-numbers.html

National Institute of Urban Affairs (NIUA). (2004). *Reforming the property tax system. Research Study Series No. 94*. New Delhi, India: NIUA Press.

National Office for the Information Economy. (2003). *E-government benefits study*. Canberra, Australia: NOIE.

Navarra, D., & Cornford, T. (2003). A policy making view of e-government innovations in public governance. *Proceedings of the Ninth Americas Conference on Information Systems (AMCIS)*, Tampa, Florida, (p. 103).

Ndou, V. D. (2004). E-government for developing countries: Opportunities and challenges. *The Electronic Journal on Information Systems in Developing Countries*, *18*(1), 1–24.

Ndubisi, N. O. (2004). Understanding the salience of cultural dimensions on relationship marketing, it's underpinnings and aftermaths. *Cross Cultural Management*, *11*(3), 70–89. doi:10.1108/13527600410797855

NetHope. (2008). *Obtained through the Internet*. Retrieved from http://www.nethope.org

Neumann, J., & Morgenstern, O. (1944). *Theory of games and economic behavior*. Princeton University Press.

New Zealand Ministry of Economic Development (MED). (2000). *E-commerce: A guide for New Zealand business*. Wellington, New Zealand: Author.

Newman, J. (Ed.). (2005). *Remaking governance: Peoples politics and the public sphere*. Bristol, UK: The Policy Press.

Nielsen, J. (1993). *Usability engineering*. San Diego, CA: Academic Press.

Nielsen, J. (1994). Heuristic evaluation. In Nielsen, J., & Mack, R. L. (Eds.), *Usability inspection methods*. New York, NY: John Wiley and Sons.

Nielsen, J. (2000). *Designing web usability: The practice of simplicity*. Indianapolis, IN: New Riders Publishing.

Nielsen, J. (2003). *Usability 101: Introduction to usability*. Jakob Nielsen's alertbox. Retrieved from http://www.useit.com/alertbox/20030825.html

Nielsen, J. (2005). *Useit.com: Jakob Nielsen's website*. Retrieved from http://www.useit.com

Nielsen, J., & Tahir, M. (2002). *Homepage usability: 50 Web sites deconstructed*. Indianapolis, IN: New Riders Publishing.

Niemi, R. G., & Herrnson, P. S. (2003). Beyond the butterfly: The complexity of U.S. ballots. *Perspectives on Politics*, *1*(2), 317–326. doi:10.1017/S1537592703000239

Nilekani, N. (2004, 25th October). Redemption in this world, this land. *The Economic Times*, Editorial Page. Retrieved from http://economictimes.indiatimes.com/articleshow/897648.cms

Noor, M. M. (2010). Assessing the impacts of rural internet center programs on quality of life in rural areas of Malaysia. *Proceedings of IPID Postgraduate Strand at ICTD* (pp. 43-48). Karlstad University Studies.

Norden, L., Creelan, J. M., Kimball, D., & Quesenbery, W. (2006). *The machinery of democracy: Usability of voting systems*. Brennan Center for Justice, New York University School of Law.

Norden, L., Kimball, D., Quesenbery, W., & Chen, M. (2008). *Better ballots*. Brennan Center for Justice, New York University School of Law.

Norman, D. A. (2002). Emotion and design: Attractive things work better. *Interaction Magazine*, *9*(4), 36–42.

Norris, P. (2001). *Digital divide: Civic engagement, information poverty and the internet worldwide*. Cambridge, UK: Cambridge University Press. doi:10.1017/CBO9781139164887

Norzaidi, M. D., Chong, S. C., Murali, R., & Salwani, M. I. (2007). Intranet usage and managers' performance in the port. *Industrial Management & Data Systems*, *107*(8), 1227–1250. doi:10.1108/02635570710822831

Norzaidi, M. D., Chong, S. C., Murali, R., & Salwani, M. I. (2009). Towards a holistic model in investigating the effects of Intranet usage on managerial performance: A study on Malaysian port industry. *Maritime Policy & Management*, *36*(3), 269–289. doi:10.1080/03088830902861235

Norzaidi, M. D., Chong, S. C., Salwani, M. I., & Lin, B. (2011). The indirect effects of Intranet functionalities on middle managers' performance – Evidence from the maritime industry. *Kybernetes*, *40*(1-2), 166–181. doi:10.1108/03684921111117988

O'Donnell, S. (2001). Analysing the internet and the public sphere: The case of Womenslink. *Javnost (Ljubljana)*, *8*(1), 39–58.

O'Donnell, S., Perley, S., Walmark, B., Burton, K., Beaton, B., & Sark, A. (2007). Community-based broadband organizations and video communications for remote and rural First Nations in Canada. In *Proceedings of Community Informatics Research Network Conference*, Prato, Italy, 2007.

O'Donnell, S., & Ramaioli, G. (2004). Sustaining an online information network for non-profit organizations: The case of community exchange. In *Proceedings of the Community Informatics Research (CIRN) Network Conference. Sustainability and Community Technology: What Does this Mean for Community Informatics?* Prato, Italy, 2004.

O'Neill, O. (2008). Transparency and the ethics of communication. In Hood, C., & Heald, D. (Eds.), *Transparency: The key for better governance?* (pp. 75–91). Oxford, UK: Oxford University Press.

Odendaal, N. (2002). ICTs in development – Who benefits? Use of geographic information systems on the Cato Manor Development Project, South Africa. *Journal of International Development, 14,* 89–100. doi:10.1002/jid.867

OECD. (n.d.). *Denmark: Efficient e-government for smarter service delivery.* OECD Publishing. Retrieved June 1, 2012, from http://dx.doi.org/10.1787/9789264087118-en

OECD. (2005). *Modernising government: The way forward.* Paris, France: OECD.

Ok, S. J., & Shon, J. H. (2006). The determinants of internet banking usage behaviour in Korea: A comparison of two theoretical models. [th December, Adelaide.]. *CollECTor, 06,* 9.

Okediran, O., Omidiora, E., Olabiyisi, S., & Ganiyu, R. (2011). A survey of remote internet voting vulnerabilities. *World of Computer Science and Information Technology Journal, 1*(7), 297–301.

Okediran, O., Omidiora, E., Olabiyisi, S., Ganiyu, R., & Alo, O. (2011). A framework for a multifaceted electronic voting system. *International Journal of Applied Science and Technology, 1*(4), 135–142.

Olowu, D. (2004). *Property taxation and democratic decentralisation in developing countries* (Working Paper Series No. 401). The Hague, The Netherlands: Institute of Social Studies.

Olsen, S. O., Tudoran, A. A., Brunson, K., & Verbeke, W. (2012). Extending the prevalent consumer loyalty modelling: The role of habit strength. *European Journal of Marketing, 47*(1), 1–30.

Olson, J. R., & Olson, G. M. (1990). The growth of cognitive modeling in human-computer interaction since GOMS. *Human-Computer Interaction, 5*(2), 221–265. doi:10.1207/s15327051hci0502&3_4

Oostveen, A., & Van den Besselaar, P. (2009). Users' experiences with e-voting: A comparative case study. *International Journal of Electronic Governance, 2*(4). doi:10.1504/IJEG.2009.030527

Orange, G. Burke, A. Elliman, T., & Kor, A. (2006). CARE: An integrated framework to support continuous, adaptable, reflective evaluation of e-government systems. *European and Mediterranean Conference on Information Systems,* Alicante, Spain, July 6–7.

Organization for Economic Cooperation and Development (OECD). (2008). *Public governance.* Retrieved from http://www.oecd.org/dataoecd/39/19/40556222.pdf

Osimo, D. (2008). *Web 2.0 in government: Why and how?* Luxembourg: Joint Research Centre. Institute for Prospective Technological Studies, Office for Official Publications of the European Communities.

Ou, C. X., & Sia, C. L. (2010). Consumer trust and distrust: An issue of website design. *International Journal of Human-Computer Studies, 68*(12), 913–934. doi:10.1016/j.ijhcs.2010.08.003

Padovani, E., & Young, D. W. (2012). *Managing local governments. Designing Management control systems that deliver value.* London, UK: Routledge.

Pallis, A. A., & Lambrou, M. (2007). Electronic markets business models to integrate ports in supply chains. *Journal of Marine Research, 4*(3), 67–85.

Pallot, J. (2001). Transparency in local government: Antipodean initiatives. *European Accounting Review, 10*(3), 645–660.

Palmer, J. W. (2002). Web site usability, design, and performance metrics. *Information Systems Research, 13*(2), 151–167. doi:10.1287/isre.13.2.151.88

Palvia, P. (2009). The role of trust in e-commerce relational exchange: A unified model. *Information & Management, 46*(4), 213–220. doi:10.1016/j.im.2009.02.003

Papadopoulou, P., Nikolaidou, M., & Martakos, D. (2010). What is trust in e-government? A proposed typology. *Proceedings of the 43rd Hawaii International Conference on System Sciences (HICSS),* (pp. 1-10).

Parks, T. (2005). *A few misconceptions about egovernment.* Retrieved from http://www.asiafoundation.org/pdf/ICT_eGov.pdf

Parthasarathy, M., & Bhattacherjee, A. (1998). Understanding post-adoption behavior in the context of online services. *Information Systems Research, 9*(4), 362–379. doi:10.1287/isre.9.4.362

Pasquier, M., & Villeneuve, J.-P. (2012). *Marketing management and communications in the public sector.* London, UK: Routledge.

Pathak, R. D., & Prasad, R. S. (2005). The role of egovernment in tackling corruption: The Indian experience. In R. Ahmad (Ed.), *The Role of Public Administration in Building a Harmonious Society, Selected Proceedings from the Annual Conference of the Network of Asia-Pacific Schools and Institutes of Public Administration and Governance (NAPSIPAG)*, December 5-7, 2005, (pp. 343 – 463).

Pathak, R. D., Singh, G., Smith, R. F. I., & Naz, R. (2008). *Contribution of information and communication technology in improving government coordination, services and accountability in Fiji.* Retrieved April 22, 2012, from http://www.napsipag.org/pdf/RAGHUVAR_PATHAK.pdf

Patil, V. (2010). Secure EVS by using blind signature and cryptography for voter's privacy & authentication. *Journal of Signal and Image Processing, 1*(1), 1–6.

Patrick, A. (2009). *Ecological validity in studies of security and human behaviour.* Key note talk: ISSNet Workshop. Retrieved April 16, 2012, from http://www.andrewpatrick.ca/cv/Andrew-Patrick-ecological-validity.pdf

Paul, S., & Shah, M. (1997). Corruption in public service delivery. In Guhan, S., & Paul, S. (Eds.), *Corruption in India: Agenda for action.* New Delhi, India: Vision Books.

Pennington, R., Wilcox, H. D., & Grover, V. (2003). The role of system trust in business-to-consumer transactions. *Journal of Management Information Systems, 20,* 197–226.

Pérez de Celis Herrero, C., Lara Alvarez, J., Cossio Aguilar, G., & Somodevilla García, M. J. (2011). An approach to art collections management and content-based recovery. *Journal of Information Processing Systems, 7*(3), 447–458. doi:10.3745/JIPS.2011.7.3.447

Petkov, D., Petkova, O., Andrew, T., & Nepal, T. (2007). Mixing multiple criteria decision making with soft system thinking techniques for decision support in complex situations. *Decision Support Systems, 43*(4), 1615–1629. doi:10.1016/j.dss.2006.03.006

Petricek, V., Escher, T., Cox, I. J., & Margetts, H. (2006). *The web structure of e-government developing a methodology for quantitative evaluation.* International World Wide Web Conference, Edinburgh, UK.

Pettigrew, A., Ferlie, E., & McKee, L. (1992). *Shaping strategic change.* London, UK: Sage.

Pettigrew, A., & Lapsley, I. (1994). Meeting the challenge: Accounting for change. *Financial Accountability and Management, 10*(2), 79–92. doi:10.1111/j.1468-0408.1994.tb00146.x

Pettigrew, A., & Whipp, R. (1991). *Managing change for competitive success.* Oxford, UK: Basil Blackwell.

PEW Research Center. (2011). *Demographics of internet users.* PEW Internet and American Life Project. Retrieved March 28, 2012, from http://pewinternet.org/Static-Pages/Trend-Data/Whos-Online.aspx.

Philippine National Red Cross. (2012). *Donate now.* Retrieved May 11, 2012, from http://www.redcross.org.ph/donatenow

Phillips, A., & Chaparro, A. (2009). Visual appeal vs. usability: Which one influences user perceptions of a Website more? *Usability News,* Vol. 11.2. Retrieved from http://www.surl.org/usabilitynews/112/aesthetic.asp

Pieterson, W., & Ebbers, W. (2008). The use of service channels by citizens in the Netherlands: Implications for multi-channel management. *International Review of Administrative Sciences, 74*(1), 95–110. doi:10.1177/0020852307085736

Pilemalm, S. (2002). *Information technology for non-profit organizations extended participatory design of an information system for trade union shop stewards.* PhD Thesis, 2002, Linköping University, Sweden.

Pini, B., Brown, K., & Previte, J. (2004). Politics and identity in cyberspace: A case study of australian women in agriculture online. In de Donk, W. (Eds.), *Cyberprotest, new media, citizens and social movements* (pp. 259–275). London, UK: Routledge.

Pipek, V. (2005). *From tailoring to appropriation support: Negotiating groupware usage.* PhD Thesis, University of Oulu, Finland.

Pipek, V., Rosson, M. B., Stevens, G., & Wulf, V. (2006). Supporting the appropriation of ICT: End-user development in civil societies. *Journal of Community Informatics, 2*(2).

Plouffe, C. R., Hulland, J. S., & Vandenbosch, M. (2001). Richness versus parsimony in modelling technology adoption decisions - Understanding merchant adoption of a smart card-based payment system. *Information Systems Research, 12*(2), 208–222. doi:10.1287/isre.12.2.208.9697

PNAC. (2006). *Report of PNAC participation in world social forum 2006 Karachi.* Retrieved from http://www.pnac.net.pk/Reports/WSF-Report.pdf

Pollitt, C. (2008). *Time, policy, management. Governing with the past.* Oxford, UK: Oxford University Press.

Pollitt, C., & Bouckaert, G. (2011). *Public management reform: A comparative analysis - New public management, governance, and the neo-Weberian state.* Oxford, UK: Oxford University Press.

Poon, W. C. (2008). Users' adoption of e-banking services: The Malaysian perspective. *Journal of Business and Industrial Marketing, 23*(1), 59–69. doi:10.1108/08858620810841498

Popoveniuc, S., & Hosp, B. (2010). An introduction to punchscan. In Chaum, D., Jakobsson, M., Rivest, R. L., Ryan, P. Y. A., & Benaloh, J. (Eds.), *Towards trustworthy elections* (pp. 242–259). Berlin, Germany: Springer-Verlag. doi:10.1007/978-3-642-12980-3_15

Porter, M. (2001). Strategy and the Internet. *Harvard Business Review,* (March): 63–78.

Prasad, H., Halderman, J., & Gonggrijp, R. (2010). Security analysis of India's electronic voting machines. *Proceeding of 17th ACM Conference on Computer and Communications Security (CCS '10),* October, (pp. 1-24).

Pratchett, L. (1999). New technologies and the modernization of local government: An analysis of biases and constraints. *Public Administration, 77*(4), 731–750. doi:10.1111/1467-9299.00177

Preece, J., Rogers, Y., Sharp, H., Benyon, D., Holland, S., & Carey, T. (1994). *Human computer interaction.* Edinburgh Gate, UK: Addison-Wesley Longman Limited.

Premkumar, G., & Bhattacherjee, A. (2008). Explaining information technology usage: A-test of competing models. *The International Journal of Management Science, 36*(1), 64–75.

Prentzas, J., & Hatzilygeroudis, I. (2011). Techniques, technologies and patents related to intelligent educational systems. In Magoulas, G. D. (Ed.), *E-infrastructures and technologies for lifelong learning: Next generation environments* (pp. 1–28). Hershey, PA: Information Science Reference.

Prosser, A., Schiessl, K., & Fleischhacker, M. (2007). E-voting: Usability and accpetance of two-stage voting procedures. In Wimmer, M. A., Scholl, H. J., & Grönlund, A. (Eds.), *EGOV 2007 (Vol. 4656,* pp. 378–387). Lecture Notes in Computer ScienceBerlin, Germany: Springer-Verlag.

Puiggali, J., & Morales-Rocha, V. (2007). Remote voting schemes: A comparative analysis. *Lecture Notes in Computer Science, 4896,* 16–28. doi:10.1007/978-3-540-77493-8_2

R.A.V. v. St. Paul. (June 22, 1992).505 U.S. 377 U.S. Supreme Court.

Rahim, R. (2011, May 14). 227,851 taxpayers successfully refunded. *The STAR. Nation (New York, N.Y.),* N3.

Ramayah, T., Ramoo, V., & Ibrahim, A. (2008). Profiling online and manual tax filers: Results from an exploratory study in Penang, Malaysia. *Labuan e-Journal of Muamalat and Society, 2,* 1-8.

Randall, D., Harper, R., & Rouncefield, M. (2007). *Fieldwork for design: Theory and practice. Computer Supported Cooperative Work.* New York, NY: Springer.

Ranganathan, S. R. (1965). The colon classification. In Artandi, S. (Ed.), *Rutgers series on systems for the intellectual organization of information (Vol. 4).* Rutgers, NJ: Rutgers University Press.

Rao, G., & Patil, S. (2011). Three dimensional virtual environment for secured and reliable authentication. *Journal of Engineering Research and Studies, 2*(2), 68–75.

Rao, N. R. (1986). *Municipal finances in India (theory and practice)*. New Delhi, India: Inter-India Publications.

Rao, V. (2003). *Property tax reforms in Bangalore*. Paper presented to the Innovations in Local Revenue Mobilisation Seminar. Retrieved from http://www1.worldbank.org/publicsector/decentralization/June2003SeminarPresentations/VasanthRao.ppt

Reddick, C. (2004). Empirical models of e-government growth in local governments. *E- Service Journal, 3*(2).

Reddick, C. (2005). Citizen interaction with e-government: From the streets to servers? *Government Information Quarterly, 22*(1), 38–57. doi:10.1016/j.giq.2004.10.003

Reddick, C. (2010). Comparing citizens' use of e-government to alternative service channels. *International Journal of Electronic Government Research, 6*(2), 54–67. doi:10.4018/jegr.2010040104

Reddy, A. (2011). A case study on Indian E.V.M.S using biometrics. *International Journal of Engineering Science & Advanced Technology, 1*(1), 40–42.

Redell, T., & Woolcock, G. (2004). From consultation to participatory governance? A critical review of citizen engagement strategies in Queensland. *The Australian Journal of Public Administration, 63*(3), 75–87. doi:10.1111/j.1467-8500.2004.00392.x

Reid, M., & Levy, Y. (2008). Integrating trust and computer self-efficacy with technology acceptance model: An empirical assessment of customers' acceptance of banking information systems (BIS) in Jamaica. *Journal of Internet Banking and Commerce, 12*(3). Retrieved 15th April, 2009, from www.arraydev.com/commerce/jibc

Remenyi, D., Money, A., Sherwood-Smith, M., & Irani, Z. (2000). *Effective measurement and management of IT costs and benefits*. Oxford, UK: Butterworth-Heinemann.

Remmert, M. (2004). Towards European standards on electronic voting. *Proceedings of the Electronic Voting in Europe Technology, Law, Politics and Society, A Workshop of the ESF TED Programme together with GI and OCG*, July 7th–9th, 2004 in Schloß Hofen/Bregenz, Lake of Constance, Austria. Retrieved from http://www.gi-ev.de/LNI

Rexha, B., Dervishi, R., & Neziri, V. (2011). Increasing the trustworthiness of e-voting systems using smart cards and digital certificates – Kosovo case. *Proceedings of the 10th WSEAS International Conference on E-Activities (E-ACTIVITIES '11)*, Jakarta, Island of Java, Indonesia, December 1-3, 2011, (pp. 208-212).

Rhodes, R. A. W. (1996). The new governance: Governing without government. *Political Studies, 44*, 652–667. doi:10.1111/j.1467-9248.1996.tb01747.x

Ribeiro, E. F. N. (2006). Urban growth and transformations in India: Issues and challenges. In *Proceedings of the National Seminar on Urban Governance in the Context of the Jawaharlal Nehru National Urban Renewal Mission*, India Habitat Centre, New Delhi 24th – 25th November 2006, (pp. 1 – 11).

Rigby, E. (2004). Usability can work for all online marketers. *Revolution (Staten Island, N.Y.)*, 56.

Rivero, M. (1999). Web objectives save time, money. *New Hampshire Business Review, 21*(27), 6.

Rizky, A., Medawati, H., & Hermana, B. (2012). Do information's richness of provincial government websites will support regional economies in Indonesia. In *Proceedings of 3rd International Conference on E-education, E-business, E-management and E-learning*, (IPDER, Vol. 27, pp. 200-204). Singapore: IACSIT Press.

Robertson, J. (2009). *What every intranet team should know*. Broadway, Australia: Step Two Designs.

Robertson, J. (2010). *Designing intranets - Creating sites that work*. Broadway, Australia: Step Two Designs.

Robins, D., & Holmes, J. (2008). Aesthetics and credibility in Web site design. *Information Processing & Management, 44*(1), 386. doi:10.1016/j.ipm.2007.02.003

Rodríguez, M. P., Alcaide, L., & López, A. M. (2010). Trends of e-government research: Contextualization and research opportunities. *International Journal of Digital Accounting Research, 10*, 87–111. doi:10.4192/1577-8517-v10_4

Roeger, K. L., Blackwood, A., & Pettijohn, S. L. (2011). *The nonprofit sector in brief: Public charities, giving, and volunteering*. Washington, DC: The Urban Institute.

Rogers, E. M. (1995). *Diffusion of innovations* (4th ed.). New York, NY: The Free Press.

Rohde, M. (2004). Find what binds: Building social capital in an Iranian NGO community system. In Huysman, M., & Wulf, V. (Eds.), *Social capital and information technology* (pp. 75–112). Cambridge, MA: MIT Press.

Ronaghan, S. A. (2002). *Benchmarking e-government: A global perspective.* The United Nations Division for Public Economics and Public Administration (DPEPA) Report.

Rose, S., Clark, M., Samouel, P., & Hair, N. (2012). Online customer experience in e-retailing: An empirical model of antecedents and outcomes. *Journal of Retailing, 88*(2), 308–322. doi:10.1016/j.jretai.2012.03.001

Rosengard, J. K. (1998). *Property tax reform in developing countries.* Boston, MA: Kluwer Academic Publications. doi:10.1007/978-1-4615-5667-1

Roth, S. K. (1998). Disenfranchised by design: Voting systems and the election process. *Information Design Journal, 9*(1), 29–38. doi:10.1075/idj.9.1.08kin

Rowley, J. (2011). E-government stakeholders – Who are they and what do they want. *International Journal of Information Management, 31*(1), 53–62. doi:10.1016/j.ijinfomgt.2010.05.005

Roy, M. C., & Bouchard, L. (1999). Developing and evaluating methods for user satisfaction measurement in practice. *Journal of Information Technology Management, 10*(3-4), 49–58.

Roy, S. (2005). *Globalisation, ICT and developing nations: Challenges in the information age.* New Delhi, India: Sage Publications.

Rubin, J., & Chisnell, D. (2008). *Handbook of usability testing: How to plan, design, and conduct effective tests* (2nd ed.). Indianapolis, IN: Wiley Publishing.

Saaty, T. L. (1980). *The analytic hierarchy process.* New York, NY: McGraw-Hill.

Saaty, T. L. (2008). Decision making with the analytic hierarchy process. *International Journal of Services Sciences, 1*(1), 83–98. doi:10.1504/IJSSCI.2008.017590

Sacco, G. M. (2000). Dynamic taxonomies: A model for large information bases. *IEEE Transactions on Knowledge and Data Engineering, 12*(2), 468–479. doi:10.1109/69.846296

Sacco, G. M. (2003). The intelligent e-sales clerk: The basic ideas. *Proceedings of INTERACT'03 -- Ninth IFIP TC13 International Conference on Human-Computer Interaction,* (pp. 876-879).

Sacco, G. M. (2005). Guided interactive diagnostic systems. *18th IEEE International Symposium on Computer-Based Medical Systems (CBMS'05),* (pp. 117-122).

Sacco, G. M. (2005). Guided interactive information access for e-citizens. In *EGOV05 – International Conference on E-Government, within the Dexa Conference Framework, Springer Lecture Notes in Computer Science 3591,* (pp. 261-268).

Sacco, G. M. (2005). No (e-)democracy without (e-) knowledge. In *E-Government: Towards Electronic Democracy, International Conference IFIP TCGOV 2005, Lecture Notes in Computer Science 3416,* Bolzano, (pp. 147-156). Springer

Sacco, G. M. (2006). Analysis and validation of information access through mono, multidimensional and dynamic taxonomies. *FQAS 2006, 7th International Conference on Flexible Query Answering Systems, Lecture Notes in Artificial Intelligence.* Springer.

Sacco, G. M. (2006). *User-centric access to e-government information: e-citizen discovery of e-services.* 2006 AAAI Spring Symposium Series, Stanford University.

Sacco, G. M. (2012). Global guided interactive diagnosis through dynamic taxonomies. *18th IEEE International Symposium on Computer-Based Medical Systems (CBMS'12),* Rome.

Sacco, G. M., Nigrelli, G., Bosio, A., Chiarle, M., & Luino, F. (2012). Dynamic taxonomies applied to a web-based relational database for geo-hydrological risk mitigation. *Computers & Geosciences, 39,* 182–187. doi:10.1016/j.cageo.2011.07.005

Sacco, G. M., & Tzitzikas, Y. (Eds.). (2009). *Dynamic taxonomies and faceted search – Theory, practice, and experience. The Information Retrieval Series* (Vol. 25). Springer.

Sachdeva, P. (1993). *Urban local government and administration in India.* Allahabad, India: Kitab Mahal.

Saeed, S., Pipek, V., Rohde, M., & Wulf, V. (2010). Managing nomadic knowledge: A case study of the European Social Forum. In *28th International Conference on Human Factors in Computing Systems,* (pp. 537-546). New York, NY: ACM.

Saeed, S., & Rohde, M. (2010). Computer enabled social movements? Usage of a collaborative web platform within the European Social Forum. In *9ᵗʰ International conference on the Design of Cooperative Systems,* (pp. 245-264). Springer.

Saeed, S., Rohde, M., & Wulf, V. (2008). A framework towards IT appropriation in voulantary organizations. *International Journal of Knowledge and Learning, 4*(5), 438–451. doi:10.1504/IJKL.2008.022062

Saeed, S., Rohde, M., & Wulf, V. (2009). Technologies within transnational social activist communities: An ethnographic study of the European Social Forum. In *Fourth international Conference on Communities and Technologies (C&T '09)*, (pp. 85-94). New York, NY: ACM.

Saeed, S., Rohde, M., & Wulf, V. (2011). Analyzing political activists' organization practices: Findings from a long term case study of the European Social Forum. *Journal of Computer Supported Collaborative Work, 20*(4-5), 265–304. doi:10.1007/s10606-011-9144-0

Saeed, S., Rohde, M., & Wulf, V. (2011). *Communicating in a transnational network of social activists: The crucial importance of mailing list usage.* In 17ᵗʰ CRIWG Conference on Collaboration and Technology. Springer.

Saeed, S., Rohde, M., & Wulf, V. (2012). Civil society organizations in knowledge society: A roadmap for ICT support in Pakistani NGOs. *International Journal of Asian Business and Information Management, 3*(2), 23–35. doi:10.4018/jabim.2012040103

Saeed, S., Rohde, M., & Wulf, V. (2012). IT for social activists: A study of World Social Forum 2006 organizing process. *International Journal of Asian Business and Information Management, 3*(2), 62–73. doi:10.4018/jabim.2012040106

Sahu, G. P., Panda, P., Gupta, P., Ayaluri, S., Bagga, R. K., & Prabhu, G. S. N. (2010). E-governance project assessment: Using analytical hierarchical process methodology. In Gupta, P., Bagga, R. K., & Ayaluri, S. (Eds.), *Enablers of change: Selected e-governance initiatives in India* (pp. 21–51). India: IUP.

Sahu, H., & Choudhray, A. (2011). Intelligent polling system using GSM technology. *International Journal of Engineering Science and Technology, 3*(7), 5641–5645.

Sakowicz, M. (2006). *How to evaluate e-government? Different methodologies and methods.* Warsaw School of Economics, Department of Public Administration. Retrieved from http://unpan1.un.org/intradoc/ groups/ public/documents/NISPAcee/UNPAN009486.pdf

Sandikkaya, M., & Orencik, B. (2006). Agent based offline electronic voting. *International Journal Of Social Sciences, 1*(4), 259–263.

Sandoval-Almazan, R., & Gil-García, J. (2012). Are government internet portals evolving towards more interaction, participation, and collaboration? Revisiting the rhetoric of e-government among municipalities. *Government Information Quarterly, 29*(2), S72–S81. doi:10.1016/j.giq.2011.09.004

Sang, S., Lee, J. D., & Lee, J. (2010). E-government adoption in ASEAN: The case of Cambodia. *Internet Research, 19*(5), 517–534. doi:10.1108/10662240910998869

Sarkar, I., Alisherov, F., Kim, T., & Bhattacharyya, D. (2010). Palm vein authentication system: A review. *International Journal of Control and Automation, 3*(1), 27–33.

Sathye, M. (1999). Adoption of internet banking by Australian consumers: An empirical investigation. *International Journal of Bank Marketing, 17*(7), 324–334. doi:10.1108/02652329910305689

Sattar, A., & Baig, R. (2001). Civil society in Pakistan: A preliminary report. *CIVICUS Index on Civil Society Occasional Paper Series, 1*(11).

Saveourvotes.org. (2008). *Cost analysis of Maryland's electronic voting system, 2008.* Retrieved from www. saveourvotes.org

Saxena, K. B. C. (2005). Towards excellence in e-governance. *Centre for Excellence in Information Management (CEXIM) Working Paper Series,* No. 2005-1, (pp. 1-15).

Scaplehorn, G. (2012). *Bringing the internet indoors: Socializing your intranet.* Retrieved July 30, 2012, from http://www.contentformula.com/articles/2010/bringing-the-internet-indoors-socializing-your-intranet

Schaupp, L. C., & Carter, L. (2010). The impact of trust, risk and optimism bias on e-file adoption. *Information Systems Frontiers, 12*(3), 299–309. doi:10.1007/s10796-008-9138-8

Schaupp, L. C., Carter, L., & McBride, M. E. (2010). E-file adoption: A study of U.S. taxpayers' intentions. *Computers in Human Behavior, 26*(4), 636–644. doi:10.1016/j.chb.2009.12.017

Schneiderman, B., & Plaisant, C. (2005). *Designing the user interface: Strategies for effective human computer-interaction* (4th ed.). Reading, MA: Addison-Wesley.

Schware, R. (2000). Useful starting points for future projects. In Bhatnagar, S., & Schware, R. (Eds.), *Information and communication technology in development: Cases from India* (pp. 206–213). Delhi, India: Sage Publications.

Sedera, D., Gable, G., & Rosemann, M. (2001). A balanced scoreboard approach to enterprise system performance measurement. *Proceedings of the Twelfth Australasian Conference on Information Systems.*

Selker, T., Hockenberry, M., Goler, J., & Sullivan, S. (2005). *Orienting graphical user interfaces reduce errors: The low error voting interface.* VTP Working Paper #23. Retrieved on August 3, 2012 from www.vote.caltech.edu.

Selker, T., Rosenzweig, E., & Pandolfo, A. (2006). A methodology for testing voting systems. *Journal of Usability Studies, 2*(1), 7–21.

Sendong Relief Operations Facebook Account. (2012). Retrieved May 10, 2012, from www.Facebook.com/pages/Sendong-Relief-Operations

Serafeimidis, V., & Smithson, S. (2000). Information systems evaluation in practice: A case study of organizational change. *Journal of Information Technology, 15*(2), 93–105. doi:10.1080/026839600344294

Shan, S., Wang, L., Wang, J., Hao, Y., & Hua, F. (2010). Research on e-government evaluation model based on the principal component analysis. *Information Technology Management, 12*, 173–185. doi:10.1007/s10799-011-0083-8

Shari, I. (2010, May 18). Govt forks out a bomb for taxpayers who shy away from e-filing. *The STAR.* Retrieved from http://thestar.com.my/news/story.asp?file=/2010/5/18/nation/6281573&sec=nation

Sharma, S., & Gupta, J. (2003). Building blocks of an e-government: A framework. *Journal of Electronic Commerce in Organizations, 1*(4), 1–15. doi:10.4018/jeco.2003100103

Sharp, H., Rogers, Y., & Preece, J. (2007). *Interaction design: Beyond human-computer interaction.* John Wiley and Sons.

Shenkman, B. O., & Jonsson, F. (2000). Aesthetics and preferences of Web pages. *Behaviour & Information Technology, 19*, 367–377. doi:10.1080/014492900750000063

Shung, S., & Seddon, P. (2000). *A comprehensive framework for classifying the benefits of ERP systems.* Paper presented at American Conference on Information Systems, Dallas, Texas.

Siau, K., & Shen, K. (2003). Building customer trust in mobile commerce. *Communications of the ACM, 46*(4), 91–94. doi:10.1145/641205.641211

Silverstone, R., & Haddon, L. (1996). Design and the domestication of information and communication technologies: Technical change and everyday life. In Silverstone, R., & Mansell, R. (Eds.), *Communication by design: The politics of information and communication technologies* (pp. 44–74). Oxford, UK: Oxford University Press.

Simon, H. A. (1996). *The sciences of the artificial* (3rd ed.). Cambridge, MA: MIT Press.

Simpson, D., & Moll, L. (1994). *The crazy quilt of government: Units of government in Cook County, 1993.* Office of Publications Services of the University of Illinois.

Singh, A. (1990). Computerisation of the Indian income tax department. *Information Technology for Development, 5*(3), 235–251. doi:10.1080/02681102.1990.9627198

Singh, N. (1996). *Governance and reform in India.* Paper presented at Indian National Economic Policy in an Era of Global Reform: An Assessment, Cornell University, March 29-30 1996.

Singh, S. S., & Misra, S. (1993). *Legislative framework of Panchayati Raj in India.* New Delhi, India: Intellectual Publishing House.

Slaughter, L., Norman, K. L., & Shneiderman, B. (1995). Assessing users' subjective satisfaction with the information system for youth services (ISYS). In *Proceedings of the Third Annual Mid-Atlantic Human Factors Conference* (pp. 164-170).

Smith, A. (2010). *Government online*. Pew Research Center. Retrieved March 20, 2012, from http://www.pewinternet.org/Reports/2010/Government-Online.aspx

Smith, K. A. (2004). Voluntary reporting performance measures to the public: A test of accounting reports from U.S. cities. *International Public Management Journal*, *7*(1), 19–48.

Smith, K. G., & Grimm, C. M. (1987). Environmental variation, strategic change and firm performance: A study of railroad deregulation. *Strategic Management Journal*, *8*(4), 363–376. doi:10.1002/smj.4250080406

Smith, R. (2002). Electronic voting: Benefit and risks. *Australian Institute of Criminology*, *224*, 1–6.

Socialbakers. (2012). *Facebook statistics*. Retrieved May 11, 2012, from http://www.socialbakers.com/facebook-statistics/philippines

SocialFresch. (n.d.). *Community manager report2012*. Retrieved June 10, 2012, from http://socialfresh.com/community-manager-report-2012/

Sodiya, A., Onashoga, S., & Adelani, D. (2011). A secure e-voting architecture. *Proceeding of 8th International Conference on Information Technology: New Generations*, Las Vegas, Nevada, USA, April 11-13, 2011, (pp. 342-347).

Solehria, S., & Jadoon, S. (2011). Cost effective online voting system for Pakistan. *International Journal of Electrical & Computer Sciences*, *11*(3), 39–47.

Sorrentino, M., Naggi, R., & Luca Agostini, P. (2009). E-government implementation evaluation: Opening the black box. *Proceedings of the 8th International Conference on Electronic Government*, Linz, Austria.

Southeastern Promotions, Ltd. V. Conrad et al. (October 17, 1974).420 U.S. 546; 95 S. Ct. 1239; 43 L. Ed. 2d 448; 1975 U.S. LEXIS 3; 1 Media L. Rep. 1140 U.S. Supreme Court.

Spot.ph. (2012). *"Its More Fun in the Philippines" meme: Top 30 fun photos on the web*. Retrieved May 25, 2012, from http://www.spot.ph/featured/50181/its-more-fun-in-the-philippines-meme-top-30-fun-photos-on-the-web

Sreenivisan, R., & Singh, A. (2009). An overview of regulatory approaches to ICTs in Asia and thoughts on best practices for the future. In Akhtar, S., & Arinto, P. (Eds.), *Digital review of Asia Pacific 2009-2010* (pp. 15–24). Haryana, India: Orbicom, IDRC and Sage Publications.

Staggers, N., & Norgio, A. F. (1993). Mental models: Concepts for human computer interaction research. *International Journal of Man-Machine Studies*, *38*(4), 587–605. doi:10.1006/imms.1993.1028

Standard Digital. (2012, July 22). Suddenly, everybody loves Maina Njenga. *Standard Digital*. Retrieved from http://www.standardmedia.co.ke/?articleID=2000003272&pageNo=2

State of Illinois. (2010). *2000 census population compared to 1990: Illinois municipalities*. State of Illinois. Retrieved November 22, 2011, from http://illinoisgis.ito.state.il.us/census2000/dplace_census.asp?theSelCnty=001

State of Illinois. (2010). *Census 2010*. State of Illinois. Retrieved April 29, 2012, from http://www2.illinois.gov/census/Pages/default.aspx

State of Illinois. (2010). Freedom of Information Act (5 ILCS 140).

Stenerud, I. S. G., & Bull, C. (2012). When reality comes knocking: Norwegian experiences with verifiable electronic voting. In M. J. Kripp, M. Volkamer, & R. Grimm. (Eds.) *5th International Conference on Electronic Voting EVOTE 2012* (pp. 21-33). Bregenz, Austria.

Stevens, G. (2009). *Understanding and designing appropriation infrastructures*. PhD Thesis University of Siegen, Germany.

Stoker, G. (1998). Governance as theory: Five propositions. *International Social Science Journal*, *50*(155), 17–28. doi:10.1111/1468-2451.00106

Storer, T., Little, L., & Duncan, I. (2006). An exploratory study of voter attitudes towards a pollsterless remote voting system. In D. Chaum, R. Rivest & P. Y. A. Ryan (Eds.), *IaVoSS Workshop on Trustworthy Elections (WOTE 06) Pre-Proceedings* (pp. 77–86).

Straub, D., Boudreau, M. C., & Gefen, D. (2004). Validation guidelines for IS positivist research. *Communications of AIS*, *13*, 380–427.

Streib, G., & Navarro, I. (2006). Citizen demand for interactive e-government: The case of Georgia consumer services. *American Review of Public Administration*, *36*(3), 288–300. doi:10.1177/0275074005283371

Streib, G. D., & Willoughby, K. G. (2005). Local governments as e-governments: Meeting the implementation challenge. *Public Administration Quarterly*, *29*(1), 78–110.

Suki, N. M., & Ramayah, T. (2010). User acceptance of the e-government services in Malaysia: Structural equation modelling approach. *Interdisciplinary Journal of Information, Knowledge, and Management*, *5*, 395–413.

Swanson, E. B. (1982). Measuring user attitudes in MIS research: A review. *Omega*, *10*(2), 157–165. doi:10.1016/0305-0483(82)90050-0

Swift, O. (2011). *Developments in new technology & implications for seafarers' welfare – Seafarers' access to WiFi and WiMax in ports.* Technical Report, International Committee on Seafarers' Welfare.

Symons, V., & Walsham, G. (1988). The evaluation of information systems: A critique. *Journal of Applied Systems Analysis*, *15*, 119–132.

Szajna, B. (1996). Empirical evaluation of the revised technology acceptance model. *Management Science*, *42*(1), 85–92. doi:10.1287/mnsc.42.1.85

Szmigin, I., & Foxall, G. (1998). Three forms of innovation resistance: The case of retail payment methods. *Technovation*, *18*(6/7), 459–468. doi:10.1016/S0166-4972(98)00030-3

Tajuddin, U. (2010). *From pilot to national initiative.* Retrieved from http://www.itu.int/ITUD/asp/CMS/ASP-CoE/2010/IRD/S4-Mr_Uzer_Tajuddin.pdf

Tan, C. W., Benbasat, I., & Cenfetelli, R. T. (2008). Building citizen trust towards e-government services: Do high quality websites matter? *Proceedings of the International Conference on System Sciences, Vancouver*, 7-10 Jan, (p. 217).

Tan, M., & Teo, T. S. H. (2000). Factors influencing the adoption of internet banking. *Journal of the Association for Information Systems*, *1*, 1–42.

Taylor, J., Lips, M., & Organ, J. (2007). Information-intensive government and the layering and sorting of citizenship. *Public Money and Management*, *27*(2), 161–164. doi:10.1111/j.1467-9302.2007.00573.x

Taylor, J., & Williams, H. (1988). *Information and communication technologies and the transformation of local government* (Working Paper 9). Centre for Urban and Regional Development Studies (Newcastle).

Taylor, S., & Todd, P. A. (1995). Understanding information technology usage: A test of competing models. *Information Systems Research*, *6*(2), 144–176. doi:10.1287/isre.6.2.144

Teo, H. H., Oh, L. B., Lui, C., & Wei, K. K. (2003). An empirical study of the effects of interactivity on Web user attitude. *International Journal of Human-Computer Studies*, *58*, 281–305. doi:10.1016/S1071-5819(03)00008-9

Teo, T. S. H., Srivastava, S. C., & Jiang, L. (2008). Trust and electronic government success: An empirical study. *Journal of Management Information Systems*, *25*(3), 99–131. doi:10.2753/MIS0742-1222250303

Ter Bogt, H. J., & Van Helden, G. J. (2000). Management control and performance measurement in Dutch local government. *Management Accounting Research*, *11*(2), 263–279. doi:10.1006/mare.2000.0132

Terry Ma, H., & Zaphiris, P. (2003). *The usability and content accessibility of the e-government in the UK.* London, UK: Centre for Human-Computer Interaction Design, City University. Retrieved from http://www.soi.city.ac.uk/~zaphiri/Papers/

The Crusader Publication Facebook Account. (2012). Retrieved June 11, 2012, from http://www.facebook.com/thecrusaderpublication

The Economic Times. (2008, June 8). Urban India gets under the digital mapping radar. *The Economic Times*, p. 14.

The eGovernments Foundation. (2003). *Street naming and property numbering guide.* Bangalore, India: Author.

The eGovernments Foundation. (2004). *The property tax information system with GIS.* Presentation document. Bangalore, India: Author.

The Government of India. (2003). *Electronic governance – A concept paper.* Retrieved from http://egov.mit.gov.in

The Government of Karnataka. (2005). *A note on the process of implementation of computerisation etc – Guidance* notes. Unpublished.

The Times of India. (2006, July 22). E-governance, GIS: New face of BMP. *The Times of India,* p.1

The Times of India. (2009, January 8). Popular debut for online tax calculator: Applicable for residential properties, citizens rue increase in net amount. *The Times of India,* p. 2

The Times of India. (2009, January 10). E-calculator spreads its wings. *The Times of India,* p. 2

The World Bank. (2004). *Building blocks of egovernment: Lessons from developing countries* (PREM Notes No. 91), August 2004.

Themistocleous, M., & Irani, Z. (2001). Benchmarking the benefits and barriers of application integration. *Benchmarking: An International Journal, 8*(4), 317–331. doi:10.1108/14635770110403828

Theodorakis, N. D., Koustelios, A., Robinson, L., & Barlas, A. (2009). Moderating role of team identification on the relationship between service quality and repurchase intentions among spectators of professional sports. *Managing Service Quality, 19*(4), 456–473. doi:10.1108/09604520910971557

Thomas, J. C. (2004). Public involvement in public administration in the information age: Speculations on the effects of technology. In Malkia, M., Anttiroiko, A., & Savolainen, R. (Eds.), *Etransformation in governance: New directions in government and politics* (pp. 67–84). Hershey, PA: Idea Group Publishing.

Thomas, J., & Streib, G. (2003). The new face of government: Citizen-initiated contacts in the era of e-government. *Journal of Public Administration: Research and Theory, 13*(1), 83–102. doi:10.1093/jpart/mug010

Tilson, R., Dong, J., Martin, S., & Kieche, E. (1998). *Factors and principles affecting the usability of four e-commerce sites. Proceedings of Human Factors and the Web, June 5, 1998.* NJ, US: Basking Ridge.

Titah, R., & Barki, H. (2006). E-government adoption and acceptance: A literature review. *International Journal of Electronic Government Research, 2*(3), 23–57. doi:10.4018/jegr.2006070102

Toh, K. K. T., Welsh, K., & Hassall, K. (2010). A collaboration service model for a global port cluster. *International Journal of Engineering Business Management, 2*(1), 29–34.

Tolbert, C. J., & Mossberger, K. (2006). The effects of e-government on trust and confidence in government. *Public Administration Review, 66*(3), 354–369. doi:10.1111/j.1540-6210.2006.00594.x

Tornatzky, L. G., & Klein, K. J. (1982). Innovation characteristics and innovation adoption implementation: A meta-analysis of findings. *IEEE Transactions on Engineering Management, 29*(1), 28–45.

Torres, L., Pina, V., & Acerete, B. (2005). E-government developments on delivering public services among E.U. cities. *Government Information Quarterly, 22*(2), 217. doi:10.1016/j.giq.2005.02.004

Tractinsky, N. (1997). Aesthetics and apparent usability: Empirically assessing cultural and methodological issues. *Proceedings of CHI 97 Conference: Looking to the Future,* March 22-27, 1997, Atlanta, Georgia, US.

Tractinsky, N., Katz, A. S., & Ikar, D. (2000). What is beautiful is usable. *Interacting with Computers, 13*(2), 127–145. doi:10.1016/S0953-5438(00)00031-X

Traugott, M. W., Hanmer, M. J., Park, W., Herrnson, P. S., Niemi, R. G., Bederson, B. B., & Conrad, F. G. (2005). *The impact of voting systems on residual votes, incomplete ballots, and other measures of voting behavior.* Paper presented at the Annual Meeting of the Midwest Political Science Association, Chicago, IL.

Trigg, R. H. (2000). From sand box to "fund box": Weaving participatory design into the fabric of a busy non-profit. In *Proceedings of the Participatory Design Conference,* Palo Alto CA, 2000, (pp. 174-183).

Trigg, R. H., Moran, T. P., & Halasz, F. G. (1987). Adaptability and tailorability in NoteCards. [North-Holland.]. *Proceedings of IFIP INTERACT, 87,* 723–728.

Trust, B. B. C. (2006). *Africa media development initiatives.* AMDI.

Tuomi, I. (2002). *Networks of innovation: Change and meaning in the age of the internet.* Oxford, UK: Oxford University Press.

Turban, E., King, D., Viehland, D., & Lee, J. (2006). *Electronic commerce – A managerial perspective.* New Jersey: Pearson Education Inc.

Turner, V. W., & Bruner, E. M. (1986). *The anthropology of experience.* Urbana, IL: University of Illinois Press.

UNCTAD. (2010). *Informational economy report.* Geneva, Switzerland: Author.

United Nations. (2008). *UN e-government survey – From e-government to connected government.* United Nations. Retrieved from http://unpan1.un.org/intradoc/groups/public/documents/UN/UNPAN028607.pdf

United Nations. (2010). *United Nations global e-government survey.* Retrieved from http://egovernments.wordpress.com/2010/04/15/united-nations-global-e-government-survey-2010/

United Nations Development Programme. (1997). *Corruption and good governance: Discussion paper 3.* New York, India: UNDP.

United Nations E-Government Survey (UNPAN). (2010). *Leveraging e-government at a time of financial and economic crisis.* Retrieved from http://unpan1.un.org/intradoc/groups/public/documents/un/unpan038851.pdf

University of South Florida Virtual Library. (2012). Retrieved from www.lib.usf.edu

Usability First. (2005). Usability in website and software design. *Usability First.* Retrieved from http://www.usabilityfirst.com/methods/index.txl

Van Der Westhuizen, D., & Fitzgerald, E. P. (2005). Defining and measuring project success. *Proceedings of the European Conference on IS Management, Leadership and Governance,* Reading, United Kingdom.

Van Deursen, A., & van Dijk, J. (2009). Improving digital skills for the use of online public information and services. *Government Information Quarterly, 26*(2), 333–340. doi:10.1016/j.giq.2008.11.002

Van Dooren, W., Bouckaert, G., & Halligan, J. (2010). *Performance management in the public sector.* London, UK: Routledge.

Van Hoof, J. J., Gosselt, J. F., & De Jong, M. D. T. (2007). *The reliability and usability of the Nedap voting machine.* University of Twente, Faculty of Behavioural Sciences. Retrieved August 3, 2012, from http://wij-vertrouwenstemcomputersniet.nl/images/c/ca/UT_rapportje_over_nedap.pdf

van Rijsbergen, C. J. (1979). *Information retrieval.* London, UK: Butterworths.

Vargo, S. L., & Lusch, R. L. (2004). Evolving to a new dominant logic for marketing. *Journal of Marketing, 68,* 1–17. doi:10.1509/jmkg.68.1.1.24036

Venkatesh, V., & Davis, F. D. (1996). A model of the antecedents of perceived ease of use: Development and test. *Decision Sciences, 27,* 451–481. doi:10.1111/j.1540-5915.1996.tb01822.x

Venkatesh, V., Morris, M. G., & Ackerman, P. L. (2000). A longitudinal field investigation of gender differences in individual technology adoption decision making processes. *Organizational Behavior and Human Decision Processes, 83,* 33–60. doi:10.1006/obhd.2000.2896

Venkatesh, V., Morris, M. G., Davis, G. B., & Davis, F. D. (2003). User acceptance of information technology: Toward a unified view. *Management Information Systems Quarterly, 27*(3), 425–478.

Victor, G. J., Panikar, A., & Kanhere, V. K. (2007). E-government projects–Importance of post completion audits. *Proceedings of the 5th International Conference of e-Government.* Retrieved from http://www.iceg.net/2007/books/1/20_308.pdf

Vijayadev, V. (2008). *Private communication.* (SAS 02-03 to 06-07 Excel spreadsheet), State Nodal Officer, Municipal Reforms Cell, Directorate of Municipal Administration.

Vijayasarathy, L. R. (2004). Predicting consumer intentions to use on-line shopping: The case for an augmented technology acceptance model. *Information & Management, 41*(6), 747–762. doi:10.1016/j.im.2003.08.011

Village of Downers Grove. (2012). *Village of Downers Grove info.* Retrieved February 1, 2012, from http://www.facebook.com/pages/Village-of-Downers-Grove-Illinois/156234227805840?sk=wall#!/pages/Village-of-Downers-Grove-Illinois/156234227805840?sk=info

Village of Grayslake. (2012). *Village of Grayslake info*. Retrieved February 2, 2012, from http://www.facebook.com/pages/Village-of-Grayslake-IL/209863695730984?sk=info

Vincent, S. (2004). A new property map for Karnataka. *IndiaTogether.org*. Retrieved from http://www.indiatogether.org/2004/mar/gov-karmapgis.htm

Virkar, S. (2011). Exploring property tax administration reform through the use of information and communication technologies: A study of e-government in Karnataka, India. In J. Steyn & S. Fahey (Eds.), *ICTs and sustainable solutions for global development: Theory, practice and the digital divide, Vol. 2: ICTs for development in Asia and the Pacific*. Hershey, PA: IGI Global.

Vredenburg, K., Isensee, S., & Righi, C. (2001). *User-centered design: An integrated approach*. USA: Prentice Hall.

Wade, R. H. (1985). The market for public office: Why the Indian state is not better at development. *World Development*, *13*(4), 467–497. doi:10.1016/0305-750X(85)90052-X

Wang, H., & Zheng, L. (2010). Measuring the performance of the information system using data envelopment analysis. *International Journal of Digital Context Technology and Applications*, *4*(8), 92–101. doi:10.4156/jdcta.vol4.issue8.10

Wang, L., Bretschneider, S., & Gant, J. (2005). Evaluating web-based e-government services with a citizen-centric approach. *Proceedings of 38th Annual Hawaii International Conference on Systems Sciences*, Big Island, Hawaii, January 3–6.

Wang, Y. S., & Liao, Y. W. (2008). Assessing eGovernment systems success: A validation of the DeLone and McLean model of information systems success. *Government Information Quarterly*, *25*, 717–733. doi:10.1016/j.giq.2007.06.002

Wangpipatwong, S., Chutimaskul, W., & Papasratorn, B. (2007). The role of technology acceptance model's beliefs and computer self-efficacy in predicting e-government website continuance intention. *WSEAS Transaction on Information Science and Applications*, *4*(6), 1212–1218.

Wangpipatwong, S., Chutimaskul, W., & Papasratorn, B. (2008). Understanding citizen's continuance intention to use e-government website: A composite view of technology acceptance model and computer self-efficacy. *The Electronic. Journal of E-Government*, *6*(1), 55–64.

Watson, R. T., Berthon, P., Pitt, L. F., & Zinkhan, G. M. (2000). *Electronic commerce: The strategic perspective*. Fort Worth, TX: Dryden Press.

Watson, R. T., & Mundy, B. (2001). A strategic perspective of electronic democracy. *Communications of the ACM*, *44*(1), 27–31. doi:10.1145/357489.357499

Weber, J., & Hengartner, U. (2009). *Usability study of the open audit voting system Helios*. Retrieved August 3, 2012, from http://www.jannaweber.com/wp-content/uploads/2009/09/858Helios.pdf

Weerakkody, V., & Choudrie, J. (2005). Exploring e-government in the UK: Challenges, issues and complexities. *Journal of Information Science and Technology*, *2*(2), 25–45.

Welch, E. (2005). Linking citizen satisfaction with e-government and trust in government. *Journal of Public Administration: Research and Theory*, *15*(3), 371–391. doi:10.1093/jopart/mui021

Welch, E. W., Hinnant, C. C., & Moon, M. J. (2005). Linking citizen satisfaction with e-government and trust in government. *Journal of Public Administration: Research and Theory*, *15*(3), 371–391. doi:10.1093/jopart/mui021

Weldemariam, K., Villafiorita, A., & Mattioli, A. (2007). Assessing procedural risks and threats in e-voting: Challenges and an approach. *Proceeding VOTE-ID'07: 1st International Conference on E-Voting and Identity*, (pp. 1-12).

West, D. (2007). *State and federal e-government in the United States*. Brown University. Retrieved from http://www.insidepolitics.org/egovt07us.pdf

West, D. M. (2005). *Digital government: Technology and public sector performance*. Princeton, NJ: Princeton University Press.

Whipp, R., Rosenfeld, R., & Pettigrew, A. (1987). Understanding strategic change process: Some preliminary British findings. In Pettigrew, A. M. (Ed.), *The management of strategic change* (pp. 14–55). Oxford, UK: Basil Blackwell.

Wijaya, S., Dwiatmoko, A., Surendro, K., & Sastramihardja, H. S. (2012). A statistical analysis of priority factors for local e-government in a developing country: Case study of Yogyakarta Local Government, Indonesia. In IRMA (Ed.), *Digital democracy: Concepts, methodologies, tools, and applications* (pp. 559-576). Hershey, PA: Information Science.

Williams, P. (2002). The learning web: The development, implementation and evaluation of internet-based undergraduate materials for the teaching of key skills. *Active Learning in Higher Education, 3*(1), 40–53. doi:10.1177/1469787402003001004

Wimmer, M., & Holler, U. (2003). Applying a holistic approach to develop user-friendly, customer-oriented e-government portal interfaces. *Lecture Notes in Computer Science, 2615*, 167–178. doi:10.1007/3-540-36572-9_13

Winch, G., & Joyce, P. (2006). Exploring the dynamics of building, and losing, consumer trust in B2C eBusiness. *International Journal of Retail and Distribution Management, 34*(7), 541–555. doi:10.1108/09590550610673617

Winckler, M., Bernhaupt, R., Palanque, P., Lundin, D., Leach, K., & Ryan, P. … Strigini, L. (2009). Assessing the usability of open verifiable e-voting systems: A trial with the system Prêt à Voter. In *Proceedings of ICE-GOV,* (pp. 281 - 296).

Wollersheim, D., & Rahayu, W. (2002). Methodology for creating a sample subset of dynamic taxonomy to use in navigating medical text databases. *Proceedings of IDEAS 2002 Conference,* (pp. 276-284).

World Bank. (2007). *The World Bank website: Report from 2007.* Retrieved May 26, 2011, from http://web.worldbank.org

World Bank. (2010). *Definition of e-government.* Retrieved from http://web.worldbank.org

World Wide Web Consortium. (W3C). (2008). *Web content accessibility guidelines (WCAG).* Retrieved August 10, 2012, from http://www.w3.org/TR/WCAG20/

Xie, H. I., & Cool, C. (2000). Ease of use versus user control: An evaluation of Web and non-Web interfaces of online databases. *Online Information Review, 24*(2), 102–115. doi:10.1108/14684520010330265

Xin, C., Ding, R., & Xie, W. (2010). Performance evaluation of e-government information service system based on balanced scorecard and rough set. *Proceedings of International Conference on Information Management, Innovation Management and Industrial Engineering.*

Yanqing, G. (2010). E-government: Definition, goals, benefits and risks. *International Conference on Management and Service Science (MASS),* 24-26 August, 2010, Wuhan, China, (pp. 1-4).

Yee, K.-P., et al. (2003). Faceted metadata for image search and browsing. Proceedings of ACM CHI 2003, (pp. 401-408).

Young, D. W. (2004). Improving operating room financial performance in a center-of-excellence. *Healthcare Financial Management, 58,* 70–74.

Young, D. W. (2008). *Management accounting in health care organizations.* San Francisco, CA: Jossey-Bass.

Young, D. W., & Saltman, R. B. (1985). *The hospital power equilibrium: Physician behavior and cost control.* Baltimore, MD: The Johns Hopkins Press.

Yousafzai, S. Y., Foxall, G. R., & Pallister, J. G. (2010). Explaining internet banking behavior: Theory of reasoned action, theory of planned behavior, or technology acceptance model? *Journal of Applied Social Psychology, 40*(5), 1172–1202. doi:10.1111/j.1559-1816.2010.00615.x

Yu, J. J., Qin, X. S., Larsen, L. C., Larsen, O., Jayasooriya, A., & Shen, X. L. (2012). A GIS-based management and publication framework for data handling of numerical model results. *Advances in Engineering Software, 45,* 360–369. doi:10.1016/j.advengsoft.2011.10.010

Yumeng, F., Liye, T., Fanbao, L., & Chong, G. (2011). Electronic voting: A review and taxonomy. *American Journal of Engineering and Technology Research, 11*(9), 1937–1946.

Zahed, A., & Sakhi, M. (2011). A novel technique for enhancing security in biometric based authentication systems. *International Journal of Computer and Electrical Engineering, 3*(4), 520–523.

Zakaria, Z., Hussin, Z., Zakaria, Z., Noordin, N. B., Sawal, M. Z. H. B. M., Saad, S. F. B. M., & Kamil, S. B. O. (2009). E-filing system practiced by inland revenue board (IRB): Perception towards malaysian taxpayers. *Cross Cultural Communication, 5*(4), 10.

Zhang, N., Guo, X., & Chen, G. (2007). Diffusion and evaluation of e-government systems: A field study in China. *Proceedings of the 11th Pacific Asia Conference on Information Systems*, (pp. 271-283).

Zickuhr, K., & Madden, M. (2012). Older adults and internet use: *For the first time, half of adults ages 65 and older are online.* Pew Research Center's Internet & American Life Project, June 6, 2012. Retrieved from http://pewinternet.org/Reports/2012/Older-adults-and-internet-use.aspx

Zickuhr, K., & Smith, A. (2012). *Digital differences.* Pew Research Center's Internet & American Life Project, April 13, 2012. Retrieved from http://pewinternet.org/Reports/2012/Digital-differences.aspx

Zmud, R. W. (1982). Diffusion of modern software practices: Influence of centralization and formalization. *Management Science, 28*(12), 1421–1431. doi:10.1287/mnsc.28.12.1421

About the Contributors

Saqib Saeed is an Assistant Professor at the Computer Science department at Bahria University Islamabad, Pakistan. He has a Ph.D.in Information Systems from University of Siegen, Germany and a Master's degree in Software Technology from Stuttgart University of Applied Sciences, Germany. He is also a certified Software Quality Engineer from American Society of Quality. His research interests lie in the areas of human centered computing, computer supported cooperative work, empirical software engineering, and ICT4D.

Christopher G. Reddick is an Associate Professor and Chair of the Department of Public Administration at the University of Texas at San Antonio, USA. Dr. Reddick's research and teaching interests are in information technology and public sector organizations. Some of his publications can be found in *Government Information Quarterly, Electronic Government*, and the *International Journal of Electronic Government Research*. Dr. Reddick recently edited the two volume book entitled Handbook of Research on Strategies for Local E-Government Adoption and Implementation: Comparative Studies. He is also author of the book Homeland Security Preparedness and Information Systems, which deals with the impact of information technology on homeland security preparedness.

* * *

Hisham Abdelsalam holds a Master's of Science and a Ph.D. in Mechanical Engineering (Old Dominion University, Norfolk, Virginia, USA). He obtained his Bachelor's degree with honors in Mechanical Engineering from Cairo University (Cairo, Egypt). Dr. Abdelsalam is an Associate Professor in the Operations Research and Decision Support Department, Faculty of Computers and Information, Cairo University. In 2009, Dr. Abdelsalam was appointed as the Director of the Decision Support and Future Studies Center in Cairo University. During the past four years, D. Abdelsalam has led several consultancy and research projects and published several scholarly articles on e-government.

Emad A. Abu-Shanab earned his PhD in Business Administration, majoring in MIS area in 2005 from Southern Illinois University – Carbondale, USA. He earned his MBA from Wilfrid Laurier University in Canada and his Bachelor in Civil Engineering from Yarmouk University (YU) in Jordan. He is an Associate Professor in MIS, where he taught courses like operations research, e-commerce, e-government, introductory and advanced courses in MIS, production information systems, and legal issues of computing. His research interests are in areas like E-government, technology acceptance, and E-learning. He has published many articles in journals and conferences and authored two books in e-government area. Dr. Emad worked as an Assistant Dean for students' affairs, quality assurance officer in Oman, and the director of Faculty Development Center at YU.

Sujana Adapa is a Lecturer in Management (Strategy & Marketing) in the UNE Business School at the University of New England (UNE), Armidale, Australia. Sujana teaches Introduction to Marketing, Marketing Strategy & Management, Services Marketing, and International Marketing units at UNE for undergraduate and postgraduate students. Her research interests relate to the adoption of technological innovations, corporate social responsibility, sales management, destination visitations, and branding. She has published research papers in reputed journals and presented her research in national and international journals.

Vanni Agnoletti earned MD, specialization in Anaesthesia and Intensive Care, University of Bologna, Italy. He has been consultant anaesthesiologist and teacher of anaesthesia. Currently, he is working at Morgagni-Pierantoni Hospital in Forlì (Italy) as tutor of anaesthesia and clinical practice. He has been a speaker and guest speaker at national and international meetings on thoracic anaesthesia-analgesia, difficult airway, operating room management, quality, and safety in medicine, and is author or co-author of 29 publications listed in PubMed.

Jaffar Ahmad Alalwan is an Assistant Professor at the Institute of Public Administration in the eastern province of Saudi Arabia. He holds a PhD from Virginia Commonwealth University in Information Systems, an MBA from the University of Scranton in Management Information Systems and Marketing, and a BS in Business Administration from King Abdul Aziz University. Alalwan has published research in the areas of strategic information systems planning, enterprise content management systems, electronic government, and the Semantic Web.

Erwin A. Alampay obtained his PhD in Development Administration and Management at the University of Manchester, and his MA in Development Studies at the Institute of Social Studies, The Hague. He is an Associate Professor at the National College of Public Administration and Governance (NCPAG) in the University of the Philippines. He is a Senior Editor for the *Electronic Journal for Information Systems in Developing Countries* (EJISDC) and recently edited the book, "Living the Information Society in Asia" (2009) ISEAS Publishing and IDRC. His research interests are in telecommunication policy, e-governance, ICT for Development, e-commerce, and the non-profit sector. He is the current Director of the Center for Leadership, Citizenship, and Democracy (CLCD) of NCPAG.

Izzat Alsmadi is an Associate Professor in the department of Computer Information Systems at Yarmouk University in Jordan. He obtained his PhD degree in Software Engineering from NDSU (USA). His second Master's in Software Engineering is from NDSU (USA), and his first Master's in CIS is from University of Phoenix (USA). He had a BSc degree in telecommunication engineering from Mutah University in Jordan. He has several published books, journals, and conference articles largely in software engineering different fields.

Upali Aparajita is a Professor in Social and Development Anthropology at Utkal University, Bhubaneswar (India). She earned her MA degree in Social Anthropology in 1985, and PhD degree in Developmental Anthropology in 1991, both from Utkal University. Her research interest is primarily on development issues in traditional societies in the framework of culture. She has published over thirty five research papers in journals of repute as well as in edited volumes including IGI Global publication, dealing with various aspects of management and development highlighting cultural dimension in the development process of various societies as well as organizations. Currently, she is working in Business Anthropology and has guided few Doctoral scholars in Anthropology and Management.

Manuel Pedro Rodríguez Bolívar is an Associate Professor in accounting at Granada University (Spain). He is a member of the European Accounting Association, the World Accounting Forum, the Spanish Association of Accounting University Teachers and the Spanish Association of Accounting and Business Management. His main research interests are focused on e-government, public sector management and international public sector accounting. He is the author of numerous articles in national and international journals, including *Public Money & Management, Public Administration and Development, American Review of Public Administration, Online Information Review, Government Information Quarterly, Administration and Society, Abacus, International Public Management Journal, International Review of Administrative Sciences,* and *Academia Revista Latinoamericana de Administración (ARLA).* He has also contributed chapters to books edited by Kluwer Academic Publishers, Springer, Routledge, NovaPublishers, and IGI Global, and he is author of books published by both the Ministry of Economy and the Ministry of the Treasury in Spain.

Matteo Buccioli graduated in Clinical Engineering, University of Trieste, Italy, is the Health Information Manager at Morgagni-Pierantoni Hospital in Forlì, Italy, and is also a specialist in data management and logistic and management in operating rooms.

Charlie E. Cabotaje obtained his MSc in Governance and Spatial Information Management at the University of Twente's Faculty of Geo-Information Science and Earth Observation (ITC) in The Netherlands. He finished his post graduate Diploma in Urban and Regional Planning at the University of the Philippines' School of Urban and Regional Planning (SURP). He currently works as a University Researcher at the Center for Leadership, Citizenship, and Democracy and as a Lecturer at the National College of Public Administration and Governance (NCPAG). He has written case studies on people's participation in the contexts of poverty alleviation and government intervention. His research interests include environment and governance, people's participation, ICTs, and policy studies.

Vincenzo Cavallo obtained his PhD in 2010 at IULM University. In 2004 he developed an eParticipation project that was awarded as best Public Communication Project at COMPA 2006. In 2007 he worked for the United Nations Department of Economic and Social Affairs. In 2008 he co-founded Cultural Video Foundation, an independent NGO based in Nairobi, working in the field of New Media and Social/Participatory-video productions. In 2009 he started the Urban Mirror project and developed an eParticipation platform using Ushahidi software to map public spaces and urban art in Nairobi. In 2010 he won the International Human Rights Film festival of Naples and in 2011 was awarded by the National Television Committee of Chile. In 2012 was invited to Oxford University to give a seminar about the "eParticipation ecologies of Kenya." He is currently working on a crowd-sourcing project to map sites of memories in Kenya.

Yu-Che Chen, Ph.D., is an Associate Professor in the Division of Public Administration at Northern Illinois University. Dr. Chen received his Master of Public Affairs and Ph.D. in Public Policy from Indiana University. His current research projects are on e-governance performance, government 2.0 and social media, and international implementation of eXtensible Business Reporting Language (XBRL). His most recent co-edited book is entitled "Electronic Governance and Cross-boundary Collaboration." His e-government research can be found in scholarly journals such as *Public Administration Review*, *American Review of Public Administration*, and *Government Information Quarterly*. His teaching interests are in electronic government and collaborative public management. He received NIU's MPA Professor of the Year Awards in 2007 and 2009. He is the Chair of the Technology Advisory Committee and the Chair of the Section on Science and Technology in Government for the American Society for Public Administration (ASPA).

Gregory Derekaris received the Diploma and MSc degrees in Computer Engineering and Informatics from the Department of Computer Engineering and Informatics, University of Patras, Greece, in 1997 and 2000, respectively. He is currently an Informatics teacher at high school. His main research interests include e-learning, web-based systems, GIS, and e-government. He has participated in national and EU research programs. He has published approximately 10 papers in international journals, edited volumes, and proceedings of conferences and workshops.

Hatem Elkadi was born in Egypt in 1960. He graduated from the Faculty of Engineering, Cairo University in 1983. He got his PhD at the University of Lille, France, in 1993. He is currently an Assistant Professor at Cairo University. He is Advisor for Strategic Projects at MSAD (Ministry of State for Administrative Development), and supervises a number of national e-government projects. He was the Director of the Egyptian eGovernment Services Delivery Program which was ranked 23th worldwide in 2010, and in 2008, he won the first prize for the All Africa Public Service Innovation Award. He is member of the "National Dispute Settlement Committee for ICT issues," as well as the steering committee for the "National ID Card project." During his carrier, he managed several successful ICT projects with the government of Egypt, private sector and NGO's, as well as consulted for National Projects in Yemen and Kuwait.

Melissa Foster is a recent Master of Public Administration graduate from Northern Illinois University with a specialization in Public Management and Leadership. She previously worked at the Village of Berkeley, Illinois where she managed the Village's electronic communications and website, as well as contributed to special projects and Village operations on a daily basis. She plans to continue her work in municipal government and to eventually return for her Ph.D. in Public Administration and continue research in the public sector. This chapter was completed with research compiled for her Capstone thesis paper while in the Public Administration Masters program.

Sara Gamal is a Research Assistant at the Decision Support and Future Studies Center, Faculty of Computers and Information – Cairo University. She also worked as Research Assistant at the Information and Decision Support Center, the Egyptian Cabinet. She studied Statistics and Economics at the Faculty of Economic and Political Science (FEPS) – Cairo University, at the moment she is preparing here Masters' degree in Statistics at FEPS.

Antonio M. López Hernández is Professor of Accounting at the University of Granada. He teaches public sector management and control. His research interests are focused on the financial information disclosures on the Web (e-government), on the management system and financial information in the federal and local governments. He has published in journals such as, *Government Information Quarterly, International Review of Administrative Science, American Review of Public Administration, International Public Management Journal, Online Information Review, Public Administration and Development, Public Money & Management*, and *Public Management Review*. He has been also the author of several book chapters published in prestigious international publishers such as Kluwer Academic Publishers, Springer, and IGI Global.

Charles C. Hinnant is an Assistant Professor in the College of Communication and Information at Florida State University. His research interests include social and organizational informatics, knowledge work in distributed teams, digital government, information management and policy, social science research methods, and applied statistics. He is particularly interested in how organizations employ Information and Communication Technology (ICT) to alter organizational processes and structures and how the use of ICT impacts institutional governance and knowledge production mechanisms. His has appeared or is forthcoming in journals such as *Journal of the American Society for Information Science and Technology, Administration and Society, Government Information Quarterly, Journal of Public Administration Research and Theory*, and *IEEE Transactions on Engineering Management*. He earned his B.S. and M.P.A. at North Carolina State University and his Ph.D. in Public Administration from the Maxwell School of Citizenship and Public Affairs at Syracuse University.

Rawan T. Khasawneh graduated from Yarmouk University with a Bachelor of Management Information Systems in 2011, and still a Master's degree student specializing in Management Information System at Yarmouk University in Jordan. Rawan has research interest in e-government topics like digital divide, security issues, and adoption of e-government. Also, she works on research projects in social media and group decision support systems.

Jisue Lee is a Doctoral student in Library and Information Studies, College of Communication and Information at Florida State University. Based on her work experiences in government sectors in Korea, her research interests include social informatics, e-government, social media, and political communication. Her recent interests focus on how citizens use social media to seek and share political information and opinion, and how government agencies/organizations utilize social media channels to interact with citizens. She presented a poster on the adoption of online technology by volunteer-based organization in 13th Annual International Conference on Digital Government Research and currently participates in projects on citizens' political communication through social network site, Twitter. She earned her B.A. in History at Ewha Womans University and her M.A. in Library and Information Science at Yonsei University in Korea.

B. K. Mangaraj is working as a Professor in Production & Operations Management Area at XLRI School of Business and Human Resources, Jamshedpur, India since 1st November, 2006. Prior to joining XLRI, he was a Professor and Head of the Department of Business Administration at Utkal University, Bhubaneswar, India. He did his MSc in Mathematics & Ph.D. in Operations Research, both from Indian

Institute of Technology, Kharagpur, India. He also holds a D. Sc degree in Development Anthropology from Utkal University, India. His areas of research interests include multi-criteria decision-making, fuzzy and stochastic optimisation, cross-cultural management, and business anthropology. Recipient of gold medals and awards for some of his research works; he has already guided nine scholars for their Ph.D. degrees in Management. Several of his publications have already appeared in reputed journals and edited volumes including IGI Global publications. He has also presented forty two research papers in various conferences and has written management teaching cases for classroom discussion. A reviewer for several edited volumes and conference proceedings, he has also been reviewing for international journals including *European Journal of Operational Research, International Journal of Shipping and Transportation Logistics, Agricultural Economics,* and *International Food and Agribusiness Marketing.* His consulting experience ranges from government to private sectors.

Lorri Mon is an Associate Professor at Florida's iSchool, the Florida State University College of Communication and Information in the School of Library and Information Studies. She teaches in the areas of social media management, government information, and reference and information services. Her research has explored the changing nature of the provision of online information services in e-government, online education, and digital libraries, including how users approach and interact with virtual information services, and how institutions and information providers have experimented with implementing information services over a wide range of digital communication technologies including chat, e-mail, instant messaging, text messaging, social media, and virtual worlds.

Robert Jeyakumar Nathan is Bachelors in Marketing and Multimedia (MMU); Masters of Philosophy (Research in Ergonomics and Internet Marketing), Multimedia University, Malaysia. He was formerly a System Analyst for Infineon Technologies AG, a semiconductor manufacturing company based in Munich, Germany, specialising in Manufacturing Statistics and Data Analysis. He has conducted statistical, data mining and enterprise document and knowledge management system trainings in various plants in Asia, Europe, and North America. Mr. Nathan is currently attached to Multimedia University, Malaysia and is active in various research projects in Malaysia, Singapore, Australia, and in the Middle East. His research interests include marketing and information technology, electronic commerce, social networks technologies, usability and ergonomics, and occupational safety & health research. Mr. Nathan is an active member of the National Institute of Occupational Safety and Health (NIOSH) Malaysia and is the Assistant Secretary General for Malaysian Academic Movement; affiliated to The Education International. He currently holds Associate Lecturer position with Newcastle Business School under the University of Newcastle Australia (Singapore Campus).

M. Maina Olembo has been a PhD student at Center for Advanced Security Research Darmstadt (CASED) since January 2011. As a member of SecUSo, and under the supervision of Prof. Dr. Melanie Volkamer, the focus of her PhD work is Usable Verifiability in Remote Electronic Voting. She holds a Bachelor's degree in Computer Science from the University of Nairobi, and a Master's in Information Technology (System Security and Audit) from Strathmore University. Prior to starting her PhD, she was an Assistant Lecturer at Strathmore University in Nairobi, Kenya. She is a professional member of the Association for Computing Machinery (ACM).

Rebecca L. Orelli, BA Business Economics, University of Bologna, Italy, Ph.D. Management of the Public Sector, University of Salerno, Italy, is visiting scholar at London School of Economics, Accounting Department, UK. She is Lecturer in Business Economics, Department of Management, University of Bologna, Italy; affiliated with EBEN; current research interests include new public management and public services changes in local governments, management accounting and management control in public sector. More info at: www.unibo.it/faculty/rebecca.orelli.

Emanuele Padovani, BA Business Economics, University of Bologna, Italy, Ph.D. Business Administration, University of Ferrara, Italy, is Associate Professor of Public Management and Accounting in the Department of Management at the University of Bologna, Italy. She has research interests in public management with specific reference to management control systems, performance measurement for management and auditing, benchmarking, and management of outsourcing. More info at: www.unibo.it/faculty/emanuele.padovani.

Carmen Caba Pérez, Ph. D. in Governmental Accounting, is Associate Professor at the Department of Accounting and Finance at the University of Almeria (Spain). She is a foundational member of the Spanish Association of Accounting University Teachers and member of European Accounting Association. She teaches Financial Accounting and Public Sector Management and control. Her research interests are focused on the financial information disclosures on the Web (e-government), on the management system and financial information in the federal and local governments. She is author of numerous articles in national and international journals indexed at the ISI/JCR including *Government Information Quarterly, Online Information Review, International Review of Administrative Sciences, American Review of Public Administration, Public Administration and Development and Public Management Review*. Also, she has written some chapters to book edited by prestigious international Editorials (Kluwer Academic Publishers, IGI Global, and Springer) on different topics related to e-government.

Jim Prentzas received the Diploma degree in Computer Engineering and Informatics and the MSc and PhD degrees from the Department of Computer Engineering and Informatics, University of Patras, Greece, in 1997, 2001, and 2003, respectively. He is currently an Assistant Professor at the Department of Education Sciences in Pre-School Age, Democritus University of Thrace, Greece. His main research interests include artificial intelligence, knowledge representation and reasoning, expert systems, e-learning, web-based systems and e-government. He has been member of Program Committees of international conferences. He has participated in a number of National and EU research programs. He has published over 60 papers in international journals, edited volumes, and proceedings of conferences and workshops.

T. Ramayah has an MBA from Universiti Sains Malaysia (USM). Currently he is a Professor at the School of Management in USM. He is an avid researcher, especially in the areas of technology management and adoption in business and education. His publications have appeared in *Computers in Human Behavior, Direct Marketing: An International Journal, Information Development, Journal of Project Management (JoPM), Management Research News (MRN), International Journal of Services and Operations Management (IJSOM), Engineering, Construction and Architectural Management (ECAM),* and *North American Journal of Psychology*. He also serves on the editorial boards and program committees of many international journals and conferences of repute. In 2006 he was awarded the "AGBA

Distinguished ASEAN Scholar" for his contribution to research and publication in the ASEAN region. He was a Visiting Professor at the National Taiwan University for a month in 2007.

Markus Rohde studied psychology and sociology at the University of Bonn and is one of the founders of the International Institute for Socio-Informatics (IISI). At the moment, Dr. Rohde is working as project manager for IISI and as research leader of Community Informatics at the Institute for Information Systems,University of Siegen Germany.

Giovanni Maria Sacco is Associate Professor of Information Systems and HCI with the Department of Informatics, University of Torino, Italy. Before that, he had worked at Purdue University, at the IBM San Jose Research Lab (in the System-R group), and at the University of Maryland, among others. Sacco's work on security with Dorothy Denning was the first attack on key distribution protocols and one of the bases of MIT's Kerberos. His work with Mario Schkolnick on buffer management for relational database systems introduced predictive buffer management. He introduced fragmentation, later known as recursive hash partitioning, the first sub-sort/merge join method, which is widely implemented in industry. Since the 80's he has been active in the area of information retrieval, in which he led research and industrial projects. He introduced dynamic taxonomies (aka faceted search) and has published over 25 papers on this topic. He also holds several US patents in information technology.

T. Santhanamery was an Operation Officer for Maybank Berhad from 1996 – 2002. Her experience in banking industry for 6 years was in the area of banking operations and handling customer service. She obtained her first degree in Economics from Universiti Utara Malaysia in the year 1996. She completed her Master's in Business Administration (MBA) from Universiti Utara Malaysia in year 2001. She started her career in lecturing with Faculty of Business Management, Universiti Teknologi MARA Malaysia in year 2002. Currently, she is pursuing her PhD by research in the area of Technology Management in Universiti Sains Malaysia under the scholarship of Ministry of Higher Education (MOHE) Malaysia. Having experience in banking and education industry for almost 14 years, she would like to collaborate and share her experience in technology management and operation management area.

Norazah Mohd Suki is an Associate Professor at the Labuan School of International Business & Finance, Universiti Malaysia Sabah. Her research interests include E-Commerce, M-Commerce, and areas related to Marketing. She actively publishes articles in international journals. She is the editor-in-chief to *Labuan e-Journal of Muamalat & Society*, and member in advisory board for several outstanding journals. She has sound experiences as speaker to public and private universities, government bodies on courses related to Structural Equation Modelling (SEM), Statistical Package for Social Sciences (SPSS), Research Methodology.

Manoj A Thomas is an Assistant Professor in the Department of Information Systems at Virginia Commonwealth University. He holds a Ph.D in Information Systems, Masters in Information Systems, Masters of Business Administration and Bachelor's in Engineering. His primary research interests focus on knowledge engineering, innovation and ICT4D. He has published in many practitioner and academic journals, and has presented at various information systems conferences. He rides and works on motorcycles when he wants to get away from the digital realm.

Athanasios Tsakalidis received a Diploma degree in Mathematics from Aristotle University of Thessaloniki, Greece. Afterwards he received the diploma, MSc and PhD degrees in Computer Science from the University of Saarland in Germany. The supervisor of his PhD Thesis was Kurt Mehlhorn who is currently the Director of Max Planck Institute for Informatics. He is currently a Professor at the Department of Computer Engineering and Informatics, University of Patras, Greece. He is Director of the Graphics, Multimedia, and GIS Laboratory. His research interests include data structures, algorithms, computational geometry, GIS, databases, information retrieval, bioinformatics, multimedia, e-learning, web-based systems, and e-government. He has supervised over twenty PhD theses. He has published many papers in international journals, edited volumes, and proceedings of conferences and workshops. He was one of the 48 authors of the "Handbook of Theoretical Computer Science" published by Elsevier Science and co-published by MIT Press. He has participated in several EU research programs.

Shefali Virkar is research student at the University of Oxford, UK, currently reading for a D.Phil. in Politics. Her doctoral research seeks to explore the growing use of Information and Communication Technologies (ICTs) to promote better governance in the developing world, with special focus on the political and institutional impacts of ICTs on local public administration reform in India. Shefali holds an M.A. in Globalisation, Governance and Development from the University of Warwick, UK. Her Master's thesis analysed the concept of the Digital Divide in a globalising world, its impact developing countries and the ensuing policy implications. At Oxford, Shefali is a member of Keble College.

Melanie Volkamer has been an Assistant Professor at the Department of Computer Science of Technische Universität Darmstadt (Germany) since February 2012. She heads the research group "SecUSo - IT Security, Usability and Society." Before this appointment, she was a postdoctoral researcher in the group of Prof. J. Buchmann at the Technische Universität Darmstadt and coordinated the Research Area "Secure Data" at the Center for Advanced Security Research Darmstadt (CASED). She has been an advisory board member of many e-voting projects and initiatives. In particular, she acted as an OSCE election observer at the first parliamentary remote electronic election in Estonia in 2007. Furthermore, she was invited by the German Federal Constitutional Court as a technical expert for e-voting in 2008. She has presented her research at numerous conferences and to many organizations like the Council of Europe. Prof. Dr. Volkamer is a co-author of two BSI-certified Common Criteria Protection Profiles for electronic voting systems: Basic set of security requirements for Online Voting Products, and Digitales Wahlstift-System (only in German). She received her PhD on: "Evaluation of Electronic Voting: Requirements and Evaluation Procedures to Support Responsible Election Authorities" (published by Springer LNBIP), from the University of Koblenz in October 2008.

Index

A

actors-influences 275
anticorruption 97
authentication methods 71-72, 75, 78, 80, 83
authoring tool 124, 130

B

best practices 7, 15, 17, 21, 23, 31, 36, 83, 129, 153, 181-182, 222, 237
biometrics 71, 76, 78-80, 84-85
blogs 16, 128, 254-255, 269, 271, 286
bookmark services 115
bulletin board service 115, 120-121, 125, 130

C

calendar services 115
change management 101, 251
change process 242, 250
citizen engagement 8, 18, 35, 56, 225-229, 236, 270, 291
citizen participation 16-18, 34, 89, 91, 135, 226-229, 232, 236-237, 325, 327
civil society-led projects 93
Clarity of Goals in Websites (CGW) 61, 64, 66
collaborative ventures 93
communication needs 50
communities of practice 53
community manager 253-259, 261-271
Community Multimedia Centers (CMCs) 277
computer supported cooperative work (CSCW) 51, 54-56
continuance intention 290-291, 294-295, 298, 309-310
Control Objectives for Information and related Technology (COBIT) 134, 144
cost reductions 95
Court Appointed Special Advocate (CASA) 203-210, 215-216, 219-222

D

dashboard 251
decision-making 2, 10, 17, 44, 51, 60, 64, 68, 83, 89, 96-100, 103, 115, 127-128, 149, 155-156, 165-168, 170-171, 227, 235, 241, 250, 256, 272
democratic e-governance 227, 235-237
developing countries 12, 94-95, 99, 101-110, 134, 142-143, 163, 269, 272, 274-275, 277-278, 282, 284, 288, 304, 307, 310, 312-313, 325-327
Diffusion of Innovations (DOI) 3, 5-6
digital divide 109-110, 149, 151, 165, 302-303, 313, 322, 325, 327
disaster response 225-226, 229-230, 232, 234-235
discourse-influence 273
discussion forum service 126, 130
Download Speed of Website (DSOW) 63-64
dynamic taxonomies 37-47

E

Ease of Web Navigation (EWN) 62, 65
ecology of games 89-90, 105-106, 108
e-commerce 11, 42, 44, 46, 59-60, 62, 64-65, 67, 69-70, 83, 117, 138, 141, 226, 254, 305-307
economic thinking 241
e-democracy 73, 83-85, 89, 105, 138, 141, 179, 197
efficiency gains 7, 95, 99, 210, 216, 301
e-filing system 290-291, 293-294, 296, 298-301, 303-306, 309-310
e-governance 11, 15-19, 35-36, 49, 88-89, 91-93, 105, 107-108, 110-111, 116, 148-156, 160-169, 225-227, 235-237, 239-240, 251, 288, 303-304, 308, 326-327
e-government 2, 6-8, 10-14, 17-18, 26, 35-38, 41, 43-47, 71-72, 74, 80-86, 88-92, 94-113, 124, 127, 131-147, 154-156, 164-170, 200, 204, 206, 222-224, 226-227, 236-237, 251, 255, 257-259, 268-271, 287, 290-291, 293-317, 319, 323-327